Solved and Unsolved Problems of Structural Chemistry

Solved and Unsolved Problems of Structural Chemistry

Milan Randić

Professor Emeritus, Drake University, Des Moines, IA

Marjana Novič

The National Institute of Chemistry, Ljubljana, Slovenia

Dejan Plavšić

The Ruđer Bošković Institute, Zagreb, Croatia

CRC Press
Taylor & Francis Group
Boca Raton London New York

CRC Press is an imprint of the
Taylor & Francis Group, an **informa** business

The characterization of aromaticity by conjugated circuits has been introduced by M. Randić almost 40 years ago (M. Randić, J. Am. Chem. Soc. 99 (1977) 444). The three Kekulé valence structures of naphthalene and their decomposition in conjugated circuits are shown over an older copy of Linus Pauling's drawing of 42 canonical structures contributing to the normal state of the naphthalene molecule. Conjugated circuits lead to generalized Hückel 4n+2 Rule: Polycyclic compounds having only 4n+2 conjugated circuits (n = 1,2,3,. . .) are aromatic.

CRC Press
Taylor & Francis Group
6000 Broken Sound Parkway NW, Suite 300
Boca Raton, FL 33487-2742

First issued in paperback 2021

© 2016 by Taylor & Francis Group, LLC
CRC Press is an imprint of Taylor & Francis Group, an Informa business

Version Date: 20160106

No claim to original U.S. Government works

ISBN 13: 978-0-367-86226-8 (pbk)
ISBN 13: 978-1-4987-1151-7 (hbk)

Contents

Foreword

Our universe consists of matter and energy, and both of these occur in discrete forms, atoms and quanta. Chemical structure is determined by the various ways atoms combine with each other. For molecules consisting of atoms that share electrons in their valence shells, the structure is expressed by chemical formulas or molecular graphs with points (vertices) symbolizing atoms, and lines (edges) symbolizing shared electron pairs. These formulas are the universal language of chemistry, and thanks to computer programs that can search huge databases, it is now possible to learn in a few minutes whether a certain structure has (or has not) been described among more than 70 million chemical substances observed experimentally or investigated theoretically.

This book written by Milan Randić, Marjana Novič, and Dejan Plavšić offers a well-documented overview of several aspects of structural chemistry that benefitted from Professor Randić's background in discrete mathematics and quantum physics and from his original contributions that have enriched the field of chemical graph theory. I take pride in having "converted" Milan Randić to graph-theoretical applications in chemistry when he was in the audience at Harvard University as I lectured on that topic in 1975, one year before the publication of a book in that field that I had edited. As an organic chemist, my kind of contribution was complementary to those brought by him, and our "cross-conjugation" resulted in several collaborative efforts. I also rejoice in having passed the torch to a worthy and faster runner, like the good teacher who is proud when one of his students overtakes him.

It is common in science that by solving one problem, one opens pathways for exploring several other, yet unsolved, problems. Application of graph theory to chemistry has, on one side, brought *novelty* and, on the other side, solved several *old problems* viewed for decades as beyond our reach. Illustrations of the former are graphs depicting degenerate rearrangements, isospectral graphs, and conjugated circuits. Illustrations of the latter are the generalization to polycyclic systems of the Hückel $4n + 2$ rule, which is valid for monocyclic systems; the numerical characterization of Clar sextet theory; and the exact solution to the protein alignment problem. There is no doubt that the current successes of chemical graph theory and graphical bioinformatics are due to the development of new tools based on discrete mathematics made available. A brief enumeration of topics that are discussed in this book, besides those mentioned above, and that bear the imprint of Dr. Randić's contributions, includes molecular connectivity; quantitative structure–property relationships using orthogonal molecular descriptors (which allows the use of highly collinear descriptors); the use of conjugated circuits for classification of polycyclic aromatic systems and for ring current calculations; novel matrices; numerical characterization of proteomic maps; numerical characterizations of graphical representations of DNA, RNA, and proteins; and novel graphical representations of DNA and proteins as well as graphical alignment of amino acid sequences. Of course, these and other topics also raise new questions about unsolved problems that still await answers.

A noteworthy side activity of Professor Randić is in developing a strong collaboration with scientific schools in parts of former Yugoslavia, particularly in his native Croatia (Nenad Trinajstić, Ante Graovac, Ivan Gutman, and Dejan Plavšić) and in neighboring Slovenia (Jure Zupan, Marjana Novič, Matevz Pompe, Marko Razinger, Marjan Vračko, and the mathematician Tomaž Pisanski). Growing contacts with similar schools then emanated in closely situated countries: Ivan Gutman in Serbia, Danail Bonchev in Bulgaria, István Lukovits in Hungary, and Mircea Diudea in Romania, contributing to making central and southeastern Europe the cradle of chemical graph theory for more than a quarter of a century.

This book opens up a wide perspective on how scientists transgress borders between discrete mathematics, quantum physics, biochemistry, genetics, organic and inorganic chemistry, bioinformatics, and computer science. It is hoped that it may inspire young people to bring their ideas and efforts to solving new problems and contribute to improving our surroundings by adding new bricks to the ever-growing tower of scientific understanding.

Alexandru T. Balaban
Texas A&M University at Galveston, USA

Preface

In absentia luci, tenebrae vincunt

The title of this book already suggests what it is about, but it does not tell how this book came about or what its background, scope, and intentions are. The above motto says, *"In the absence of light, darkens prevails."* We decided, therefore, to bring some light to a number of solved and unsolved problems in chemistry because even some problems believed to have been solved have left a number of details of the solutions in the dark. Let us mention here at the beginning that the book is unusual and unconventional, and just as it has been an adventure for us to write some of its sections, we will not be surprised if readers find reading it also to be adventurous, loaded with unexpected insights, novelty, and clarifications of uncertainties. It may appear to some risky to challenge some established views and opinions, but it would be more risky not to. After all, if we are not right, they have nothing to lose. But let those who have nothing to win in maintaining the *status quo* decide on their own who is right. Underneath, in a way concealed, this book speaks about a number of overlooked, misunderstood, unrecognized, or unaccepted and unknown observations and findings in science. As will be seen, solving a problem has been, in a number of cases, not the end but the beginning of the end as it has taken or may take additional time and effort to have the solutions accepted.

The book speaks also of scientific results that could have been found or solved decades earlier but were not; it speaks of scientific results that have been found or solved decades earlier but, at best, few chemists have been aware of them, and most researchers, many of whom could benefit from such overlooked "details," remain unaware or uninformed.

Science is not a race; it is about important findings and novel insights, and it occasionally offers new views on old things, whether found accidentally, guessed, or well thought out. All these one can find in some solved and unsolved problems that we have selected to elaborate on. This book started as a report on an invited lecture at the international conference about Chemometrics and Quantitative Structure–Activity Relationship (QSAR), in Sopron, Hungary, in the summer of 2013, organized by Karoly Herberger. The report turned out to be too long for the proceedings of the conference. As we expanded further into the neighboring areas of mathematical chemistry, including also topics involving aromaticity, which is one of the central themes of chemistry, and topics from bioinformatics, which focused on comparative studies of DNA and proteins, we considered submitting the review to *Chemical Reviews*. This is the leading journal on chemistry, in which we have previously reviewed both aromaticity (in 2003) and graphical bioinformatics (in 2011). However *Chemical Reviews* viewed the scope of the proposed material not *comprehensive enough*—and, in our view, to be comprehensive enough on the theme of solved and unsolved problems in chemistry would require at least that we write a book having three volumes. We decided to write a single volume because we limited ourselves to the domain to which we have contributed and with which we have been familiar.

Surely, there are a number of interesting and intriguing solved and unsolved problems in chemistry that require the attention of chemists but that we leave to the attention of those in whose area of research such problems belong!

That this book is somewhat unusual may already be seen from the inclusion of a table in the preface, which one usually does not see. In the table, we list 12 of the problems that receive greater visibility in our book and, in a few words, indicate what they are about. These are not the only topics covered in the book, and the book is not only about solved and unsolved problems. There are also problems somewhere in between, some of which are problems assumed by those involved to have been solved, but they represent, at best, partial solutions, and there are problems, the solutions of which are in *statu nascendi*. An illustration of the former is the single equation fallacy, the presumption that in multivariate regression analysis (MRA) a single equation allows one to interpret the role of the descriptors used, and an illustration of the latter is the ring closure matrix, which represents a generalization of matrices involving only distances between terminal vertices of acyclic graphs, which fully characterizes such graphs to cyclic systems. The ring closure matrix, if it eventually is found to be general enough, would illustrate also a case of *statu nascendi* of a novel *tool* in chemistry. Toolmaking, which has been one of our longtime interests and a longtime misunderstood activity, is unfortunately not well appreciated among tool users! One can almost extend C. P. Snow's theme of "two cultures," a lack of intercommunication between natural sciences and the arts, to "two subcultures" in science, toolmakers and tool users, with occasional giants, such as Galileo Galilei, who excelled in both!

It is needless to say how much unwarranted hardship this hostile attitude of tool users has brought to the misunderstood minority of the toolmakers. Fortunately, there have been scientists who have appreciated our work, which was a welcome encouragement at the time of struggle for survival. Support by a single outstanding scientist would suffice, but we have been fortunate to have a dozen, among whom we are proud to mention Robert S. Hansen, Roald Hoffmann, Alan Katritzky, Harry Kroto, Per Olov Lowdin, Robert G. Parr, and E. Bright Wilson. Although the writing of this book took less than a year, the work about which we write is mostly part of our many years of research. This started with our interest in the application of discrete mathematics, and graph theory in particular, to chemistry since the mid-1970s and the extension of this to bioinformatics in early 2000.

We hope this book will open the eyes of some who may have not been aware of the "riches" of almost "lost" science and that it may awaken some of those in the "sleeping" communities of chemists. This book is not for those who have been working all of their lives on a single problem, and it is not for those who have been ignoring the work outside their own narrow focus of research attention. We don't welcome their comments; let them live happily in darkness.

Some of the Topics of This Book Reflecting on Logic, Elegance, and Beauty in Chemistry	
Problem	**About**
1 Periodic table of isomers	Regularities in properties of isomers
2 Chemical shift sums	Regularities beyond atomic properties

(Continued)

Some of the Topics of This Book Reflecting on Logic, Elegance, and Beauty in Chemistry	
Problem	**About**
3 Highly collinear descriptors	The problem that is not
4 The variable molecular descriptors	Efficient and optimal molecular descriptors
5 Highly discriminative descriptors	Molecular similarity descriptors
6 Generalized Hückel $4n + 2$ rule	Going beyond a single cycle
7 Graphical alignment of proteins	No approximations, but use of trial-and-error
8 Sequential nearest neighbors	Simple binary codes for maps
9 Ring currents via conjugated circuits	Discreteness versus continuum
10 Renormalization in chemistry	Self-similarity in chemistry
11 Aromatic sextet revisited	Undiscovered beauty
12 Exact solution to protein alignment	Sleeping giant

We would like to acknowledge almost 20 years of continuing financial support from the Science Authorities in Slovenia and Croatia, which made our scientific collaboration, coming from two continents and three countries, possible and productive.

Milan Randić
Ames, Iowa, USA
mrandic@msn.com

Marjana Novič
Ljubljana, Slovenia
marjana.novic@ki.si

Dejan Plavšić
Zagreb, Croatia
dplavsic@irb.hr

MATLAB® is a registered trademark of The MathWorks, Inc. For product information, please contact:

The MathWorks, Inc.
3 Apple Hill Drive
Natick, MA 01760-2098 USA
Tel: 508 647 7000
Fax: 508-647-7001
E-mail: info@mathworks.com
Web: www.mathworks.com

Acknowledgments

We thank Alexandru T. Balaban and Douglas Klein from Texas A&M University at Galveston, TX, and Ovidiu Ivanciuc from the Medical School, University of Texas at Galveston, TX, for their many helpful comments. We also thank J. Gasteiger for taking time to examine the manuscript of the book prior to being sent to the publisher. Ivan Gutman has read the manuscript with great care and found, among others, a number of incomplete and incorrect references, which have been corrected. We also received useful suggestions on literature from Matthias Dehmer, Vienna, Austria, and Guillermo Restrepo, Pamplona, Columbia. Last but not least, authors would like to thank Ms. Adel Rosario, CRC Project Manager, Manila Typesetting Company, Philippines for her patience and advices on improving the manuscript. This work has been supported by the Ministry of Science, Education of the Republic of Slovenia under Research Grant P1017 and by the Ministry of Science, Education and Sport of the Republic of Croatia under the Project 098-0982929-2917. The authors decided not to entertain discussions about who has written which part or which section of the book. But the senior author MR is responsible and is to be blamed for any inaccuracy or error that may have escaped our attention.

Authors

Milan Randić was born in 1930 in Belgrade, Serbia, his family originally from the Croatian Adriatic coast. He has been living in the United States since 1971 and is a citizen of the United States and Croatia. Dr. Randić studied theoretical physics from 1949 to 1953 at the University of Zagreb under Professor Ivan Supek, a student of Werner Heisenberg. Directed by Professor Supek to do research in theoretical chemistry, he got his PhD degree in 1958 at the University of Cambridge, England. Returning to Zagreb in 1960, he founded the Theoretical Chemistry group at the Institute Rudjer Bošković. In 1965, Dr. Randić joined the department of chemistry at the University of Zagreb as a professor; however, early in 1971, he left for the United States. From 1971 to 1980, Dr. Randić visited several universities, including the Johns Hopkins University (R. G. Parr) and Harvard University (E. Bright Wilson). In 1973, he started research on an application of discrete mathematics to chemistry. Since 1980, Dr. Randić has been in the department of mathematics and computer science at Drake University, Des Moines, IA, retiring in 1999 as a distinguished professor. In 1988, he received the Governor of Iowa annual science award. In 1996, he received the Skolnik Award of the American Chemical Society. Since 1995, he has been making annual visits to the Laboratory of Chemometrics, National Institute of Chemistry, Ljubljana, Slovenia. In 1997, Dr. Randić became a member of the Croatian Academy of Sciences and Arts. In 2008, he became an honorary member of the Slovenian Chemical Society and, in 2009, became an honorary member of the National Institute of Chemistry, Ljubljana. In 2010, he received the Grand Pregel Award of the National Institute of Chemistry of Slovenia. In 2014, Dr. Randić reported the *exact* solution of the protein alignment problem. The protein alignment problem has been around for more than 45 years, believed by many not to have an exact solution—but it has!

Marjana Novič was born in Ljubljana, Slovenia, where she took studies in physical chemistry at the University of Ljubljana during 1974 to 1979. She obtained her PhD in 1985 at Faculty of Chemistry and Chemical Technology, University of Ljubljana, with the thesis "Hierarchical Clustering and Recognition of Chemical Structures and Structural Fragments on the Basis of ^{13}C NMR spectra." She started her career at the National Institute of Chemistry in Ljubljana, initially developing automated information systems for infrared and NMR spectroscopy. Her post-doctoral specialization was carried out at the University of Lausanne, Switzerland (1986–1987) in the field of automated pattern recognition in 2-D NMR spectra. In 1989, she was given the "Boris Kidrič" national award for scientific achievement. Later, she visited several research laboratories (in Tarragona and Cordoba, Spain, and in Hobart, Tasmania) while being employed at the Laboratory of Chemometrics at the National Institute of Chemistry, Ljubljana, Slovenia. Currently she is the head of the laboratory. Her expertise includes the development of chemometrics methods, Quantitative Structure Activity Relationship (QSAR) and Artificial Neural Network (ANN) modeling, structural elucidation of transmembrane segments of membrane proteins, and innovative merging of chemometrics methods with molecular modeling, which facilitates effective drug design. She is also teaching chemometrics at the University of Ljubljana.

 **Dejan Plavšić** was born in Zagreb, the Republic of Croatia, where he received his primary and secondary education. After obtaining his BSc degree in chemistry from the Faculty of Science, University of Zagreb, he joined the department of physical chemistry at the Ruđer Bošković Institute in Zagreb. He earned his MSc degree in chemistry from the University of Zagreb in 1980, and received his PhD degree in chemistry in 1984 from the same university under the supervision of Professor Jaroslav Koutecký, Institut für Physikalische Chemie, Freie Universität Berlin, Germany, and Professor Leo Klasinc, Ruđer Bošković Institute. In 2005, Dr. Plavšić joined the NMR Center at the Ruđer Bošković Institute, where he is a senior research scientist. His research interests are in the areas of mathematical chemistry, chemical graph theory and its applications, metal clusters, organometallic compounds, and catalysis. Since 2003, he has become interested in extending graphical and numerical characterizations to DNA, proteins, and proteome maps using graph theoretical tools, entering thus into bioinformatics. Two of his papers in this area have received recognition as being among the 50 Most Cited Papers in Chemical Physics Letters during the 2003–2007 period. In 2005, when the International Academy of Mathematical Chemistry was founded, he was one of the initial founding members of the Academy. In 2006, Dr. Plavšić was the first recipient of the International Award Latium between Europe and the Mediterranean for Medicine, Physics, or Chemistry presented in Rome, Italy.

1 Introduction

The main theme of this book is the structure–property relationship, which is the relationship between the molecular structure and its properties, including *mathematical* properties of chemical structures. The list of physicochemical properties of molecules has been fairly standard over a relatively long time even though, with advances in technology of instruments, which is to continue, the list is not likely to remain unchanged. The same is even more expected about molecular activities, which are to continue to diversify. Study of *mathematical properties of molecules* has emerged more recently, and it has not been recognized by many for a while to be the subject of chemistry. The role of mathematics in chemistry has been apparent for over a century ever since the establishment of physical chemistry. The term "physical chemistry" appeared first as a title of a course due to M. V. Lomonosov (1711–1765) in 1752 at St. Petersburg University. It took another hundred years until J. W. Gibbs (1839–1903) in 1876 published the seminal work on application of thermodynamics, and in 1882, H. L. F. von Helmholtz (1821–1894) published fundamental work on thermodynamics, which was essential for transforming the qualitative physical chemistry of that time into rigorous scientific discipline. Gibbs, who was a professor of mathematical physics at Yale University, together with J. C. Maxwell (1831–1879), a Scottish mathematical physicist and L. E. Boltzmann (1844–1906), an Austrian physicist and philosopher, created statistical mechanics (the term that he coined), which explains the laws of thermodynamics as the results of statistical properties of bulk (large assembly of particles). Physical chemistry, which today in a broad sense includes quantum chemistry, quantitative structure–activity relationship (QSAR), and mathematical chemistry, was firmly established by the works of F. W. Ostwald (1853–1932), J. H. van't Hoff (1852–1911), and S. A. Arrhenius (1859–1927). Ostwald and van't Hoff already in 1887 founded *Zeitschrift für Physikalische Chemie*, which established itself as the standard journal in the field.

The role of mathematics became apparent with the development of quantum chemistry in works of L. Pauling (1901–1994) and others [1], and more recently discrete mathematics has started to play a role in solving diverse chemical problems even though this is, by far, less apparent. One of the tasks of this book is to bring to the attention of a wider circle of chemists the definite role that *discrete mathematics* plays in chemistry. In contrast to "traditional" mathematics, discrete mathematics is concerned with studies of properties of objects having discrete components, such as atoms in a molecule. It is therefore quite surprising that it took such a long time to recognize the relevance of discrete mathematics in chemistry, which facilitated the growth of mathematical chemistry [2–5]. The early indication to the importance of enumerations in structural chemistry, if we consider the enumeration of isomers in 1876 of Flawitsky [6] as a contribution to stereochemistry (which, of course, is a part of structural chemistry, too), was drawn by G. N. Lewis (1875–1946), who in 1923 wrote about his octet rule [7], a fundamental contribution to the understanding of the covalent bond, years before the emergence of quantum chemistry.

In this book, we review several solved and unsolved problems in structure–property–activity studies, in theoretical chemistry, and in bioinformatics, which are related to the structure–property relationship in general. We have elected to discuss and elaborate on a number of problems, whether solved or unsolved, which appear to us to have not been widely or well known. Among the selected problems we have included, in particular, some problems that relate to the following four subjects:

i. The multiple regression analysis (MRA)
ii. The use of partial ordering, one of the general methods of discrete mathematics, which is apparently not receiving enough attention in chemistry
iii. The often overlooked role of Kekulé valence structures in chemistry
iv. One of the central problems of bioinformatics, the problem of protein and DNA alignments

As will be seen, the misunderstandings and the tendency of many chemists to overlook the approaches coming from the "nontraditional" directions of chemistry have not been isolated solely to QSAR studies, but have been seen in several other areas of theoretical chemistry.

In view that the main tool to be used here is mathematics, it seems appropriate as an introduction to review a selection of solved and unsolved problems in mathematics. We do not assume familiarity of readers with details of mathematics, discrete or not, and what we need we will clearly explain. We will start with a few problems that have been considered in antiquity, followed with a few so-called "famous" problems, the problems that can be easily understood even by laymen, but the solving of which may cause even outstanding professionals to have difficulties. In addition, we have selected a few "illustrious" problems. According to *Webster's Dictionary of Synonyms & Antonyms* [8], "illustrious" stands for something *"enduring and merited honor and glory."* We feel that the selected problems deserve such a title.

In reviewing these problems, it will be seen that a number of problems remain, not because there were not enough capable and clever people around trying to solve them, but because, at the time, there were not suitable tools available for solving them. One should keep this in mind because some of the unsolved problems of today, in mathematics, computer science, physics, or chemistry, remain unsolved and may continue to be unsolved for the same reason, lack of suitable tools. However, tool making appears not to be appreciated and understood, even today, in some circles of the scientific "establishment" as has been reflected from time to time in the comments of reviewers and the practice of editorial boards of some scientific journals (see Appendices 1–3) [9].

Another important message of this book is that one should be aware of unsolved problems in a wider area of his or her own research interests, so that if one comes across some novel tool or some tool of which he or she was not aware before, it may be worthwhile to revisit earlier unsolved problems and find out if the novel tool could facilitate finding their solution.

1.1 PROBLEMS OF ANTIQUITY

The three well-known Greek problems of antiquity are known as the following:

1. Doubling a cube
2. Trisection of an angle
3. Squaring a circle

The solution to these geometrical problems was sought using only a straightedge (unmarked ruler) and compass. The later development of mathematics and novel mathematical tools enabled solutions of these three problems. All three problems were transformed into algebraic problems involving constructible numbers. A constructible number is one that can be obtained by a *finite* number of *additions, subtractions, multiplications, divisions*, and *finite square root* extractions of integers. "Constructible numbers" correspond to *line segments*, which can be constructed using only *straightedge* and *compass.*

1.1.1 DOUBLING A CUBE

Solving the first problem requires the solution of the equation $x^3 = 2$, which is equivalent to the geometrical construction of the number $\sqrt[3]{2}$. The construction of $\sqrt{2}$ is simple; one has just to construct a square, the side of which is one unit, and draw its diagonal. Unlike $\sqrt{2}$ which is constructible, the cube root, the number $\sqrt[3]{2}$ is not constructible, and therefore, the doubling the cube problem could not be solved in antiquity by using only a compass and straightedge. As is known today, one can with a compass and straightedge construct only numbers expressed as square roots and repeated square roots. A remarkable illustration of this is a result of Karl Friedrich Gauss (1777–1855), one of the greatest mathematicians of all time, who in 1796 (at the age of 19), showed that it is possible to construct a heptadecagon, a regular polyhedron having 17 sides, using only a compass and straightedge [10]. The significance of this result is reflected in the statement [11], "*This proof represented the first progress in regular polygon construction in over 2000 years.*" Constructing a regular heptadecagon involves finding the cosine of angle $2\pi/17$ in terms of square roots, which involves an equation of degree 17. The solution in modern notation is

$$16\cos\frac{2\pi}{17} = -1 + \sqrt{17} + \sqrt{34 - 2\sqrt{17}} + 2\sqrt{17 + 3\sqrt{17} - \sqrt{34 - 2\sqrt{17}} - 2\sqrt{34 + 2\sqrt{17}}}.$$

So much on this topic.

1.1.2 TRISECTION OF AN ANGLE

The second problem of antiquity requires dividing an arbitrary angle into three equal angles by using only straightedge and compass. Dividing an arbitrary angle into two equal parts is easy: (i) By compass, draw about the vertex V of the angle a circular arc intersecting the sides of the angle in two points A and B; then draw about the points A and B two further circular arcs of equal radius that intersect in point C. (ii) By straightedge, draw the segment $\overline{VC}$ bisecting the angle. This process can be repeated many times allowing the construction of 1/4, 1/8, 1/16 ... of an arbitrary angle. It is known that the infinite sum in which every term is four times smaller gives as result 1/3:

$$1/4 + 1/16 + 1/64 + 1/256 + 1/1024 + ... = 1/3$$

The above sum in decimal form becomes

$$0.25000 + 0.06250 + 0.015625 + 0.003906 + 0.000977 + 0.000244 + 0.000061 + ...$$

giving for the partial sums the sequence

$$0.25000, 0.31250, 0.32813, 0.33203, 0.33308, 0.33325, 0.33313...$$

with the limit 0.333333 or 1/3. But the process of construction never ends, so 1/3 of an arbitrary angle cannot be constructed! In 1837, P. L. Wantzel (1814–1848) published a rigorous algebraic proof that 1/3 of an arbitrary angle cannot be constructed by using only a straightedge and compass. The problem of trisection of an arbitrary angle is equivalent to constructing a segment whose length is the root of the equation

$$\cos\theta = 4\cos^3(\theta/3) - 3\cos(\theta/3),$$

which is a third-degree equation, and its root is not constructible.

1.1.3 SQUARING A CIRCLE

The third problem of antiquity states that for a given circle construct a square with the same area as the circle using only a *compass* and *straightedge*. The solution of the problem requires the construction of the number $\sqrt{\pi}$ because the formula for the area of a circle or radius r is $r^2\pi$. Because π is transcendental number, which means that π is not a solution of any algebraic equation (a polynomial equation in one variable having rational coefficients), the construction of π by compass and straightedge is impossible. In 1882, F. Lindemann (1852–1939) proved the transcendence of π.

In Figure 1.1, we illustrate as a curiosity how J. Kepler (1571–1630), more than 400 years ago, found the formula for area of a circle. It was known that the circumference of a circle, which one can always measure, is $2r\pi$, but the area of a circle could not be measured! Kepler started dividing a circle into a dozen sections, just as one does with a large pizza. The sections can then be arranged in a sequence of "up" and "down" oriented cuts as illustrated in Figure 1.1. In the next step, one can double the number of cuts and continue the process of orienting cuts in the up–down sequence. When one continues to increase the number of cuts, the resulting pattern more and more resembles a rectangle, the height of which is r and the base $r\pi$, because the top and the bottom of a rectangle together add to the circle circumference $2r\pi$. Therefore the area of a circle is $r^2\pi$.

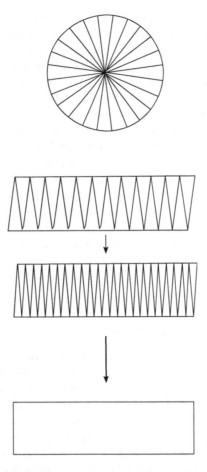

FIGURE 1.1 Kepler's method for finding the area of a circle by transforming it into a rectangle of height r and length $r\pi$ (1/2 of circumference $2r\pi$).

1.2 FAMOUS PROBLEMS

The so-called "famous problems" are mathematical problems that can be understood by almost everybody and solved by almost nobody! One can even characterize the famous problems by stating that these are the problems that will make those who solve them famous! We will briefly mention two to illustrate the scope of the mathematics they cover.

1.2.1 THE FOUR-COLOR PROBLEM

One of the most famous of these problems is the four-color problem, which states that any geographical map can be colored by using only four colors so that no adjacent countries (regions) of the map have the same color. According to W. W. Rouse Ball [12], the author of a well-known book on recreational mathematics, A. F. Möbius (1790–1868), one of the distinguished German mathematicians, mentioned this problem in his lectures in 1840, but it came to wider visibility when, in 1852, Francis Guthrie (who was an earlier student of Scottish mathematician A. De Morgan) asked his brother Frederick, a current student of De Morgan, to inquire about it. De Morgan considered the claim that any map can be colored with four colors a hypothesis that needed to be proved. In 1878, Alfred Kempe published his proof, and the following year, P. G. Tait also published another proof. However, some 10 years later, Percy J. Heawood found an error in the proof of Kempe's, and Julius Petersen found an error in the proof of Tait. Not much happened with the four-color problem until 1976 when Kenneth Appel and Wolfgang Haken announced a proof of the four-color problem obtained by use of a computer after an analysis of almost 2000 structurally unrelated maps (which took about 1000 hours of computing). Later, the number of essential maps to be checked was reduced to about 1500, but this hardly changes the complexity and the nature of the proof. This may be the first mathematical problem the proof of which was obtained by extensive assistance of a computer. Not surprisingly the situation has aroused controversy and doubts in some mathematical circles about such computer-assisted proofs [13]. Perhaps we have to wait to see if another, different, computer-assisted proof will confirm that the four-color problem was solved.

1.2.2   GRAPH RECONSTRUCTION PROBLEM

The graph reconstruction problem appeared in a book on a collection of mathematical problems by S. M. Ulam [14] and a paper of P. J. Kelly [15], a student of Ulam. This problem has considerable relevance to chemistry and mathematical chemistry by being related to the graph isomorphism problem, which considers if two graphs (that may represent complex molecular structures) are identical or not. We will outline the problem of graph reconstruction as essentially formulated by F. Harary [16]. In Figure 1.2, we have illustrated a "deck" of seven "cards" on each of which is depicted one subgraph of a graph obtained by erasing a single vertex and incident edges. These subgraphs have been referred to as *Ulam subgraphs*. The question is can one from a given set of Ulam subgraphs (i.e., given the deck of cards) reconstruct the graphs that produced the seven subgraphs?

A related problem is that of verification if a given "deck of cards" of Ulam subgraphs is legitimate, that is, it has not been tampered with or changed. Suppose that one of the cards in a set has been deliberately changed. Is there a way to find this out? In Appendix 4, we show that if one card, accidentally or deliberately, has been altered, this can be detected and corrected.

There is an additional interesting property of Ulam subgraphs reported by F. H. Clarke [17] and V. Levenshtein et al. [18], which we will outline in the section considering the characteristic polynomial of graphs. Let us end this section on graph reconstruction by mentioning that if the reconstruction hypothesis is to be proven, the abovementioned Clarke theorem would offer a relatively simple isomorphism test for graphs based on the characteristic polynomials of graphs and their Ulam subgraphs. For now, this simple approach at least holds for isomorphism testing of acyclic graphs because, for them, the reconstruction problem has been known to hold.

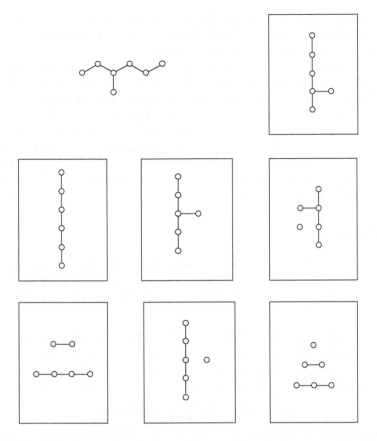

FIGURE 1.2 Molecular graph of carbon skeleton of 3-methylhexane and the "deck" of cards with its seven Ulam subgraphs obtained by deleting one carbon atom and its incident CC bonds at the time.

1.3 ILLUSTRIOUS PROBLEMS

The number of interesting problems in mathematics that can be understood by a layman is not so small. There are, for example, a number of such problems involving prime numbers. Consider the question of whether the number of twin prime numbers is finite or infinite. Twin prime numbers are pairs of prime numbers that differ by 2, such as (3, 5), (5, 7), (11, 13), (17, 19), (29, 31), (41, 43), (59, 61), and (71, 73) so and on. Twin prime numbers are not so uncommon and continue even for very large prime numbers. The question of whether the number of twin prime numbers is finite or infinite is considered one of great unsolved problems of number theory, a branch of mathematics studying properties of numbers. Significant progress toward solving this problem was reported in 2013 by Chinese-American mathematician Y. Zhang of the University of New Hampshire [19]. Incidentally, the eight pairs of twin primes listed above are the only twin primes smaller than 100. There are an additional 27 twin primes smaller than 1000. The frequency of twin primes becomes smaller as the numbers increase, and the question of whether the number is finite or infinite appears legitimate. This question has been attributed to Euclid (ca. 325 BC–ca. 265 BC), which, if true, would make this question the oldest unsolved conjecture in mathematics being around for about 2300 years.

Although proving theorems is generally a difficult task even for highly professional mathematicians, posing questions apparently may not be so difficult. However, posing interesting questions and questions that may result in interesting answers and lead to new directions in research, is again, a difficult task. A question may appear simple, but it need not be, and a question may appear difficult and need not be. Only time will tell which questions are interesting and important and which are not. But worse than asking questions that may turn out not to be so interesting is not to pose questions. In that spirit, we will pose a question relating to prime numbers, more specifically relating to twin prime numbers:

Problem 1

Looking at a list of smaller twin numbers, we noticed that prime number 5 appears to be the only prime number that appears in two twin prime pairs: (3, 5) and (5, 7). Is this the only prime number that is a member of two twin pairs, or in other words, is the triplet (3, 5, 7) the only twin prime number triplet?

As far as we are aware, this question has not been previously posed. There is a question relating to Fibonacci numbers that bears some similarity. The Fibonacci numbers [20] we will meet later when discussing the count of Kekulé valence structures for a special class of benzenoid hydrocarbons. The initial Fibonacci numbers are

$$1, 1, 2, 3, 5, 8, 13, 21, 34, 55, 89, 144...$$

They satisfy the simple recursion $F_n = F_{n-1} + F_{n-2}$ with seed values $F_1 = 1$ and $F_2 = 1$. Clearly, each number F_n for $n \geq 2$ is given by adding two preceding numbers in the sequence. Thus, $1 + 1 = 2$, $1 + 2 = 3$, $2 + 3 = 5$, $3 + 5 = 8$, and so on. It can be shown that $\lim_{n \to \infty} \frac{F_{n+1}}{F_n} = (1+\sqrt{5})/2 = 1.618...$ is the so-called *golden ratio*.

Leonardo of Pisa, known as Fibonacci (ca. 1170–ca. 1250), introduced these numbers in his book *Liber abaci* ("Book of Calculation") in a story of rabbit multiplication. One starts with a pair of rabbits, and it takes a month for newborn rabbits to grow to maturity. So the initial count is 1, 1. Next month, the mature pair breeds, and one has two pairs of rabbits. A month after, the mature pair breeds again, producing an additional young pair, and the young pair of the previous month is grown to maturity, so we have three pairs. The next month, we have two mature pairs that breed and produce two additional pairs, which added to the three of the previous month give the count of rabbits of five pairs, and so on. It is interesting that Fibonacci numbers appear often in nature, in particular in some plants, their flowers and seeds (cones) in counting petals in various flowers, of spirals in cones, and even in the count of branches in some trees as mentioned by Hugo Steinhaus (1887–1972), a Polish mathematician of Jewish roots [21]. In Figure 1.3, we have illustrated the growth of branches in such trees. If this figure is inverted, it would illustrate the count of breeding rabbits mentioned in the book of Fibonacci.

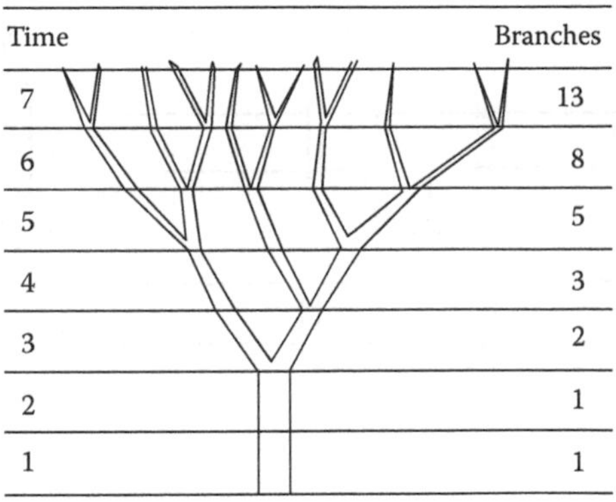

Time		Branches
7		13
6		8
5		5
4		3
3		2
2		1
1		1

FIGURE 1.3 Fibonacci numbers illustrated on the count of growth of annual branches of some trees.

Here is a mathematical question raised and still not answered: Is 144, the 12th member in the sequence, the only Fibonacci number that is a square of a number, 144 being 12^2? As far as we know, this is an unsolved problem. We have not included this problem in *our* list of problems as we could not trace its origin.

Clearly, the above two questions are for mathematicians, not chemists, and have been mentioned merely to illustrate the basic distinction between mathematics and chemistry. This difference has been well summarized by the late R. B. Woodward (1917–1979), the leading organic chemist of recent time, considered by many to be the preeminent organic chemist of the 20th century. In his Cope Lecture, Woodward spoke of the special challenges of chemistry compared to mathematics [22]:

> While in mathematics, presumably one's imagination may run riot without limit, in chemistry one's ideas, however beautiful, logical, elegant, imaginative they may be in their own right, are simply without value unless they are actually applicable to the one physical environment we have—in short, they are only good if they work!

In view of the above, we therefore selected for discussion in this section only two problems:

  i. The Problem of Seven Bridges of Königsberg
  ii. Halmos Handshake Problem

The first problem was solved by Swiss mathematician Leonhard Euler in 1736, and it signifies the beginning of graph theory as a novel mathematical discipline. The second problem, which is of a more recent time, will illustrate how a *novel tool*, here graph theory, can facilitate solving problems, which, at first, appear unsolvable.

1.3.1 THE SEVEN BRIDGES OF KÖNIGSBERG

The river Pregel in East Prussia divides the city into four parts, which are connected with seven bridges as illustrated in Figure 1.4. The inhabitants of the city were

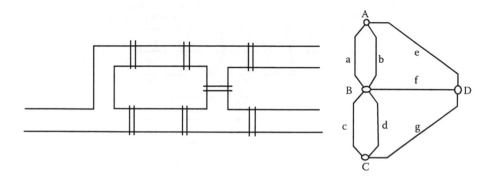

FIGURE 1.4 Schematic map of the city of Königsberg and its seven bridges.

wondering if it is possible to make a walk through the city crossing each bridge just once and return to the place where the walk started. Leonhard Euler (1707–1783), one of the greatest mathematicians of all time, was born in Basel, Switzerland, but spent most of his time in St. Petersburg, Russia. He published more than 900 works, and he demonstrated that such a walk across the Königsberg bridges is not possible [23].

Euler solved the problem, algebraically. If one labels the four regions of the city with letters A, B, C, and D and the seven bridges with a, b, c, d, e, f, and g, then any walk through the city over the bridges that is possible could be listed as a sequence of capital and small letters. One such walk, which crosses six bridges, would be, for example, BaAbBcCdBfDe, and another one is AaBcCgDfBbAe; each misses one bridge and does not return to the part of the town where the walk started. Euler concluded that in order to satisfy the conditions it is not essential in which order the bridges are crossed, but that every area to be visited is connected by an even number of bridges, so that one bridge is used to enter the area and one to leave it. Because the four areas of Königsberg all have an odd number of bridges, Euler concluded that it is not possible to complete a walk crossing each bridge once and return to the starting point. Even if one is to relax the condition and only require that one crosses all the bridges but need not return to the place where the walk started, such a walk is not possible in Königsberg or in any city, which could have many bridges, unless all sections of such cities have an *even* number of connections except for the starting and end points of the walk.

One can simplify the maps of such cities by drawing a graph in which different areas of a city are represented as vertices and connections (bridges) by edges as shown for the case of Königsberg in Figure 1.4. To state the Euler theorem, which applies to any graph regardless of the number of vertices and edges, we have to define the Euler path and the Euler circuit:

Definition 1: A Euler path is a path in a graph G that traverses each edge of G exactly once.

Definition 2: A Euler circuit is a circuit in a graph G that contains each edge of G exactly once.

We can now state the Euler theorem, which resolved the problem of the Königsberg bridges.

Euler theorem: An undirected connected graph G has at least one Euler path if and only if G has zero or two vertices of odd degree. If each vertex in G has even degree, then G has a Euler circuit.

A closely related problem is that of searching for a shortest closed path connecting all vertices. Such closed paths, which are generally difficult to find, are known as Hamilton circuits. The problem of determining whether a Hamilton path or a Hamilton circuit exists in a graph is of considerable interest not only in graph theory, but in its many applications, including chemistry. It has been shown that this problem is *NP* complete. Interested readers should consult general literature on graph theory, in which the problem has also been known to be closely related to the so-called "traveling salesman problem," which asks what route a salesman should take in a round trip through a given set of cities to minimize the total distance traveled. We have listed limited literature on Hamiltonian circuits, which relates to some chemistry problems [24–30].

1.3.2 HALMOS HANDSHAKE PROBLEM

The next problem that we selected for consideration was outlined in one of the articles of P. R. Halmos (1916–2006), a distinguished American-Hungarian mathematician. It was published in 1982 in *The College Mathematics Journal* of the Mathematical Association of America [31]. Halmos stated several mathematicians considered that this is an ill-defined, invalid problem, which has no solution and cannot be solved. At the end of the paper, he gives the solution although he did not describe how the solution was obtained, presumably not to spoil the joy of those who may try to solve it. Despite its appearance as an ill-defined impossible problem, it is a legitimate problem with a single unique solution. In Appendix 5, again not to spoil the joy of those who may try to solve the problem, we have outlined the solution of the problem. One of the reasons we have selected this problem is because we also wanted to illustrate to the readers how a new tool, when appropriate, can facilitate finding solutions to problems that appear at first sight unsolvable.

According to Halmos's article, he and his wife were invited to a party, to which in all were invited five couples, 10 persons. Halmos noticed that on arrival some people were shaking many hands and some few. He had been intrigued by this, and when the party was over, he was standing at the exit and asked each person how many hands he or she had shaken. He got the following nine answers: 0, 1, 2, 3, 4, 5, 6, 7, 8. It is assumed that any person does not shake hands with his or her spouse. The question is: How many hands did Mrs. Halmos shake?

It does look like an impossible problem that is based on insufficient information, but be assured that the answers that Halmos received are sufficient to solve the problem. One of the present authors, in his lectures to the special class of a dozen outstanding students on graph theory, mentioned this problem at the beginning of one of his lectures. At the end of the class, one student came to tell his solution, and he was right. Not knowing how much time it took him to find the solution and assuming that it took some time, after congratulating him he was told that it was very nice that he solved the problem, but at the same time, he was not listening to the lecture while solving the problem. This anecdote illustrates that although most of the time in schools students learn from professors, occasionally also professors can learn from students: The lesson for this professor was that in the future he should not tell problems at the *beginning* of the class, but at the *end*!

Before we leave this "impossible" problem of Halmos, let us briefly mention one of its modifications that we came across in writing this book. Suppose that Halmos knew all the people at the party by their first names or if not knowing asked each person as they were leaving not only how many hands he or she shook but also their first name. Then having the nine answers: 0, 1, 2, 3, 4, 5, 6, 7, 8 and the corresponding names, say, Ann, Barbara, Carmen, Doris, Elisabeth, Francis, George, Henry, and Ivan (in short, the nine names A, B, C, D, E, F, G, H, and I), one of which is an assumed name for Mrs. Halmos, one can ask two questions:

1. How many hands did Mrs. Halmos shake?
2. What is the (here assumed) name of Mrs. Halmos?
3. Identify the five couples by their names.

We had to change the real name of Mrs. Halmos to an assumed name in order not to give an unfair advantage to those who knew or could find out the real name of Mrs. Halmos. The answers to the additional two questions as well as the answer to the question of how many hands Halmos shook can be found in Appendix 5.

1.4 HILBERT PROBLEMS

We will end this excursion in mathematics by mentioning that there are additional lists of unsolved problems in mathematics and closely related fields, which have played an important role in the development of current mathematics. At the turn of the 20th century, at the Second International Congress of Mathematicians in Paris in 1900, David Hilbert (1862–1943), very distinguished and recognized as influential and one of the last universal mathematicians, delivered a lecture in which he outlined 23 challenging and outstanding problems in mathematics as he saw them. Only very recently from his notes it was found that his list had 24 problems, but just before the lecture, Hilbert decided to eliminate one of them. The Hilbert lecture was not merely listing the 23 problems, but it was about the mathematics behind them and is generally considered by many to be *"the most influential speech ever given to mathematicians, given by a mathematician, or given about mathematics"* [32].

With the following words, David Hilbert opened the Second International Congress of Mathematicians in Paris in the year 1900:

Who of us would not be glad to lift the veil behind which the future lies hidden; to cast a glance at the next advances of our science and at the secrets of its development during future centuries? What particular goals will there be toward which the leading mathematical spirits of coming generations will strive? What new methods and new facts in the wide and rich field of mathematical thought will the new centuries disclose?

We will, but only briefly, comment on five of the 23 Hilbert problems, just to throw some "light" on a part of mathematics of which we in chemistry rarely have the opportunity to hear about, yet nevertheless, we can appreciate their content to a degree. These are problems 3, 6, 7, 8, and 18. Let start with 3:

Hilbert Problem 3: *Given any two tetrahedra T_1 and T_2 with equal base area and equal height (and therefore equal volume), is it always possible to find a finite number of tetrahedra, so that when these tetrahedra are glued in some way to T_1 and also glued to T_2, the resulting polyhedra are scissors congruent?*

Hilbert's problem 3 is clearly easy to understand and has been mentioned just to remind everyone that rules, formulas, and theorems that we all use and take for granted, someone at some time had to discover and prove. Hilbert apparently expected that the answer would be negative. This is a problem of three-dimensional space, which had already been known in two-dimensons to Euclid. Two plane polygons of the same area are related by "scissors congruence." This means that one can always cut one of them up into polygonal pieces that can be reassembled to give the other. The "scissors congruence" process is reminiscent of games for children in reassembling different objects from a set of cut pieces of different shapes. Hilbert's problem asks if a similar "cutting" approach could be used to reassemble one tetrahedron to another of the same volume, having the same height and equal base. The answer is no (as Hilbert suspected) and was proven already the next year by M. Dehn [33].

Hilbert Problem 6: *Mathematical treatment of the axioms of physics.*

Hilbert's problem 6 is really more about theoretical physics than mathematics. Fully stated, it is as follows:

> The investigations on the foundations of geometry suggest the problem: To treat in the same manner, by means of axioms, those physical sciences in which already today mathematics plays an important part; in the first rank are the theory of probabilities and mechanics.

This is the only problem of Hilbert in which he himself has been continually engaged over almost 40 years (from 1894 to 1932). Hilbert starts by mentioning the axiomatics of geometry, which goes back to Euclid. In Appendix 6, we have listed the first page of the first book of Euclid to illustrate the basic rule of mathematics: Start any work by first giving definitions so that others know exactly what your words stand for. This more than 2000-year-old tradition is still highly respected in mathematical literature. In contrast, in chemistry, words are used to represent vague concepts, which to different people stand for different things, and many continue to remain unclear for a long time as illustrated, for example, by such common terms as aromaticity, resonance energy, molecular surface, reaction coordinate, etc. to mention just few. Nevertheless, it is very true that they are and will remain very useful, but would probably be even more useful if they would have been defined rigorously!

The German philosopher Immanuel Kant (1724–1804) more than 300 years ago made a significant observation relating to the nature of chemistry [34]:

> In any special doctrine of nature there can only be as much proper science as there is mathematics therein. And since chemistry fails to satisfy this condition, chemistry can be nothing more than a systematic art or experimental doctrine, but never a proper science.

For a full quote, see Appendix 7. It is clear that, at that time and for the following almost 200 years, chemistry was generally perceived to be "beyond" mathematics. But less that 100 years later, A. Crum Brown (1838–1922), who, after finishing study of medicine in Edinburgh, went to visit Bunsen in Heidelberg and continued with his interest in chemistry, came with a more optimistic reference, stating [35],

> ...chemistry will become a branch of applied mathematics; but it will not cease to be an experimental science. Mathematics may enable us retrospectively to justify results obtained by experiment, may point out useful lines of research and even sometimes predict entirely novel discoveries. We do not know when the change will take place, or whether it will be gradual or sudden...

In 1864, Crum Brown developed a diagrammatic representation of chemical compounds [36], which was an influential contribution to the systematic representation of chemical structures. It is interesting to mention that almost 100 years later another "student" of medicine became interested in problems of chemical nomenclature. Joshua Lederberg (1925–2008) devised the DENDRAL computer program [37] (in collaboration with computer scientists Edward Feigenbaum and Bruce Buchanan and chemist Carl Djerassi) for the determination of the molecular structure of organic compounds based on the information from known groups of related compounds. In particular, Lederberg worked on a notational algorithm to represent three-dimensional molecular structures in a form that computers could understand [29]. In 1958, when only 33 years old, Lederberg won the Nobel Prize in Physiology or Medicine for discovering that bacteria can mate and exchange genes.

With the development of quantum chemistry and statistical mechanics in the mid-1950s, the chemical graph theory in the mid-1970s, the use of computers in retro-synthesis, and graphical bioinformatics since the year 2000, the characterization of Kant of chemistry as "*a systematic art or experimental doctrine,*" although still, in part, true, holds only for parts of chemistry. Just as probabilities and mechanics started to transform parts of physics 100 years ago into mathematical physics, so in more recent times quantum chemistry, statistical mechanics, chemical graph theory, the use of computers in retro-synthesis, and graphical bioinformatics started to transform parts of chemistry into mathematical chemistry.

Thus, today, it is understandable that it would be desirable to have axiomatic approaches, which are so characteristic of mathematics, also in other sciences that use mathematics, which includes chemistry. Hilbert himself made significant and fundamental contributions to theoretical physics, which was at that time the nearest candidate for more rigorous formulation. Chemistry has been moving gradually in that direction. In the mid-1950s, chemistry was still viewed as an empirical science. This is not currently the case. In those earlier days, there were rumors among physicists that chemistry was

too difficult for chemists and that physicists had to enter chemistry. Few physicists know
that around 1900 Hilbert was known for saying: *"Physics is too hard for physicists."*

Hilbert Problem 7: *Irrationality and transcendence of certain numbers.*

As is known, irrational numbers, numbers that cannot be expressed as the ratio of
two integers, which were known in the time of Pythagoras (ca. 570 BC–ca. 495 BC), were
not well received by some at that time. Apparently, some Pythagoreans must have
been (irrationally) against irrational numbers and disturbed when it was found that
$\sqrt{2}$ is irrational. According to historical records (although details are of some disput-
able nature), this cost Hippasus of Metapontum (5th century BC), a Pythagorean philosopher, sometimes credited for discovery of irrational numbers, his life as he was
found drowned at a sea with an accepted explanation that this must have been a pun-
ishment of the gods. Algebraic numbers, of which $\sqrt{2}$ is an illustration, are numbers
that are solutions of polynomial equations with rational coefficients. Transcendental
numbers, illustrated with π and e, the base of natural logarithms, are numbers that
cannot arise as solutions of polynomial equations with rational coefficients. This
problem of Hilbert asks about the nature of numbers such as $2^{\sqrt{2}}$ and e^{π}. Are they
transcendental or not? From the fundamental Euler identity that connects the five
basic mathematical constants: 0, 1, π, e, and i (imaginary number $\sqrt{-1}$):

$$e^{i\pi} + 1 = 0$$

it is obvious that $e^{i\pi} = -1$, but what kind of number is e^{π} is the question.

Hilbert Problem 8: *Problems (with the distribution) of prime numbers.*

We mentioned the prime numbers not only because our first problem in this book
relates to prime numbers, but we will later describe some patterns and properties
of sequences of prime numbers, which are one-dimensional mathematical objects
(written as sequences of numbers) as two-dimensional objects, illustrated as maps.
The same idea has been used already in graphical bioinformatics to represent DNA
and protein sequences as maps [38–41].

Hilbert Problem 18: *Building up space from congruent polyhedra.*

This problem relates to infinite networks of polyhedral forms and is related to
topics of crystallography. We will discuss in one of the coming sections the Kepler
conjecture, which stated that in three-dimensional Euclidean space the most dense
packing of spheres of equal size is cubic and hexagonal close packing. As we will
see, this *Kepler conjecture* has been more recently solved, but the solution is based
on a *computer-assisted proof*, which is a novelty, considered by some even to be to
some extent controversial because of its lack of verifiability by a human reader in a

reasonable time. Be this as it may, this problem as well as several other Hilbert problems, will continue to keep the attention of some mathematician for a while.

We decided to offer a collection of problems in this book as problems may stimulate further interest in the topics discussed in this book. Science books without figures tend to be unnecessarily tedious, and science books without problems tend to be unnecessarily stagnant. Our problems are of two kinds: (i) They may broaden the topics considered and strengthen one's interest in continuing to explore particular topics further, and (ii) they may challenge some readers to try to solve some important problems and directly contribute to the advancement of the current science of structural chemistry, which has a number of open pending problems.

It may have been accidental that Hilbert had 23 problems, just the same as the number of definitions with which Euclid started his first book! Be this as it may, we decided, to "adorn" our book with 23 problems—this is not to imply any pretense about the importance of our problems, but a sign of respect for such illustrious persons such as Euclid and Hilbert.

1.5   P VERSUS NP

Let us end the introductory chapter by mentioning an outstanding unsolved problem known as the *P versus NP* problem, considered by most as the most important unsolved problem in computer science. The problem was introduced by S. Cook in 1971 in his seminal paper "The complexity of theorem proving procedures" [42]. *P* and *NP* are two classes of mathematical problems. Problems of class *P* give solutions in polynomial time. For *NP* problems, there is no known algorithm that will give a solution in polynomial time. If, however, a solution is given, it may be possible to verify in polynomial time whether the solution is correct or not. In other words, *P* problems are quickly solvable with the help of a computer, and *NP* problems are ones whose possible solutions can be quickly checked with the help of a computer but that may require an impossibly long time to be solved [43]. The *P* versus *NP* problem is one of the seven Millennium Prize Problems in mathematics that were stated by the Clay Mathematics Institute in 2000. The solution to any of the problems results in a million dollar prize. The solution of this problem would also have some bearing on chemistry because in chemistry also arise some *NP* problems.

That verification for many specific problems is simpler than solving the problems is to be expected in simple cases, but for complex problems, this need not be evident. For example, to determine the symmetry of the graph in Figure 1.5 is an *NP* problem, and this problem was solved using a particular canonical numbering of vertices (to be elaborated later in this book).

When the solution was submitted to the prestigious *Journal of Chemical Physics*, it was rejected on the basis of an absurd recommendation of a referee, for whom the problem reminds him of what is his 10-year-old is doing! In order to challenge such an editorial decision, the author has sent the figure of the graph and his solution to a dozen distinguished researchers in this area (see Appendix 3). This included Professor Coxeter, the recognized leading authority on geometry. All the answers

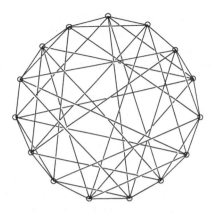

FIGURE 1.5 Regular graph on $n = 15$ vertices of degree 6. What is its symmetry? (Reproduced from "On history of the Randić index and emerging hostility toward chemical graph theory," *MATCH Commun. Math. Comput. Chem.* (2008).)

were supportive, including the letter from Professor Coxeter, reproduced below in complete:

> Dear Dr. Randić,
>   Thanks for your letter about the 15-point graph. I find it interesting and consider the referee to be incompetent; he should not have called your conclusion 'obvious.' It took me about ten minutes to see how to name the 15 vertices with the pairs of 6 symbols so that two of them are joined if they have no common symbol in common.
>   Best regards,
>   H. S. M. Coxeter

In summary, if verification of a problem, which is P, takes the leading expert 10 minutes, clearly solving the problem, which is NP, is not what 10-year-olds are doing. For more details on this story, interested readers may examine Appendix 3 and Ref. [44].

1.6 ON UNSOLVED PROBLEMS

Speaking of unsolved problems, there are many problems that one can think of, but recall E. Bright Wilson's comment [45]: *"Many scientists owe their greatness not to their skill in solving problems but to their wisdom in choosing them."* Hence, before trying to solve a problem one should try to get an idea of how important the problem is. Consider the problem of ordering variables in the construction of the orthogonal basis for a structure space of a particular MRA study. If one is interested in using regression in a search for the most active compound, the ordering of descriptors is not important at all because whatever order one takes, the final regression equations (and their accompanied statistics) are the same. If one is interested in *modeling and interpretation* of the results of a regression analysis, then the ordering of descriptors may be of considerable importance. But the question for a researcher remains: Is this particular modeling so important for me, or are there some other problems that should have higher priority?

The answer depends on the person considering such questions. We recall an anecdote of asking our colleague organic chemist [46], Professor S. Borčić, after he has briefly explained what he has been doing, why it is so important. His answer was, "*If I know of something more important I would do that.*" That ought to be the rule in science, not the exception.

So do we know of something more important than searching for additional molecular descriptors, improving the interpretation of our models, and similar questions? Are there some "giant" problems in QSAR that deserve our attention? And if there are, what are these problems? One may expect that there are at least several highly important problems in any "alive" discipline of science because, if not, the discipline would be "dead" and not "alive." To find them may not, however, be easy.

What we may add to this is that in order to search and find important problems it may be for theoreticians very useful to speak to experimental scientists rather than only theoretical as the former ones tend to be closer to novelty. And the opposite is also true: For experimental scientists, it may be very useful to speak to theoreticians, at least to become aware of the capabilities of the latest theoretical approaches. Be it experimental or theoretical, it is important to talk to other scientists rather than continue to solve the problems that one *knows* how to solve!

Let us end this "philosophical" debate by mentioning one problem in QSAR that we consider that deserves wide attention even if it may not qualify for the title of being a "giant" problem. The problem relates to the structure–property rather than the structure–activity area of QSAR. In structure–activity, one expects that many, if not most, scientists will agree that the search for pharmacophore and its characterization is a very important problem. In the literature there are many illustrations, starting with the work of L. B. Kier in 1967 [47,48] on quantitative characterization of the geometry of pharmacophores or the use of graph theoretical characterizations of pharmacophores reported in the more recent past [49–54]. There have been some controversial views on the history of the origin of the concept of pharmacophores, and interested readers should consult [55,56] and earlier contributions on this topic. Part of the problem is not only who first proposed a particular "definition," but also which particular contribution made an impact and was followed.

The essence of the problem is identification of the molecular fragment whether a connection of disconnected atoms or atomic groups, which can be viewed as critical in determining molecular activity.

An "unsolved problem" that we would like to mention and that we view as being rather important is to explore and find, if it exists, a counterpart of the "pharmacophore" in structure–property correlations. Let us refer to this counterpart as the "physicophore," in the view that the structure–property relationship relates mostly to the physical properties of molecular physicochemical properties. Let us propose this formally as

Problem 2

The question is whether there are some molecular fragments or sets of atoms and atomic groups, at least in the case of some physical properties of ensembles of molecules, which are critical structural factors for the considered molecular property and which could define such fragments as the "physicophore."

We suspect that there may be some such fragments, at least in some cases; otherwise, all physical molecular properties would parallel each other, and it is known that they do not.

REFERENCES AND NOTES

1. E. B. Wilson, Fifty years of quantum chemistry, *Pure Appl. Chem.* 47 (1976) 41–47.
2. A. T. Balaban (Eds.), *Chemical Applications of Graph Theory*, Academic Press, London (1976).
3. N. Trinajstić and I. Gutman, Mathematical chemistry, *Croat. Chem. Acta* 75 (2002) 329–356.
4. A. T. Balaban, Reflections about mathematical chemistry, *Found. Chem.* 7 (2005) 289–306.
5. G. Restrepo and J. L. Villaveces, Mathematical thinking in chemistry, *HYLE* 18 (2012) 3–22.
6. F. Flavitsky, The method of computing the number of saturated monohydroxy alcohols isomers, *J. Russ. Chem. Soc.* 3 (1871) 160–167.
7. G. N. Lewis, *Valence and the Structure of Atoms and Molecules*, Dover Publications, Inc., New York, 1966.
8. *Editors of Merriam-Webster, Webster's Dictionary of Synonyms & Antonyms,* Barnes & Noble Books, New York, 2003.
9. For illustration, see the correspondence with the editor of *J. Chem. Phys.* (Appendix 1), the statement published in the *Journal of Chemical Information and Computer Sciences* (Appendix 2), and a "Letter to Colleagues" (Appendix 3).
10. C. F. Gauss, *Disquisitiones Arithmeticae*, Gerh. Fleischer, Leipzig, 1801.
11. A. Jones, S. A. Morris, and K. R. Pearson, *Abstract Algebra and Famous Impossibilities*, Springer Verlag, New York, 1991, p. 178.
12. W. W. Rouse Ball and H. S. M. Coxeter, *Mathematical Recreations and Essays*, 13th ed., Dover Publications, Inc., New York, 1987.
13. H. S. M. Coxeter, private correspondence.
14. S. M. Ulam, *A Collection of Mathematical Problems*, Wiley, New York, 1980.
15. P. J. Kelly, A congruence theorem for trees, *Pacific J. Math.* 7 (1957) 961–968.
16. F. Harary, On the reconstruction of a graph from a collection of subgraphs, in: *Proceedings of the Symposium Theory of Graphs and its Applications, Prague*, M. Fiedler (Ed.); reprinted by Academic Press, New York, 1964, pp. 47–52.
17. F. H. Clarke, A graph polynomial and its applications, *Discrete Math.* 3 (1971) 305–313.
18. V. Levenshtein, E. Konstantinova, E. Konstantinov, and S. Molodtsov, Reconstruction of a graph from 2-vicinities of its vertices, *Discrete Appl. Math.* 156 (2008) 1399–1406.
19. Y. Zhang, Bounded gaps between primes, *Ann. Math.* 179 (2014) 1121–1174.
20. L. Pisano, *Fibonacci's Liber Abaci: A Translation into Modern English of the Book of Calculation*, Sources and Studies in the History of Mathematics and Physical Sciences, (translated by L. E. Siegler), Springer, New York, 2002.
21. H. Steinhaus, *Mathematic Snapshots (Dover Recreational Mathematics)*, Dover, Mineola, NY (2011).
22. Woodward on the difference between mathematics and chemistry, *Wavefunction*, February 17, 2011. http://wavefunction.fieldofscience.com/2011/02/woodward-on -difference-between.html.
23. L. Euler, Solutio problematis ad geometriam situs pertinentis. *Comment. Acad. Sci. U. Petrop.* 8 (1736) 128–140, reprinted in *Opera Omnia Series Prima*, Vol. 7, pp. 1–10, 1766.
24. A. T. Balaban, D. Babić, and D. J. Klein. W. R. Hamilton: His genius, his circuits, and the IUPAC nomenclature for fullerenes, *J. Chem. Educ.* 72 (1995) 693–698.

25. D. Babić, A. T. Balaban, and D. J. Klein, Nomenclature and coding of fullerenes, *J. Chem. Inf. Comp. Sci.* 35 (1995) 515–526.
26. P. Hansen, Hamiltonian circuits, Hamiltonian paths and branching graphs of benzenoid systems, *J. Math. Chem.* 17 (1995) 15–33.
27. J. Lederberg, Hamiltonian circuits of convex trivalent polyhedral (up to 18 vertices), *Am. Math. Monthly* 74 (1967) 522–527.
28. J. Lederberg, Topological mapping of organic molecules, *Proc. Natl. Acad. Sci. U.S.A.* 53 (1965) 134–139.
29. J. Lederberg, Topology of molecules, in: *The Mathematical Sciences, A Collection of Essays*, National Research Council on Support of Research in the Mathematical Sciences (Ed.), The MIT Press, Cambridge, MA, 1969, pp. 37–51.
30. R. J. Gould, Hamiltonian graphs, Ch. 4.5, in: *Handbook of Graph Theory*, 2nd ed., *Discrete Mathematics and Its Applications*, J. L. Gross, J. Yellen, and P. Zhanf (Eds.) (Series Editor: K. H. Rosen), CRC Press, Boca Raton, FL, 2014, pp. 314–335.
31. P. R. Halmos, The thrills of abstraction, *Two-Year Coll. Math. J.* 13 (1982) 243–251.
32. Opening sentence of an article "Mathematical Problems of David Hilbert," available at http://www.cmi.ac.in/~smahanta/hilbert.html (accessed August 20, 2015).
33. M. Dehn, Uber den Rauminhalt, *Math. Ann.* 55 (1901) 465–478.
34. I. Kant, *Metaphysiche Anfangsgründe der Naturwissenschaft*, Hartkoch Verlag, Riga, 1786.
35. A. Cayley, On the analytical forms called trees, with applications to the theory of chemical combinations, *Rep. Brit. Assoc. Adv. Sci.* 45 (1875), 257–305.
36. A. Crum Brown, On the theory of isomeric compounds, *Trans. Roy. Soc. Edinburgh* 23 (1864) 707–719.
37. J. Lederberg, DENDRAL-64, a system for computer, construction, enumeration and notation of organic molecules as tree structures (NASA CR 57029).
38. M. Randić, J. Zupan, A. T. Balaban, D. Vikić-Topić, and D. Plavšić, Graphical representation of proteins, *Chem. Rev.* 111 (2011) 790–862.
39. M. Randić, M. Novič, and D. Plavšić, Milestones in graphical bioinformatics, *Int. J. Quantum Chem.* 113 (2013) 2413–2446.
40. M. Randić, Graphical representation of DNA as 2-D map, *Chem. Phys. Lett.* 386 (2004) 468–471.
41. M. Randić, N. Lerš, D. Plavšić, S. C. Basak, and A. T. Balaban, Four-color map representation of DNA or RNA sequences and their numerical characterization, *Chem. Phys. Lett.* 407 (2005) 205–208.
42. S. Cook, The complexity of theorem proving procedures, in: *Proc. Third Annual ACM Symposium on Theory of Computing*, 1971, pp. 151–158.
43. M. R. Garey and D. S. Johnson, *Computers and Intractability (A Guide to the Theory of NP-Completeness)*, Freeman, San Francisco, 1979.
44. M. Randić, On the history of the Randić index and emerging hostility towards chemical graph theory, *MATCH—Commun. Math. Comput. Chem.* 59 (2008) 5–124.
45. E. B. Wilson, *Introduction to Scientific Research*, McGraw-Hill, New York, 1952.
46. S. Borčić (1931–1994) Professor at the University of Zagreb, School of Pharmacy, Zagreb, Croatia.
47. L. B. Kier, Molecular orbital calculation of preferred conformations of acetylcholine, muscarine, and muscarone, *Mol. Pharmacol.* 3 (1967) 487–494.
48. L. B. Kier, *Molecular Orbital Theory in Drug Research*, Academic Press, Boston, 1971, pp. 164–169.
49. M. Randić, S. C. Grossman, B. Jerman-Blažič, D. H. Rouvray, and S. El-Basil, An approach to modeling the mutagenicity of nitroarenes, *Math. Comput. Model.* 11 (1988) 837–842.

50. M. Randić, G. A. Kraus, and B. Jerman-Blažič, Ordering graphs as an approach to structure-activity studies, in: *Chemical Applications of Topology and Graph Theory*, R. B. King (Ed.), Elsevier, Amsterdam, *Studies Phys. Theor. Chem.* 28 (1983) 192–205.
51. S. C. Grossman, B. Jerman-Blažič Džonova, and M. Randić, A graph theoretical approach to quantitative structure-activity relationship, *Int. J. Quantum Chem.: Quantum Biol. Symp.* 12 (1985) 123–139.
52. M. Randić, B. Jerman-Blažič, S. C. Grossman, and D. H. Rouvray, A rational approach to the optimal drug design, *Math. Modelling* 8 (1986) 571–582.
53. M. Randić, On characterization of pharmacophore, *Acta Chim. Slov.* 47 (2000) 143–151.
54. M. Randić, Design of molecules with desired properties. A molecular similarity approach to property optimization, in: *Concepts and Applications of Molecular Similarity*, M. A. Johnson and G. Maggiora (Eds.), John Wiley & Sons, New York, 1990, pp. 77–145.
55. O. F. Güner, History and evolution of the pharmacophore concept in computer-aided drug design, *Curr. Top. Med. Chem.* 2 (2002) 1321–1332.
56. J. H. van Drie, Monty Kier and the origin of the pharmacophore concept, *Internet Electron. J. Mol. Des.* 6 (2007) 271–279.

2 Mathematical Chemistry

This book is considering several solved and unsolved problems in chemistry in the structure–property and the structure–activity areas in particular. A number of problems that we will review relate to multiple regression analysis, the oldest statistical approach in sciences. After numerous illustrations relating to QSAR, we will end by mentioning a few solved and unsolved problems in theoretical chemistry in general. The problems that we have selected appear not to be so well known or, if known, have not attracted due attention or may have been even misrepresented. As we will see, the misunderstandings and the tendency to overlook approaches coming from different "schools" of thought have not been confined solely to QSAR studies. They have been also observed in several other areas of theoretical chemistry over the past decades. As illustrations, among others, let us mention the unenthusiastic acceptance of quantum chemistry among physical chemists in the 1930s. This is well reflected in a letter of Professor R. G. Parr to M. Randić relating to initiation of the *Journal of Chemical Physics* (JCP):

> ...In the very early 1930s the leading young American physical chemists got very upset by the disdain for quantum mechanics, molecular spectroscopy, and the like, shown toward such subjects by the editor of the then Journal of Physical and Colloid Chemistry (later JPC), a chap named Bancroft, and the fact that the chemical leadership in the country would not force a change. They {Harold Urey, Joe Mayer, John Kirkwood, Robert Mulliken, others} went to the American Physical Society, and there soon followed the founding of JCP. Take a look at issue 1, volume 1, of JCP, and you will amazed by the high quality. It took forty years or so for JPC to recover (perhaps one should say "be born again")....

Similar lack of appreciation followed the initial lack of acceptance of the exact solution of the inverse problem to X-ray diffraction in late 1950, the hostility toward mathematical molecular descriptors (the so-called topological indices), the hostility to chemical graph theory in general in the 1980s, and the widespread suspicion toward the density functional theory (DFT) as an alternative to the standard quantum chemical calculations, and a lack of interest in Clar's Aromatic Sextet theory by most quantum chemists in the 1990s. We will not be concerned with and will not speculate about why these cases of initiation of novel directions in the research appear to have been overlooked, ignored, or considered unimportant. Instead, in the "Concluding Remarks" of the book, we speculate about the chances that the "Sleeping Beauties" and "Sleeping Giants" of QSAR, theoretical chemistry, chemical graph theory, and bioinformatics be awakened sooner than later in part due to the potential visibility that this book may offer. We also hope that the attention the current book may motivate other chemists to report on solved and unsolved problems in areas of their research and that they seek and try to identify the "Sleeping Beauties" and possibly even the "Sleeping Giants" in areas of chemistry of their interest.

2.1 DISCRETE MATHEMATICS AND CHEMISTRY

The traditional mathematics, the central part of which is calculus, has, soon after its conception in the works of Isaac Newton (1642–1727) and Gottfried Leibnitz (1646–1716), become the major tool for study of physics and has gradually entered physical chemistry and biology in more recent times. The discrete mathematics, a branch of mathematics considering the mathematical properties of sets of discrete objects, has remained mostly unknown among students of physics and chemistry until most recent times. With the exception of the followers of computer science, in which the importance of discrete mathematics has been recognized from the start of the discipline, discrete mathematics, even today, remains little known among the general body of physicists and chemists. This is despite that all physicists and many chemists are familiar with the notable early contributions due to Kirchhoff, who formulated the laws of currents [1], and that many chemists may have known of the early enumerations of isomers even if they were not aware of Flawitsky [2,3]. For the early applications of graph theory to chemistry, see a review of Randić and Trinajstić on lesser known early contributions to chemical graph theory [4].

We may mention that number theory, the branch of pure mathematic concerned with the study of integers, and graph theory, a part of discrete mathematics dealing with binary relations among discrete objects, are currently two of the most researched areas of mathematics. But while number theory has been receiving wider attention, in part due to its association with prime numbers, graph theory is generally less known even among mathematicians, despite that Paul Erdös (1913–1996), one of the rare mathematicians of more recent times who, in addition to contributing to approximation theory, set theory, and probability theory, was particularly heavily involved in graph theory.

Bollobás and Erdös in one of their publications considered the upper bounds on the Randić index [5], which has been known in chemistry as the connectivity index (χ) [6]. Undoubtedly, the work Bollobás and Erdös attracted the interest of many mathematicians in graph theory to the study of the mathematical properties of the Randić index [7,8]. More recently, such work extended toward the study of several additional topological indices. The Randić index or the connectivity index χ is well known in QSAR. It is one of the most often used molecular descriptors in the structure–property–activity studies. As of the middle of 2014, the seminal publication of the connectivity index has passed 2500 citations, in part possibly also due to the continuing interest of mathematicians in the mathematical properties of molecular descriptors.

In this book, we will illustrate the use of two branches of discrete mathematics: the graph theory and partial ordering (including lattices, which are a special case of partially ordered sets). Unfortunately, topics of partial ordering are widely unknown among chemists. This may be due to a lack of courses on graph theory and discrete mathematics for students of chemistry at most universities. We will introduce the elementary concepts, which ought to suffice for chemical topics that we will cover later. We assume no prior knowledge of readers of graph theory [9,10], partial ordering theory [11–13], or the lattice theory [14,15]. The indicated references [9–15] are very useful for elementary introduction, but any introductory textbooks on discrete mathematics will have a section on the topics of partial ordering and lattice theory for those who want to pursue the subject on their own.

The two books on lattice theory that we are recommending have been characterized as "Elementary texts recommended for those with limited *mathematical maturity*." It seems therefore appropriate to clarify somewhat ambiguous notion of "mathematical maturity." What is mathematical maturity? In searching for an answer, we came across a collection of "qualities," presented in the list below (most of which we have extracted from [16,17]). On one side, the list below characterizes mathematicians, and on the other side, it allows the rest of us to get an impression of how many "mathematical" qualities we may possess! We expect many readers may be surprised to find that they may have several hidden mathematical "abilities" themselves, which they have not recognized as such.

Mathematical maturity:
The capacity to generalize from a specific example to broad concept
The capacity to handle increasingly abstract ideas
To learning through understanding
The capacity to separate the key ideas from the less significant
The ability to link a geometrical representation with an analytic representation
The ability to translate verbal problems into mathematical problems
The ability to recognize mathematical patterns
Make and use connections with other problems and other disciplines
Fill in missing details
Recognize and appreciate elegance
Draw a line between what you know and what you don't know

2.2 PARTIAL ORDER

In mathematics, a set is a finite or infinite unordered collection of distinct objects called the elements of the set. Often the order of elements in a set is important. Binary relations are often used for ordering some or all of the elements of sets. A binary relation on a set is a partial order or partial ordering if the relation is reflexive, antisymmetric, and transitive. A set together with a partial ordering is called a partially ordered set or poset. A simple illustration of partial order is ordering objects (people included) with respect to *two* characteristics. Consider ordering of a group of people with respect to their height and weight, allowing that two persons may have the same height or weight, but not both. To be specific, we will consider the following eight persons, denoted by A–H, whose heights (in centimeters) are between 200 and 170, and weights (in kilograms) between 100 and 70:

A (200, 100), B (200, 80), C (195, 90), D (180, 100), E (195, 80), F (185, 85), G (180, 95), H (170, 70)

When we order them with respect to their height, the order is

$$A = B > C = E > F > D = G > H.$$

When we order the same group of people with respect to their weight, then we have

$$A = D > G > C > F > B = E > H.$$

To obtain the partial order for this set (group of people), we have to combine the information of two orderings. As one can see, A dominates the other seven persons, which means A has greater weight *and* height than any other person. Looking at B and C, we see for height B > C, but for weight C > B. So neither B dominates C nor does C dominate B. In the terminology of partial ordering, B and C are incomparable. The same is true for the pairs BD and CD. Similarly, one finds that EF, EG, and FG are also noncomparable. However, B dominates E, C dominates F, and D dominates G because B has greater height *and* weight than E, and C has greater height *and* weight than F, and D has greater height *and* weight than G. Similarly E, F, and G dominate H.

All this can be graphically represented by a Hasse diagram [18] shown in Figure 2.1. Helmut Hasse (1898–1979) was a German mathematician who used such diagrams extensively. In a Hasse diagram, the unconnected vertices (elements) are mutually incomparable, and elements above dominate all those elements below to which they are connected. This particular poset in Figure 2.1 represents a lattice, which is an ordered set having one element at the top that dominates every other element and one element at the bottom, which is dominated by all elements.

In mathematics, the lattice is defined as follows [19]:

A *lattice* is a partially ordered set in which every two elements have the greatest least lower bounds (referred to as *infinum*) and the least lower bounds (referred to as *supremum*).

In a less formal way, one may refer to the top and the bottom vertices as the *master* and the *slave*, respectively.

As another illustration, consider the group of eight combinations of elements of the set {x, y, z}:

A (x, y, z), B (x, y), C (x, z), D (y, z), E (x), F (y), G (z), H (Ø)

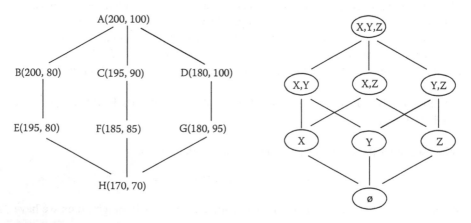

FIGURE 2.1   Hasse diagrams for two partially ordered sets. At left, the set {A–H} and, at right, the set based on combinations of {x, y, z}.

Now we take as the binary relation inclusion. For example A dominates B, C, and D; B dominates E; F dominates H; and so on. We show the related Hasse diagram in Figure 2.1. As one can see, partial order diagrams are of a *qualitative* nature; they show the relative dominance or lack of dominance for a pair of elements, but do not involve numerical values. Nevertheless, they offer useful insights. Let us return to the eight persons of different weight and height and consider their potential as basketball players. One can bet that players B and E have the best natural advantages, being tall and not having excessive weight. Or let us go back in history for 250 years and mention that Maria Theresa (1717–1780)—the only female ruler of the Habsburg royal family, the emperor of Austria, Hungary, Croatia, Bohemia, Mantua, Milan, Lodomeria and Galicia, the Austrian Netherlands, and Parma—naturally, as does every emperor or king, had her royal guard. But to be enrolled as a royal guard of Emperor Maria Theresa, a person had to be 2 m or more in height. As we see from the Hasse diagram in Figure 2.1, only A and B of our eight persons would qualify for this royal duty, so the others, being inferior (dominated by A and B), could, at best, look for some inferior royal duties!

Partial order can be used to relate molecules that are structurally similar, such as is the case with benzene derivatives, the partial ordering of which is illustrated in Figure 2.2. At the top of Figure 2.2 is benzene, which is followed by one mono-substituted benzene. Additional substitution gives three di-substituted benzenes as shown in the third row of the diagram. Adding another substituent results again

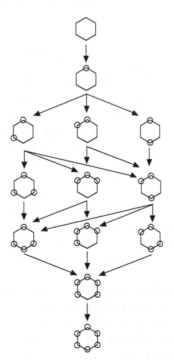

FIGURE 2.2 Partial ordering of the set of substituted benzenes. The circles indicate carbon atoms at which substitution occurs.

in three isomers of tri-substituted derivatives of benzene. At this point, one comes to the middle of the diagram with all isomers having an equal number of substituted and not substituted carbon atoms. The remaining part of the diagram is centrosymmetric, ending with fully substituted benzene. The arrows in the diagram indicate possible synthetic routes.

Just as has been the case with drawing graphs, similarly, drawing a Hasse diagram for a partially ordered set is not unique. As a rule, one has to respect the dominance relationships, placing the dominant elements above the dominated. Sometimes, partially ordered sets are drawn so that the dominant elements are at left of those being dominated (for illustration, see Ref. [20]).

In the next sections, we see additional illustrations of partial ordering, including partial ordering of isomers, partial ordering of smaller benzenoid hydrocarbons, partial ordering of Kekulé valence structures, and use of partial ordering in the search for the phamacophore, the active part of bioactive molecules. Finally, we also illustrate the use of partial ordering for numerical characterization of proteome maps [20].

2.3   ENUMERATION OF KEKULÉ VALENCE STRUCTURES

2.3.1   CATA-CONDENSED BENZENOID HYDROCARBONS

Gordon and Davison described an elegant algorithm for finding the number of Kekulé valence structures K for cata-condensed benzenoid hydrocarbons [21], which can be illustrated on naphthalene, anthracene, and phenanthrene. According to Gordon and Davison, one writes in the first benzene hexagon number 2 because benzene has two Kekulé valence structures. In the next cata-condensed benzene ring, one writes number 3 because naphthalene has three Kekulé valence structures. One continues adding 1 to each successive benzene ring linearly fused until one reaches a "kink" ring. The "kink" is a ring at which the direction of adding fused rings changes. At the "kink" rings, one adds the number in the preceding benzene ring and continues adding that number until the next "kink" ring appears, and so on. This is illustrated in Figure 2.3a on an arbitrary cata-condensed benzenoid hydrocarbon.

The algorithm of Gordon and Davison is well known and admired for its simplicity and elegance. However, if one considers a lengthy cata-condensed system, its application starts to become less attractive because to lengthy repetitive additions at each step of calculation. Can this simple algorithm be further simplified for application to lengthy cata-condensed benzenoid hydrocarbons? Surprisingly, this question was not considered for a full 50 years as it surfaced "accidentally" when one of the present authors was writing a lengthy review on aromaticity in polycyclic conjugated hydrocarbons [22].

The modified Gordon and Davison algorithm is illustrated in Figure 2.3b and c. The modification consists of starting construction by selecting one of the "kink" rings in the molecular interior part. The number of Kekulé valence structures is obtained by summing two contributions: The first term is obtained by applying the Gordon and Davison algorithm to the two terminal segments of the cata-condensed benzenoid and multiplying the values in the rings adjacent to the selected initiation ring. For the case of Figure 2.3 this gives $10 \times 17 = 170$. The second term is obtained

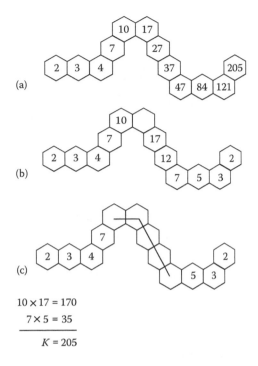

(a)

(b)

(c)

$$10 \times 17 = 170$$
$$7 \times 5 = 35$$
$$K = 205$$

FIGURE 2.3 (a) Illustration of the algorithm of Gordon and Davison for counting Kekulé valence structures in cata-condensed benzenoid. (b and c) A modification of the Gordon-Davison algorithm to give K for cata-condensed benzenoids by selecting a ring in the interior of a molecule as a starting point for calculating K.

by crossing out all benzene rings, which are linearly fused to the initiation ring as shown in Figure 2.3c and multiplying the values in the rings adjacent to the linearly fused rings to the initiation ring. This contribution is $7 \times 5 = 35$, which, combined with the first term, gives $K = 205$.

2.3.1.1 Branched Cata-Condensed Benzenoids

Unbranched cata-condensed benzenoid hydrocarbons are structurally the simplest benzenoid hydrocarbons, for which holds the elegant algorithm of Gordon and Davison for enumeration of Kekulé valence structures. But can the algorithm of Gordon and Davison be extended to branched cata-condensed benzenoid hydrocarbons and peri-condensed benzenoid hydrocarbons?

This question received limited attention in the literature. Here, we want to mention generalization of the Gordon and Davison algorithm to branched cata-condensed benzenoids. The problem was considered by Gordon and Davison, but their solution lacked the elegance of their basic algorithm for cata-condensed benzenoid systems. In Figure 2.4, we illustrate generalization of enumeration of Kekulé valence structures for branched cata-condensed benzenoids, which is simple and closely related to the algorithm of Gordon and Davison for nonbranched cata-condensed benzenoid hydrocarbons. This approach has been mentioned also in the review article on aromaticity [23].

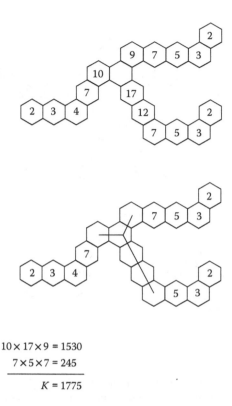

$$10 \times 17 \times 9 = 1530$$
$$7 \times 5 \times 7 = 245$$
$$K = 1775$$

FIGURE 2.4 Generalization of the Gordon-Davison algorithm to give K for branched cata-condensed benzenoids by selecting the branching ring of the molecule as the starting point for calculating K.

The generalization of the algorithm of Gordon and Davison for branched cata-condensed benzenoids has similarity with the just mentioned modified enumeration of K for cata-condensed benzenoids, which was initiated by selecting a "kink" ring in the molecular interior. As illustrated in Figure 2.4, there are two terms that contribute to K. The number of Kekulé valence structures is obtained by summing the contribution obtained by applying the Gordon and Davison algorithm to the three terminal segments of the branched cata-condensed benzenoid and multiplying the values in the rings adjacent to the branching ring. For the case of Figure 2.4, this gives $10 \times 17 \times 9 = 1530$. The second contribution is obtained by crossing out all benzene rings that are linearly fused to the branching ring, as shown in the lower part of Figure 2.4, and multiplying the values in the rings adjacent to the branching ring. This contribution is $7 \times 5 \times 7 = 245$, which, combined with the first term, gives $K = 1775$.

Let us return to Figure 2.3a. Observe that if one stops at the third ring one obtains the number of Kekulé structures of anthracene $C_{14}H_{10}$ ($K = 4$); if one stops at the fourth ring, one obtains the number of Kekulé structures of benzanthracene $C_{18}H_{12}$ ($K = 7$); if one stops at the fifth ring, one obtains the number of Kekulé structures of pentacene $C_{22}H_{14}$ ($K = 10$); and so on. In Figure 2.5, we have illustrated several five-ring cata-condensed benzenoids, and we inserted in their rings the successive counts of their

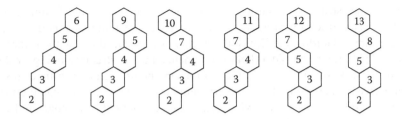

FIGURE 2.5 The count of Kekulé valence structures in smaller benzenoid hydrocarbons, depicted as lattices, based on the algorithm of Gordon and Davison.

Kekulé valence structures. One can easily see that the number of Kekulé valence structures depends on the positioning of "kink" rings and increases with the number of "kink" rings. Of the six benzenoids of Figure 2.5, the largest number of Kekulé valence structures has picene, the last benzenoid of Figure 2.5.

Observe that every internal benzene ring of picene is a "kink" ring, which results in a sequence of Fibonacci numbers: 2, 3, 5, 8, 13 [24,25]. Such benzenoids, the smallest of which are phenanthrene and chrysene, have been referred to as fibonacenes, which are among the most stable cata-condensed benzenoids. It has been pointed out in Ref. [26] there are two different spellings for these compounds: fibonaccenes and fibonacenes, with a suggestion from A. T. Balaban:

> We are aware that *fibonaccenes* would be etymologically more suitable but we suppressed one c for the simplicity and for similarity with the established name "acenes."

However, the name fibonaccenes has also been suggested [27] and used [28].

2.3.2 BENZENOID LATTICES

The four-ring pyrene, as usually drawn, represents the smallest benzenoid lattice if we exclude cata-condensed benzenoid hydrocarbons. The peri-condensed benzenoids shown in Figure 2.6 illustrate additional smaller benzenoid lattices, which we use to introduce a generalization of the Gordon and Davison algorithm for counting

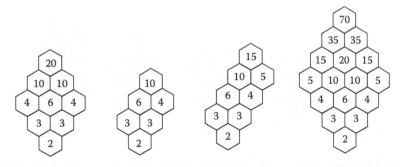

FIGURE 2.6 The count of Kekulé valence structures in smaller "convex" benzenoid lattices by extending the algorithm for cata-condensed benzenoids of Gordon and Davison.

Kekulé valence structures in lattices of peri-condensed benzenoids. Enumeration of Kekulé valence structures in benzenoid lattice structures requires a modification of the algorithm of Gordon and Davison, which was introduced in chemical literature by Randić [22]. Let us consider dibenzocoronene, the first structure in Figure 2.6.

To arrive at K, one inscribes in the bottom ring number 2 and continues inscribing in adjacent rings along the periphery, numbers 3 and 4, as one would do if the periphery of dibenzocoronene is viewed as a cata-condensed benzenoid. In the next step, one starts to fill in blank rings, again starting from the bottom, by adding the K values of two rings below them. The process is reminiscent of the construction of the Pascal triangle, and it can be shown that so obtained entries represent the count of paths from the master to the slave vertex. So one obtains for dibenzocoronene $K = 20$. The same procedure applies also to the remaining benzenoids of Figure 2.6, giving for their count $K = 10$, $K = 15$, and $K = 70$. This procedure applies to all benzenoid lattices that can be characterized as "convex," by which we understand that the inner duals of such benzenoid lattices are convex polygons. The "inner" duals, which were introduced by Balaban and Harary in their work on enumeration of benzenoid hydrocarbons [26], are constructed by replacing each benzene hexagon by a vertex and connecting adjacent vertices, but ignoring the outside region, which in the construction of proper geometrical duals has to be considered.

In the case of the lattices shown in Figure 2.7, the above procedure cannot be completed because when arriving at a "bay" region, one is missing the entry on the ring above the bay. One thus cannot continue with the Pascal-like procedure. So, in order to continue in these cases, we will introduce the rule for the "kink" rings on the periphery of peri-condensed benzenoids: One assigns to the "kink" ring the sum of two preceding rings, just as has been the case in cata-condensed benzenoids. Thus, for the first two lattices of Figure 2.7, the "kink" ring is assigned the value of $2 + 3 = 5$, and for the next two lattices the kink ring becomes $3 + 6 = 9$, which applies to both left and right "bay" regions. The same is again in the fourth lattice, and in the last lattice, the left ring becomes $4 + 10 = 14$, and the ring becomes $6 + 10 = 16$. After assignment of the peripheral kink, ring counting continues applying the Pascal rule. To complete the assignment of the counting numbers to the remaining rings of such lattices, complete the count of K for these lattices as illustrated in Figure 2.8.

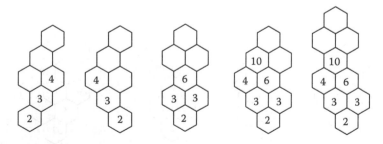

FIGURE 2.7 "Non-convex" peri-condensed benzenoid lattices, for which the count of Kekulé valence structures cannot be completed because of the presence of bay regions on the molecular periphery.

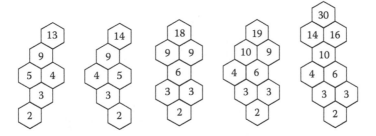

FIGURE 2.8 The completion of the count of Kekulé valence structures for peri-condensed benzenoid lattices of Figure 2.6.

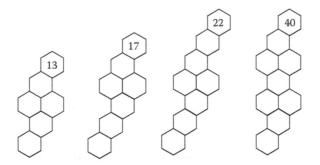

FIGURE 2.9 The Kekulé valence structure counts for several smaller peri-condensed benzenoid non-convex lattices considered by Clar in his booklet *The Aromatic Sextet*.

The procedure outlined applies to all "nonconvex" lattices, the "inner" dual of which is a nonconvex polygon. In Figure 2.9, we have illustrated the Kekulé valence structure counts for peri-condensed benzenoid lattices considered by Clar in his booklet *The Aromatic Sextet* [29]. This time, we only show the result in the top benzene ring leaving it to the readers to fill in the empty benzene rings and verify, as an exercise, that the count shown is correct. In the next section, we will briefly outline the count of Kekulé valence structures for a more general class of benzenoid hydrocarbons.

2.3.3 JOHN AND SACKS ALGORITHM

In this section, we will consider the count of Kekulé valence structures for a large class of more general benzenoid hydrocarbons, such as the benzenoids illustrated in Figures 2.10 through 2.12. John and Sachs arrived at an elegant algorithm for finding the count of Kekulé valence structures for these more general benzenoids [30,31]. One may refer to these more general benzenoids as "multi-lattice" benzenoids because they can be viewed as obtained by overlap of two or more lattice benzenoids. The algorithm of John and Sacks can be viewed as an algebraic extension of the just outlined approach for finding K for lattices, illustrated in Figures 2.6 through 2.9.

The systems considered have to have the same number of vertices at the top and the bottom. In the case of benzo[*ghi*]perylene, the benzenoid shown in Figure 2.10,

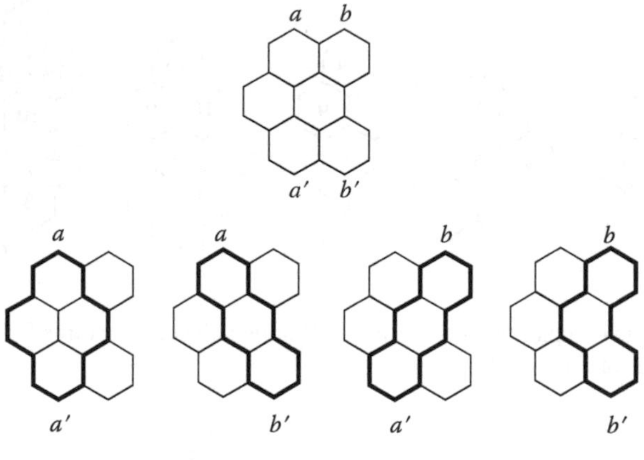

$$\begin{vmatrix} aa' & ba' \\ ab' & bb' \end{vmatrix} = \begin{vmatrix} 6 & 4 \\ 4 & 5 \end{vmatrix} = 6 \times 5 - 4 \times 4 = 14$$

FIGURE 2.10 Calculation of Kekulé valence structures for benzo[*ghi*]perylene using the method of John and Sachs.

there are two top and two bottom vertices, which have labels a, b and a', b', respectively. To find K in this benzenoid, one has to construct a 2×2 determinant, the elements of which are the number of shortest paths from a to a' and b' and from b to a' and b'. As already mentioned, the count of paths in lattices built from fused benzene rings is the same as the number of Kekulé valence structures of respective benzenoids. The construction of the 2×2 determinant of dibenzo[*ghi*]perylene is illustrated in Figure 2.10, giving $K = 14$. In Figure 2.11 are shown the 2×2 determinants for the collection of benzenoids discussed in Clar's *The Aromatic Sextet* book.

In Figure 2.12, we have illustrated determinants for an additional half a dozen benzenoid hydrocarbons that have three top and three bottom vertices. They have also been mentioned in Clar's booklet except for the last structure, which has no Kekulé valence structure and illustrates a hypothetical system, referred to as "concealed" non-Kekuléan benzenoids [32] because, in some cases, it is not immediately apparent that they have no F. A. Kekulé (1829–1896) structures. For some of the benzenoid hydrocarbons shown in Figures 2.11 and 2.12 later in the chapter, there is an even simpler way of finding K, but the approach of John and Sachs is quite general and is often the simplest way of calculating K.

2.4 GRAPH THEORY

Graph theory started with L. Euler and his paper in which he considers the problem of the seven bridges of Königsberg (then East Prussia, now Kaliningrad, Russia) [33]. Already in 1877, only 12 years after F. A. Kekulé (1829–1896) originated his cyclic valence structure of benzene, J. J. Sylvester (1814–1897), the first professor of mathematics at the Johns Hopkins University in Baltimore, recognized the potential

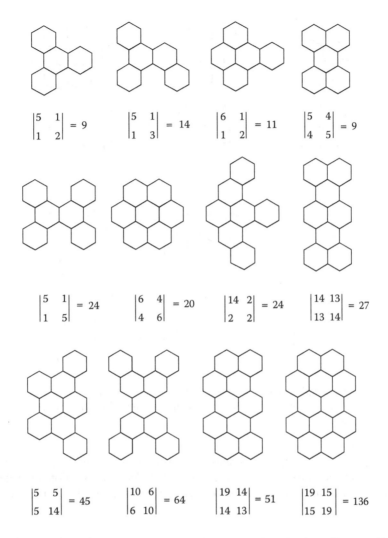

FIGURE 2.11 Calculation of Kekulé valence structures for selection of benzenoid hydrocarbons from the booklet of Clar on aromatic sextets.

of graph theory in chemistry [34,35]. For a useful introduction to the origins of chemical graph theory, we recommend a chapter written by Rouvray in the book on chemical graph theory edited by D. Bonchev and D. H. Rouvray [36]. For a useful introduction to the elements of graph theory, especially written for chemists, we recommend a chapter written by O. E. Polansky [37], a chemist, in the same book.

2.4.1 GRAPHS

A *graph* is defined as a set of discrete elements (objects) V_i, referred to as vertices, and a binary relationship $R_{i,j}$, which indicates whether a pair of elements (vertices) are related or not. Pictorially, vertices are represented by small circles and

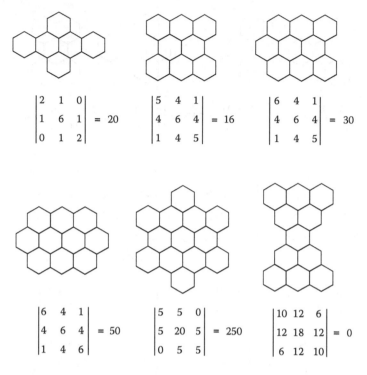

$$\begin{vmatrix} 2 & 1 & 0 \\ 1 & 6 & 1 \\ 0 & 1 & 2 \end{vmatrix} = 20 \qquad \begin{vmatrix} 5 & 4 & 1 \\ 4 & 6 & 4 \\ 1 & 4 & 5 \end{vmatrix} = 16 \qquad \begin{vmatrix} 6 & 4 & 1 \\ 4 & 6 & 4 \\ 1 & 4 & 5 \end{vmatrix} = 30$$

$$\begin{vmatrix} 6 & 4 & 1 \\ 4 & 6 & 4 \\ 1 & 4 & 6 \end{vmatrix} = 50 \qquad \begin{vmatrix} 5 & 5 & 0 \\ 5 & 20 & 5 \\ 0 & 5 & 5 \end{vmatrix} = 250 \qquad \begin{vmatrix} 10 & 12 & 6 \\ 12 & 18 & 12 \\ 6 & 12 & 10 \end{vmatrix} = 0$$

FIGURE 2.12 Calculation of Kekulé valence structures for several benzenoid hydrocarbons from the booklet of Clar on aromatic sextets having three top and bottom vertices requiring 3×3 determinates.

relationships by connecting lines (edges). The absence of a connecting line indicates that the considered pairs of objects are not related. As an illustration, consider the list below. We assume that (i, j) implies (j, i), so there is no need to list (8, 1), (9, 1), (10, 1), etc.

> Hydrogen suppressed molecular graph of adamantane defined by the list of
> vertices and the list of edges:
> V = 1, 2, 3, 4, 5, 6, 7, 8, 9, 10
> R = (1, 8), (1, 9), (1, 10), (2, 6), (2, 7), (2, 10), (3, 5), (3, 7), (3, 9), (4, 5), (4, 6),
> (4, 8)

Here we have a graph on 10 vertices and 12 edges, which we illustrated at the top of Figure 2.13 in two different ways. At left, we have arranged vertices 1 through 10 in a circle and drawn the connecting lines, and at right, the same graph is drawn more symmetrically. Finally, in Figure 2.14, we show the same graph depicted even more symmetrically as an object in 3-D space. It represents a carbon skeleton of adamantane $C_{10}H_{16}$, which can be viewed as a unit part of an infinite diamond lattice. Incidentally, adamantane was synthesized for the first time in Zagreb, Croatia, in 1941 by Prelog and Seiwerth [38].

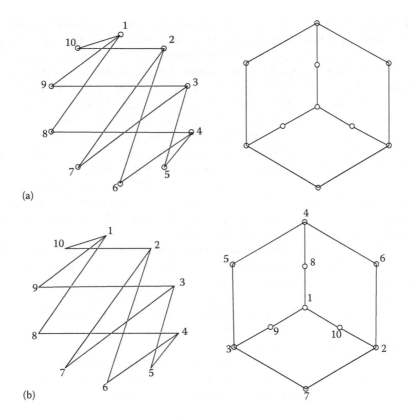

(a)

(b)

FIGURE 2.13 (a) Two graphs having $n = 10$ vertices. It is not obvious that the graph on the left has the same connectivity as the graph on right; hence, it represents the same graph depicted in two different ways. (b) The same two graphs with assigned labels that show all vertices have the same neighbors in both graphs.

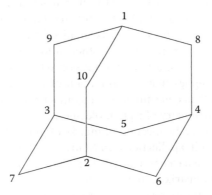

FIGURE 2.14 The graph of Figure 2.13 depicted as a 3-D structure of adamantane.

The graph of Figure 2.13 is not complicated; thus, one can verify that the two different drawings of the graph represent the same graph. In the language of mathematics, the two graphs of Figure 2.13 are isomorphic. One way to demonstrate this is to assign labels 1 through 10 to the unlabeled graph so that the resulting list of neighbors in both graphs is the same. If, for example, one assigns the labels 1, 2, 3, and 4 to the four vertices of degree three as is shown in the lower part of Figure 2.13, one can easily assign the remaining labels for the vertices in the second graph to agree with list of neighbors of the first graph. Thus, the unlabeled vertex between 1 and 2 should have label 10 because in the list of neighbors vertex 10 has neighbors 1 and 2. Similarly, the unlabeled vertex between 1 and 3 should have label 9, and so on. Observe also that the suggested procedure shows that the graph at right in Figure 2.13 has higher symmetry than its geometrical representation because the four vertices of degree three are not equivalent in Figure 2.13, but in fact, they are fully equivalent as one can see from Figure 2.14.

The graph isomorphism problem, also referred to as "the graph isomorphism disease" [39,40], in the view that researchers "infected" with this problem not only had a hard time solving the problem, but apparently also had a hard time quitting solving the problem. The problem appears to be on the border between P and NP (polynomial and nonpolynomial) problems, which may grow exponentially with the increase in the size of the graphs considered. What is known for certain is that the subgraph isomorphism problem is NP complete.

Isomorphism testing consists of demonstrating whether the two graphs, like those shown at the top of Figure 2.13, are isomorphic or not. To each graph, one can associate an adjacency matrix A, the element $[A]_{ij}$ of which is defined by

$$[A]_{ij} = \begin{cases} 1 & \text{if vertices } i \text{ and } j \text{ are adjacent} \\ 0 & \text{otherwise.} \end{cases}$$

The problem is to find common vertex labels such that the two adjacency matrices are equal.

Observe two important aspects of graphs: (i) They can be drawn arbitrarily, and (ii) vertices can be labeled arbitrarily. Thus, the adjacency matrix of a graph of adamantine shown in Matrices 2.1 can be written in many different forms. If a graph on n vertices represents an object without symmetry, there will be $n!$ different forms for the adjacency matrix, but this number is smaller for graphs having symmetry. It is extremely difficult in a general case to establish for a given two matrices of the same size if they belong to two different graphs or to the same graph just numbered differently. This also implies that determination of the symmetry of a graph is a more difficult problem than determination of symmetry of objects having a definite geometry, such as the symmetry of rigid molecules.

The adjacency matrix is the most basic matrix of graph, but there are additional matrices that arise often in applications. In Matrices 2.2, we have listed for norbornane, in addition to the adjacency matrix, also the distance matrix D, the valence matrix V, the Laplacian matrix L, the adjacency matrix squared (A^2) and raised to third power (A^3), the centrality matrix CV, and the canonical adjacency matrix (A^*).

Different forms of the adjacency matrix of adamantine or tricyclo[3. 3. 1. (3,7)] decane

| Standard Chemical Labeling | Canonical Numbering of Carbon Atoms |

$$
\begin{bmatrix}
0 & 1 & 0 & 0 & 0 & 1 & 0 & 1 & 0 & 0 \\
1 & 0 & 1 & 0 & 0 & 0 & 0 & 0 & 0 & 0 \\
0 & 1 & 0 & 1 & 0 & 0 & 0 & 0 & 1 & 0 \\
0 & 0 & 1 & 0 & 1 & 0 & 0 & 0 & 0 & 0 \\
0 & 0 & 0 & 1 & 0 & 1 & 0 & 0 & 0 & 1 \\
1 & 0 & 0 & 0 & 1 & 0 & 0 & 0 & 0 & 0 \\
0 & 0 & 0 & 0 & 0 & 0 & 0 & 1 & 1 & 1 \\
1 & 0 & 0 & 0 & 0 & 0 & 1 & 0 & 0 & 0 \\
0 & 0 & 1 & 0 & 0 & 0 & 1 & 0 & 0 & 0 \\
0 & 0 & 0 & 0 & 1 & 0 & 1 & 0 & 0 & 0
\end{bmatrix}
\begin{bmatrix}
0 & 0 & 0 & 0 & 0 & 0 & 0 & 1 & 1 & 1 \\
0 & 0 & 0 & 0 & 0 & 1 & 1 & 0 & 0 & 1 \\
0 & 0 & 0 & 0 & 1 & 0 & 1 & 0 & 1 & 0 \\
0 & 0 & 0 & 0 & 1 & 1 & 0 & 1 & 0 & 0 \\
0 & 0 & 1 & 1 & 0 & 0 & 0 & 0 & 0 & 0 \\
0 & 1 & 0 & 1 & 0 & 0 & 0 & 0 & 0 & 0 \\
0 & 1 & 1 & 0 & 0 & 0 & 0 & 0 & 0 & 0 \\
1 & 0 & 0 & 1 & 0 & 0 & 0 & 0 & 0 & 0 \\
1 & 0 & 0 & 0 & 0 & 0 & 0 & 0 & 0 & 0 \\
1 & 1 & 0 & 0 & 0 & 0 & 0 & 0 & 0 & 0
\end{bmatrix}
$$

$$(2.1)$$

All these matrices will be described in more detail later in this book. The elements of the distance matrix list the number of edges between two vertices; the valence matrix is diagonal, the elements of which show the valence of vertices: the Laplacian matrix is given by $L = V - A$. The centrality matrix counts for each edge pair of *common vertices* (CV) at the same distances from the edge. Finally, the canonical adjacency matrix is the adjacency matrix based on specially selected vertex labels. Vertex labels are chosen to produce for the adjacency matrix the smallest binary number if its rows are concatenated and read from left to right and from top to bottom [41,42]. The standard numbering of carbon atoms in norborane is shown in Figure 2.15a, and the canonical labels for carbon atoms are shown in Figure 2.15b. Before elaborating on the canonical labeling of vertices, their properties, advantages, and applications in structural chemistry, we will briefly review the topic of chemical nomenclature.

2.5 CHEMICAL NOMENCLATURE

For an introduction to the problems and the early developments of chemical nomenclature, interested readers can examine an article of Goodson [43], which starts with historical comments and continues with chemical line notations and application of graph theory to chemical notation. The importance of chemical notation became acutely recognized with the use of computers in structure and substructure searching. One problem to resolve was setting rules for the assignment of labels to atoms (vertices of molecular graphs). An interesting early solution to this was the Morgan algorithm [44]. Morgan introduced the notion of the augmented vertex valence, a simple method to assign to vertices different numerical values. This is an iterative procedure that starts so that one replaces the vertex valence by the sum of the valences of adjacent vertices. At each following step, one replaces the current vertex values by the sum of the values of its adjacent vertices. The process continues until all symmetry nonequivalent vertices in the molecular structure receive different numerical values. At that stage, one can assign unique labels $1 - n$ (where n is the number of vertices in the molecular graph) to atoms in a molecule except for symmetry equivalent atoms.

Norbornane matrices: adjacency matrix and six additional matrices, which arise in various applications of chemical graph theory

Adjacency Matrix A

$$
\begin{bmatrix}
0 & 1 & 0 & 0 & 0 & 1 & 0 \\
1 & 0 & 1 & 0 & 0 & 0 & 0 \\
0 & 1 & 0 & 1 & 0 & 0 & 1 \\
0 & 0 & 1 & 0 & 1 & 0 & 0 \\
0 & 0 & 0 & 1 & 0 & 1 & 0 \\
1 & 0 & 0 & 0 & 1 & 0 & 1 \\
0 & 0 & 1 & 0 & 0 & 1 & 0
\end{bmatrix}
$$

Distance Matrix D

$$
\begin{bmatrix}
0 & 1 & 2 & 3 & 2 & 1 & 2 \\
1 & 0 & 1 & 2 & 3 & 2 & 2 \\
2 & 1 & 0 & 1 & 2 & 2 & 1 \\
3 & 2 & 1 & 0 & 1 & 2 & 2 \\
2 & 3 & 2 & 1 & 0 & 1 & 2 \\
1 & 2 & 2 & 2 & 1 & 0 & 1 \\
2 & 2 & 1 & 2 & 2 & 1 & 0
\end{bmatrix}
$$

Valence (Degree) Matrix V

$$
\begin{bmatrix}
2 & 0 & 0 & 0 & 0 & 0 & 0 \\
0 & 2 & 0 & 0 & 0 & 0 & 0 \\
0 & 0 & 3 & 0 & 0 & 0 & 0 \\
0 & 0 & 0 & 2 & 0 & 0 & 0 \\
0 & 0 & 0 & 0 & 2 & 0 & 0 \\
0 & 0 & 0 & 0 & 0 & 3 & 0 \\
0 & 0 & 0 & 0 & 0 & 0 & 2
\end{bmatrix}
$$

Laplacian Matrix L

$$
\begin{bmatrix}
2 & -1 & 0 & 0 & 0 & -1 & 0 \\
-1 & 2 & -1 & 0 & 0 & 0 & 0 \\
0 & -1 & 3 & -1 & 0 & 0 & -1 \\
0 & 0 & -1 & 2 & -1 & 0 & 0 \\
0 & 0 & 0 & -1 & 2 & -1 & 0 \\
-1 & 0 & 0 & 0 & -1 & 3 & -1 \\
0 & 0 & -1 & 0 & 0 & -1 & 2
\end{bmatrix}
$$

A^2

$$
\begin{bmatrix}
2 & 0 & 1 & 0 & 1 & 0 & 1 \\
0 & 2 & 0 & 1 & 0 & 1 & 1 \\
1 & 0 & 3 & 0 & 1 & 1 & 0 \\
0 & 1 & 0 & 2 & 0 & 1 & 1 \\
1 & 0 & 1 & 0 & 2 & 0 & 1 \\
0 & 1 & 1 & 1 & 0 & 3 & 0 \\
1 & 1 & 0 & 1 & 1 & 0 & 2
\end{bmatrix}
$$

A^3

$$
\begin{bmatrix}
0 & 3 & 1 & 2 & 0 & 4 & 1 \\
3 & 0 & 4 & 0 & 2 & 1 & 1 \\
1 & 4 & 0 & 4 & 1 & 2 & 4 \\
2 & 0 & 4 & 0 & 3 & 1 & 1 \\
0 & 2 & 1 & 3 & 0 & 4 & 1 \\
4 & 1 & 2 & 1 & 4 & 0 & 4 \\
1 & 1 & 4 & 1 & 1 & 4 & 0
\end{bmatrix}
$$

Canonical Adjacency Matrix A *

$$
\begin{bmatrix}
0 & 0 & 0 & 0 & 0 & 1 & 1 \\
0 & 0 & 0 & 0 & 1 & 0 & 1 \\
0 & 0 & 0 & 1 & 0 & 0 & 1 \\
0 & 0 & 1 & 0 & 1 & 0 & 0 \\
0 & 1 & 0 & 1 & 0 & 1 & 0 \\
1 & 0 & 0 & 0 & 1 & 0 & 0 \\
0 & 0 & 1 & 0 & 0 & 1 & 0
\end{bmatrix}
$$

(2.2)

A somewhat distantly related procedure was introduced by Randić and Plavšić for their characterization of molecular complexity [45]. Complexity, just as with several other notions on molecules, such as *beauty*, is not easy to define and tends to depend on the eye of the beholder. It is therefore not surprising that different authors introduced different numerical characterizations of complexity as its definition remains

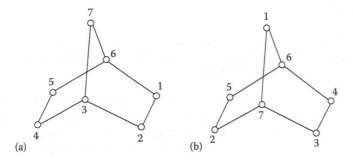

FIGURE 2.15 Standard numbering of atoms in norbornane (a) and the canonical numbering of atoms that produces the smallest binary code for norbornane (b).

elusive. In Figure 2.16 and Table 2.1, we show evaluation of the *complexity* index of Randić and Plavšić on two small graphs having six vertices [45]. The proposed index of complexity is atom additive quantity, which is obtained by dividing the sums of valences at increasing distances from the vertex considered by powers of 2. Observe how this *simple* idea discriminates between different vertices of a graph and allows one to speak of atom complexity, the sum of which gives molecular complexity. As an unsolved problem or a project, we would like to mention:

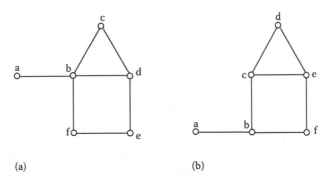

FIGURE 2.16 Two graphs on six vertices whose augmented valences are listed in Table 2.1.

TABLE 2.1
Construction of Augmented Valences for Vertices of Two Graphs Depicted in Figure 2.16

Vertex	Graph A	Graph B
a	$1 + 4/2 + 7/4 + 2/8 = 5$	$1 + 3/2 + 5/4 + 5/8 = 4.375$
b	$4 + 8/2 + 2/4 = 8.5$	$3 + 6/2 + 5/4 = 7.25$
c	$2 + 7/2 + 5/4 = 6.75$	$3 + 8/2 + 3/4 = 7.75$
d	$3 + 8/2 + 3/4 = 7.75$	$2 + 6/2 + 5/4 + 1/8 = 6.375$
e	$2 + 5/2 + 6/4 + 1/8 = 6.125$	$3 + 7/2 + 3/4 + 1/8 = 7.375$
f	$2 + 6/2 + 6/4 = 6.5$	$2 + 6/2 + 6/4 = 6.5$

Problem 3

Explore if the complexity algorithm mentioned above holds for selected classes of molecules, such as (i) acyclic alkane graphs, (ii) cata-condensed benzenoid graphs, or (iii) smaller polycylic graphs.

Clearly all vertices in vertex and edge transitive graphs will have the same vertex "complexity," so calculation of vertex complexities in irregular graphs may offer some measure of deviations, the degree of irregularity of irregular graphs.

Wiswesser line notation [46–48] is another important historical development in chemical notation that offered useful service in early searches for chemical compounds. More recent developments were more mathematical, based on suitable modifications of graph theoretical concepts. In the case of structurally related molecules, simple nomenclatures have been developed for benzenoid hydrocarbons by Balaban based on the concept of *inner duals*. In this approach, benzene rings are represented by single vertices that are connected if benzene rings have a common CC bond [26,49]. Balaban and Schleyer extended the dualist graph principle to polymantane, which is built from adamantane units superimposed on a diamond lattice [50,51].

In the case of less regular families of compounds of interest, consider extension of the concept of the center of the graph. The concept of the center of graphs was introduced by Jordan and Sylvester independently [52,53]. This concept was generalized to cyclic graphs, thus arriving at numerical indices that indicate the closeness of the selected vertex from the center of the graph [54–61]. The notion of the centrality of vertices has received considerable attention in more recent studies of complex networks. One of the latest additions in the literature is a structural approach in which the notion of centrality is based on the count of pairs of neighbors at the same distance from each vertex. It has been shown that vertices having a high count of neighbors at the same distance are closer to the center of the network [62]. Here we may mention work of Dehmer and coworkers [63–65] who introduced a novel class of graph entropies associated with measures based on assigning probability to each vertex of a graph. Another graph-based approach to chemical nomenclature is the nodal nomenclature of Lozac'h, Goodson, and coworkers [66–69]. Even mathematicians had interest and considered development of chemical nomenclatures as has been demonstrated by R. C. Read [70].

Finally, we should mention the "simplified molecular-input line-entry system" (SMILES), a line notation for chemical molecules using short ASCII strings, which was initiated by Weininger in 1980 [71–73]. ASCII is abbreviation for the American Standard Code for Information Interchange. Originally, the code included the numbers 0–9, the letters a–z and A–Z of the English alphabet, some basic punctuation symbols, a blank space, and some control codes. SMILE is a graph-based computational procedure, which, as is often the case in chemical graph theory, to remove hydrogen atoms from considerations. In the case of cyclic systems, SMILES considers spanning trees, which are acyclic subgraphs of cyclic graphs that include all vertices but not all edges (see the next section for more detailed explanation). Thus, this nomenclature is based on the nomenclature of acyclic systems, and in cyclic systems, all cycles have been broken and labels are used to indicate the connected nodes. In Chapter 8 of this book, we will outline the ring closure matrix, a matrix of reduced size, which contains all information on cyclic systems and allows their reconstruction. This illustrates that nomenclature based on spanning trees can be further condensed without loss of information.

2.5.1 ON UNIFORM REPRESENTATIONS OF MOLECULES

As we have seen, there have been diverse approaches to the design of molecular codes, some of which had different specific goals in mind. Here we will outline one such approach which is aimed at a uniform representation of molecules. Good structure representations for modeling should fulfill the following demands [74]:

1. Each compound should have one and only one code, with different codes for different structures (uniqueness).
2. Each compound should be represented by the same number and type of variables (uniformity).
3. One should be able to retrieve the structure from the code (reversibility).
4. Code should be independent of translation or rotation of a structure (translational and rotational invariance).

Although it is not difficult to satisfy, individually, any of the four demands, it is a difficult and unsolved problem to simultaneously satisfy all the four conditions. One of the problems is that one wants to represent molecules of *different* size by *the same number* of variables. For general cases, structure representation should not depend on the choice of coordinate origin. We will briefly illustrate an application of the proposed approach to model biological activity. We have selected a well-defined group of 28 flavonoid analogues (Table 2.2) considered by Novič and Vračko [75], who applied as modeling techniques the counter-propagation artificial neural network and multiple linear regression. The main skeleton and chemical structure of 28 flavonoid derivatives considered acting as inhibitors of the enzyme p56lck protein tyrosine kinase by binding on the enzyme receptor active site [76–79] are shown in Figure 2.17. From the modeling results, it follows that the suggested representation is well suited for QSAR studies. However, in view of the presence of heteroatoms (oxygen and nitrogen) in additional to pure geometry, electronic factors of the compounds must be included in the representation of molecular structures because they play an important role in the reaction mechanism inhibition of the enzyme p56lck.

Descriptors used in the present study are calculated from the information about the connections between the atoms, from atomic 3-D coordinates, and from information about atomic electronic properties. In this work were used (i) graph theoretical (topological descriptors), (ii) geometric descriptors, (iii) electrostatic descriptors, (iv) 3-D descriptors for spectrum-like representation, and (v) density of states spectra. We elaborate on topological descriptors adopted in this particular work in view of a multitude of topological indices available and, in this way,

TABLE 2.2

Main Skeleton and Chemical Structure of 28 Flavonoid Derivatives

ID Number	Substituents	ID Number	Substituents
1	6-OH,5,7,4'-NH$_2$	15	7-OH,6,8,4'-NH$_2$
2	6-OH,5,7,3'-NH$_2$	16	7-OH,6,8,4'-NO$_2$
3	6-OMe,8,3'-NH$_2$	17	5-OMe,8,4'-NH$_2$
4	6,4'-NH$_2$	18	7-OH,8,4'-NO$_2$
5	6,8,4'-NH$_2$	19	6,4'-NO$_2$
6	6-OH,8,4'-NH$_2$	20	8,4'-NO$_2$
7	8,4'-NH$_2$	21	7-OH,6,4'-NO$_2$
8	7-OH,6,4'-NH$_2$	22	5-OH,8,4'-NO$_2$
9	6,3'-NH$_2$	23	6-OMe,8,4'-NO$_2$
10	5-OH,6,4'-NH$_2$	24	5-OMe,8,4'-NO$_2$
11	5-OH,8,4'-NH$_2$	25	6,8,4'-NO$_2$
12	7-OH,8,4'-NH$_2$	26	6-OH,8,4'-NO$_2$
13	6-OMe,8,4'-NH$_2$	27	5-OH,6,4'-NO$_2$
14	7-OH,6,3'-NH$_2$	28	6-OH,5,7,4'-NO$_2$

point to descriptors that we found useful. We also describe construction of 3-D descriptors for spectrum-like representation and thus illustrate their uniqueness and uniformity.

2.6 MOLECULAR DESCRIPTORS

Molecular descriptors are derived from the connectivity characteristics of molecular graphs and describe the atomic connectivity in the molecule [80–95]. Graphs are represented by the adjacency matrix and the distance matrix, in which the distances are given by the count of edges separating vertices. The topological descriptors used as components of structure representation vectors reflect structural features, such as size, shape, symmetry, branching, and cyclicity. Some descriptors constitute families, the members of which are of different order. *Order* is determined by the length of the paths between the vertices considered. In the case of the first-order connectivity index proposed by Randić [6], atomic factors are determined by the reciprocal of the square root of the degree of vertices.

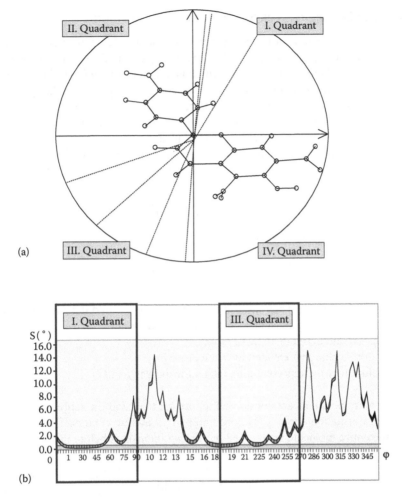

FIGURE 2.17 Spectrum-like structure representation (b) of the 6-OH,5,7,4'-NO flavonoid derivative as obtained from the XY projection of oriented 3-D structure (a).

The Wiener index [86] can be expressed in terms of the distance matrix [87] and equals the half-sum of all distance matrix entries. Randić [88] and Kier and Hall indices of order 0–3 [89] are calculated from coordination numbers of atoms or from values of atomic connectivity. The Kier shape index (order 1–3) [90] depends on the number of skeletal atoms, molecular branching, and the ratio of the atomic radius and the radius of the carbon atom in the sp^3 hybridization state. The Kier flexibility index [90] is derived from the Kier shape index. The Balaban index depends on the row sums of the entries of the distance matrix and the cyclomatic number [92,93]. The information content index and its derivatives (order 0–2) are based on the Shannon information theory [95]. Modifications of the information content index are structural information content, complementary information content, and bond information content [96].

2.6.1   3-D DESCRIPTORS FOR SPECTRUM-LIKE REPRESENTATION

3-D descriptors for spectrum-like representation of a molecular structure, defined by 3-D coordinates of its atoms, are obtained by a projection of all atomic centers of a molecule onto a sphere in the center of which is the molecule [97,98]. The projection ray from the central point of the sphere defines a pattern of points on the sphere, and each point represents a particular atom. Then, the center of each atom is represented by a "bell-shaped" function, the intensity of which is determined by the distance between the coordinate origin and a particular atom. The Mulliken charge of the particular atom is an additional parameter incorporated into this function. As the "bell-shaped" function of atoms has been taken Lorentzian curve with the form:

$$s_i(\varphi_i,\vartheta_1) = \rho_i / \left\{(\varphi_j - \varphi_i)^2 + \sigma_i^2\right\} + \rho_i / \left\{(\vartheta_1 - \vartheta_i)^2 + \sigma_i^2\right\}$$

$$\varphi_j = 2\pi/k,\ldots 2\pi j/k,\ldots 2\pi; \qquad j = 1, k \tag{2.3}$$

$$\varphi_1 = \pi/(k/2),\ldots \pi l/(k/2),\ldots \pi; \quad 1 = 1, k/2$$

where $s_i(\varphi_i,\vartheta_1)$ is the "spectrum intensity" related to atom i, and the parameters are
  ρ_i = distance between the center of the sphere and atom i
  φ_i,ϑ_1 = polar and azimuth angle of atom
  i; σ_i = atomic charge (extended by 1) on atom i
  k = resolution of the representation (steps for indices j and l)

If only atom positions are considered, σ_i are set to 1, and the representation contains only information about 3-D coordinates of all atomic centers. The total intensity for the entire molecule is obtained by adding intensities belonging to individual atoms:

$$S(\varphi_i,\vartheta_1) = \Sigma s_i(\varphi_i,\vartheta_1) \tag{2.4}$$

This representation fulfills the following demands:

1. Uniqueness (unique and distinct code for different structures).
2. Uniformity, that is, the number, type, and domain of variables are the same for all structures regardless of the number of atoms they have.
3. Reversibility, which means that by proper initial orientation of the structures, atomic coordinates can be obtained back from the code.

In practice, rather than using the projection on the entire sphere, the projections on three perpendicular equatorial trajectories have been considered. Thus, the structure is represented by three "spectra," each defined on the interval (0, 2). By selection of k equidistant points in this interval, the structure is represented with three k-dimensional vectors. The larger the number k, the better the resolution. If k is smaller than the number of atoms in the largest molecule in the set considered, some information will be lost. Because the largest part of the skeletons of the molecules considered in that particular in study is planar, it suffices to consider only the projection on one trajectory (X–Z plane). If Mulliken charges on atoms i are incorporated, then recovering the atom positions from the code is more computer-intensive.

In Figure 2.17b, we show the spectrum-like structure representation of 6-OH, 5,7,4X-NO flavonoid derivative, obtained from the XY projection of oriented 3-D structure (Figure 2.17a). Three dotted lines in the first quadrant (having three atoms) and four dotted lines in the third quadrant, connecting the center of the circle (having four atomic centers), mark the appearance of peaks in the "spectrum" (Figure 2.17b). If the azimuth angles of the two atoms are approximately equal, two peaks are merged into one. At positions $\varphi = 84$ and $\varphi = 85$, two atoms are represented by one peak; however, the intensity is the sum of contributions of both atoms.

With this, we will end this section on uniform representations of molecules. As one can see, the spectral representation may vary with increased resolution. In Figure 2.17, one can easily detect 20 peaks and possibly two shoulders, but the total number of atoms in this molecule is 33, so about a dozen atomic centers overlap or are too close to the same projection lines to be sufficiently discriminated, but as already mentioned, molecular structures can always be reconstructed from the spectral representation. This is one of the important features showing that there is no loss of information accompanying spectral-like uniform representations of molecules. For more on spectral-like uniform representations of molecules one should consult references listed in this section.

2.7 CANONICAL LABELING OF VERTICES

In this section, we illustrate the general approach to finding if two graphs, which may appear different, are different or illustrate the same graph. We have labeled vertices in the two graphs of Figure 2.18 so that readers can easily verify that the two pictorial representations depict the same graph as they both produce the same list of adjacent neighbors.

We have selected the particular labeling of vertices producing the adjacency matrix, the rows of which, when read from left to right and from top to bottom, give the smallest binary code for this graph if compared with other symmetry-unrelated

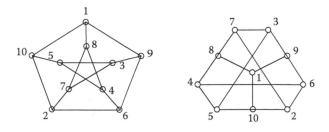

FIGURE 2.18 Two alternative forms of the Petersen graph, which represents trigonal bypiramidal XY_5 axial to equatorial rearrangements. Observe that the labels have the same neighbors in both graphs, demonstrating their isomorphism.

numberings of this graph [99,100]. We refer to this labeling of vertices of a graph producing the smallest binary code for a graph as the canonical labeling. The labeling of diamantane in Figure 2.19 has also been selected so that the adjacency matrix produces the smallest binary code for diamantane.

We know that all vertices in the Petersen graph are equivalent, which will help in the construction of the canonical labeling and also allows one to determine the symmetry of the graph [101]. Clearly, label 1 can be placed at 10 places, label 10 has three possible (equivalent) sites adjacent to 1, label 9 only two, and there is but one site for label 8. Thus far, for the canonical assignment of labels 1, 10, 9, and 8 there are $10 \times 3 \times 2 = 60$ possibilities. The next smallest label 2 has to be adjacent to 10, which gives two additional possibilities. However, as one can verify, all the successive labels have only one site available. For example, labels 6 and 7 have to be adjacent to vertex 2, but only if 7 is placed as shown, one will make it possible for label 3 to have vertices with the largest labels as neighbor, which secures the smallest binary number for the third row. Thus, it follows that the symmetry group of the Petersen graph has 120 symmetry operations [102].

For most graphs of chemical interest, finding the canonical labeling, illustrated with the Petersen graph in Figure 2.18, need not be difficult. However, finding the canonical labeling for general graphs remains a difficult problem and may not turn out to be as practical as it has been for molecular graphs. Also the problem of determining the symmetry of a graph may take more time for graphs representing degenerate rearrangements, such as the already mentioned Petersen graph and the graph of Figure 2.20, representing the rearrangement of $CH_3-CH_2^+$ constructed by Balaban and coworkers in 1966, which was the first rearrangement graph in chemical literature [103].

If one wishes to get some experience with finding labels for molecular graphs, one may try to find canonical labels for the 18 isomers of octane, which is a fair exercise in the search for canonical labels of more complex graphs. Eight factorials is 40,320, but for more than half of the isomers of octane one can find canonical labels in a single construction! This is the case with n-octane, 2M-C_7, 2,2MM-C_6, 2,5MM-C_6, 3,3MM-C_6, 3E,2M-C_5, 2,2,4MMM-C_5, 2,3,3MMM-C_5, 2,3,4MMM-C_5,

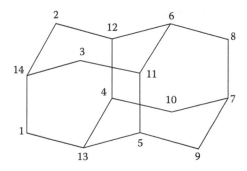

FIGURE 2.19   3-D carbon skeleton of the diamantane $C_{14}H_{20}$ molecule, which consists of two fused adamantane units with canonical labels.

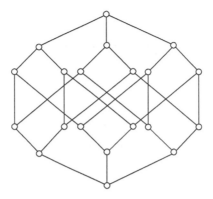

FIGURE 2.20 The Desargues–Levi graph, which represents degenerate rearrangeents of CH_3–CH_2^+, the first graph constructed by Balaban and coworkers to represent degenerate rearrangements in chemistry.

and 2,2,3,4MMMM-C_4. It appears that the most "difficult" isomer of octanes, which requires testing a dozen alternatives, is 3,4MM-C_6, and both 2M-C_7 and 3E,3M-C_5 requires testing half a dozen alternatives. The canonical labels for the 18 octane isomers are shown in Figure 2.21.

The degenerate rearrangement graphs can be unusually large, such as the "monster graph" of the rearrangements of bullvalene, tricyclic $C_{10}H_{10}$, the molecule in which 10 carbon atoms continually change connectivity. Bullvalene was conceived by Doering [104,105] and already the next year was synthesized by Schröder [106–109]. The rearrangement graph of bullvalene has 10!/3 or 1,209,600 vertices [110,111]. The rearrangement of structurally related P_7^{3-} [112], in which seven phosphorous atoms continually change their connectivity, is represented by the "baby" monster graph, which has 7!/3 or 1680 vertices [113].

2.7.1 Use of Canonical Labeling for Determining Symmetry of Graphs

In Figure 2.19, we have illustrated a 3-D carbon skeleton of diamantane $C_{14}H_{20}$, a molecule that consists of two fused adamantane units. We have assumed canonical labels, the finding of which was not difficult. For a graph having 14 vertices, there are 14! = 87,178,291,200, that is, more than 87 billion ways of placing 14 labels. Even if one takes the symmetry of diamantane into consideration, the number of cases to be considered appears beyond practicality, but one may be surprised that it took less than dozen tests to find the canonical labels shown in Figure 2.19. It is not difficult to find that there are six locations having vertices of degree 2 for label 1, which must have 14 and 13 as adjacent, in order that the first row of the adjacency matrix is the smallest possible number, which is for a 14 × 14 binary matrix:

$$0\,0\,0\,0\,0\,0\,0\,0\,0\,0\,0\,0\,1\,1$$

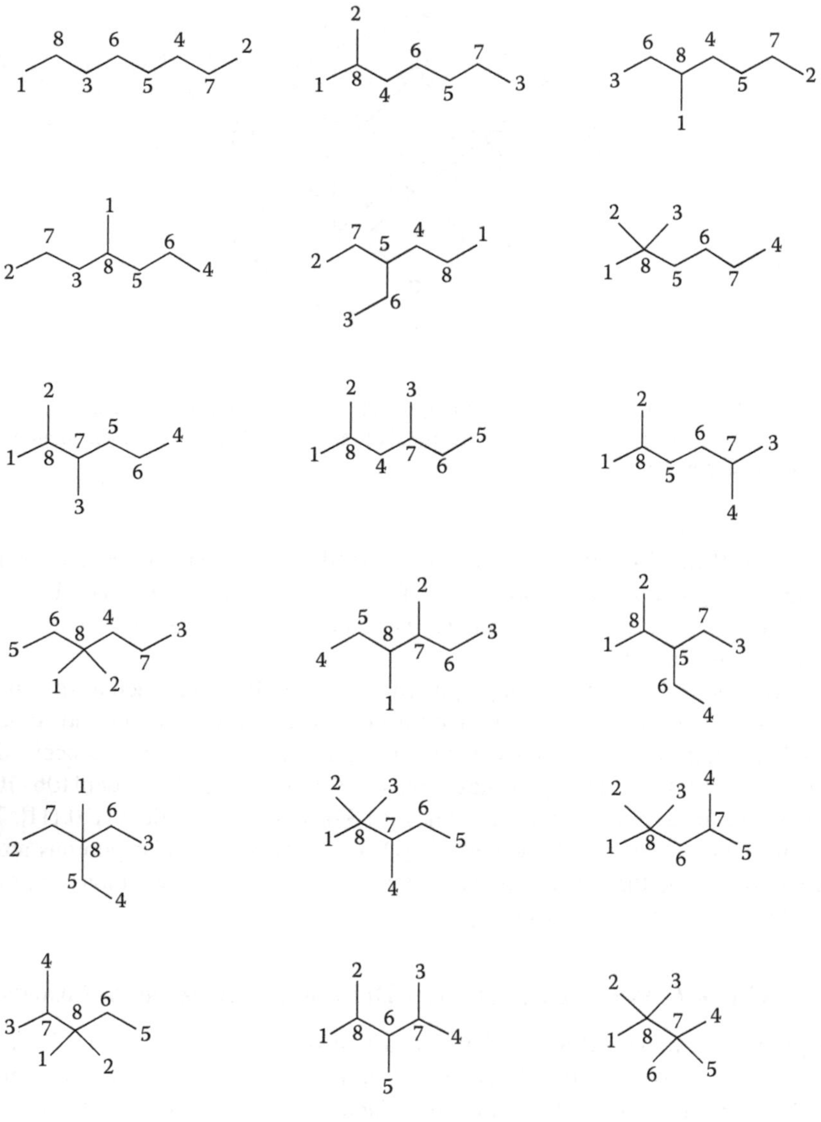

FIGURE 2.21   The canonical labeling of octane isomers.

In continuation, as outlined in Appendix 8, label 14 has only one site where it must go, which is also the case with label 13. The next smallest labels have two possibilities, but all the remaining labels have unique locations and could not be assigned to alternative neighboring locations. Thus, one finds that the total number of symmetry operations for diamantane, which is equal to the number of alternative labeling of vertices that will produce the same adjacency matrix, is $6 \times 2 = 12$. In Appendix 9, we show the 12 possible canonical labels for diamantane, and in Table 2.3, we have listed all 12 permutations of labels that leave the adjacency matrix invariant. As one

TABLE 2.3
Symmetry Elements of Diamantane

	Symmetry Elements	Partition
AA	(1) (2) (3) (4) (5) (6) (7) (8) (9) (10) (11) (12) (13) (14)	1^{14}
AB	(1) (2, 3) (4, 5) (6) (7) (8) (9, 10) (11,12) (13) (14)	$1^6\, 2^4$
AC	(1, 2) (3) (4) (5, 6) (7) (8, 9) (10) (11) (12, 13) (14)	$1^6\, 2^4$
AD	(1, 3, 2) (8, 10, 9) (4, 5, 6) (11, 12, 13) (7) (14)	$1^2\, 3^4$
AE	(1, 3) (2) (4, 6) (5) (7) (8, 10) (11, 13) (12) (14)	$1^6\, 2^4$
AF	(1, 2, 3) (8, 9, 10) (4, 6, 5) (11, 13, 12) (7) (14)	$1^2\, 3^4$
AG	(1, 8) (2, 9) (3, 10) (4, 11) (5, 12) (6, 13) (7, 14)	2^7
AH	(1, 8) (2, 10) (3, 9) (4, 12) (5, 11) (6, 13) (7, 14)	2^7
AI	(1, 9) (2, 8) (3, 10) (4, 11) (5, 13) (6, 12) (7, 14)	2^7
AJ	(1, 10, 2, 8, 3, 9) (4, 12, 6, 11, 5, 13) (7, 14)	$2^1\, 6^2$
AK	(1, 10) (2, 9) (3, 8) (4, 13) (5, 12) (6, 11) (7, 14)	2^7
AL	(1, 9, 3, 8, 2, 10) (4, 13, 5, 11, 6, 12) (7, 14)	$2^1\, 6^2$

Note: Labeling of carbon atom vertices is shown in Figure 2.19, and structures A–L are shown in Appendix 9.

can see the symmetry elements belong to the following five classes: 1^{14}, $1^6 2^4$, $1^2 3^4$, 2^7, $2^1 6^2$, which represent the identity, the reflection planes involving opposite external CC bonds, rotation for 120° around the axis passing through atoms 7 and 14, reflection through the center of the molecule, and rotation by 180° followed by reflection through the plane of atoms 4, 5, 11, and 12.

2.8 COUNTING OF PATHS

Another difficult problem in chemical graph theory, which to the uninitiated may not appear so difficult, is counting paths in a graph. In 1947, J. R. Platt [114] suggested that counts of paths of different lengths in a molecule could be useful molecular descriptors. Paths received greater attention in chemical graph theory around 1980 [115,116] as we elaborate in the next chapter of the book. Paths of length one are edges, paths of length two are two consecutive edges, paths of length three are three consecutive edges, and so on. In acyclic graphs, any two vertices (i, j) are separated by a single path of length d_{ij}, which is given by the number of edges between them. The count of paths is not difficult for acyclic graphs, but it can be very difficult even for relatively small polycyclic graphs.

In Figure 2.22, we have illustrated a way to count of paths in the hydrogen-suppressed molecular graph of norbornane, C_7H_{12}. We have selected three symmetry nonequivalent vertices 1, 3, and 7 and have, for each of these three vertices separately, constructed acyclic graphs by listing in succession vertices forming various branches, each branch having seven different labels, representing a spanning tree of norbornane molecular graph. The spanning tree of the polycyclic graph is an acyclic subgraph covering all vertices. The total number of spanning trees of

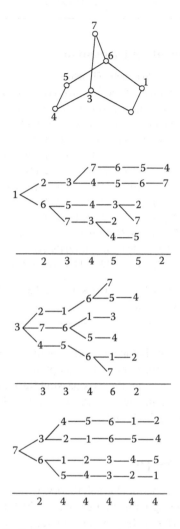

FIGURE 2.22 Count of paths for three symmetry nonequivalent carbon atoms of norbornane.

a polycyclic graph can be obtained from the determinant of the Laplacian matrix [117]. The Laplacian matrix is defined as V − A, where V is the valence matrix, having all elements zero except for the elements on the main diagonal, which have the value of the valence (degree) of the corresponding vertex.

From Figure 2.22, one sees that the three symmetry nonequivalent vertices of norbornane have the following path counts:

Vertex 1   2, 3, 4, 5, 5, 2

Vertex 3   3, 3, 4, 6, 2

Vertex 7   2, 4, 4, 4, 4, 4

When the above are multiplied by 4, 2, and 1, respectively, so as to include the contributions from symmetry equivalent vertices, and added, one obtains the path count for the norbornane molecule, which is 16, 22, 28, 36, 28, 12. This has to be divided by 2, because each path has been counted twice, which gives 8, 11, 14, 18, 14, 6. This gives for the total number of paths in norbornane 71. The path numbers 8, 11, 14, 18, 14, 6, and the total number of paths, 71, clearly do not depend on how one labels the vertices of norbornane or how one depicts the carbon skeleton of norbornane and thus represent *invariants*. Graph invariants can be used as molecular descriptors or as parameters in quantitative comparative studies of molecules. Whether they will remain mere mathematical constructions or they will become chemically useful descriptors, can only be established *after* they have been tested on a number of structure–property correlations. We will address this topic in more detail in the following chapters and will examine closely a collection of more widely used ("popular") molecular descriptors.

In Table 2.2, we have shown, in addition to the adjacency matrix A, the distance matrix D, also the valence matrix V, the Laplacian matrix L, and the powers A^2 and A^3 of the adjacency matrix, all for norbornane. Distance matrix D of a graph, another useful matrix in chemical graph theory, was introduced by F. Harary [118]. The element $[D]_{ij}$ of the D matrix are defined as

$$[D]_{ij} = \begin{cases} d_{ij} & \text{if } i \neq j \\ 0 & \text{otherwise,} \end{cases}$$

where d_{ij} denotes the topological distance (the length of the shortest path connecting two vertices in a graph) between the vertices i and j.

The element $[L]_{ij}$ of the Laplacian matrix is

$$[L]_{ij} = \begin{cases} -1 & \text{if } i \neq j \text{ and vertices } i \text{ and } j \text{ are adjacent} \\ d_i & \text{if } i = 1 \\ 0 & \text{otherwise,} \end{cases}$$

where d_i didenotes the valence (degree) of the vertex i.

The elements $[A^2]_{ij}$ of the A^2 matrix gives the number of walks of length two between the vertices i and j; the diagonal elements give the count of self-returning walks of length two. The element $[A^3]_{ij}$ of the A^3 matrix gives the number of walks of length three between the vertex i and the vertex j. The nth power of the adjacency matrix similarly would give the count of walks between vertices i and j of length n. The diagonal elements, which give the count of self-returning walks, are necessarily zero for all odd powers of the adjacency matrix.

2.9 EXPERIMENTAL MATHEMATICS

With the availability of computers, it has been recognized that not only can some problems (which up to that time were beyond the possibility of being numerically solved) be reexamined and solved, but also new problems can emerge, be discovered, considered, and even solved, which were unknown in pre-computer time. Let us briefly mention two such problems that have some novelty, both of which have some "connection" with chemistry: (i) the proof or "almost a proof" of the Kepler conjecture about dense packing of spheres and (ii) the problem relating to Ulam's spiral.

We will end with the outline of a new problem of experimental mathematics that arose in the process of writing this book and relates to isospectral graphs.

2.9.1 On Kepler's Conjecture on Dense Packing of Spheres

The Johannes Kepler conjecture (1611) stated that the most dense packing of spheres of equal size is obtained by placing spheres in layers on a regular arrangement as schematically suggested at the top of Figure 2.23, which shows the first layer of dense packing of spheres. In considering placing the following layer, let us focus on the centers of balls and the small "triangles" between three adjacent spheres. In the second layer, balls can be placed in one way only, but this creates two possibilities for the third layer: (i) The centers of balls can be placed above triangles in the second layer, which are above triangles in the first layer, and (ii) the centers of balls placed above the triangles in the second layer can be placed above the centers of the spheres of the first layer. If such a rule is followed for succeeding layers, one obtains either the hexagonal close packing or the cubic close packing. Both have the same maximal density of $\pi/(3\sqrt{2}) = 0.74048$.

The Kepler conjecture received attention from outstanding mathematicians in the past, including Gauss and Hilbert, and was only relatively recently solved. Thomas Hales [119] used computer to exhaustively check various possible, very large but finite, numbers of different arrangements (it was shown by the Hungarian mathematician Fejes Tóth in 1953 that the problem can be so reduced [120]). The proof of Hales took

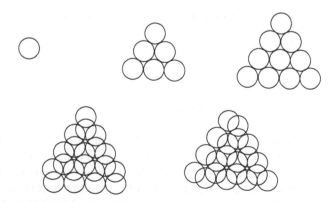

FIGURE 2.23 Packing of spheres of the same size in layers. Shown are the first two layers. In the third layer at left, each ball is above the center of three adjacent balls, and at right, each ball is above a single center of the first level.

some 100 pages and involved minimizing a function with 150 variables. The mathematical tool was linear programming and the simplex method (of George Dantzig, *vide infra*), which involved more than 5000 different configurations of spheres. In 1998, Hales announced that the proof was found; it consisted of 250 pages and three gigabytes of computer program, which was published in the respectable *Annals of Mathematics* [121]. The panel of referees of the publication reported that they consider the proof "99% certain," which is a novelty of a kind in mathematics that may become a rule and not an exception for computer proofs of the near future.

Some elder chemists may remember the time when chemical calculations were made by use of logarithmic tables to be supplemented later by electric calculating machines. These were then replaced by early computers with very modest memory capacities of a dozen kilobytes, which, with time, grew to the current pocket and laptop calculators having gigabyte capacities. Speaking of logarithms, let us mention that at the University of Chicago, Enrico Fermi (1901–1954) and Leo Szilárd (1898–1964) created the first artificial controlled nuclear chain reaction on December 2, 1942, there. This is a historic day of a special kind regardless of the later use and misuse of atomic energy. In comparison with the energies involved in a chemical reaction, a single nuclear reaction releases more than a million times more energy than a single chemical reaction. At the University of Chicago, there is a memorial room remembering this historic experiment in which there is a large painting of Enrico Fermi in which he is standing and keeping something, like a ruler, in his hand. On the occasion when one of the authors visited Chicago and the Enrico Fermi memorial room, he was puzzled by what Enrico Fermi has in his right hand. It took about 5 minutes to realize that this is a logarithmic calculator (slide ruler)! If it took 5 minutes to recognize a logarithmic calculator for a person who, as a student used a logarithmic calculator daily, the younger generation who have never seen logarithmic calculators may never realize that significant results were obtained with such a "primitive" tool and people were still able to do some complex calculations at the time when electronic calculators were unknown.

Before leaving the subject of the Kepler conjecture, we would briefly comment on linear programming, a mathematical discipline developed during World War II, which had important application in various sciences, starting with decision science and including economics. Mathematically, it consists of minimizing a linear function of n variables subject to m linear inequalities and/or equality constraints. In applications, one searches for optimal numerical solutions of larger sets of inequalities. The methodology of how this is to be done has been and was mainly developed by George B. Danzig (1914–2005), who introduced the simplex algorithm or simplex method for finding a solution [122]. Let us quote the opening introductory two paragraphs from *Linear Programming, "The Story about How It Began: Some legends, a little about its historical significance, and comments about where its many mathematical programming extensions may be headed"* [123]. According to Dantzig,

> Linear programming can be viewed as a part of a great revolutionary development which has given mankind the ability to state general goals and to lay out a path of detailed decisions to take in order to "best" achieve its goals when faced with practical situations of great complexity. Our tools for doing this are ways to formulate real-world problems in detailed mathematical terms (models), techniques for solving the models (algorithms), and engines for executing the steps of algorithms (computers and software).

This ability began in 1947, shortly after World War II, and has been keeping pace ever since with the extraordinary growth of computing power. So rapid has been the advance in decision science that few remember the contributions of the great pioneers that started it all. Some of their names are von Neumann, Kantorovich, Leontief, and Koopmans. The first two were famous mathematicians. The last three received the Nobel Prize in economics for their work.

2.9.2 ULAM'S SPIRAL

Another illustration of "experimental" mathematics is the Ulam spiral, which is constructed by writing prime numbers in a spiral form over the square network. Stanislaw Ulam (1909–1984) is a renowned Polish-American mathematician, who, among other accomplishments, invented the Monte Carlo probability methodology— named after the well-known gambling city of Europe in the Principality of Monaco— in view that this method has something in common with card players in casinos (if they want to be successful), keeping a record of previous events (games) so as to be able to estimate the probability of high and low cards remaining in the deck. During the World War II, Ulam was invited to participate in America's "Manhattan Project" and was involved in computations relating to nuclear armament.

The Ulam spiral is a result of Ulam attending an uninteresting lecture, during which he started sketching his prime number spiral. This somewhat unusual format for writing sequences offers a two-dimensional pattern for a one-dimensional mathematical object, here, the sequence of prime numbers [124]. In Figure 2.24, we have illustrated Ulam's spiral for prime numbers smaller than 100.

This small spiral only illustrates its construction, and it is too small to display interesting properties of Ulam's spiral. In Figure 2.25, we show the Ulam spiral constructed on a 32 × 32 grid, which contains all prime number smaller than 1000. Here, one can see that in such representation of prime numbers, many prime numbers are located along various diagonal lines. Observe also that some diagonals are "rich" in prime numbers, and some are not, which is something that also caught the attention of Ulam and lead to a better generalizing formula for prime numbers.

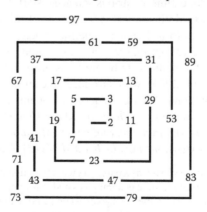

FIGURE 2.24 Ulam's spiral showing locations of prime numbers smaller than 100, which tend to be lined along various diagonals.

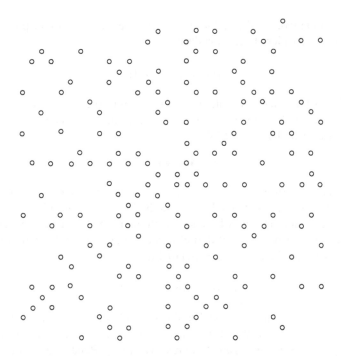

FIGURE 2.25 Ulam's spiral for the first 1000 prime numbers.

Formulas that produce prime numbers, known as *generating formulas*, have been known for a while. In 1772, Euler found the quadratic polynomial: $P(n) = n^2 + n + 41$, which, for *all* values of $n = 0, 1, 2, 3, \dots 39$ gives prime numbers, 40 primes in total. These are a selection of prime numbers that start with 41, 43, 47, 53, 61, 71, 83, ... for $n = 0$ to $n = 6$ and so on and end with 1601, for $n = 39$. For $n = 40$, one obtains for P(n):

$$P(40) = 40^2 + 40 + 41 = 40 (40 + 1) + 41 = 40 \times 41 + 41 = 41 \times 41, \text{ or } 41^2$$

Clearly, the result is not a prime number. Observe that the differences between two successive prime numbers generated by the Euler quadratic polynomial are 2, 4, 6, 8, 10, 12, etc. This simple regularity for prime numbers may look to the uninformed surprising, but such regularity is a property of quadratic functions, for which holds the rule that the differences between equidistant values for the variable grow linearly. For example, the simple parabola: $y = x^2$, for $x = 1, 2, 3, 4, 5, \dots$ takes the values 1, 4, 9, 16, 25, ..., respectively, the differences between which are odd numbers 3, 5, 7, 9... Of course, the 40 prime numbers between 41 and 1601 generated by Euler's formula are not consecutive, but what is important is that all the numbers produced by the formula for n in the interval (0, 39) are prime numbers. There are formulas that will produce prime numbers, but occasionally some of their outputs are not prime numbers, and such formulas are of limited interest in mathematics. There are additional generating functions of considerable interest in current mathematics, which involve polynomials of higher degrees and many variables, but we leave this topic to those with special interest in prime numbers.

Let us mention two quadratic forms that are related to the Euler polynomial. In 1798, Legendre considered the polynomial:

$$P(n) = n^2 - n + 41 \tag{2.5}$$

which differs from Euler's formula only in the sign of the linear term, which gives the same prime numbers for $n = 0, 1, 2, 3, \ldots 40$. Another related formula, shown below, is of Hardy and Wright [125]:

$$P(n) = (n - 40)^2 + (n - 40) + 41 \tag{2.6}$$

This formula, for n running from 0 to 79 gives 80 prime numbers and gives the same prime numbers given by the Euler (and the Legendre) formulas; however, each number is given twice. This may be a good formula for bad students, who do not give enough attention to what they are doing! It is to be expected (although this was not indicated in the material at our disposal) that the formula

$$P(n) = (n - 40)^2 - (n - 40) + 41 \tag{2.7}$$

will do the same for n running from 1 to 80. The only novelty, besides the mentioned polynomials of many variables and other higher polynomials, is that there are also linear functions $L(n) = an + b$, where a and b are *relative* primes. Relative primes are the numbers that have no common divisor, such as 4 and 9, none of which is a prime number. For illustration, here is one such *linear prime number generator* of more recent date [126]:

$$43142746595714191 + 5283234035979900 \times n \tag{2.8}$$

This formula gives 26 prime numbers, for all n from 0 to 25.

We have mentioned Ulam's spiral as an illustration of a "tool" for representation of sequences as 2-D maps. Independently, the same idea has been used in graphical bioinformatics for representing DNA sequences as 2-D maps, which then allows selection of various map invariants as descriptors to be used to estimate the degrees of similarity between different DNA sequences. In view of DNA sequences being of very long lengths, map representation of DNA has an important advantage of being rather compact. In a later section, we outline construction of DNA maps based on the spiral representation of DNA.

An alternative graphical representation of DNA as maps is based on the Chaos Game of Barnsley [127], also to be examined more closely later in the book in the section on graphical bioinformatics. In passing, we add that Michael Barnsley is a British mathematician, who got his PhD in at the University of Wisconsin, Madison in theoretical chemistry.

Before leaving the Ulam spiral, we should mention that there have been a few modifications of Ulam's spiral. For example, R. Sacks constructed the Archimedean spiral by plotting integers uniformly on the spiral, and when composite numbers have been deleted, one obtains what is known as the Sacks prime number spiral [128]. On this spiral, prime numbers that are obtained from Euler's prime number generator $x^2 - x + 41$ are clearly seen on a line approaching the left horizontal axis. There are modifications of the

Archimedean prime number spiral. We may also mention that quadratic formulas with coefficient $a = 4$ generate a relatively large number of primes in comparison with other quadratic formulas. In particular, the formula $4n^2 - 2n + 41$ is one of such prime number generators. Hardy and Littlewood [129] proposed a conjecture that quadratic polynomials satisfying some simple requirements can generate infinite numbers of prime numbers. Moreover, they found that asymptotically the number of primes is proportional to A, a number that only depends on the coefficient of the quadratic formula (a, b, c). The larger value of A indicates the diagonals in Ulam's spiral having more prime numbers. It was found that for $4n^2 - 2n + 41$ the A value is 6.6, which means that the number of prime numbers it generates are close to seven times larger than the number of prime numbers that one can expect in examining the same number of random numbers. Let us end this excursion in the prime number territory by reporting that the highest currently known value of A associated with quadratic polynomials appears to be about 11.3, which was discovered by Jacobson and Williams in 2003 [130]. By now, we all know enough of the spirit of mathematicians to expect that the "race" for even higher A will continue.

We thought that we might add a novelty to the topic of prime number spirals by considering only odd numbers in the construction of the spiral. In Figure 2.26, we have illustrated a section of a so-modified Ulam's spiral on a 24 × 24 Cartesian grid,

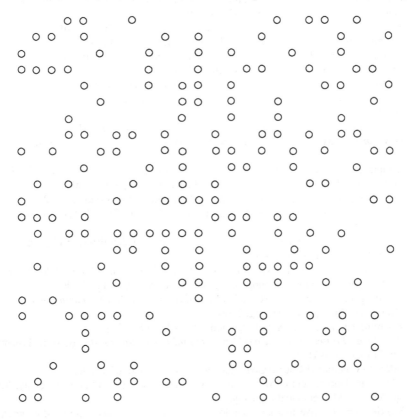

FIGURE 2.26 Patterns of the smallest 190 odd prime numbers on the Ulam spiral constructed on a 24 × 24 Cartesian grid for odd numbers only.

on which we plotted the sites of the odd prime numbers, ending with the prime number 1129, and excluding, of course, number 2, the only even prime number. The revisited spiral shows numerous horizontal and vertical lines as well as many diagonals that carry a high population of primes. The novel Ulam spiral is clearly more dense as it excludes even numbers. Whether it will offer additional information that may be currently hidden in the standard Ulam spiral has yet to be seen. Observe, however, that the pattern of dots representing prime numbers in Figure 2.26 shows sections of horizontal and vertical lines and double lines that appear not to bear primes. It will be of interest to see whether the same continues and for how long in still larger spirals. This feature, as far as we are aware, has not yet been mentioned or noticed in Ulam's spiral except for the trivial case of even number diagonals, which remain unpopulated except for the two diagonals meeting at number 2, the only even prime number.

REFERENCES AND NOTES

1. G. Kirchhoff, Über die Auflösung der Gleichungen, auf welche man bei der untersuchung der linearen verteilung galvanischer Ströme geführt wird, *Ann. Phys. Chem.* 72 (1847) 497–508.
2. F. Flavitsky, The method of computing the number of saturated monohydroxy alcohols isomers, *J. Russ. Chem. Soc.* 3 (1871) 160–167.
3. F. Flavitsky, Bemerkung zu der Abhandlung des Herrn Hugo Schiff: Zur Statistik Chemischer Verbindungen, *Ber. Dtsch. Chem. Ges.* 9 (1876) 267.
4. M. Randić and N. Trinajstić, Notes on some less known early contributions to chemical graph-theory, *Croat. Chem. Acta* 67 (1994) 1–35.
5. B. Bollobás and P. Erdös, Graphs of extremal weights, *Ars Combin.* 50 (1998) 225–233.
6. M. Randić, Characterization of molecular branching, *J. Am. Chem. Soc.* 97 (1975) 6609–6615.
7. X. Li and I. Gutman, *Mathematical Aspects of Randić-type Molecular Structure Descriptors*, Mathematical Chemistry Monographs No. 1, Kragujevac, Serbia, 2006.
8. I. Gutman and B. Furtula (Eds.), *Recent Results in the Theory of Randić Index*, Mathematical Chemistry Monographs No. 6, Kragujevac, Serbia, 2008.
9. R. Wilson, *Introduction to Graph Theory*, Oliver and Boyd, London, 1972.
10. N. L. Biggs, E. K. Lloyd, and R. J. Wilson, *Graph Theory 1736–1936*, Clarendon Press, Oxford, 1968.
11. D. J. Klein and J. Brickmann (Eds.), Partial orderings in chemistry, *MATCH Commun. Math. Comput. Chem.* 42 (2000) 1–290.
12. P. R. Duchowicz and E. A. Castro, *The Order Theory in QSPR—QSAR Studies*, Mathematical Chemistry Monographs No. 7, Kragujevac, Serbia, 2006.
13. R. Brüggemann and L. Carlsen, (Eds.), *Partial Order in Environmental Sciences and Chemistry*, Springer, Heidelberg, 2006.
14. T. Donnellen, *Lattice Theory*, Pergamon, Oxford, 1968.
15. G. Grätzer, *Lattice Theory: First Concepts and Distributive Lattices*, W. H. Freeman, San Francisco, 1971.
16. LBS 119 Calculus II Course Goals, Lyman Briggs School of Science.
17. K. Suman, Department of Mathematics and Statistics, Winona State University, Jump up, A Set of Mathematical Equivoques.
18. S. Skiena (Ed.), Hasse diagrams, Section 5.4.2, in: *Implementing Discrete Mathematics: Combinatorics and Graph Theory with Mathematica*, Addison-Wesley, Reading, MA, 1990.

19. K. H. Rosen, *Discrete Mathematics and Its Applications*, 7th ed., McGraw-Hill, New York, 2011.
20. M. Randić, A graph theoretical characterization of proteomics maps, *Int. J. Quantum Chem.* 90 (2002) 848–858.
21. M. Gordon and W. H. T. Davison, Theory of resonance topology of fully aromatic hydrocarbons, *J. Chem. Phys.* 20 (1952) 428–435.
22. M. Randić, On enumeration of complete matchings in hexagonal lattices, in: *Graph Theory, Combinatorics and Applications*, Vol. 2, Y. Alavi, G. Chartrand, O. R. Ollermann, and A. J. Schwenk (Eds.), J. Wiley and Sons, New York, 1991, pp. 1001–1008. Gordon and Breach Sci. Publ., New York, 1991.
23. M. Randić, Aromaticity of polycyclic conjugated hydrocarbons, *Chem. Rev.* 103 (2003) 3449–3605.
24. I. Gutman and S. Klavžar, Chemical graph theory of fibonacenes, *MATCH Commun. Math. Comput. Chem.* 55 (2006) 39–54.
25. A. T. Balaban, Chemical graphs. 50. Symmetry and enumeration of fibonacenes (unbranched catacondensed benzenoids isoarithmic with helicenes and zigzag cata-fusenes), *MATCH Commun. Math. Comput. Chem.* 24 (1989) 29–38.
26. E. Clar, *The Aromatic Sextet*, John Wiley & Sons, London, 1972.
27. A. T. Balaban and F. Harary, Chemical graphs V. Enumeration and proposed nomenclature of benzenoid cata-condensed polycyclic aromatic hydrocarbons, *Tetrahedron* 24 (1968) 2505–2516.
28. P. G. Anderson, Fibonaccene, in: *Fibonacci Numbers and Their Applications*, A. N. Philippou, G. E. Bergum, and A. F. Horadam (Eds.), Reidel, Dordrecht, 1986, p. 2.
29. S. El-Basil and D. J. Klein, Fibonacci numbers in the topological theory of benzenoid hydrocarbons and related graphs, *J. Math. Chem.* 3 (1989) 1–23.
30. P. John and H. Sachs, Wegesysteme und Linear Faktoren in Hexagonaler und Quadratischen Systemen, in: *Graphen in Forschung und Unterricht*, R. Bodendiek, H. Schumacher, and G. Walther (Eds.), Franzbecker Verlag, Bad Salzdetfurth, 1985, pp. 85–101.
31. P. John and H. Sachs, Calculating the number of perfect matchings and of spanning trees, Pauling's orders, the characteristic polynomial, and the eigenvectors of benzenoid systems, in: *Advances in the Theory of Benzenoid Hydrocarbons, Topics in Current Chemistry*, I. Gutman, S. J. Cyvin (Eds.), Vol. 153, 1990, pp. 145–179.
32. S. J. Cyvin and I. Gutman, *Kekulé structures in benzenoid hydrocarbons*, Springer, Berlin, New York, Paris, 1988.
33. L. Euler, Solutio problematis ad geometriam situs pertinentis. *Comment. Acad. Sci. U. Petrop.* 8 (1736) 128–140, reprinted in *Opera Omnia Series Prima*, Vol. 7, pp. 1–10, 1766.
34. J. J. Sylvester, Chemistry and algebra, *Nature* 17 (1877–1878) 284.
35. J. J. Sylvester, On an application of the new atomic theory to the graphical representation of the invariants and covariants of binary quantities, *Am. J. Math.* 1 (1878) 64–125.
36. D. H. Rouvray, The origin of chemical graph theory, in: *Chemical Graph Theory, Introduction and Fundamentals*, D. Bonchev and D. H. Rouvray (Eds.), Abacus Press, New York, 1991.
37. O. E. Polansky, Elements of graph theory for chemists, in: *Chemical Graph Theory, Introduction and Fundamentals*, D. Bonchev and D. H. Rouvray (Eds.), Abacus Press, Gordon and Breach Sci. Publ., New York, 1991, pp. 41–96.
38. V. Prelog and R. Seiwerth, Über die Synthese des Adamantans, *Berichte der Deutschen Chemischen Gesellschaft (A and B Series)* 74 (1941) 1644–1648.
39. R. C. Read and D. G. Corneil, The graph isomorphism disease, *J. Graph Theory* 1 (1977) 339–336.
40. G. Gati, Further annotated bibliography on the isomorphism disease, *J. Graph Theory* 3 (1979) 95–109.

41. M. Randić, On the recognition of identical graphs representing molecular topology, *J. Chem. Phys.* 60 (1974) 3920–3928.
42. M. Randić, On rearrangement of the connectivity matrix of a graph, *J. Chem. Phys.* 62 (1975) 309–310.
43. A. L. Goodson, Nomenclature of chemical compounds, in: *Chemical Graph Theory, Introduction and Fundamentals*, D. Bonchev and D. H. Rouvray (Eds.), Abacus Press, Gordon and Breach Sci. Publ., New York, 1991, pp. 97–132.
44. H. L. Morgan, The generation of a unique machine description for chemical structures—A technique developed at Chemical Abstracts Service, *J. Chem. Doc.* 5 (1965) 107–113.
45. M. Randić and D. Plavšić, On the concept of molecular complexity, *Croat. Chem. Acta* 75 (2002) 107–116.
46. W. J. Wiswesser, *A Line-Formula Chemical Notation*, Thomas Y. Crowell Co., New York, 1954.
47. E. G. Smith and P. A. Baker, *The Wiswesser Line-Formula Chemical Notation (WLN)*, 3rd ed., Chemical Information Management, Cherry Hill, NJ, 1976.
48. W. J. Wiswesser, 107 Years of line-formula notation (1861–1968), *J. Chem. Doc.* 8 (1968) 46–150.
49. A. T. Balaban, Chemical graphs VII. Proposed nomenclature of branched cata-condensed benzenoid polycyclic hydrocarbons, *Tetrahedron* 25 (1969) 2949–2656.
50. A. T. Balaban, Enumeration of catafusenes, diamondoid hydrocarbons, and staggered alkane C-rotamers, *MATCH—Commun. Math. Comput. Chem.* 2 (1976) 51–61.
51. A. T. Balaban and P. v. R. Schleyer, Systematic classification and nomenclature of diamond hydrocarbons. I. Graph-theoretical enumeration of polymantanes, *Tetrahedron* 34 (1978) 3599–3609.
52. C. Jordan, Sur les assemblages de lignes, *J. Reine Angew. Math* 70 (1896) 185.
53. J. J. Sylvester, A question in the geometry of situation, *Quart. J. Math.* 1 (1857) 79.
54. D. Bonchev, A. T. Balaban, and M. Randić, The graph center concept for polycyclic graphs, *Int. J. Quantum Chem.* 19 (1981) 61–82.
55. D. Bonchev, A. T. Balaban, and O. Mekenyan, Generalization of the graph center concept, and derived topological centric indexes, *J. Chem. Inf. Comp. Sci.* 20 (1980) 106.
56. D. Bonchev, O. Mekenyan, and A. T. Balaban, Iterative procedure for the generalized graph center in polycyclic graphs, *J. Chem. Inf. Comp. Sci.* 29 (1989) 91.
57. A. T. Balaban, D. Bonchev, and W. A. Seitz, Topological/chemical distances and graph centers in molecular graphs with multiple bonds, *Mol. Struct. (Theochem.)* 280 (1993) 253.
58. D. Bonchev, A. T. Balaban, and M. Randić, The graph center concept for polycyclic graphs, *Int. J. Quantum Chem.* 19 (1981) 61–82.
59. D. Bonchev and A. T. Balaban, Topological centric coding and nomenclature of polycylic hydrocarbons, *J. Chem. Inf. Comp. Sci.* 21 (1981) 223–229.
60. D. Bonchev, The concept for the centre of a chemical structure and its application, *J. Mol. Struct. (Theochem.)* 185 (1989) 155–168.
61. A. T. Balaban, S. Bertelsen, and S. C. Basak, New centric topological indices for acyclic molecules (trees) and substituents (rooted trees), and coding of rooted trees, *MATCH Commun. Math. Comput. Chem.* 30 (1994) 55–72.
62. M. Randić, M. Novič, M. Vračko, and D. Plavšić, On the centrality of vertices of molecular graphs, *J. Comput. Chem.* 34 (2013) 1409–1419.
63. M. Dehmer, S. Borgert, and F. Emmert-Streib, Entropy bounds for Hierarchical molecular networks, *PLoS One* 3 (2008) e3079.
64. M. Dehmer, Information-theoretic concepts for the analysis of complex networks, *Appl. Artif. Intell.* 22 (2008) 684–706.
65. M. Dehmer, Information processing in complex networks: Graph entropy and information functionals, *Appl. Math. Comput.* 201 (2008) 82–94.

66. N. Lozac'h, A. L. Goodson, and W. H. Powell, Nodal nomenclature. I. General principles, *Angew. Chem.* 91 (1979) 951–964.
67. A. L. Goodson, Graph-based chemical nomenclature. 1. Historical background and discussion, *J. Chem. Inf. Comp. Sci.* 20 (1980) 167–172.
68. A. L. Goodson, Application of graph based chemical nomenclature to theoretical and preparative chemistry, *Croat. Chem. Acta* 56 (1983) 319–328.
69. A. Goodson, Graph based chemical nomenclature. 2. Incorporation of graph theoretical principles into Taylor's nomenclature proposal, *J. Chem. Inf. Comp. Sci.* 20 (1980) 172–176.
70. R. C. Read, A new system for the designation of chemical compounds: 1. Theoretical preliminaries and the coding of acyclic compounds, *J. Chem. Inf. Comp. Sci.* 23 (1983) 135–149.
71. D. Weininger, SMILES, a chemical language and information system. 1. Introduction to methodology and encoding rules, *J. Chem. Inf. Modeling* 28 (1988) 31–36.
72. D. Weininger, A. Weininger, and J. l. Weininger, SMILES. 2. Algorithm for generation of unique SMILES notation, *J. Chem. Inf. Modeling* 29 (1989) 97–101.
73. D. Weininger, SMILES. 3. DEPICT. Graphical depiction of chemical structures, *J. Chem. Inf. Modeling* 30 (1990) 237–243.
74. J. Zupan, M. Vračko, and M. Novič, New uniform and reversible representation of 3D chemical structures, *Acta Chim. Slov.* 47 (2000) 19–37.
75. M. Novič and M. Vračko, Comparison of spectrum-like representation of 3D chemical structure with other representations when used for modeling biological activity, *Chemom. Intell. Lab. Syst.* 59 (2001) 33–44
76. M. Cushman, D. Nagarathnam, and L. R. Geahlen, Synthesis and evaluation of hydroxylated flavones and related compounds as potential inhibitors of the protein–tyrosine kinase, *J. Nat. Prod.* 54 (1991) 1345–1352.
77. M. Cushman, D. Nagarathnam, L. D. Burg, and L. R. Geahlen, Synthesis and protein–tyrosine kinase inhibitory activities of flavonoid analogues, *J. Med. Chem.* 34 (1991) 798–806.
78. M. Cushman, H. Zhu, L. R. Geahlen, and J. A. Kraker, Synthesis and biochemical evaluation of a series of aminoflavones as potential inhibitors of protein–tyrosine kinases p56, EGFr, *J. Med. Chem.* 37 (1994) 3353–3362.
79. M. Novič, Z. Nikolovska-Coleska, and T. Šolmajer, Quantitative structure–activity relationship of flavonoid p56lck protein tyrosine kinase inhibitors: A neural network approach, *J. Chem. Inf. Comput. Sci.* 37 (1997) 990–998.
80. A. T. Balaban (Ed.), *From Chemical Topology to Three-Dimensional Geometry*, Plenum, New York, 1997, pp. 73–116.
81. M. Randić and M. Razinger, On characterization of three-dimensional molecular structure, in: *From Chemical Topology to Three-Dimensional Geometry*, A. T. Balaban (Ed.), Plenum Press, New York, 1966, pp. 159–236.
82. L. B. Kier and L. H. Hall, *Molecular Connectivity in Structure-Activity Analysis*, Research Studies Press, Letchworth, England, 1986.
83. J. Stankevich, I. V. Stankevich, and N. S. Zefirov, Topological indexes in organic chemistry, *Usp. Khim.* 57 (1988) 337–366.
84. S. El-Basil and M. Randić, Equivalence of mathematical objects of interest in chemistry and physics, *Adv. Quantum Chem.* 24 (1992) 239–290.
85. K. Varmuza and P. Filzmoser, *Introduction to Multivariate Statistical Analysis in Chemometrics*, CRC Press, Taylor & Francis Group, Boca Raton, FL, 2009.
86. H. Wiener, Structural determination of paraffin boiling points, *J. Am. Chem. Soc.* 69 (1947) 17–20.
87. H. Hosoya, Topological index. A newly proposed quantity characterizing topological nature of structural isomers of saturated hydrocarbons, *Bull. Chem. Soc. Jpn.* 44 (1971) 2332–2339.

88. M. Randić, Topological indices. in: *The Encyclopedia of Computational, Chemistry,* P. R. Schleyer, N. L. Allinger, T. Clark, J. Gasteiger, P. A. Kollman, H. F. Schaefer III, and P. R. Schreiner (Eds.), Wiley, Chichester, 1999, pp. 3018–3032.

89. L. B. Kier and L. H. Hall, *Molecular Connectivity in Chemistry and Drug Research,* Academic Press, New York, 1976.

90. L. B. Kier, Indexes of molecular shape from chemical graphs, in: *Computational Chemical Graph Theory,* D. H. Rouvray (Ed.), Nova Science Publishers, New York, 1990, pp. 151–174.

91. A. T. Balaban, Chemical graphs. 48. Topological index *J* for heteroatom-containing molecules taking into account periodicities of element properties, *MATCH Commun. Math. Comput. Chem.* 21 (1986) 115–122.

92. A. T. Balaban, Topological indices based on topological distances in molecular graphs, *Pure Appl. Chem.* 55 (1983) 199–206.

93. A. T. Balaban, Highly discriminating distance–based topological index, *Chem. Phys. Lett.* 89 (1982) 399–404.

94. L. B. Kier, Use of molecular negentropy to encode structure governing biological activity, *J. Pharm. Sci.* 69 (1980) 807–810.

95. D. Bonchev, *Information Theoretic Indices for Characterization of Chemical Structure,* Wiley-Interscience, New York, 1983.

96. S. C. Basak, D. K. Harriss, and V. R. Magnuson, Comparative study of lipophilicity versus topological molecular descriptors in biological correlations, *J. Pharm. Sci.* 73 (1984) 429–437.

97. M. Novič and J. Zupan, A new general and uniform structure representation, in: J. Gasteiger (Ed.). *Software-Entwicklung in der Chemie,* Vol. 10, Gesellschaft Deutscher Chemiker, Frankfurt am Main, 1996, pp. 47–58.

98. J. Zupan and M. Novič, General type of a uniform and reversible representation of chemical structures, *Anal. Chim. Acta* 348 (1997) 409–418.

99. M. Randić, On the recognition of identical graphs representing molecular topology, *J. Chem. Phys.* 60 (1974) 3920–3928.

100. M. Randić, On canonical numbering of atoms in a molecule and graph isomorphism, *J. Chem. Inf. Comput. Sci.* 17 (1977) 171–180.

101. M. Randić, On discerning symmetry properties of graphs, *Chem. Phys. Lett.* 42 (1976) 283–287.

102. M. Randić, A systematic study of symmetry properties of graphs. I. Petersen graph, *Croat. Chem. Acta* 49 (1977) 643–655.

103. A. T. Balaban, D. Fărcaşiu, and R. Bănică, Graphs of multiple 1, 2-shifts in carbonium ions and related systems, *Rev. Roum. Chim.* 11 (1966) 1205.

104. W. von E. Doering and W. R. Roth, A rapidly reversible degenerate Cope rearrangement: Bicyclo[5.1.0]octa-2,5-diene, *Tetrahedron* 19 (1963) 715–737.

105. A. Ault, The bullvalene story. The conception of bullvalene, a molecule that has no permanent structure, *J. Chem. Educ.* 78 (2001) 924.

106. G. Schröder, Als Mitteilung I gilt: Synthese und Eigenschaften von Bullvalen, *Angew. Chem.* 75 (1963) 722.

107. G. Schröder, Synthese und Eigenschaften von Tricyclo[3.3.2.04.6]decatrien-(2.7.9)2.3) (Bullvalen), *Chemische Berichte* 97 (1964) 3140.

108. G. Schröder and J. F. M. Oth, Recent chemistry of bullvalene, *Angew. Chem. Int. Ed.* 6 (1967) 414–423.

109. J. F. M. Oth, K. Müllen, J.-M. Giles, and G. Schröder, Comparison of ^{13}C- and ^{1}H-magnetic resonance spectroscopy of technique for the quantitative investigation of dynamic processes. The Cope rearrangement in bullvalene, *Helv. Chim. Acta* 57 (1974) 1415.

110. J. Brocas, The reaction graph of the Cope rearrangement in bulvallene, *J. Math. Chem.* 15 (1994) 389–395.
111. T. Živković, Bullvalene reaction graph, *Croat. Chem. Acta* 69 (1996) 215–222.
112. M. Baudler, Chain and ring phosphorus compounds-analogies between phosphorus and carbon chemistry. *Angew. Chem. Int. Ed.* 21 (1982) 492.
113. M. Randić, D. O. Oakland, and D. J. Klein, Symmetry properties of chemical graphs. IX. The valence tautomerism in the P_7^{3-} skeleton, *J. Comput. Chem.* 7 (1986) 35–54.
114. J. R. Platt, Influence of neighbor bonds on additive bond properties in paraffins, *J. Chem. Phys.* 15 (1947) 419–420.
115. M. Randić, Characterization of atoms, molecules, and classes of molecules based on path enumerations, *MATCH—Commun. Math. Comput. Chem.* 5 (1979) 3–60.
116. M. Randić, On representation of molecular graphs by basic graphs, *J. Chem. Inf. Comput. Sci.* 32 (1992) 57–69.
117. B. Mohar, Laplace eigenvalues of graphs—A survey, *Discrete Math.* 109 (1992) 171–183.
118. F. Harary, *Graph Theory*, Addison-Wesley, Reading, MA, 1969.
119. T. C. Hales, A computer verification of the Kepler conjecture, in: *Proc. Int. Congress Mathematicians, Vol. II. Invited lectures. Held in Beijing, August 20–28, 2002*, T. Li (Ed.), Higher Education Press, Beijing, China, 2002, pp. 795–804.
120. L. Fejes Tóth, *Lagerungen in der Ebene, auf der Kugel und im Raum*, Die Grundlehren der Mathematischen Wissenschaften in Einzeldarstellungen mit besonderer Berücksichtigung der Anwendungsgebiete, Band LXV, Berlin, Springer-Verlag, New York, 1953.
121. T. C. Hales, A proof of the Kepler conjecture, *Ann. Math.* 162 (2005) 1065–1185.
122. G. B. Dantzig, *Linear Programming and Extensions*, Princeton University Press, Princeton, 1963.
123. G. B. Dantzig, Linear programming, *Oper. Res.* 50 (2002) 42–47.
124. See: M. Gardner, Mathematical games: The remarkable lore of the prime number, *Sci. Am.* 210 (1964) 120–128.
125. G. H. Hardy and E. M. Wright, *An Introduction to the Theory of Numbers*, 5th ed., Clarendon Press, Oxford, England, 1979.
126. B. Perichon, *A World Record AP26 (Arithmetic Progression of 26 primes)*, The AP26 is listed in "Jens Kruse Andersen's Primes in Arithmetic Progression Records page," 2010 (Available on Internet).
127. M. Barnsley, Chaos game, in: *Fractals Everywhere*, M. Barnsley (Ed.), Academic Press, Cambridge, 1993.
128. R. Sacks, Number spiral, 2003. http://www.numberspiral.com/.
129. G. H. Hardy and J. E. Littlewood, Some problems of "partitio numerorum": III: On the expression of a number as a sum of primes, *Acta Math.* 44 (1923) 1–70, reprinted in *Collected Papers of G. H. Hardy*, Vol. I, pp. 561–630, Clarendon Press, Oxford, 1966.
130. M. J. Jacobson Jr. and H. C. Williams, New quadratic polynomials with high densities of prime values, *Math. Comput.* 72 (2003) 499–519.

3 Graph Theory and Chemistry

In the early days of quantum chemistry, two distinct theoretical approaches emerged: the molecular orbital (MO) theory and the valence bond (VB) theory. The former is an extension of the notion of atomic orbitals from atoms to molecules, which led to the Hückel molecular orbital (HMO) calculations, which were applied to benzene and similar conjugated systems by Hückel and others. The representative book on this is E. Hückel's *Fundamentals of Theory of Unsaturated Aromatic Compounds*, published in 1938 [1]. The second theory is a generalization of the calculations of Heitler and London on the H_2 molecule extended to conjugated systems of organic chemistry. The representative book on this is Pauling's *The Nature of the Chemical Bond and the Structure of Molecules and Crystals*, to be published in 1939, the next year [2]. Interestingly, although conceptually and computationally different, both methods, the MO and VB, found early contact with graph theory. As already mentioned, only 12 years after Kekulé introduced a cyclic valence structure for benzene, Sylvester recognized the graph theoretical content of Kekulé valence structures [3,4]. The graph theoretical content of HMO is, in this respect, of a more recent time, the 1970s, which is almost 100 years later, although in 1956 Günthard and Primas published a paper on the connection between the theory of graphs and the MO theory of molecules with conjugated bond. However, by mid-1970, graph theory was confused and identified by many with HMO although HMO, at best, parallels certain sections of spectral graph theory.

In Table 3.1, we have listed the titles of 10 chapters of an outstanding introductory book on graph theory by Biggs, Lloyd, and Wilson: *Graph Theory 1736–1936* [5]. This book covers 200 years of the development of graph theory from the days of Euler and the problem the seven bridges of Königsberg [6], which marks the beginning of graph theory, until the emergence of the first book on graph theory by D. König in 1936 [7]. As one can see from Table 3.1, application of graph theory to chemistry received some attention during the 200 years of graph theory. However, this was not about HMO but about Kekulé valence structures.

3.1 UNFAMILIARITY WITH GRAPH THEORY

Pictorial representations of graphs are so easily intelligible that chemists are often satisfied with inspecting and discussing them without paying too much attention to their algebraic aspects, but it is evident that some familiarity with the theory of graphs is necessary for deeper understanding of their properties.

V. Prelog [8]

TABLE 3.1

Ten Chapters of a Book about 200 Years of *Graph Theory 1736–1936*

1. Paths
   The problem of the Königsberg bridges
   Diagram-tracing puzzles
   Mazes and labyrinths
2. Circuits
   The knight's tour
   Kirkman and polyhedra
   The Icosian Game
3. Trees
   The first studies of trees
   Counting unrooted trees
   Counting labeled trees
4. Chemical Graphs
   Graphic formulae in chemistry
   Isomerism
   Clifford, Sylvester, and the term "graph"
   Enumeration, from Cayley to Pólya
5. Euler's Polyhedral Formula
   The history of polyhedral
   Planar graphs and maps
   Generalizations of Euler's formula
6. The Four Color Problem – Early History
   The origin of the four-color problem
   The "proof"
   Heawood and the five-color theorem
7. Coloring Maps on Surfaces
   The chromatic number of a surface
   Neighboring regions
   One-sided surfaces
8. Ideas from Algebra and Topology
   The algebra of circuits
   Planar graphs
   Planarity and Whitney duality
9. The Four Color Problem – to 1936
   The first attempt to reformulate the problem
   Reducibility
   Birkhoff, Whitney, and chromatic polynomials
10. The Factorization of Graphs
    Regular graphs and their factors
    Petersen's theorem on trivalent graphs
    Alternative view: correspondences

The degree of unfamiliarity of many chemists with graph theory appears widespread and can be illustrated with the lack of awareness that (i) graphs, despite that they can be drawn in an arbitrary manner, nevertheless have *metric*; (ii) graph identification is a nontrivial problem; (iii) graph symmetry cannot be determined from their pictorial representations because graphs have no fixed geometry and can be represented in widely different ways.

In order to illustrate the last point, consider two graphs of Figure 2.18, which look different and display different symmetry but are, in fact, two different graphical representations of the same graph. This graph is known in mathematical literature as the Petersen graph [9]. It is a *cubic* graph, which means that all vertices have the degree 3. The Petersen graph is named after Julius Petersen (1839–1910), a Danish mathematician, who, in 1898, constructed it to be the smallest bridgeless cubic graph with no three-*edge*-coloring, that is, the Petersen graph is the smallest cubic graph with the property that if one colors its *vertices* so that adjacent vertices have different colors then one needs *three* colors. "Bridgeless" means that the graph has no edge that when erased breaks the graph in two disconnected parts. If one is to color its *edges* so that no edges of the same color meet at any vertex, it needs *four* colors. The Petersen graph has been not only of considerable interest in mathematics, but also of interest in chemistry. It represents degenerate isomerization of the trigonal bipyramidal complex XY_5 when the vertices represent the XY_5 complex and the edges relate to an exchange between equatorial and axial ligands [10]. The contributions of Peterson to mathematics led to a new mathematical discipline, graph theory, and he has been considered the father of modern graph theory [11].

3.2 ON HMO, THE HÜCKEL MOLECULAR ORBITAL THEORY

The fundamental laws necessary for the mathematical treatment of a large part of physics and the whole of chemistry are thus completely known, and the difficulty lies only in the fact that application of these laws leads to equations that are too complex to be solved.

P. A. M. Dirac [12]

The above profound quote of Dirac, well and widely known, clearly professes and, by extension, predicts the rise of quantum chemistry as the science of the future, which, as we know, is a major part of theoretical chemistry of today. The acceptance of novelty then, as is also the case today, was by no means always straightforward and welcome. However, two major events of early quantum chemistry could not be overlooked even by the most conservative circles skeptical of novelty: (i) The so-called "chemical forces," forces that in contrast to gravitational "classical" forces show (1) short range, (2) directional properties, and (3) saturation, are the consequence of the new laws of physics, quantum mechanics, as was demonstrated by Heitler and London in 1927 [13] by solving the Schrödinger equation for the H_2 molecule. (ii) Erich Hückel (1896–1980), a physicist (just as were Heitler and London), applied the novel quantum mechanics to benzene C_6H_6 by solving the problem, focusing attention solely to the six π electrons for which he developed a simplified computational model, and was able to show why benzene C_6H_6 and cyclooctatetraene C_8H_8 are so different.

70 Solved and Unsolved Problems of Structural Chemistry

This model is today known as the Hückel molecular orbital (HMO) method [1]. Less known is that Hückel adopted the so-called "hard ball" potential of Bloch, used in a simplified theory of metals [14]. As mentioned, Hückel was able to explain the fundamental difference between the aromaticity of benzene and the lack of aromatic character in the behavior of eight π electrons of cyclooctatetraene, C_8H_8, known today as the Hückel $4n + 2$ rule. This has been one of the most outstanding and significant accomplishments for early quantum chemistry that follows from *the Hückel $4n + 2$ rule:*

If a cyclic, planar molecule has $4n + 2$ π electrons, it is considered aromatic.

We will relate to the Hückel $4n + 2$ rule in more detail in Chapter 11, which is concerned with aromaticity.

Observe, withstanding Dirac's *"the difficulty lies only in the fact that application of these laws leads to equations that are too complex to be solved,"* that the above two, probably most outstanding early theoretical chemistry results have been based on very *approximate* solutions, proving the correctness of another statement of Dirac: *"I consider that I understand an equation when I can predict the properties of its solutions, without actually solving it."*

Clearly, Heitler and London, as well as Hückel, understood that quantum chemistry is the way to solve the problems of chemistry. It took the time and talents of people such as Linus Pauling and other early pioneers of quantum chemistry before quantum chemistry has been accepted by a wider circle of chemists.

Although Hückel framed his approach in the language of quantum mechanics, the approximations that he used, reduced the Hamiltonian of HMO to the adjacency matrix of a molecular graph of the carbon skeleton of benzene. This was realized already in the mid-1950s [15,16] and associated later HMO with the spectral graph theory [17]. By 1970, however, HMO was followed by more elaborate Hamiltonians, which took into account some electron–electron repulsions, such as the Hamiltonian of Pariser and Parr [18,19] and Pople [20], known as the PPP method [21]. The calculated spectra from the PPP method correlated fairly well with UV experimental spectra of benzenoid hydrocarbons, which made it viable and, at the same time, made HMO an obsolete model for discussing molecular spectra. A number of chemists developing chemical graph theory, notably Gutman, Trinajstić, and coworkers [22–24], who made sizable literature contributions to the chemical graph theory at that time, continued to explore some aspects of HMO.

An already mentioned outstanding introductory book on graph theory, *Graph Theory 1736–1936*, covers historical development of graph theory from the days of Euler to the appearance of the first monograph on graph theory by Hungarian mathematician D. König (1884–1044): *Theorie der endlichen und unedndlichen Graphen*, Akademische Verlagsgesellschaft, Leipzig, 1936 (reprinted: Chelsea, New York, 1950 [25]). Denis König, to evade persecution from the Nazis as a Hungarian Jew, apparently committed suicide in 1944.

According to Biggs, Lloyd, and Wilson, notable graph theorists, application of graph theory to chemistry, today's chemical graph theory, received some visibility during the 200 years of graph theory. But this was because of Kekulé and Kekulé valence structures. Although the spectral graph theory is a visible part of graph theory today [26], it was not part of the theory of graphs during 1736–1936. We will discuss briefly some aspects of the spectral graph theory in the following section.

Criticisms directed toward spectral graph theory that came at that time have overlooked the fact that in addition to UV spectra molecules have additional properties on which the HMO model may offer some information. Apparently, some may be forgetting that the famous simple Hückel $4n + 2$ rule of aromaticity, which illustrates a special case of spectral graph theory, has been based on HMO! Simple does not necessarily mean simplistic.

Let us mention two simple theoretical models that have brought significant insight into inorganic and organic chemistry, respectively.

3.2.1 The Crystal Field Model

This model describes the breaking of degeneracies of d-electron orbitals and f-electron orbitals in complexes due to an electric field produced by surrounding charged ligands. It was used to describe optical spectra of transition metal coordination complexes, some magnetic properties, and colors. It was developed by Bethe and van Vleck [27] in the early 1930s. The crystal field theory was later combined with molecular orbital theory to form the complex ligand field theory [28], which gave insight into the chemical bonding in transition metal complexes.

3.2.2 Extended Hückel Molecular Orbital Method

The extended Hückel method is the semi-empirical quantum chemistry method, developed by Roald Hoffmann in 1963 [29–32]. It extends the Hückel method, which only considered π electrons, to saturated molecules having single CC bonds, referring to the resulting molecular orbitals as sigma orbitals. The extended Hückel method can determine the relative energy of different geometrical configurations. It does not include calculations of the electronic repulsions directly, and the total energy is just a sum of terms for each electron in the molecule. The off-diagonal Hamiltonian matrix elements are assumed to be proportional to the average of the corresponding diagonal elements and the overlap: $H_{ij} \approx S_{ij} (H_{ii} + H_{jj}/2)$. This approximation was due to Wolfsberg and Helmholz [33].

The method was first used by Roald Hoffmann, who developed, with R. B. Woodward, the rules for elucidating reaction mechanisms (the Woodward–Hoffmann rules). These rules explain the stereospecifity of electrocyclic reactions under thermal and photochemical conditions. The rules illustrated elegantly and simply the power of molecular orbital theory to experimental chemists.

The Woodward–Hoffmann rules [32,34–37] have been widely used ever since their introduction, but they have not been often explicitly stated, so we felt that, due to their inherent simplicity, elegance, and importance, they deserve better publicity. Below we have reproduced them as published in [29]:

> In an open-chain system containing 4n-*electrons*, the *orbital symmetry* of the *highest occupied molecule orbital* is such that a bonding interaction between the termini must involve overlap between orbital envelopes on opposite faces of the system and this can only be achieved in a *conrotatory* process.

In open systems containing $4n + 2$ *electrons*, terminal bonding interaction within ground-state molecules requires overlap of orbital envelopes on the same face of the system, attainable only by *disrotatory* displacements.

In a *photochemical reaction* an electron in the *HOMO* of the reactant is promoted to an *excited state* leading to a reversal of terminal symmetry relationships and reversal of stereospecificity.

HOMO stands for the highest occupied molecular orbital. In 1981, Hoffmann was awarded the 1981 Nobel Prize in Chemistry for this work (shared with Kenichi Fukui who developed a related model using frontier molecular orbital theory). Because Woodward had died in 1979, he was not eligible to win what would have been his second Nobel Prize for Chemistry.

Let us return to the spectral graph theory and mention five areas in which spectral graph theory made visible contributions:

1. The story of isospectral graphs
2. Experimental mathematics
3. On the difference between HMO and PPP bond orders
4. On numerical characterizations of folded chains
5. Graphene edges

Some readers may be surprised to see as one of the five areas mentioned "experimental mathematics" because mathematics has been generally perceived as a prototype of theoretical studies. But historically, mathematics started as an experimental science, based on geometrical measurements and constructions, and only as late as 1700 has mathematics been transformed into a formal theoretical discipline based on axioms, proposing conjectures, and searching for proofs. Current interest in experimental mathematics emerged about 1970, when it was recognized that novel insights in mathematics could be obtained by using computers in the search for numerical solutions of problems that previously appeared intractable. The new tool, computers, opened novel routes for studying complex patterns and even finding proofs that are too complex to digest in the "old-fashioned" traditional ways. According to Borwein and Bailey [38]:

> Experimental Mathematics is an approach to mathematics in which numerical computation is used to investigate mathematical objects and identify properties and patterns.

We will introduce a few problems and a few solutions relating to experimental mathematics in the section following the story of isospectral graphs.

3.3 THE STORY OF ISOSPECTRAL GRAPHS

A pair of graphs that has all their eigenvalues the same is called *isospectral* or *cospectral* graphs. The first pair of such graphs, having eight vertices, illustrated in Figure 3.1 (top), was reported by Collatz in 1957 [39] but were discovered before 1943.

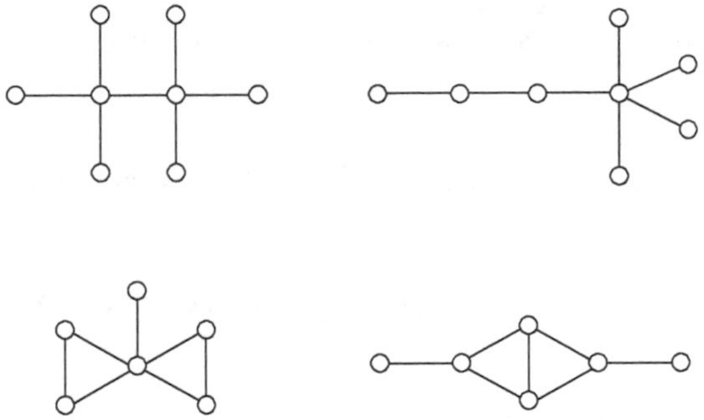

FIGURE 3.1 The smallest pair of acyclic (eight vertices) and cyclic (six vertices) isospectral graphs.

The graphs have the characteristic polynomial

$$x^8 - 7x^6 + 9x^4.$$

They were found by a PhD student of Collatz, U. Sinogowitz, who perished in World War II in 1943. In the early 1960s, when the work of Collatz and Sinogowitz was generally unknown, it was speculated that the characteristic polynomial of the adjacency matrix were uniquely related to graphs [40,41]. However, by the early 1970s, Harary et al. [42,43] demonstrated that this speculation is false by illustrating the smallest pair of isospectral graphs, the pair of graphs on six vertices, shown at the bottom of Figure 3.1 with the characteristic polynomial

$$x^6 - 7x^4 - 4x^3 + 7x^2 + 4x - 1.$$

Initially, very few such graphs have been found, but as time was passing, the number of reported isospectral graphs grew, and eventually, by the 1980s, it was found that they were not as uncommon as was initially believed [44–53]. The first pair of isospectral graphs that represent molecules of chemical interest is the pair of graphs illustrated in Figure 3.2, representing two isomers of dodecane: 4-ethyl-2,4,6-trimethylheptane and 4-isopropyl-2,6-dimethyheptane, reported by Balaban and Harary et al. [43] in 1971.

The next year, Mowshowitz reported the characteristic polynomials of all acyclic graphs having 10 and fewer vertices, from which, in addition to the pair of graphs having $n = 8$ vertices of Figure 3.1, one finds four pairs of isospectral graphs on

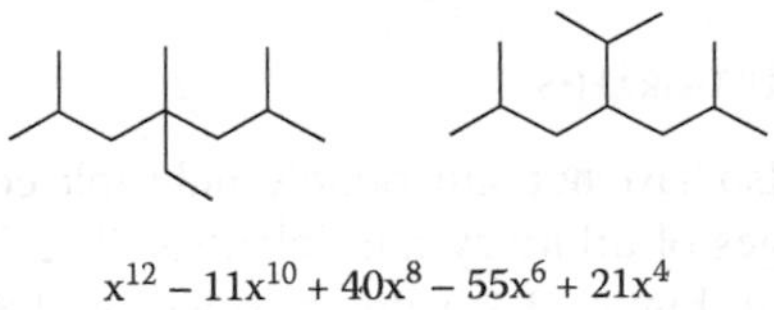

$$x^{12} - 11x^{10} + 40x^8 - 55x^6 + 21x^4$$

FIGURE 3.2 Isospectral pair 4-ethyl-2,4,6-trimethylheptane and 4-isopropyl-2,6-dimethy heptane, reported by Harary et al. [42], and their characteristic polynomials.

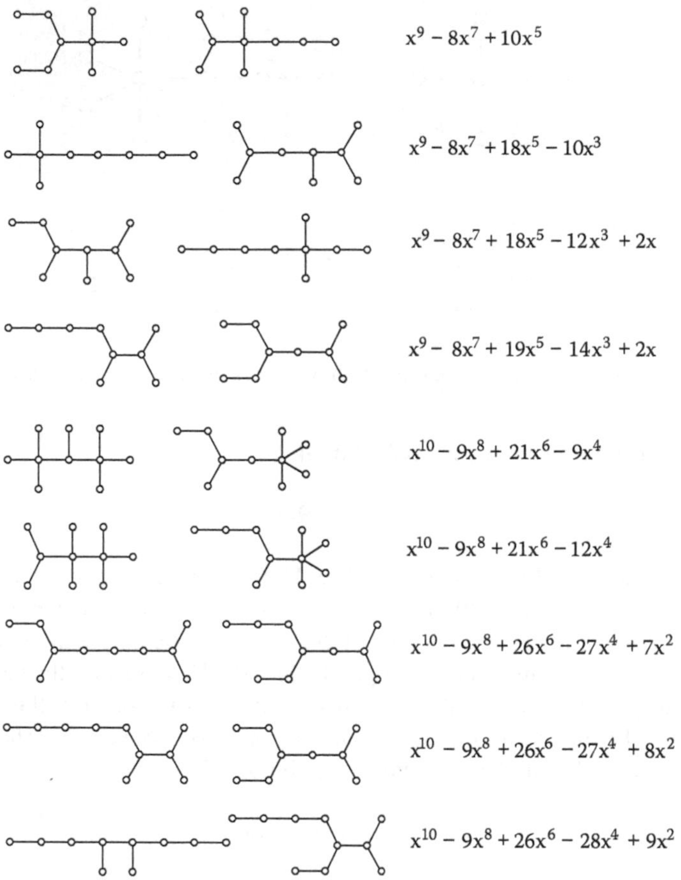

$$x^9 - 8x^7 + 10x^5$$

$$x^9 - 8x^7 + 18x^5 - 10x^3$$

$$x^9 - 8x^7 + 18x^5 - 12x^3 + 2x$$

$$x^9 - 8x^7 + 19x^5 - 14x^3 + 2x$$

$$x^{10} - 9x^8 + 21x^6 - 9x^4$$

$$x^{10} - 9x^8 + 21x^6 - 12x^4$$

$$x^{10} - 9x^8 + 26x^6 - 27x^4 + 7x^2$$

$$x^{10} - 9x^8 + 26x^6 - 27x^4 + 8x^2$$

$$x^{10} - 9x^8 + 26x^6 - 28x^4 + 9x^2$$

FIGURE 3.3 Isospectral acyclic graphs having $n = 9$ and $n = 10$ vertices.

$n = 9$ vertices and five pairs of isospectral graphs on $n = 10$ vertices, all illustrated in Figure 3.3. Except for the first two pairs of graphs on 10 vertices, the remaining seven pairs of graphs are nonane or decane isomers. All the isospectral graphs shown in Figure 3.3 have been detected by a brute-force method, simply by calculating characteristic polynomials for all 47 graphs on $n = 9$ vertices and all 106 graphs on $n = 10$ vertices. The corresponding number of alkanes (for which the maximal valence is four) is 35 and 75, respectively.

3.3.1 ISOSPECTRAL MULTIGRAPHS

Isospectral graphs can also have not only double and triple edges, which are of interest in chemistry, but edges of arbitrary multiplicities. The first pairs of isospectral multigraphs, illustrated in Figure 3.4, were reported in 1988 [53]. It is interesting that although isospectral graphs have received a fair visibility both in chemical and mathematical literature, this is not the case with isospectral multigraphs. So we will

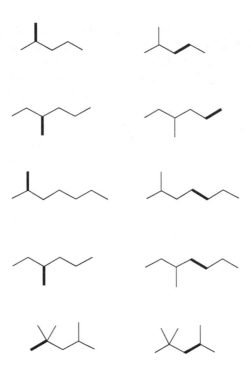

FIGURE 3.4 The smallest isospectral multigraphs having from six to eight vertices.

here present few basic results regarding isospectral multi-trees. All cases considered involve just one multi-edge, which can have arbitrary multiplicity k. The cases with $k = 2$ and $k = 3$ will be of more interest in organic chemistry. The isospectral multi-graphs of Figure 3.4 have been found in a systematic search confined to acyclic graphs having less than nine vertices. Using the "ultimate pruning" procedure, which consists of selecting a single edge and constructing a special 2×2 determinant, which gives the characteristic polynomial of the graph considered; some 300 graphs have been examined, and for $n = 6$, the pair of Figure 3.4 was found, and for $n = 7$ again, one pair is found; for $n = 8$, three isospectral pairs are found, and for $n = 9$ five isospectral pairs are found.

The "ultimate pruning" procedure is described in the section on factoring of the characteristic polynomial that shortly follows, which expressed the characteristic polynomials of general graphs in terms of polynomials of path (linear) graphs. Observe Figure 3.4, in which are depicted the isospectral multi-trees for $n = 6$, $n = 7$, and $n = 8$, and each pair of isospectral multigraphs when, for the multiplicity of the edge, is considered the special case, $k = 1$, reduce to the same graph. The same is the case also for the five isospectral multigraphs having $n = 9$ vertices (see Figure 5 in Ref. [53]). The question can therefore be raised, *Are there isospectral multigraphs that for $k = 1$ do not reduce to the same graph?*

The answer to this question is hitherto unknown although it will not be surprising that there are isospectral multi-trees that for $k = 1$ do not reduce to the same graph.

In Table 3.2, we show the coefficients of the 15 characteristic polynomials of all multigraphs for $n = 6$, expressed in terms of the path polynomial L_n: $L_6 + c_4L_4 + c_2L_2 + c_0L_0$. As one can see from Table 3.2, the coefficient c_4 for all multi-trees having $n = 6$ vertices is equal $(1 - k)$. This means that all isospectral multi-trees must have the same k, the same multiplicity of the edge. The path polynomials L_n appear very suitable for computations of the characteristic polynomials of graphs, and it is useful to have coefficients of x^n powers expanded in L_n. In Table 3.3 at left, we show the coefficients of L_n polynomials when the x^n powers are expanded in L_n.

TABLE 3.2

The Coefficients of Characteristic Polynomials in Chebyshev Expansion for Isomers of Hexene

Multigraph	L_4	L_2	L_0
	$1-k$		
	$1-k$	$1-k$	
	$1-k$	$1-k$	$1-k$
	$1-k$	$-k$	
	$1-k$	$1-2k$	$1-k$
	$1-k$	$-k$	
	$1-k$	-1	$-1+k$
	$1-k$	-1	-1
	$1-k$	$1-2k$	$-k$
	$1-k$	$-k$	$-k$
	$1-k$	$-1-k$	-1
	$1-k$	$1-3k$	$1-2k$
	$1-k$	$1-k$	$1-k$
	$1-k$	$1-k$	$1-k$
	$1-k$	$1-k$	$1-k$

TABLE 3.3

Truncated Pascal Triangle Giving the Coefficients in Expansion of x^n in Terms of L_n

Coefficients						Polynomials
					1	$x = L_1$
				1	1	$x^2 = L_2 + 1$
			1	2		$x^3 = L_3 + 2L_1$
		1	3	2		$x^4 = L_4 + 3L_2 + 2$
		1	4	5		$x^5 = L_5 + 4L_3 + 5L_1$
	1	5	9	5		$x^6 = L_6 + 5L_4 + 9L_2 + 5$
	1	6	14	14		$x^7 = L_7 + 6L_5 + 14L_3 + 14L_1$
1	7	20	28	14		$x^8 = L_8 + 7L_6 + 20L_4 + 28L_2 + 14$
1	8	27	48	42		$x^9 = L_9 + 8L_7 + 27L_5 + 48L_3 + 42L_1$
1	9	35	73	90	42	$x^{10} = L_{10} + 9L_8 + 35L_6 + 75L_4 + 90L_2 + 42$

The pattern of numbers in the left part of Table 3.3 has a close relationship with the Pascal triangle and can be viewed as a vertically truncated Pascal-like triangle. The numbers appearing in the last two columns (which are identical) are known as the Catalan numbers:

$$1, 2, 5, 14, 42, 132$$

given by the expression

$$C(n) = (2n)!/[n!(n + 1)!].$$

They were introduced in mathematics by Belgian mathematician Eugène C. Catalan (1814–1894). In 1838, Catalan considered the following problem [54]: *How many different ways can a product of* n *different factors be calculated by pairs?*

Here "calculated by pairs" counts every product of two factors. Thus product of *a b c d* leads to the following five paired multiplications:

$$[(a\ b)\ c]\ d \quad [a\ (b\ c)]\ d \quad (a\ b)(c\ d) \quad a\ [(b\ c)\ d] \quad a\ [b\ (c\ d)].$$

and arise in various geometry-related and combinatorial enumerations. For example, consider the count of the number of ways brackets can be combined. For the case of two and three pairs of brackets, one has two and five possibilities shown below:

$$( )( ) \quad (( ))$$

$$( )( )( ) \quad (( ))( ) \quad ( )(( )) \quad (( )( )) \quad ((( ))),$$

which are equal to C(2) and C(3).

As is not uncommon in science and mathematics problems thought to be new, may have been considered before in a somewhat different context. Thus, in 1751, the numbers

1, 2, 5, 14, 42, 132, 429

were known to Euler who posed the problem: *In how many ways can a (plane convex) polygon of* n *sides be divided into triangles by a diagonal?*

Obviously, a triangle has a single triangulation, and a square has two because it has two different diagonals. That a pentagon has five triangulations is fairly obvious because it has five vertices, from each two internal edges to complete the triangulation. The case of a hexagon (benzene ring) is not obvious, but it is not difficult to find. In Figure 3.5, we have illustrated all different patterns of triangulations of a hexagon. When rotated the first pattern appears 5 times, the next two appear twice, giving additional 9 triangulation, totaling with five shown 14.

The difficulties start with larger polygons, but Euler found the number of triangulations for polygons having 7, 8, and 9 sides, and on the basis of these initial numbers was able to develop the formula:

$$E_n = [2 \times 6 \times 10 \times \ldots (4n - 10)]/(n - 1)!$$

Here E_n is the number of triangulations of the n-polygon. Several years later, in 1758, Euler sent a letter to Segner sending him his first numbers, 1, 2, 5, 14, 42, 132, and 429 and his recursion formula shown above. Segner considered a polygon having $m + n$ sides as being made of two polygons, having m and $n - 1$ sides, respectively, fused to a triangle. This has led to the recurrence formula for the number of triangulations of polygons having n-sides:

$$E_n = E_2 \times E_{n-1} + E_3 \times E_{n-2} + \ldots + E_{n-1} \times E_2$$

the factor E_2 has the value 1. For example,

$$E_6 = E_2 \times E_5 + E_3 \times E_4 + E_4 \times E_3 + E_5 \times E_2 = 14$$

$$E_7 = E_2 \times E_6 + E_3 \times E_5 + E_4 \times E_4 + E_5 \times E_3 + E_6 \times E_2 = 42$$

And so on. In addition, it is nice looking, and being useful for relatively small E_n numbers, the formula becomes tedious and impractical for calculation in the case of

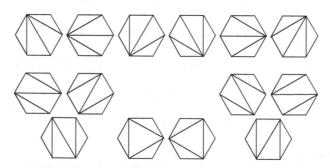

FIGURE 3.5 The 14 possible triangulations of a hexagon.

large n. It was only relatively recently (in 1941, that is, 190 years after Euler), that H. Urban [55] proposed the simple recurrence expression:

$$E_n = [(4n - 10)/(n - 1)]E_{n-1}$$

This appears to be an elegant contraction of the initial Euler's recurrence formula. For more on this topic, interested readers can consult pp. 21–21 of Ref. [56].

In addition to mathematics, Catalan numbers have found application in chemistry and biology. For discussion of Catalan numbers in chemistry, see a paper by Okhami and Hosoya [57]. Randić et al. [58] reported the count of the number of canonical "excited" valence structures for polycyclic conjugated hydrocarbons having $(n + 1)$ CC double bonds to be $C(n)$. In molecular biology, Manfred Eigen, a German biophysical chemist who shared the Nobel Prize in chemistry in 1967 with R. G. W. Norrish and G. Porter for their study of extremely fast reactions, came in his studies to the "Origin of Life" across the vertically truncated Pascal triangle [59]. An introductory article by D. M. Campbell on the computation of Catalan numbers can be found in *Mathematical Magazine* [60].

3.3.2 Vertex Neighbor Sum Rule

Although literature on isospectral graphs is abundant and much has been learned about isospectral graphs and their construction, apparently not all questions relating to isospectral graphs have been answered. For example, 20 years ago, Randić, Guo, and Kleiner proposed the ring enlargement process for construction of subspectral graphs [49]. Ring enlargement consists of constructing a graph double the size from a smaller graph such that it has the same "vertex neighbor sum rule." The "vertex neighbor sum rule" consists of a list of the sum of coefficients of adjacent vertices for all vertices. In a later section on subspectral graphs, we illustrate such graphs. For subspectral graphs, holds the following:

Let g be a smaller graph whose vertices have labels a, b, c... If vertices of a larger graph G satisfy the neighbor sums of a smaller graph g, or linear combinations of the same sum rules, then necessarily the eigenvalues of G include the eigenvalues of g.

What Randić, Guo, and Kleiner noticed is that if two graphs g and h are bipartite and isospectral, so will be graphs G and H obtained by the ring enlargement procedure. From this follows:

Conjecture: If graph g yields non-isomorphic bipartite enlargements G_1, G_2, etc., the derived graphs are not only subspectral to the parent graph g but are also mutually isospectral.

As was noticed, at that time, some isospectral graphs g and h generate isospectral pairs G, H via the ring enlargement procedure, and some do not. Why? The answer is not known, so we have an open problem:

Problem 4

Under what conditions will a pair of isospectral graphs g, h upon enlargement produce isospectral pairs G and H?

Liu and Jiang [50] have considered the problem discussed above and came up with some generalizations that involve Möbius molecules. They also extended the Sachs theorem on the graphical content of the coefficients of the characteristic polynomials to Möbius molecules. Möbius molecules are molecules in which the overlap between the adjacent π electrons can be negative. The Sachs theorem [61,62] gives expressions for the coefficients of the characteristic polynomial of a graph G in terms of the set of subgraphs of G, which are either isolated edges and/or disjointed rings. An isolated edge is a formally complete graph on two vertices (designated as K_2) and disjointed rings are isolated cycles (designated as C_n, where n is the size of the cycle). We would strongly recommend to those readers who may become interested to look at the above topic to consult the very informative publication of Liu and Jiang [50]. We also recommend an introductory review on characteristic polynomials of a chemical graph by Trinajstić [62]. For a brief review on the impact of the Sachs theorem on theoretical chemistry, see the testimony of Gutman [63].

3.3.3   ISOSPECTRAL BENZENOID DERIVATIVES

The first pair of isospectral graphs among benzenoid molecules relating to π electrons is the pair 1,4-divynilbenzene and 2-phenylbutadiene, shown in the lower part of Figure 3.6, reported in 1973 by T. Živković [64]. It is interesting to learn how this pair of graphs was discovered. In 1965, the time when the HMO method was widely applied to benzenoid hydrocarbons, Coulson and Streitweiser published their *"Dictionary of Pi-Electron Calculations"* [65]. In this book were listed HMO results, which include all eigenvalues and eigenvectors of a collection of benzenoid compounds and their derivatives. Apparently, the book was well received and followed with an expanded two-volume extension. The *Dictionary of Pi-Electron Calculations* included the eigenvalues and eigenvectors of 1,4-divinylbenzene and 2-phenylbutadiene of Figure 3.6, just a few pages

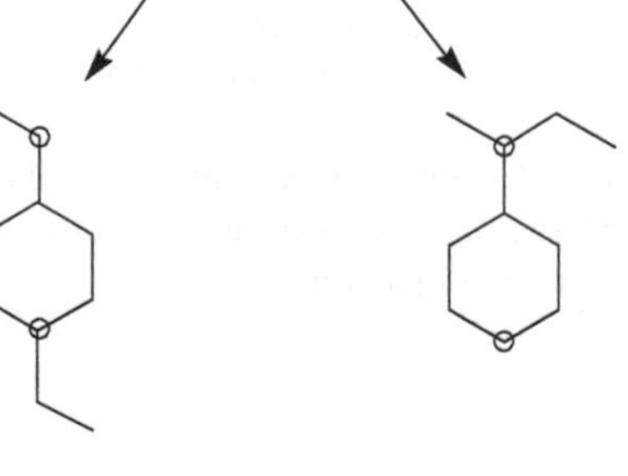

FIGURE 3.6   Isospectral pair 1,4-divynilbenzene and 2-phenylbutadiene have common fragment vinylbenzene, the two endospectral vertices are shown as small circles.

apart. However, nobody observed that the two molecules have *all* eigenvalues identical until 8 years later when T. Živković, browsing through the pages of the *Dictionary* of Coulson and Streitweiser, found that the abovementioned two molecules are isospectral; that is, they have the same set of eigenvalues. Apparently, HMO quantum chemists of that time, probably including Coulson and Streitweiser, did not expect that different molecules could have identical eigenvalues and so were not looking to find them even though isospectral graphs were known at that time in mathematics and physics.

A close examination of the graphs of 1,4-divynilbenzene and 2-phenylbutadiene, including also their eigenvectors, has led to the notion of *endospectral graphs* [66,67]. Endospectral graphs (*endo* being Greek for inside, within) are graphs that have *a pair of symmetry nonequivalent* vertices that generate a pair of isospectral graphs when carrying the same substitution fragment. They be can identified by searching for symmetry nonequivalent vertices that have the same coefficients in all their eigenvectors. A consequence of this is that at such vertices, one may attach an edge or any larger combination of edges, and the two resulting graphs will be isospectral. The endospectral vertices have to be symmetry nonequivalent because if they were equivalent, by attaching a single edge or any fragments to such vertices, one would produce a pair of isomorphic graphs.

The graph of vinylbenzene (shown at the top of Figure 3.6) is the smallest benzenoid endospectral graph having Kekulé valence structures and having a pair of nonequivalent vertices having the same coefficients in all their eigenvalues. Hence, when an ethyl group is attached to two of its endospectral vertices (vertices indicated by small circles in Figure 3.6), it produces the isospectral pair 1,4-divinylbenzene and 2-phenylbutadiene. In Figure 3.7, we show several of the smallest endospectral acyclic graphs, which can produce many larger acyclic isospectral pairs. The graphs of Figure 3.7 have been introduced in references [46,48,51,68–72] during 1973–1989.

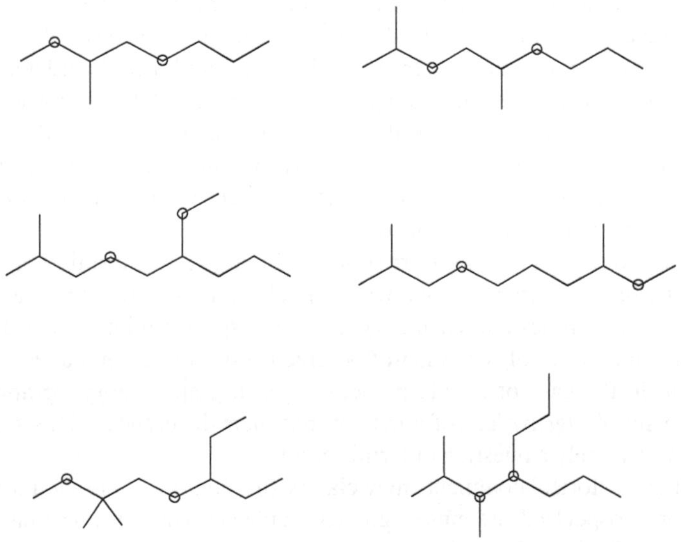

FIGURE 3.7   Endospectral graphs having from $n = 9$ to $n = 12$ vertices, having one, two, and three branching vertices. The endsopectral vertices are indicated by small circles.

Thus, one should not be surprised to find out that isospectral graphs are not so uncommon among larger graphs. Mathematician Schwenk has proven a theorem that shows that as n, the size of graphs, tends to infinity, the ratio of isospectral acyclic graphs to all acyclic graphs tends toward 1. This would justify the provocative title of his article, "*Almost all trees are cospectral*" [68]. More recently, however, this was questioned, at least when one considers random graphs [48].

For us in chemistry, the important question is, are isospectral graphs relevant for chemistry? Just because they were not discovered when the HMO "ruled the waves" of quantum chemistry, this does not make them intrinsically important. Heilbronner and Jones in their note to the editor of the *Journal of the American Chemical Society* [73], one of the leading journals of chemistry in the world, went on to claim that isospectral graphs are not useful because HMO eigenvalues do not offer good spectral results. This was published when it was generally accepted that HMO is no longer useful in chemistry. They based their argument on comparing the *actual* molecular UV spectra of 1,4-divinylbenzene and 2-phenylbutadiene. As could have been expected, the UV spectra corresponding to transitions involving π electrons of the two isospectral molecules were found not to be the same and not even similar to experimental results.

To several researchers, including us, the note from Heilbronner and Jones appeared as an unwarranted "dramatization" and misrepresentation of the "irrelevance" of graph spectra in chemistry. Thus, one of the present authors has sent to the editor of the *Journal of the American Chemical Society* a comment on the note by Heilbronner and Jones, titled "*Spectral differences between 'isospectral' molecules.*" This comment essentially stated that if Heilbronner and Jones wanted to show that HMO is not good for predicting UV spectra of benzenoid molecules, there was no need to consider spectra of *two* molecules; *one* molecule, isospectral or not, would be enough! After submitting this note to the *Journal of the American Chemical Society*, a letter from the editor came with the decision of rejection. The basis of rejection was the view of a reviewer that the note from Heilbronner and Jones should not have been accepted! But it has been accepted and printed! Of course, all this is not so important and was only mentioned here as an illustration that some people in the "establishment" prefer the status quo and try to block any dissent. Is this in the interest of the general public and chemistry as science? Unfortunately, similar incidents may not be so isolated as one would like to expect.

Let us remind readers that there are some differences in the terminology of chemical graph theory and graph theory when speaking of spectra. In contrast to the term "spectra" in physics and chemistry, in which spectra relate to the differences between the eigenstates of the system (the eigenvalues of the interaction matrix or Hamiltonian in the case of quantum chemistry), in graph theory, "graph spectra" relate to the set of *eigenvalues* of a matrix, not their differences. This need not be confusing; it is merely a question of terminology.

A few short historical comments may clarify the situation. One should recall that the notion of "isospectral" systems originated in physics with the question: "Can one hear the shape of a drum?" which was the title of a paper by Mark Kac [74] in 1966 in the *American Mathematical Monthly* of the Mathematical Association of America. Kac had been invited to submit an article that brought interesting mathematical topics to a wide audience. The next year, Kac was awarded two prizes for this article, in which he had tried to show what can be inferred about the shape of a drum from the knowledge of all its eigenvalues (tones) and how this problem is of interest in mathematics and physics. The problem, and even its name, has been known for some time before Kac addressed it. The problem has been traced back to Hermann Weyl (1885–1955), an outstanding German mathematician who made great contributions to mathematical physics. However, it was the article of Kac that drew wider attention to this interesting question, which, in simple language, can be transformed into a question: Can two drums (membranes) of different shapes produced exactly the same tones? The literature and current interest in various aspects of this problem started initially in physics [75,76]. The subject is still alive in physics as one can verify by visiting the Internet. The spectral theory of graphs [26,77] and isospectral graphs (known in mathematics as cospectral graphs) continue to remain of considerable interest in mathematics [78–80].

In chemistry, isospectral graphs were a short-lived episode mostly associated with the search and construction of such graphs that may represent molecules [68,69,81–87]. Unfortunately, this was mostly at the time when HMO was viewed by most quantum chemists to be outdated. But the adjacency matrix and its eigenvalues (the HMO model) continue to be of interest in other areas of chemistry, including the characterization of highly folded chains (to model folding of proteins) [88] and the characterization of fullerenes [89]. So we may end this topic on isospectral graphs by paraphrasing the traditional proclamation made following the accession of a new monarch in various countries, exclaiming: *"The King is dead; long live the King"* by exclaiming, *"HMO is dead; long live HMO."* Clearly as far as discussing molecular spectra, HMO is dead, but HMO founded novel applications, so it is not dead!

Before leaving HMO, let us mention an important theorem about HMO energies for alternant systems (π electron systems having no odd rings) due to Coulson and Rushbrooke [90], known as the *Coulson-Rushbrooke pairing theorem*:

> For every HMO energy $\alpha + x\beta$ in an alternant hydrocarbon, there exists another energy $\alpha - x\beta$; that is, the roots of the HMO secular determinant occur in pairs, which are equal in magnitude and opposite in sign. Furthermore, the coefficients of paired MOs are the same except that the algebraic sign at every atom is opposite.

Interestingly, this theorem was first found in theoretical chemistry and only later was rediscovered in mathematics. In the case of π electron systems having no odd rings but having an odd number of carbon atoms (such as is the case of allyl radicals) for which the Coulson-Rushbrooke pairing theorem also holds, one of the eigenvalues is $x = 0$. The corresponding molecular orbitals are known as nonbonding molecular orbitals. Longuet-Higgins made extensive studies of HMO molecular orbitals, including nonbonding MOs, which allows one to find the number of Kekulé valence structures of π electron systems of interest [91,92]. This is a useful illustration of the hidden connections between MO and VB methods of quantum chemistry, which reveals an intriguing association between the continuum (calculus) and the discreteness (enumeration).

3.4 THE DIFFERENCE IN HMO AND PPP BOND ORDERS

This section illustrates that HMO is not quite dead just because the HMO has no future in the prediction of molecular spectra. In that respect, HMO is certainly dead, but as previously mentioned, there are *other* potential applications of the adjacency matrix (which, in the case of π electron systems, corresponds to the Hückel matrix) that relate to the molecular connectivity. One such application is the subject of this section, which will show that HMO calculations may represent a useful standard in comparing the results of various quantum chemical methods. There are two dominant variants of the bond orders of conjugated molecular systems. The first was introduced by Pauling and is based on the consideration of Kekulé valence structures [2], and the second was introduced by Coulson and is based on the coefficients of molecular orbitals [93]. Although this has not been stressed in the past, it is worth observing that Pauling bond orders are evaluated by *counting*, and Coulson bond orders result from the application of *calculus*, implying *continuum*. Thus, we have a somewhat rare instance of overlap of approaches based on discrete mathematics and calculus. Later, we will see another such case arising in the calculation of the ring currents in benzenoid and non-benzenoid hydrocarbons.

Bond orders offer some insights into the relative lengths of CC bonds in conjugated systems, and although there are some differences between Pauling and Coulson bond orders, they both offer a fair correlation with the known experimental CC bond lengths. The approach of Coulson has an advantage over the approach of Pauling in that his idea of bond orders can be generalized beyond the HMO to other MO calculations. In principle, the Pauling bond orders also can be generalized by introducing different weights for different Kekulé valence structures although it appears that not much work was done in this direction.

The Pariser–Parr–Pople molecular orbital calculations, as already mentioned earlier in Chapter 2, consider contributions to MO arising from electron repulsions. The resulting calculations produced more reliable π electron molecular orbitals in comparison to the Hückel MO approach, which completely ignore electron–electron repulsions. The Coulson bond orders, based on PPP MO wavefunctions, are thus expected to be different *and* better than Coulson bond orders based on HMO. A comparison between the two MO bond order models could have been done in 1953 when PPP was introduced, but it waited almost 10 years for its first consideration [94,95]. The comparison is of interest because all CC bonds need not be equally affected, and the comparison may point to changes in molecular geometry due to the influence of the π electron–electron repulsion.

The comparison of bond orders based on the HMO and PPP methods has shown some interesting results on benzenoid hydrocarbons illustrated in Figure 3.8. In this figure, are shown selected Kekulé valence structures of 10 smaller benzenoid hydrocarbons. In these structures, the bonds shown as C=C are the bonds for which Coulson bond orders based on the PPP MO have increased, suggesting an increase in CC double bond character. It is very intriguing that the results represent individual Kekulé valence structures. Due to symmetry, in the case of anthracene, pyrene, and ovalene, there are two symmetry equivalent Kekulé structures similarly affected; both are shown in Figure 3.8. A close look at the valence structures of Figure 3.8 shows that these are

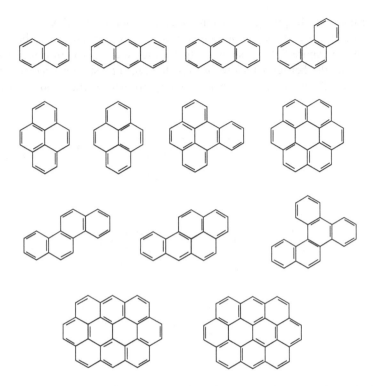

FIGURE 3.8   Fries structures for a selection of smaller benzenoid hydrocarbons. Anthracene, pyrene, and ovalene have two Fries structures.

the Kekulé valence structures having the largest possible number of rings with three C=C bonds, same as the rings in the two Kekulé valence structures of benzene. These Kekulé structures are known as Fries structures [96]. Fries came with an empirical rule stating that those Kekulé structures that have the largest number of benzene rings, each having three C=C bonds, represent the most stable Kekulé structures.

This result, which even today appears not to be widely known, is very interesting and instructive. Unexpectedly, a comparison of two distinct MO calculations offers information of considerable interest for the VB model. In VB calculations, the starting point is the assumption that all Kekulé valence structures have the same weight. The Fries rule suggests that some Kekulé valence structures appear to be more important than others.

We end this short excursion into the topic on bond orders, which has received some attention in the literature [97], by considering the possibility of exploring different weighting models for Kekulé valence structures as an "Unsolved Problem." In the past, almost always, all Kekulé valence structures have been assumed to have the same weight. But only after exploring the effects of an alternative weighting scheme, one could find out if there are better Pauling "bond order–bond length" regressions for CC bonds of benzenoid hydrocarbons.

A simple new weighting scheme, for example, could be to take the count of rings having three C=C bonds as the weight of Kekulé valence structures. The model can be further improved by giving to benzene rings having three C=C weight 3, to those having two and one C=C bonds in their Kekulé valence structure weights 2 and 1, respectively. In Figure 3.9, we show such Kekulé valence structure weights for a selection of smaller benzenoid hydrocarbons (having at most four benzene rings). In Figure 3.10, we show, for the same benzenoid hydrocarbons, the bond contributions

FIGURE 3.9   The weights of Kekulé valence structures of smaller benzenoid hydrocarbons based on the count of benzene rings with three C=C bonds.

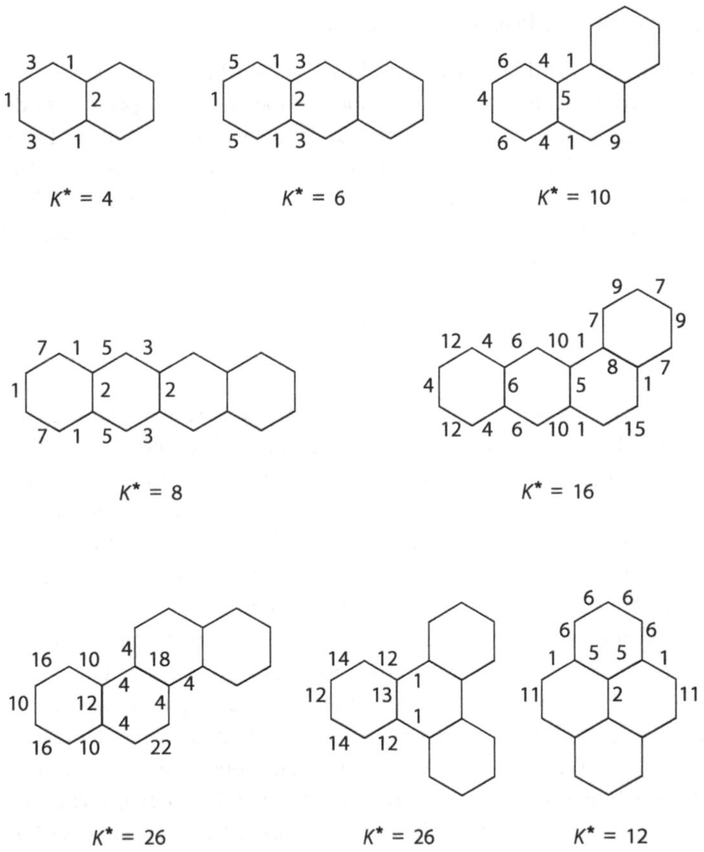

FIGURE 3.10 The bond contributions to the CC bond orders based on the weighted Kekulé valence structures. Bond orders are obtained by dividing the bond contributions by K^*.

to the CC bond orders based on so-weighted Kekulé valence structures. Because now the sum of bond weights exceeds K, the number of Kekulé valence structures, we have to introduce a virtual number of Kekulé valence structures, designated as K^*, given by the sum of bond weights of all Kekulé valence structures. Bond orders are obtained by dividing the bond contributions by K^*.

Another approach of weighting could be based on the count of conjugated circuits (to be discussed at some length later), which are circuits within individual Kekulé valence structures having alternating C=C and C–C bonds. In benzenoid systems, all conjugated circuits are of size $4n + 2$; benzene rings having three C=C represent the smallest conjugated circuits R_1, for $n = 1$ [98]. Yet another possibility is to explore the "degree of freedom of Kekulé valence structures" [99,100], which may also be of interest in arriving at more general VB bond orders.

In following this, of course, one is running a risk of being "accused" of engaging in the development of "approximate" methods at the time of *ab initio* MO calculations. But if such "approximate" explorations bring novel insights into molecular structures considered, the effort and the time spent would be justified.

88	Solved and Unsolved Problems of Structural Chemistry

3.4.1 On Modeling Folding of Proteins

Shahknovich and Gutin [88] have introduced a 3-D lattice model of protein folding in which amino acids are placed inside a cube in which protein is depicted as
a compact self-avoiding chain structure, like one shown in Figure 3.11, which was
considered by Šali, Shahknovich, and Karplus [101]. As outlined by Šali et al., for
such protein models, one can calculate the total energy E of conformation, which is
given by the sum of the contact energies B_{ij} between the non-bonded adjacent amino
acids within the $3 \times 3 \times 3$ lattice:

$$E = \sum_{i<j} \Delta(r_i, r_j) B_{ij}$$

Here, $\Delta(r_i, r_j)$ is one or zero, depending if amino acids are in "contact," that is, non-
bonded but in adjacent positions in the Cartesian grid, or not, respectively. If one
further assumes that all B_{ij} are equal, mathematically, the problem is reduced to a
HMO-type problem, but the corresponding matrix does not relate to a π electron
system but a protein embedded on lattices. The above simplification offers an alternative approach to the problem of the total energy of conformation to the folded protein, which, instead of summing the contact contributions, considers for interactions
the "hard ball" potential of Bloch, adopted by Hückel for his molecular orbital calculations of benzene and other π electron systems [1,102–105]. The simplifications
may be drastic for real proteins, but nevertheless, one might get some novel insight
into the old problem of *"How does a protein fold?"* (which is incidentally the title of
the publication of Šali et al. [101]). As already mentioned, the more realistic model of
Šali, Shahknovich, and Karplus, instead of using the adjacency matrix for the amino
acid contact interaction considered here, represents the close contacts of amino acid
interactions by a weighted "contact adjacency" matrix, in which the weight B_{ij} correspond to the contact energies between amino acids in real proteins.

Characterization of the folding of chain-like structures (not proteins) of *definite*
geometry in 2-D and 3-D space was considered some time ago by Randić et al.
[106]. They have introduced the D/D matrix, the elements of which are given as
the quotients of the distance between two vertices through space and their distance
measured along the connecting bonds. It was found that the leading eigenvalue of the

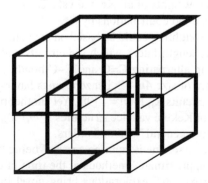

FIGURE 3.11   The lattice protein of Šali et al. [100]. Thin lines indicate the "contact" term.

D/D matrix offers good characterization of the degree of folding of such structures. The element (i, j) of the D/D matrix is defined as

$(D/D)_{i,j}$ = the length of the shortest (Euclidian) distance(s) divided by the length of the distance between vertices i and j along the connecting bonds

$$(D/D)_{i,j} = 0$$

That is, the matrix elements are given by the quotient of the "through space" and the "through bond" distances between the vertices i and j. That the leading eigenvalue of the D/D matrix is a measure of the degree of molecular folding should not be surprising. This follows from the well-known theorem of algebra, which states that the upper and the lower bounds on the leading eigenvalue of a symmetric matrix are the largest and the smallest row (or column) sums of the matrix [107,108]. Clearly, as a chain becomes more bent, the through-space distances between vertices decrease, and through-bond distances remain the same. This results in a decrease of the row sums of the adjacency matrix, which thus decreases the bounds for the leading eigenvalue of D/D matrix. The approach also has been tested on linear polymers, including fractals [109], lattice proteins [110], and proteins [111]. One can find more on D/D matrices in several reviews that discuss this subject in more detail [112–115].

For the protein of Figure 3.11, the leading eigenvalue of the D/D matrix is $\lambda_1 =$ 9.2459 and the total energy $E = 17.256$. In Figure 3.12, we have illustrated nine additional folded proteins. The proteins $(A–E)$ were considered by Li, Helling, Tang, and Wingreen [116], Kardar [117], Borman [118], and Randić [119]. Proteins F and G have been considered more recently by Tang and Zheng [120], and the protein H was mentioned by Fink in his dissertation [121]. The last lattice protein of Figure 3.12 has been constructed so to fill the $3 \times 3 \times 3$ cube with the maximal number of straight segments in order to explore if so-folded protein represents one of the least folded proteins.

The leading eigenvalues and the total energies of the 10 folded proteins of Figure 3.12 are listed in the first line of Table 3.4. Observe from Table 3.4, that there is a significant variation in the listed leading eigenvalues among the 10 folded lattice proteins. The most folded is the protein of Šali, Shahknovich, and Karplus (Figure 3.11) and the protein A of Li, Helling, Tang, and Wingreen. The least folded is the protein F, which has 10 straight line local segments, and the last protein I, having also 10 line segments, is found to be only "moderately" folded. Observe also that there is no simple correlation between the "degree of folding" and the total energy, suggesting that additional factors play a role in the eignvalues of such matrices. One such factor could be the end-to-end distance of the proteins, but even this alone cannot account for all apparent variations in the degree of folding among the 10 lattice proteins considered.

A better understanding of the folding patterns and the possible role of the leading eigenvalue and the total energy (which here we took to be the sum of all positive eigenvalues) for characterization of folded structures remains one of the unsolved problems in this area. Two factors have been considered to explain the stability of model proteins: Li, Helling, Tang, and Wingreen [116] thought of importance the

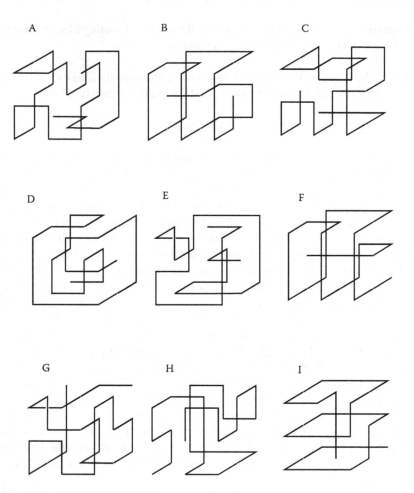

FIGURE 3.12   Additional folded proteins on a $3 \times 3 \times 3$ lattice.

TABLE 3.4
The Leading Eigenvalues (λ_1) and Total Energies (E) for Protein Folding Models of Figure 3.12

	A	B	C	D	E	F	G	H	I
λ_1	2.5699	2.8051	2.7006	2.7719	2.6868	2.7327	2.6266	2.7101	2.7881
E	15.63	16.47	16.64	16.50	16.87	16.14	16.96	16.53	17.12

number of sequences that produce the same folding pattern, and Šali, Shahknovich, and Karplus [101] considered the *speed* at which proteins fold to their unique native state. According to Levinthal, local variations in folding may also be of some relevance [122]. At the second line in Table 3.4, we have listed the total energies of conformations of the 10 lattice proteins based on the constant "contact" energies.

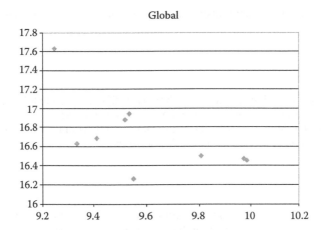

Global

FIGURE 3.13 Plot of the total energy of conformation against the degree of foldedness of proteins.

In Figure 3.13, we correlated the total conformation energies with the degrees of folding of lattice proteins. The preliminary study shows that among the proteins considered, the protein of Šali, Shahknovich, and Karplus is found to be among the *most folded* and having *the largest total energy* of conformation. Observe in Figure 3.13 that most proteins lie close to the diagonal descending from the top left corner of the diagram to the right bottom corner. From this, it follows that the "contact energy" decreases as the degree of folding decreases (or the leading eigenvalue of D/D matrix increases). There are three outliers that correspond, from the left to the right, proteins *A*, *G*, and *H*. It is not clear at this moment why these proteins deviate and have visibly smaller "contact energy" in relation to their degree of folding.

It is not easy to visualize the differences between the proteins A–I from Figure 3.12. If, however, we view the corresponding folding matrix as the adjacency matrix of a graph, one can associate an alternative pictorial representation with each protein, which we have illustrated on two of the proteins considered in Figure 3.14. These "contact interaction" graphs show an apparent lack of similarities between the considered proteins more clearly. Nevertheless, it appears that when analyzing a larger set of proteins embedded in such, "contact interaction" graphs and their invariants may offer alternative representations of proteins to be of use in their comparative study.

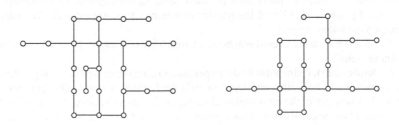

FIGURE 3.14 Illustration of "contact interaction" graphs for model folded proteins.

Let us end this section with raising some questions and proposing a few unsolved problems.

Problem 5

Is there isospectral folding? That is, are there two different protein folding that will produce the same eigenvalues of the binary "folding" matrix?

Observe that here isospectrality is associated with "contact interactions" and has nothing to do with spectral properties of proteins.

Problem 6

How many different folding patterns exist for the $3 \times 3 \times 3$ lattice? This is of some interest if one contemplates the search for "isospectral" proteins.

Problem 7

Can one find the most "stable" folding pattern, and/or find structural conditions required for the most stable folding patterns?

We are not characterizing these problems as important, but rather as possibly interesting. As of today, some of them may turn out to be of more interest to mathematicians than chemists. However, one day, it may be found that these problems are more interesting than previously thought! So if one invests a *little* time in investigating these problems, it may turn out not to be bad investment. But as a rule for tempting problems, such as these may be to some, for which it is not known how important they may be, it may be prudent to spend some time—but not too much time!

3.5　GRAPHENE EDGE

Graphene, an infinite layer of fused benzene rings, became reality on October 22, 2004, just about 10 years ago. This is how graphene was described in the editorial of Ref. [123]:

> In 2003, one ingenious physicist took a block of graphite, some Scotch tape and a lot of patience and persistence and produced a magnificent new wonder material that is a million times thinner than paper, stronger than diamond, more conductive than copper. It is called graphene, and it took the physics community by storm when the first paper appeared the following year.
>
> The man who first discovered graphene, along with his colleague, Kostya Novoselov, is Andre Geim...
>
> In October 2004, Geim published a paper announcing the achievement of graphene sheets in Science magazine, entitled "Electric field effect in atomically thin carbon films." It is now one of the most highly cited papers in materials physics, and by 2005, researchers had succeeded in isolating grapheme sheets. Graphene is a mere one atom

thick—perhaps the thinnest material in the universe—and forms a high quality crystal lattice, with no vacancies or dislocations in the structure…

Not surprisingly, in 2010, The Royal Swedish Academy of Sciences decided to award the Nobel Prize in Physics to Andre Geim and Konstantin Novoselov, both at the University of Manchester, UK "…*for groundbreaking experiments regarding the two-dimensional material graphene*." The reference of their seminal paper is [124]. The popularity of graphene grew exponentially, and for overview of the expansion of the new field, readers may consult the overview publications of Geim and Novoselov [125,126].

Graphene was theoretically studied before 2004 and continues to be studied after 2004. We are here to mention just one aspect of such studies concerned with graphene edges. The reason for selecting this topic is to illustrate that Hückel is not dead! The topic of the edges of graphene received some attention [127]. Three types of edges have been considered possible for graphite layers and isolated graphene: the zigzag, the armchair, and the reconstructed zigzag, which are illustrated in Figure 3.15. The zigzag edge is at Figure 3.15a, and the armchair edge is at Figure 3.15b. In Figure 3.15c is shown the reconstructed zigzag edge obtained by regular alternation of a five member rings and seven member rings. Just as two six-member (benzene) rings give naphthalene, so-fused five- and seven-member rings give azulene, a stable aromatic compound. The smallest benzenoid having the "armchair" edges is phenenthrene, which is obtained by fusing three benzene rings in a nonlinear fashion. Hence, instead

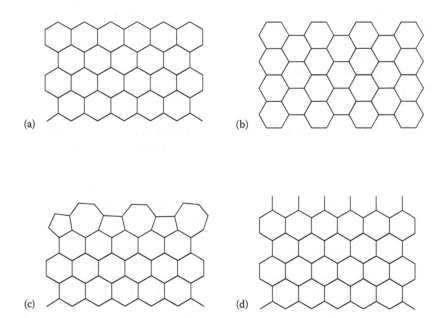

(a)   (b)   (c)   (d)

FIGURE 3.15 (a and b) Zigzag and armchair border edges; (c and d) reconstructed zigzag and extended Klein edge.

of speaking of the zigzag, the armchair, and the reconstructed zigzag, one could speak of naphthalene-, phenanthrene-, and azulene-type edges of graphene.

This was until 20 years ago, when D. J. Klein, using HMO, suggested a fourth possibility, illustrated in Figure 3.15d, which according to HMO calculations, is relatively stable. Of course, we all know that HMO is not sufficiently accurate for calculation of molecular spectra, but it has been sufficiently accurate to establish the difference between benzene stability and cyclooctatetraene, which has led to the famous Hückel $4n + 2$ rule. From the work of Klein [128] and Klein and Bytautas [129], we are assured that HMO works well for edges of graphene. As we mentioned already, Hückel was not dead, but the extended Klein edge—for short, the EK edge—was dead for 20 years until late in 2014, when it was experimentally observed by a collaboration of scientists in Oxford, England, and Seoul, Korea [130].

The work of Klein illustrates that theoretical chemistry can be occasionally ahead of experimental and apparently "premature" so to say. But this is not because sometimes experimentalists objected to the theoretical work as being too speculative (and there are such!) but because an experimental tool for direct detection of chemical bonds was not available at the time. Let's end the story on extended Klein bonds by reproducing the abstract of the paper on extended Klein edges in graphene [131], in which TEM stands for transmission electron microscopy:

> Graphene has three experimentally confirmed periodic edge terminations, zigzag, reconstructed 5–7, and arm-chair. Theory predicts a fourth periodic edge of graphene called the extended Klein (EK) edge, which consists of a series of single C atoms protruding from a zigzag edge. Here, we confirm the existence of EK edges in both graphene nanoribbons and on the edge of bulk grapheme using atomic resolution imaging by aberration-corrected transmission electron microscopy. The formation of the EK edge stems from sputtering and reconstruction of the zigzag edge. Density functional theory reveals minimal energy for EK edge reconstruction and bond distortion both in and out of plane, supporting our TEM observations. The EK edge can now be included as the fourth member of observed periodic edge structures in graphene.

Is any additional plausible edge type possible for graphene? Speaking of speculations, let us mention two that we have illustrated in Figure 3.16. At Figure 3.16a is

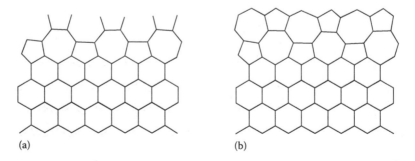

(a)   (b)

FIGURE 3.16 Speculative possibilities for additional graphene edges. (a) Extended edges on reconstructed 5–7 model zigzag. (b) Double layer of reconstructed 5–7 edges.

shown the extended edge on a reconstructed 5–7 zigzag model. This, of course is speculation because even though azulene (the 5–7 ring fusion) is aromatic, it is visibly less aromatic than benzene, which has $2R_1$ conjugated circuits while azulene has $2R_2$ conjugated circuits, which contribute less to molecular stability.

This speculation can be tested with HMO calculation, which leads to

Problem 8

Calculate using HMO the plausibility of the extended edges on a reconstructed 5–7 zigzag model.

The other speculative graphene edge is illustrated at Figure 3.16b which has a double layer of reconstructed 5–7 edges. The boundary edges here are again the CC bonds of five-member and seven-member rings, but they are different: Now we have two CC bonds from five-member rings and two CC bonds from seven-member rings, and before we had graphene edges of one CC bond from five-member rings and three CC bonds from seven-member rings. Observe that now the graphene edges of the double layer of the reconstructed 5–7 edge look fairly similar to the zigzag graphene edge (at least geometrically), just as the reconstructed 5–7 edges (of Figure 3.16a) look fairly similar to the armchair (again geometrically speaking). One can go on! So one can and add extended Klein edges to the graphene double layer of reconstructed 5–7 edges (Figure 3.16b). But it is good to recollect the message of a recent paper of Hoffmann, P. v. R. Schleyer, and Schaffer [131]: *Predicting Molecules—More Realism, Please!* So we better stop now and go to the next chapter.

REFERENCES AND NOTES

1. E. Hückel, *Grundzuge der Theorie ungesattiger und aromatischer Verbindugen*, Verlag Chemie, Berlin, 1938.
2. L. Pauling, *The Nature of the Chemical Bond and the Structure of Molecules and Crystals*, Cornell University Press, Ithaca, NY, 1939.
3. J. J. Sylvester, Chemistry and algebra, *Nature* 17 (1877–1878) 284.
4. J. J. Sylvester, On an application of the new atomic theory to the graphical representation of the invariants and covariants of binary quantities, *Am. J. Math.* 1 (1878) 64–125.
5. N. L. Biggs, E. K. Lloyd, and R. J. Wilson, *Graph Theory 1736–1936*, Clarendon Press, Oxford, 1968.
6. L. Euler, Solutio problematis ad geometriam situs pertinentis, *Comment. Acad. Sci. U. Petrop.* 8 (1736) 128–140, reprinted in *Opera Omnia Series Prima*, Vol. 7, pp. 1–10, 1766.
7. D. König, Theorie Der Endlichen Und Unendlichen Graphen: Kombinatorische Topologie Der Streckenkomplexe, Chelsea Publishing Company, New York, 1950.
8. V. Prelog, from the foreword to the book: *Chemical Applications of Graph Theory*, A. T. Balaban (Ed.), Academic Press, London, 1976.
9. J. Petersen, Sur la théorème de Tait. *L'Intermédiare des Math.* 5 (1898) 225–227.
10. T. Pisanski and M. Randić, Bridges between geometry and graph theory, in: *Geometry at Work*, C. A. Gorini (Ed.), Math. Assoc. America, Washington, DC, 2000, pp. 174–194.

11. J. Lützen, G. Sabidussi, and B. Toft, Julius Petersen 1839–1910 a biography, *Discrete Math.* 100 (1992) 9–82.
12. P. A. M. Dirac, *The Principles of Quantum Mechanics*, Clarendon Press, Oxford, 1958.
13. W. Heitler and F. London, Wechelwirkung neutraler Atome und homöopolare Bindung nach der Quantenmechanik, *Zeit. Phys.* 44 (1927) 619.
14. F. Bloch, Über die Quantummechanik der Elektronen in Kristallgittern, *Zeit. Phys.* 52 (1928) 555–600.
15. Hs. H. Günthard and H. Primas, Zusammenhang von Graphentheorie und MO-Theorie von Molekeln mit Systemen konjugierter Bindungen, *Helv. Chim. Acta* 39 (1956) 1645–1653.
16. K. Ruedenberg, Quantum mechanics of mobile electrons in conjugated bond systems. III. Topological matrix as generatrix of bond orders, *J. Chem. Phys.* 34 (1961) 1884–1892.
17. I. Gutman and N. Trinajstić, Graph theory and molecular orbitals, *Topics Curr. Chem.* 42 (1973) 49–93.
18. R. Pariser and R. G. Parr, A semi-empirical theory of the electronic spectra and electronic structure of complex unsaturated molecules. I, *J. Chem. Phys.* 21 (1953) 466–471.
19. R. Pariser and R. G. Parr, A semi-empirical theory of the electronic spectra and electronic structure of complex unsaturated molecules. II, *J. Chem. Phys.* 21 (1953) 767–776.
20. J. A. Pople, Electron interaction in unsaturated hydrocarbons, *Trans. Faraday Soc.* 49 (1953) 1375–1385.
21. J. N. Murrell, S. F. A. Kettle, and J. M. Tedder, *Valence Theory*, John Wiley & Sons Ltd., London, 1970.
22. I. Gutman and N. Trinajstić, Graph theory and molecular orbitals. Total π-electron energy of alternant hydrocarbons, *Chem. Phys. Lett.* 17 (1972) 535–538.
23. I. Gutman, M. Milun, and N. Trinajstić, Graph theory and molecular orbitals. 19. Nonparametric resonance energies of arbitrary conjugated systems, *J. Am. Chem. Soc.* 99 (1977) 1692–1704.
24. I. Gutman, B. Ruščić, N. Trinajstić, and C. F. Wilcox Jr., Graph theory and molecular orbitals. XII. Acyclic polyenes, *J. Chem. Phys.* 62 (1975) 3399.
25. D. König, Theorie der endlichen und unendlichen Graphen, Akademische Verlagsgesellschaft, Leipzig, 1936 (repreinted: Chedlsea, New York, 1950).
26. D. M. Cvetković, M. Doob, and H. Sachs, *Spectra of Graphs: Theory and application*, 3rd ed., Johann Ambrosius Barth Verlag, Heidelberg–Leipzig, 1995.
27. J. van Vleck, Theory of the variations in paramagnetic anisotropy among different salts of the iron group, *Phys. Rev.* 41 (1932) 208.
28. H. L. Schläfer and G. Gliemann, *Basic Principles of Ligand Field Theory*, Wiley Interscience, New York, 1969.
29. R. Hoffmann, An extended Hückel theory. I. Hydrocarbons, *J. Chem. Phys.* 39 (1963) 1397–1412.
30. R. Hoffmann, Extended Hückel theory. II. Sigma orbitals in the azines, *J. Chem. Phys.* 40 (1964) 2745.
31. R. Hoffmann, Extended Hückel theory. III. Compounds of boron and nitrogen, *J. Chem. Phys.* 40 (1964) 2474–2480.
32. R. Hoffmann, Extended Hückel theory. IV. Carbonium ions. *J. Chem. Phys.* 40 (1964) 2480–2488.
33. M. Wolfsberg and L. J. Helmholz, The spectra and electronic structure of the tetrahedral ions MnO_4^-, CrO_4^-, and ClO_4^-, *J. Chem. Phys.* 20 (1952) 837.
34. R. B. Woodward and R. Hoffmann, Stereochemistry of electrocyclic reactions, *J. Am. Chem. Soc.* 87 (1965) 395.
35. R. Hoffmann and R. B. Woodward, Selection rules for concerted cycloaddition reactions, *J. Am. Chem. Soc.* 87 (1965) 2046.

36. R. B. Woodward and R. Hoffmann, The conservation of orbital symmetry, *Acc. Chem. Res.* 1 (1968) 17–22.
37. R. B. Woodward and R. Hoffmann, The conservation of orbital symmetry, *Angew. Chem. Int. Ed.* 8 (1969) 781–853.
38. J. Borwein and D. Bailey, *Mathematics by Experiment: Plausible Reasoning in the 21st Century*, A. K. Peters, Wellesley, MA, 2003.
39. L. Collatz and U. Sinogowitz, Spektren endlicher Grafen, *Abh. Math. Sem. Univ. Hamburg* 21 (1957) 63–77.
40. F. Harary, The determinant of the adjacency matrix of a graph, *Soc. Ind. Appl. Math. Rev.* 4 (1962) 202–210.
41. L. Spialter, The atom connectivity matrix (ACM) and its characteristic polynomial (ACMCP): A new computer-oriented chemical nomenclature, *J. Am. Chem. Soc.* 85 (1963) 2012–2013.
42. F. Harary, C. King, A. Mowshowitz, and R. C. Read, Cospectral graphs and digraphs, *Bull. London Math. Soc.* 3 (1971) 321–328.
43. A. T. Balaban and F. Harary, The characterstic polynomial does not uniquely determine the topology of a molecule, *J. Chem. Doc.* 11 (1971) 258–259.
44. J. V. Knop, W. R. Müller, K. Szymanski, N. Trinajstić, A. F. Kleiner, and M. Randić, On irreducible endospectral graphs, *J. Math. Phys.* 27 (1986) 2601–2612.
45. M. Randić, M. Barysz, J. Nowakowski, S. Nikolić, and N. Trinajstić, Isospectral graphs revisited, *THEOCHEM—J. Mol. Struct.* 54 (1989) 95–121.
46. M. Randić and A. F. Kleiner, On the construction of endospectral graphs, *Ann. N. Y. Acad. Sci.* 555 (1989) 320–331.
47. D. Babić and I. Gutman, On isospectral benzenoid graphs, *J. Math. Chem.* 9 (1992) 261–278.
48. O. Ivanciuc and A. T. Balaban, Nonisomorphic graphs with identical atomic count of self returning walks: Isocodal graphs, *J. Math. Chem.* 11 (1992) 155–167.
49. M. Randić, X. Guo, and A. F. Kleiner, On subspectral graphs, *Congr. Numerantium* 96 (1993) 143–156.
50. J. Liu and Y. Jiang, New method for constructing isospectral graphs, *J. Chem. Inf. Comput. Sci.* 34 (1994) 1267–1272.
51. Y. Jiang and C. Liang, On endospectral bipartite graphs, *Croat. Chem. Acta* 68 (1995) 343.
52. K. Balasubramanian and S. C. Basak, Characterization of isospectral graphs using graph invariants and derived orthogonal parameters, *J. Chem. Inf. Comput. Sci.* 38 (1998) 367–373.
53. M. Randić and B. Baker, Isospectral multitrees, *J. Math. Chem.* 2 (1988) 249–265.
54. E. C. Catalan, Note sur une équation aux différences finies, *J. Math. Pure Appl.* 3 (1838) 508–516.
55. H. Urban, Zu Eulers Problem der Polygonzerlegung, *Zeitschrift für Math. Naturw. Unterricht* 4 (1941) 118–119.
56. H. Dörrie, *100 Great Problems of Elementary Mathematics. Their History and Solutions*, Dover Publ. Inc., New York, 1982.
57. N. Okhami and H. Hosoya, Wheland polynomial, *Bull. Chem. Soc. Japan* 52 (1979) 1624–1633.
58. M. Randić, H. Hosoya, N. Ohkami, and N. Trinajstić, The generalized Wheland polynomial, *J. Math. Chem.* 1 (1987) 97–122.
59. M. Eigen, Selforganization of matter and the evolution of biological macromolecules, *Naturwissenschaften* 58 (1971) 465–523.
60. D. M. Campbell, The computation of Catalan numbers, *Math. Mag.* 57 (1984) 195–208.
61. H. Sachs, Beziehungen zwischen den in einem Graphen enthaltenen Kreisen und seinem charakteristischen Polynom, *Publ. Math. (Debrecen)* 11 (1964) 119–134.

62. N. Trinajstić, Computing the characteristic polynomial of a conjugated system using Sachs theorem, *Croat. Chem. Acta* 49 (1977) 539–633.
63. I. Gutman, Impact of the Sachs theorem on theoretical chemistry: A participant's testimony, *MATCH Commun. Math. Comp. Chem.* 48 (2003) 17–34.
64. T. Živković, reported at the Quantum Chemistry School of USSR Academy of Science, Repino, USSR, December 1973.
65. C. A. Coulson and A. Streitweiser, *Dictionary of Pi-electron Calculations*, W. H. Freeman Company, San Francisco, 1965.
66. T. Živković, N. Trinajstić, and M. Randić, On conjugated molecules with identical topological spectra, *Mol. Phys.* 30 (1975) 517–532.
67. M. Randić, N. Trinajstić, and T. Živković, Molecular graphs having identical spectra, *J. Chem. Soc.—Faraday Trans. II* 72 (1976) 244–256.
68. A. Schwenk, Almost all trees are cospectral, in: *New Directions in the Theory of Graphs*, (F. Harary, Ed.), Academic Press, New York, 1973, pp. 275–307.
69. C. Godsil and B. McKay, Some computational results on the spectra of graphs, *Lecture Notes Math.* 769 (1976) 72.
70. M. Randić, Random walk and their diagnostic value for characterization of atomic environment, *J. Comput. Chem.* 1 (1980) 386–399.
71. Y. Jiang, Problems on isospectral molecules, *Sci. Sin. (Ser. B)* 27 (1984) 236.
72. M. Randić, B. Horvat, and T. Pisanski, Some graphs are more strongly-isospectral than others, *MATCH Commun. Math. Comp. Chem.* 63 (2003) 737–750.
73. E. Heilbronner and T. B. Jones, Spectral differences between "isospectral" molecules, *J. Am. Chem. Soc.* 100 (1978) 6506–6507.
74. M. Kac, Can one hear the shape of a drum?, *Am. Math. Monthly* 73 (1966) 1–23.
75. M. Fisher, On hearing the shape of a drum, *J. Comb. Theory* 1 (1966) 105–125.
76. G. A. Baker, Drum shapes and isospectral graphs, *J. Math, Phys.* 7 (1966) 2238–2242.
77. D. M. Cvetković, M. Doob, H. Sachs, and A. Torgašev, Recent results in the theory of graph spectra, *Ann. Discrete Math.* 36 (1988) 1–306.
78. F. Chung, *Spectral Graph Theory, CBMS Regional Conference Series in Mathematics, 92*, American Mathematical Society, Providence, RI, 1997.
79. I. Oren and R. Band, Isospectral graphs with identical nodal counts, arXiv:1110.0158 [math-ph] (Mathematical Physics), 2012.
80. L. Halbeisen and N. Hungerühler, Generation of isospectral graphs, user.math.uzh.ch /halbeisen/publications/pdf/spec.pdf.
81. Lecture of C. Godsil in 2007: "Are almost all graphs cospectral?" mentioned by S. Butler, *Cospectral Graphs*, University of California, San Diego, CA, April 21, 2011, available on Internet.
82. W. C. Herndon, Isospectral molecules, *Tetrahedron Lett.* 15 (1974) 671–674.
83. W. C. Herndon and M. L. Ellzey, Jr., Isospectral graphs and molecules, *Tetrahedron* 31 (1975) 99–107.
84. T. Živković, N. Trinajstić, and M. Randić, On topological spectra of composite molecular systems, *Croat. Chem. Acta* 49 (1977) 89–100.
85. E. Heilbronner, Some comments on cospectral graphs, *MATCH Commun. Math. Comp. Chem.* 5 (1979) 105–133.
86. J. P. Love and M. V. Davis, Isospectral points and edges in graph theory, *J. Math. Chem.* 5 (1990) 275–285.
87. C. Rücker and G. Rücker, Understanding the properties of isospectral points and pairs in graphs: The concept of orthogonal relation, *J. Math. Chem.* 9 (1992) 207–238.
88. E. Shahknovich and A. Gutin, Enumeration of all compact conformations of copolymers with random sequence of links, *J. Chem. Phys.* 93 (1990) 5967–5971.

89. A. T. Balaban, X. Liu, D. J. Klein, D. Babić, T. G. Schmalz, W. A. Seitz and M. Randić, Graph invariants for fullerenes, J. Chem. Inf. Comput. Sci. 35 (1995) 396–404.
90. C. A. Coulson and G. S. Rushbrooke, Note on the method of molecular orbitals, Proc. Cambridge Phil. Soc. 36 (1940) 193–200.
91. H. C. Longuet-Higgins, Some studies in molecular orbital theory I. Resonance structures and molecular orbitals in unsaturated hydrocarbons, J. Chem. Phys. 18 (1950) 265–274.
92. H. C. Longuet-Higgins, Some studies in molecular orbital theory II. xxx. . . J. Chem. Phys. 18 (1950) 283–291.
93. C. A. Coulson, Valence, 2nd ed., Oxford University Press, Oxford, 1961.
94. M. Randić, Comment on difference between bond orders calculated by SCF MO and simple MO method, J. Chem. Phys. 34 (1962) 693–694.
95. M. Randić, D. Plavšić, and N. Trinajstić, On the difference in bond orders between HMO and PPP methods, Int. J. Quantum Chem. 37 (1990) 437–448.
96. K. Fries, Über bicyclische Verbindungen und ihren Vergleich mit dem Naphtalin, Justus Liebigs Ann. Chem. 454 (1927) 121–324.
97. M. Randić and X. F. Guo, Generalized bond orders, Int. J. Quantum Chem. 49 (1994) 215–237.
98. M. Randić, Local aromatic properties of benzenoid hydrocarbons, Pure Appl. Chem. 52 (1980) 1587–1596.
99. M. Randić and D. J. Klein, Kekulé valence structures revisited. Innate degree of freedom of pi-electron coupling, in: Mathematical and Computational Concepts in Chemistry, N. Trinajstić (Ed.), Ellis Horwood, New York, 1985, pp. 274–282.
100. D. J. Klein and M. Randić, Innate degree of freedom of a graph, J. Comput. Chem. 8 (1987) 516–521.
101. A. Šali, E. Shahknovich, and M. Karplus, How does a protein fold? Nature 369 (1994) 248–251.
102. E. Hückel, Zur Quantentheorie der Doppelbindung, Zeit. Phys. 60 (1930) 423–456.
103. E. Hückel, Quantumtheoretische Beitrage zum Benzolproblem, I. Die Elektronenekonfiguration des Benzols und verwandter Verbindungen, Zeit. Phys. 71 (1931) 204.
104. E. Hückel, Quanstentheoretische Beiträge zum Benzolproblem II. Quantentheorie der induzierten Polaritäten, Zeit. Phys. 72 (1931) 310–337.
105. E. Hückel, Quantumtheoretische Beitrage zum der atomatischen und ungesattigen Verbindungen III, Zeit. Phys. 76 (1932) 628–648.
106. M. Randić, L. M. DeAlba, and A. F. Kleiner, Distance/Distance matrices, J. Chem. Inf. Comput Sci. 34 (1994) 277–286.
107. R. A. Horn and C. R. Johnson, Matrix Analysis, Cambridge University Press, Cambridge, 1990, Chapter 8.
108. The theorem is known as the Perron-Frobenius theorem and holds for any matrix, the bounds being given by the largest row sum or column sum and the smallest row sum or column sum. For a symmetric $n \times n$ matrix A with non-negative entries, the following holds: (i) The spectral radius of A is an eigenvalue, (ii) there is an eigenvalue with non-negative entries, (iii) the eigenvalue estimate is $\min_i \Sigma_j a_{ij} \leq r \leq \max_i \Sigma_j a_{ij}$.
109. L. Bytautas, D. J. Klein, M. Randić, and T. Pisanski, Use of path matrices for a characterization of molecular structures, DIMACS Ser. Discrete Math. Theor. Comput. Sci. 51 (2000) 39–61.
110. M. Randić and G. Krilov, On a characterization of the folding of proteins Int. J. Quantum Chem. 75 (1999) 1017–1026.
111. G. Krilov and M. Randić, Quantitative characterization of protein structure: Application to a novel a/b fold, New J. Chem. 28 (2004) 1608–1614.

112. M. Randić and M. Razinger, On characterization of three-dimensional molecular structure, in: *From Chemical Topology to Three-Dimensional Geometry*, A. T. Balaban, (Ed.), Plenum Press, New York, 1966, pp. 159–236.

113. M. Randić, M. Vračko, M. Novič, and S. C. Basak, On ordering of folded structures, *MATCH, Commun. Math. Comput. Chem.* 42 (2000) 181–231.

114. M. Randić, M. Vračko, and M. Novič, Eigenvalues as molecular descriptors, in: *QSPR/QSAR Studies by Molecular Descriptors*, M. V. Diudea (Ed.), Nova Sci. Publ., Huntington, 2001, pp. 147–211.

115. M. Randić, Similarity methods of interest in chemistry, in: *Mathematical Methods in Contemporary Chemistry*, S. I. Kuchanov (Eds.), Gordon & Breach Publ., Amsterdam, 1996, pp. 1–100.

116. H. Li, R. Helling, C. Tang, and N. Wingreen, Emergence of preferred structures in a simple model of protein folding, *Science* 273 (1966) 666.

117. K. Mehran, Which came first, Protein sequence or structure? *Science* 273 (1996) 610.

118. S. Borman, Protein folding model focuses on 'designability,' *Chem. Eng. News*, August 12, 1996, p. 36.

119. M. Randić, On characterization of chemical structure, *J. Chem. Inf. Comput. Sci.* 37 (1997) 672–687.

120. C. Tang, Simple models of the protein folding problem, *Physica A* 288 (2000) 31–48.

121. T. M. A. Fink, *Inverse Protein Folding, Hierarchical Optimisation and Tie Knots*, Dissertation, University of Cambridge, Cambridge, 1998.

122. C. Levinthal, How to fold graciously, in: *Mossbauer Spectroscopy in Biological Systems*, P. Debrunner, J. C. M. Tsibris, and E. Münck (Eds.), Univ. of Illinois Press, Urbana, 1969.

123. A. Chodos, Editorial: This month in physics history—October 22, 2004: Discovery of Graphene, *Am. Phys. Soc. News*, October 2009, Vol. 18, No. 9, page 2.

124. K. S. Novoselov, A. K. Geim, S. V. Morozov, D. Jiang, Y. Zhang, S. V. Dubonos, I. V. Grigorieva, and A. A. Firsov, Electric field effect in atomically thin carbon films, *Science* 306 (2004) 666–669.

125. A. K. Geim and K. S. Novoselov, The rise of grapheme, *Nat. Mater.* 6 (2007) 183–191.

126. A. K. Geim, Graphene: Status and prospects, *Science* 324 (2009) 1530–1534.

127. M. Acik and Y. J. Chabal, Nature of graphene edges: A review, *Jpn. J. Appl. Phys.* 50 (2011) 070101-1-070101-16.

128. D. J. Klein, Graphitic polymer strips with edge states, *Chem. Phys. Lett.* 217 (1994) 261–265.

129. D. J. Klein and L. Bytautas, Graphitic edges and unpaired π-electron spins, *J. Phys. Chem. A* 103 (1999) 5196–5210.

130. K. He, A. W. Robertson, S. Lee, E. Yoon, G.-D. Lee, and J. H. Warner, Extended Klein edges in graphene, *ACS Nano* 8 (2014) 12272–12279.

131. R. Hoffmann, P. v. R. Schleyer and H. F. Schaefer, III, Predicting molecules—More realism, please!, *Angew. Chem. Int. Ed.* 47 (2008) 7164–7167.

4 Characteristic Polynomial

4.1 ON CHARACTERISTIC POLYNOMIAL

We will start this section on mathematical chemistry by briefly outlining the construction and several properties of the characteristic polynomial, one of the basic attributes of graphs. There are several reasons that we decided to elaborate on the characteristic polynomial. The characteristic polynomial emerges in a number of topics that overlap with problems we discuss in this book:

1. Isospectral graphs that we have already mentioned have the same characteristic polynomial.
2. The coefficients of the characteristic polynomial offer a set of useful graph invariants.
3. The characteristic polynomial is related to another important polynomial of interest in physics and chemistry, the matching polynomial, the coefficients of which count independent sets of edges in a graph [1]. When added, these counts give the Hosoya index Z [2], one of the early molecular descriptors for structure–property regressions.
4. A relationship between the characteristic polynomial and its derivatives lead to Clarke's theorem [3], which is of interest as a possible alternative route to solving graph reconstruction problems.
5. It has been found that the square root of the last coefficient of the characteristic polynomial in Hückel molecular orbital theory [4], which can be computed as the determinant of the HMO matrix, gives the number of Kekulé valence structures of benzenoid hydrocarbons. This is one of a few very intriguing relationships connecting the valence bond (VB) method and molecular orbital (MO) method of quantum chemistry [5,6].
6. The coefficients of the characteristic polynomial of trees can be obtained in terms of connectivity parameters introduced for evaluations of moments. Graph theoretical analysis reduces the coefficients of the characteristic polynomial to linear combinations of binomial factors, which are uniquely determined from the structure of the molecular graph.

4.2 ON CONSTRUCTION OF THE CHARACTERISTIC POLYNOMIAL

Calculating the characteristic polynomial for larger systems could be tedious, but with current computer software, such as MATLAB® [7], this is no longer an issue. There are very useful recursion formulas for the characteristic polynomial of acyclic graphs due to Heilbronner [8], the simplest of which is

$$\mathrm{Ch}(G) = x\mathrm{Ch}(G - t) - \mathrm{Ch}(G - t - a) \qquad (4.1)$$

Here, Ch(G) is the characteristic polynomial of a graph G, Ch($G - t$) is the characteristic polynomial of a graph G obtained from G by deleting terminal edge t, and Ch($G - t - a$) is the characteristic polynomial of a graph G obtained from G by deleting the terminal edge and adjacent edge a. The above is a special case of a more general recursion valid for acyclic graphs and cyclic graphs having bridge edges (which, when erased, disconnect the graph into two disconnected components):

$$Ch(G) = Ch(G - uv) - Ch(G - u - v) \qquad (4.2)$$

Here edge uv is a bridge with u, v end vertices. Ch($G - uv$) is the characteristic polynomial of a graph G obtained from G by deleting the bridge uv, and Ch($G - u - v$) is the characteristic polynomial of a graph G obtained from G by deleting the bridge uv and its terminal vertices (with all incident edges). This recursion is illustrated in Figure 4.1 on a graph of 3-methylhexane.

Both recursions also hold for cyclic systems with pending bonds as long as one considers "bridge" bonds (a special case of which is terminal bonds). This can be easily verified on smaller graphs with pending bonds and smaller graphs with "bridge" bonds. Heilbronner has also extended the above recursion formula to cyclic graphs, but according to Rosenfeld and Gutman [9], that important generalization appears to have been overlooked by many and was rediscovered in mathematics by Schwenk more than 20 years later [10]. The recursion formula for cyclic graphs is

$$Ch(G) = Ch(G - uv) - Ch(G - u - v) - 2\Sigma Ch(G - C) \qquad (4.3)$$

Here Ch($G - C$) is the characteristic polynomial of a graph G obtained from G by deleting the cycle C and the summation goes over all cycles of the graph G containing the edge uv. Below, we illustrate application of the above recursion on naphthalene (as reported in [11]):

$$Ch\,(naphthalene) = Ch(P_{10}) - Ch(P_8) - Ch(P_4)^2 - 4Ch(P_4) - 2 \qquad (4.4)$$

where P_{10}, P_8, and P_4 denote path graphs on 10, 8, and 4 vertices, respectively. When the expressions for Ch(P_{10}), Ch(P_8), and Ch(P_4), which are listed in Table 4.1, are substituted in the above recursion, one obtains

$$Ch\,(naphthalene) = x^{10} - 11x^8 + 41x^6 - 65x^4 + 43x^2 - 8. \qquad (4.5)$$

Some may be skeptical, not about the accuracy of the results but about, in this time of computers, "wasting" time and energy on something that you can get easily

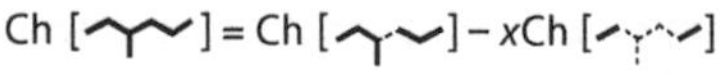

$$Ch\,[\curlyvee] = Ch\,[\curlyvee] - xCh\,[\curlyvee]$$

FIGURE 4.1 Heilbronner's general recursion expression for the characteristic polynomial of a tree illustrated on a graph of 3-methylhexane. The recursion applies also to the bridge bond of polycyclic graphs.

TABLE 4.1

Characteristic Polynomials of Carbon Skeletons of *n*-Alkanes for *n* = 1 to *n* = 12 (Path Graphs of Lengths 1–12)

$L_1 = x$	$Ch(P_1)$
$L_2 = x^2 - 1$	$Ch(P_2)$
$L_3 = x^3 - 2x$	$Ch(P_3)$
$L_4 = x^4 - 3x^2 + 1$	$Ch(P_4)$
$L_5 = x^5 - 4x^3 + 3x$	$Ch(P_5)$
$L_6 = x^6 - 5x^4 + 6x^2 - 1$	$Ch(P_6)$
$L_7 = x^7 - 6x^5 + 10x^3 - 4x$	$Ch(P_7)$
$L_8 = x^8 - 7x^6 + 15x^4 - 10x^2 + 1$	$Ch(P_8)$
$L_9 = x^9 - 8x^7 + 21x^5 - 20x^3 + 5x$	$Ch(P_9)$
$L_{10} = x^{10} - 9x^8 + 28x^6 - 35x^4 + 15x^2 - 1$	$Ch(P_{10})$
$L_{11} = x^{11} - 10x^9 + 36x^7 - 546x^5 + 35x^3 - 6x$	$Ch(P_{11})$
$L_{12} = x^{12} - 11x^{10} + 45x^8 - 84x^6 + 70x^4 - 21x^2 + 1$	$Ch(P_{12})$

---

from MATLAB or other computer software. Why bother? Well, it is true that one can get the characteristic polynomial of naphthalene in MATLAB just by typing Poly(A), where A is the adjacency matrix of naphthalene, but the alternative representation in terms of $Ch(P_n)$ has also some advantages as one can easily see the structure of the contributing paths. For illustration, let us also consider azulene, for which one could immediately obtain the characteristic polynomial:

$$Ch\ (azulene) = x^{10} - 11x^8 + 41x^6 - 2x^5 - 61x^4 + 6x^3 + 31x^2 - 2x - 41 \quad (4.6)$$

If one uses the Heilbronner's modified recursion for cyclic systems, one would obtain:

$$Ch\ (azulene) = Ch(P_{10}) - Ch(P_8) - Ch(P_3)(ChP_5) - 2Ch(P_3) - 2Ch(P_5) - 2 \quad (4.7)$$

It is worth observing the difference between the two representations of the characteristic polynomial. The standard representation (in powers of x) offers little insight into the similarity of naphthalene and azulene; the latter, the representation in terms of the characteristic polynomials of paths P_n, shows that the great structural similarity between the two molecules is also reflected in their characteristic polynomials.

In Table 4.1, we have listed the characteristic polynomials of the first dozen linear alkanes (paths of length 1 to paths of length 12). Observe that the coefficients of these characteristic polynomials are the binomial coefficients (the same numbers forming the Pascal triangle). It is a simple exercise to verify that if one selects the central edge in path P_{12} the recursion formula becomes $Ch(P_{12}) = [Ch(P_6)]^2 - [Ch(P_5)]^2$ or $Ch(P_{12}) = [Ch(P_6) - Ch(P_5)] \times [Ch(P_6) + Ch(P_5)]$, that is,

$$(x^6 - 5x^4 + 6x^2 - 3)^2 - (x^5 - 4x^3 + 3x)^2. \quad (4.8)$$

By squaring the two polynomials, one obtains

$$x^{12} - 10x^{10} + 37x^8 - 62x^6 + 46x^4 - 12x^2 + 1 \quad \text{and} \quad x^{10} + 8x^8 - 22x^6 + 24x^4 - 9x^2, \tag{4.9}$$

which when subtracted gives

$$x^{12} - 11x^{10} + 45x^8 - 84x^6 + 70x^4 - 21x^2 + 1, \tag{4.10}$$

which is the characteristic polynomial of path (P_{12}).

Of some interest has also been factorization of Ch(x). By knowing the factors, the roots of the characteristic polynomials can be obtained by solving polynomial equations of lower degree. The alternative route for Ch(P_{12}) shown above represents one such factoring of Ch(P_{12}) into two polynomials of degree six.

4.2.1 COULSON GRAPHICAL CONSTRUCTION OF Ch(x)

The Sachs theorem [12] is elegant and of a compact form representing the coefficients of the characteristic polynomial, but due to explosive growth of the contributing terms, it is not practical in actual calculations. In contrast to Sachs theorem, Coulson considered graphical expressions for the coefficients of the characteristic polynomial that are built from a relatively smaller number of the contributing subgraphs [13]. In Figure 4.2, we have illustrated the expressions for the leading coefficients C_1–C_5 of characteristic polynomials as presented by Coulson, which are obtained by counting bonds, distinct neighbor triangles, pairs of disjoint bonds, distinct neighbor squares, distinct tetrahedra, and so on.

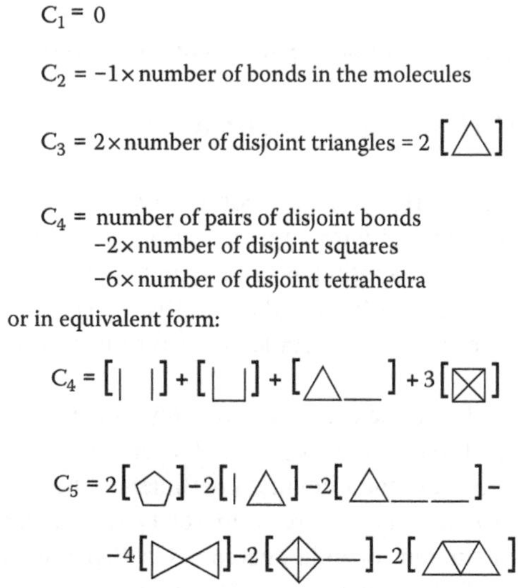

$C_1 = 0$

$C_2 = -1 \times$ number of bonds in the molecules

$C_3 = 2 \times$ number of disjoint triangles $= 2\ [\triangle]$

$C_4 =$ number of pairs of disjoint bonds
$\quad -2 \times$ number of disjoint squares
$\quad -6 \times$ number of disjoint tetrahedra

or in equivalent form:

$C_4 = [|\ \ |] + [\sqcup] + [\triangle\_\_] + 3[\boxtimes]$

$C_5 = 2[\hexagon] - 2[|\ \triangle] - 2[\triangle\_\_] -$
$\quad -4[\boxtimes\!\!\triangleleft] - 2[\diamondsuit\!\!-] - 2[\triangle\!\!\triangle]$

FIGURE 4.2 Graphical expressions for the leading coefficients of the characteristic polynomial due to Coulson.

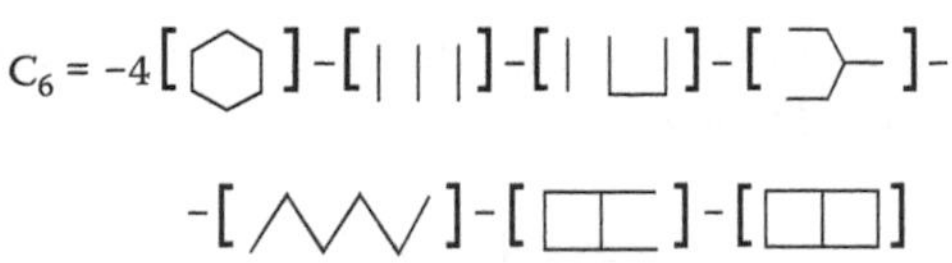

$$C_6 = -4\left[\bigcirc\right] - \left[\,|\,|\,|\,\right] - \left[\,|\,\sqcup\,\right] - \left[\,\rangle\!-\,\right] -$$

$$-\left[\wedge\!\!\wedge\!\!\wedge\right] - \left[\,\square\!\square\,\right] - \left[\,\square\!\!\square\,\right]$$

FIGURE 4.3 The graphical expression due to Coulson for the coefficients C_6 of the characteristic polynomial of *alternant* systems (i.e., systems having no odd rings).

According to Coulson, *"The calculation of C_6 in this manner is too involved,"* so he has only illustrated C_6 for alternant (even cycle) hydrocarbons, which are given by

$C_6 = -1 \times$ total number of sets of three disjointed bonds

$\quad -2 \times$ number of benzene rings (hexagon)

$\quad +2 \times$ number of double squares

$\qquad$ (i.e., two squares having one common edge)

$\quad +2 \times$ number of combinations of one square

$\qquad$ and a disjointed (noncontiguous) bond

In the case of biphenylene (shown in Figure 4.5), we show coefficient C_6 in Figure 4.3.

From Coulson's article, it follows that he arrived at the selected subgraphs for C_n coefficients by considering non-vanishing contributions arising from all minor determinants having n rows and columns. Their number increase as n increases, but is still significantly smaller than are the number of contributing terms in the Sachs theorem. Coulson has illustrated construction of coefficients C_2, C_4, and C_6 for biphenylene, which agree with the correct characteristic polynomial, which is

$$x^{12} - 14x^{10} + 69x^8 - 154x^6 + 162x^4 - 72x^2 + 9 \qquad (4.11)$$

(Due to a printing error in Coulson's paper, the C_8 term, which was not graphically calculated, was incorrectly reported as $162x$.)

It is not clear if there are some rules for construction of graphical expression for initial coefficients of the characteristic polynomial of Figures 4.2 and 4.3. Apparently Coulson knew the rules but did not hint in the margin of his publication "that he has solved the problem but there was not enough room to write down the solution," as Fermat did about his last theorem [14].

4.3 FERMAT'S LAST THEOREM

The theorem known as Fermat's last theorem could also be referred to as Fermat's lost theorem because what was lost was the proof for the theorem that Fermat claimed to have. A mathematical statement without a proof is not a theorem but a conjecture awaiting proof or disproof. To introduce Fermat's last theorem, consider the equation

$$a^2 + b^2 = c^2, \qquad (4.12)$$

which geometrically asks for a square (c^2) that can be represented by two smaller squares $(a^2$ and $b^2)$. One solution is

$$3^2 + 4^2 = 5^2. \tag{4.13}$$

The numbers 3, 4, and 5 are known as Pythagorean triplets. The name is derived from the Pythagorean theorem about right-angle triangles, which states that sides $(a$ and $b)$ of every right triangle satisfy the formula:

$$a^2 + b^2 = c^2. \tag{4.14}$$

where c is hypotenuse. Fermat's last theorem claims the equation

$$a^n + b^n = c^n \tag{4.15}$$

has solutions only for $n = 2$. He has written in a margin of a book, "*It is impossible to separate a cube into two cubes, or a fourth power into two fourth powers, or in general, any power higher than the second, into two like powers. I have discovered a truly marvelous proof of this, which this margin is too narrow to contain.*"

Despite numerous attempts, Fermat's last theorem remained unsolved until recent times. In 1995, English mathematician Wiles and his former student Taylor published the proof. Although this result many will characterize as great, it cannot be characterized as *marvelous*. After its announcement in 1994, some flaw was detected, and it took a year to be corrected. However, in the view that Fermat's last theorem was lost, we will probably never know what *truly marvelous proofs* look like.

Clearly, if the proof of Fermat's last theorem is correct, and nobody appears to challenge it, then the equation

$$a^3 + b^3 = c^3 \tag{4.16}$$

has no solution. But how about the equation

$$a^3 + b^3 + c^3 = d^3. \tag{4.17}$$

Well, there is at least one solution for that equation, already known to Euler:

$$3^3 + 4^3 + 5^3 = 6^3. \tag{4.18}$$

There could be additional nontrivial solutions, that is, solutions that are not ka, kb, kc, kd, with k being any real number, and some may try to find another nontrivial solution to the equation $a^3 + b^3 + c^3 = d^3$. One can even ask: What is the next smallest set of four numbers that satisfies the above equation. This kind of problem is for computer-oriented people, for whom this may be simpler than searching the scattered literature to find a solution, if available. That number would be the next smallest Plato number, of which not much is known. Some speculate that 6^3, which equals 216, could be the mysterious Plato's number.

Euler considered these expressions:

$$a^4 + b^4 + c^4 = d^4$$
$$a^5 + b^5 + c^5 + d^5 = e^5 \qquad (4.19)$$
$$a^6 + b^6 + c^6 + d^6 + e^6 = f^6$$

and so on ... as a generalization of the Fermat problem and conjectured that they will not have a solution, for any *higher* powers just as $a^3 + b^3 = c^3$ has no solution (which is part of the famous last theorem of Fermat). For almost 200 years, mathematicians believed that this may be true, but then, in 1966, the Euler's sum of powers conjecture, as the above is known, was shown to be wrong. Two mathematicians, Lander and Parkin, found that

$$27^5 + 84^5 + 110^5 + 133^5 = 144^5, \qquad (4.20)$$

which is the middle equation shown above, that disproves Euler's conjecture. Since then, several solutions have been reported for exponents $k = 4$, $k = 5$, $k = 7$, and $k = 8$, but a solution for $k = 6$ still is missing, and, of course, all higher values of k. It appears to be a question of time and money (computer time has a cost) before more is known on this topic.

There is a caveat to this story: Although computers can be used to find the solution to any of the power sum equations for different exponent values by searching for counter-examples, a computer can never settle the question, if one considers the set of power equations individually: Is there any value of k for which Euler's conjecture holds? At least as long as one does not find an inductive proof for showing that a solution for k can be extended to $k + 1$.

4.4 COMMON EIGENVALUES

Hückel molecular orbital calculations, of course, today are mostly of historical and educational significance, but they should not be completely dismissed as not being any more relevant in chemistry. HMO continues to be of some interest in a few specific problems, one of them being the "packing" interaction of folded model protein, outlined in the previous section. The characteristic polynomial and the eigenvalues of the characteristic polynomial thus continue to be of some interest as a tool for comparative study of structurally related systems.

The eigenvalues of the adjacency matrix have received considerable attention in the literature, including chemical literature, and the eigenvectors have received visibly lesser attention although they also offer some useful information. Here, we may mention use of eigenvectors determining geometry of carbon toroidal molecules [15], characterizing graph drawing with eigenvectors [16], and construction of "nice graph" representations, obtained by a suitable computer program [17]. In several publications [18–20], Dias considers use of symmetrical eigenvectors and mirror-plane fragments to obtain eigenvectors for larger benzenoid molecules and for complementary structures. For those interested in the use of eigenvectors in chemical graph theory besides publications already mentioned, which provide a useful introduction, they should also consult [21–23].

Several researchers noticed that certain eigenvalues, such as $0, \pm 1, \pm\sqrt{2}, \pm\sqrt{3}$, $(1+\sqrt{5})/2$ (golden ratio), etc., occur often in different molecules. It was not difficult to realize that the underlying causes for repeated occurrence of selective eigenvalues are the presence of zero coefficients in the corresponding HMO eigenvectors. For example, the eigenvalues $\pm(1+\sqrt{5})/2$ and $\pm(-1+\sqrt{5})/2$ occur as the eigenvalues of the molecular graphs shown in Figure 4.4. What have all these graphs in common? It is not difficult to find (due to symmetry) that all these graphs have eigenvectors with nodal boundaries passing through vertices, which, when excised, leave butadiene fragments. It is known that $\pm(1+\sqrt{5})/2$ and $\pm(-1+\sqrt{5})/2$ are the eigenvalues of the adjacency matrix of the butadiene graph. In the case of the first two molecules (fulvene and fulvalene), only eigenvalues +0.6780 and –1.6180 occur, which are the eigenvalues of butadiene antisymmetric eigenvectors [24–27]. Notice that these are the only molecules in which the nodal plane bisects the butadiene skeleton allowing only the antisymmetric MO solutions.

The value of the golden ratio $\varphi = (1+\sqrt{5})/2$ is

$$\varphi \approx 1.61880\ 33988\ 74989\ 48482\ 045... \tag{4.21}$$

that goes indefinitely, but we stopped after all digits have appeared at least once. This happens when the golden ratio is given to 23 decimal places. Incidentally, we have seen the number 23 before at the number of definitions in Euclid's first book and the number of Hilbert problems. This book has also 23 problems. The former two are just coincidences; the last is a deliberate choice as a way to acknowledge our respect for Euclid and Hilbert.

In comparison to the golden ration φ, the number π, if written so that all digits appear at least once, is

$$\pi \approx 3.14159\ 26535\ 89793\ 23846\ 26433\ 83279\ 50... \tag{4.22}$$

which gives π to 32 decimal places (which is 23 in reverse). If one is to apply and pursue the above topic in any depth, this would turn out to be a very shallow project, illustrating numerology. Numerology is any belief in a divine or mystical relationship between numbers and events. In early science, there were numerous illustrations of

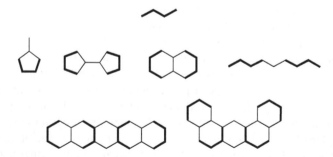

FIGURE 4.4 Graphs having eigenvalues ±1.6180 and ±0.6180, the eigenvalues of butadiene (top), except for the first two graphs, which have only the eigenvalues +0.6180 and –1.6180, eigenvalues of the antisymmetrical eigenvectors.

numerology. For example, it is well known that Kepler believed numbers and geometry to have special significance and could be used to explain natural phenomena. Thus in his first book *Mysterium Cosmographicum* published in (1596), he described the solar system, which, at that time, consisted of six planets: Mercury, Venus, Earth, Mars, Jupiter, and Saturn, to move in orbits defined by circumspheres and inspheres of Platonic solids: cube, tetrahedron, octahedron, icosahedron, and dodecahedron. Today, of course, numerology has joined astrology and alchemy, which illustrate pseudosciences that some (science illiterates) still cherish. In science, here and there, one may come across numerology in articles in which authors are unaware and give credence to numerically insignificant results. That this is a problem even today has been illustrated by Golbraikh and Tropsha in an article titled *"Beware of q^2!"* [28]. In it, the authors demonstrated that the high values of "leave-one-out" cross validation q^2 appears to be a necessary but not the *sufficient* condition for a model to have a high prediction power. They argue that this is the *general* property of QSAR models developed using "leave-one-out" cross validation, emphasizing that the external validation is the *only* way to establish a reliable QSAR model—the rest, we may say, is numerology—giving credence to insignificant numbers!

MRA is not the only area of chemical science in which numerology may be detected. Theoretical chemistry, and in particular quantum chemical calculations, have also been vulnerable in this respect. As Hoffmann et al. pointed out in an appeal, "Predicting Molecules—More Realism, Please!" [29], people often report results of calculations on an unnecessarily larger number of digits than are warranted. This apparently applies to Gaussian *ab initio* MO calculations [30] and even more so to density functional theory (DFT) [31].

In the next section, we will consider a less transparent case of common eigenvalues between different molecular graphs, which cannot be related to the nodal zero coefficients because, in such cases, the nodal planes are passing through bonds and not vertices. Hence, in such a situation, one cannot decompose graphs into smaller molecular fragments to be considered as independent entities. This problem, despite that HMO appears to be of limited relevance to modern chemistry, nevertheless may be of some interest to mathematicians, being about some unresolved mathematical issues, relating to graph spectral properties.

There appears to be growing interest in some mathematical circles toward mathematical properties of some problems of chemistry. Mathematical chemistry, after slow growth in the mid-20th century has seen considerable growth in more recent times, and its presence and identity is not likely to continue to be questioned. Nevertheless, even superficial browsing through some journals covering mathematical chemistry, such as the *Journal of Mathematical Chemistry* and *MATCH Communication in Mathematical and in Computer Chemistry*, may show that a fraction of contributions, even if not large, has much to do with mathematics and little with chemistry. There is nothing wrong with that because one never knows when some mathematical, apparently nonchemical contribution, may turn out to be important for chemistry. However, before that happens, we feel that it would be more correct to classify such contributions as contributions to *Chemical Mathematics* rather than *Mathematical Chemistry*. With this in mind, we will continue elaborating on common eigenvalues of different graphs in the next section, "Chemical Mathematics."

4.5   CHEMICAL MATHEMATICS

In chemical mathematics, we view mathematical problems that emerge from problems in chemistry, which then lead to considerations of some aspects of such problems that have some mathematical interest and have no apparent direct connection to issues of chemistry. In addition, there have been several problems of interest to mathematicians, initiated by mathematicians, and having connection to chemistry, but being, at best, of marginal relevance to problems that keep chemistry growing. For example, Siemion Fajtlowicz, a Polish mathematician, currently a professor at the University of Houston, has constructed a computer program that is making conjectures [32,33]. A conjecture is a mathematical statement that could be true or false and, hence, waits to be proven or found incorrect. It is up to the mathematical community at large to consider these conjectures and try to prove them or find a counter-example that demonstrates that a given conjecture is false. It has been interesting to see that among an initial two dozen conjectures constructed by the Graffiti program, most, if not all, are related to graph theory, and more than half of the generated conjectures were about the Randić index, the connectivity index χ. The connectivity index apparently continues to be of interest in mathematics. This is, undoubtedly, in great part due a paper by Béla Bolobás and Paul Erdös, "Graphs of Extreme Weights" [34], in which the upper and the lower bounds on the connectivity index (in which the exponent $-1/2$ has been maintained), and the generalized connectivity index (in which the exponent can be varied) has been considered for various graphs.

An illustration of an early result of "chemical mathematics" is an observation that the eigenvalues of *alternant* molecular graphs come in pairs, positive and negative. This was known to Coulson and Streitwieser [35] before the same was rediscovered in mathematics. Another illustration of chemical mathematics is Heilbronner's recursion for the characteristic polynomial of graphs containing cycles, mentioned earlier, which was rediscovered in mathematics 20 years later. An additional illustration of chemical mathematics is the abovementioned paper of Bollobás and Erdös, which is about mathematical properties (of upper and lower bounds of the generalized Randić index, which are of limited interest to chemists).

4.5.1   AN ILLUSTRATION OF CHEMICAL MATHEMATICS: SUBSPECTRAL GRAPHS

At the top of Figure 4.5 is the molecular graph of hexatriene with inscribed general coefficients for the symmetric and antisymmetric eigenvectors. Below are also shown molecular graphs of pyrene and biphenylene, both with inscribed coefficients of symmetric and antisymmetric eigenvectors. In the case of pyrene, the reflection plane passes through the four central carbon atoms, which thus necessarily have zero coefficients for antisymmetric eigenvectors as has been the case with the graphs shown in Figure 4.4 having common eigenvalues with those of butadiene. In the case of pyrene, the symmetry of reflection in a plane bisecting the molecule passing through the two central CC bonds results in two hexatriene fragments. Thus, one finds common eigenvalues between pyrene and hexatriene. Observe that pyrene vertices with zero coefficients have for the sum of the coefficients of adjacent vertices

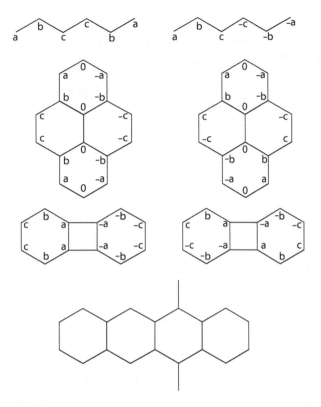

FIGURE 4.5 The molecular graph of hexatriene with coefficients for the symmetric and antisymmetric eigenvectors inscribed on pyrene and biphenylene graphs illustrating the same sum rule for their coefficients.

zero. Because of this, the sum of neighbor coefficients for the remaining vertices of pyrene are the same as the sum of coefficients for vertices of hexateriene, which are for symmetrical eigenvectors:

Vertex	Sum of Coefficients of Adjacent Vertices
a	b
b	$a+c$
c	$b+c$

and for antisymmetrical eigenvalues:

Vertex	Sum of Coefficients of Adjacent Vertices
a	b
b	$a+c$
c	$b-c$

The sum rule given above also holds for vertices of biphenylene symmetrical and antisymmetrical eigenvectors even though, in this case, two hexateriene subgraphs are not separated by the zero coefficient vertices. It appears that the above sum rule has not been recognized as a tool for studying subspectral graphs. There was an algorithm originating with Longuet-Higgins for determining the un-normalized eigenvector coefficients for a given eigenvalue λ_i:

$$-\lambda_i c_i(r, s, t) + c_i(r) + c_i(s) + c_i(t) = 0. \qquad (4.23)$$

where $c_i(r)$, $c_i(s)$, and $c_i(t)$ are the coefficients of vertices adjacent to the vertex (r, s, t) [36,37]. This λ_i weighted sum rule relates the eigenvector coefficients of adjacent vertices for different eigenvalues. In contrast, our sum rule does not involve eigenvalues but only shows how related the coefficients of adjacent vertices are without actually knowing what their values are as will be seen from the illustration presented here.

We can pose the following question: *Why are the eigenvalues of hexatriene also contained among the eigenvalues of the last molecule of Figure 4.5, a benzene derivative having four fused benzene rings?*

This question was mentioned in refs. [38,39] with a hint that those familiar with chemical graph theory may have an advantage when considering this question. The answer to the question is that the vertices in this molecule—although this is not obvious as it was in the case of pyrene and biphenylene graphs—also satisfy the sum rules for the coefficients of hexatriene. This can be verified for symmetrical eigenvectors by close examination of Figure 4.5. As one can see the sums of the coefficients of symmetrical eigenvectors of hexatriene are also satisfied for the four-ring benzene derivatives, but they require one to consider and find proper combinations of the coefficients a, b, and c for each vertex that satisfies the same rules.

The sum rules of the coefficients are more general than one may have suspected. As long as for all vertices the coefficients can be expressed as linear combinations in terms of the coefficients a, b, and c as is the case for the four-ring benzene derivative of Figure 4.5, one may be assured that such molecules will have the eigenvalues of hexatriene (shown in Appendix 10). Here we show verification that indeed for the structures of Figure 4.5 the sum rule of the hexatriene symmetric eigenvectors (shown in the Appendix 10, Table A10.1 for λ_1, λ_3, and λ_5) are satisfied also for vertices of the benzene derivative of Figure 4.6:

Vertex	Sum of Coefficients of Adjacent Vertices
$-a$	$-b$
$-a - b$	$-a - b - c$
$a - c$	$-c$
$2a + b - 2c$	$a - c$
$-a + c$	c
$a + b$	$a + b + c$

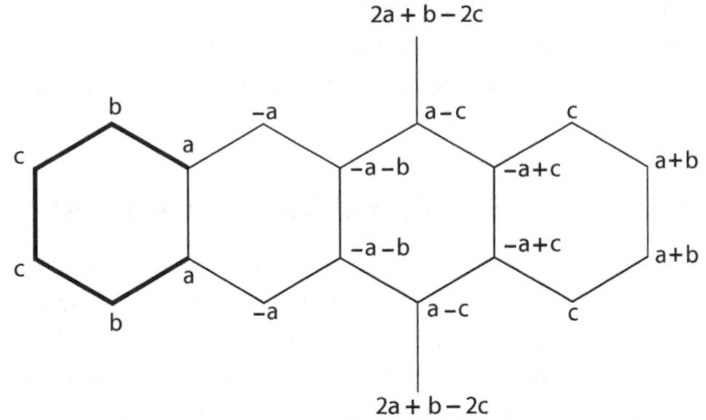

FIGURE 4.6 The coefficients of the leading eigenvector of hexatriene (a, b, c) and the coefficients of tetracene derivative, which satisfy the same sum rules.

Similar sum rules can be written for the coefficients of antisymmetrical eigenvectors (illustrated in Figure A11.1 in Appendix 11) and can be verified to be satisfied. Thus, in the case of the benzenoid of Figure 4.6, all eigenvalues of hexatriene can be found in its spectrum. In Appendix 10 (Table A10.2), we show that the sum rule is satisfied not only for the leading eigenvalue of hexatriene, but for the remaining positive eigenvalues that occur in the graph of Figure 4.6, but it can be shown to hold for all eigenvalues of hexatriene. In Appendix 11, Figure A11.1, we show that the sum rule is satisfied not only for symmetrical eigenvectors of hexatriene but also for antisymmetrical eigenvectors of hexatriene, which occur in the graph of Figure 4.6.

To the best of our knowledge so far, this is the first and the only larger molecular graphs containing all the eigenvalues of hexatriene: ±1.8019, ±1.2470, ±0.4450, in addition to graphs having vertices with zero coefficients or having nodal lines crossing in the symmetrical fashion edges of a graph. So we can pose

Problem 9

Find additional molecular structures containing the eigenvalues of hexatriene: ±1.8019, ±1.2470, ±0.4450, but having no nodes on atoms.

Problem 10

As an exercise, write down the sum rules for antisymmetric eigenvectors of hexatriene and show that the four-ring benzene derivative of Figure 4.6 satisfies such rules, which assures that this benzenoid also contains the eigenvalues of antisymmetric eigenvectors of hexatriene.

For the solution, see Figure A11.1 in Appendix 11.

Problem 11

Select an arbitrary smaller graph. Find additional graphs that satisfy the same sum rules for their coefficients, which will then contain the eigenvalues of the smaller graph.

For additional illustrations of subspectral conjugated molecular systems, see a paper by Dias [40] on a series of molecular graphs having a preponderance of common eigenvalues in which, for a collection of subspectral structures, their eigenvalues are tabulated.

4.6 REVISITING THE CHARACTERISTIC POLYNOMIAL

Before we return to the factoring of the characteristic polynomial (in the next section), let us mention that in addition to the methods described for representing the characteristic polynomial—the theorem of Sachs, elegant and beautiful, but hardly practical except for small molecules, and the Coulson approach, again of limited practical uses except for smaller molecules—there are additional approaches. In 1986, Kiang and Tang outlined a graphical evaluation of characteristic polynomials of Hückel trees (acyclic carbon chains) [41], in which formulas of coefficients of the characteristic polynomial are obtained in terms of the connectivities. A graph theoretical analysis is given in which the coefficient c_k is expressed as a linear combination of binomial factors specified by a set of graphs having $k/2$ edges. In 1971, Hosoya came up with his graphical approach, which is practical for graphs of intermediate size and particularly suitable for acyclic graphs, and it has its own recursion formulas to be used on larger graphs [42,43]. In this approach, calculation of the coefficients of the characteristic polynomial is based on the count of disjointed edges in a graph, which give number $p(G, k)$, where G is tree (acyclic graph), and k is the number of edges considered simultaneously. In Figure 4.7, we show the count of Hosoya's nonadjacent edges for 3-methylhexane. The first two rows show the count of a *single* disconnected edge, giving $p(G, 1) = 6$; the next three rows count *two* disjointed edges, giving $p(G, 2) = 9$; and the last row counts *three* disjointed edges giving $p(G, 3) = 3$. When these numbers are combined with alternating signs, one obtains the characteristic polynomial for 3-methylhexane:

$$x^7 - 6x^5 + 9x^3 - 3x. \tag{4.24}$$

To find the characteristic polynomial of polycyclic systems is more difficult because there are no simple approaches that will break them into two or more components. However, if one constructs Ulam subgraphs, one will obtain simpler graphs, the characteristic polynomial of which could be available or easier to find. But that would be of no use were it not for the Clarke theorem [3], which states that the sum of characteristic polynomials of Ulam subgraphs is the derivative of the characteristic polynomials of the parent graph. Once we know the derivative of the characteristic polynomials, upon integration, one can find the characteristic polynomials of the graph considered up to a constant term. For benzenoid hydrocarbons, the constant term can be found from the determinant of the adjacency matrix, the square root of which gives the number of Kekule valence structures. Longuet-Higgins has shown a simple way to find the number of Kekulé valence structures of π electron systems of interest using nonbonding MOs [36,37]. Gutman has pointed out that it suffices to know a single eigenvalue of a graph, which often may be the case, to find the constant [44]. So finding the constant term is not a problem. Figure 4.8 shows the four symmetry unrelated Ulam's graphs for benzocyclobutadiene.

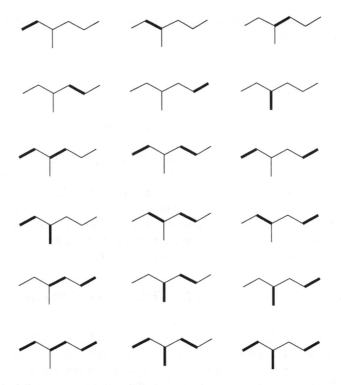

FIGURE 4.7 Count of Hosoya's nonadjacent edges for 3-methylhexane. The first two rows show the count of a *single* disconnected edge, p(G, 1) = 6; the next three rows count *two* disjointed edges p(G, 2) = 9; the last row counts *three* disjointed edges p(G, 3) = 3, which give for the characteristic polynomial: $x^7 - 6x^5 + 9x^3 - 3x$.

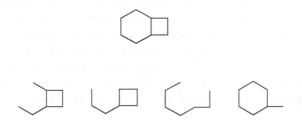

FIGURE 4.8 The four symmetry unrelated Ulam's graphs for benzocyclobutadiene.

Their characteristic polynomials are [45]

$$x^7 - 7x^5 + 11x^3 - 2x$$
$$x^7 - 7x^5 + 10x^3 - 3x$$
$$x^7 - 6x^5 + 10x^3 - 4x$$
$$x^7 - 7x^5 + 13x^3 - 7x$$

(4.25)

When these are added and multiplied by two, one obtains for the derivative of the characteristic polynomials

$$8x^7 - 54x^5 + 88x^3 - 32x \tag{4.26}$$

This upon integration gives

$$x^8 - 9x^6 + 22x^4 - 16x^2 + \text{const.} \tag{4.27}$$

Because benzocyclobutadiene is not a benzenoid molecule (having a four-member ring), one has to find the so-called algebraic Kekulé structures introduced by Wilcox [46,47]. It is based on generalizing the concept of parity of Kekulé valence structures of Dewar and Longuet-Higgins [48] to nonbenzenoid systems, such as benzocyclobutadiene. We are not to elaborate on the parity concept here, and interested readers should consult literature. It turns out that the algebraic Kekulé structures for benzocyclobutadiene equal 1, and the constant in the above expression is 1.

4.7 ON FACTORING OF CHARACTERISTIC POLYNOMIAL

Factoring $Ch(x)$ of acyclic graphs can be often accomplished by expressing their characteristic polynomials in terms of the characteristic polynomials of the n-alkanes illustrated in Table 4.1, for which we will here use the notation L_n. For illustrations of characteristic polynomials of several families of branched alkanes, including characteristic polynomials of 35 isomers of n-nonane as well as characteristic polynomials of 20 monocyclic structures with pending bonds, see ref. [45]. The multiplication table of L_n polynomials (see Table 4.2) [49], which facilitates finding factors of characteristic polynomials when expressed in terms of L_n, is very simple.

Before we describe an efficient approach to factoring the characteristic polynomial of an acyclic graph, which is based on the use of 2×2 determinants, we should outline how this approach was developed. In 1982, Balasubramanian introduced a method in which, by stepwise pruning of pending bonds of acyclic structures, construction of the characteristic polynomials of larger systems with pending bonds is simplified [50]. The same year Balasubramanian and Randić introduced *simultaneous* pruning of all terminal vertices in a graph, a novelty in construction

---

TABLE 4.2
Multiplication Table of L_n Polynomials

	L_1	L_2	L_3	L_4	L_5
L_1	$L_2 + 1$	$L_3 + L_1$	$L_4 + L_2$	$L_5 + L_3$	$L_6 + L_4$
L_2		$L_4 + L_2 + 1$	$L_5 + L_3 + L_1$	$L_6 + L_4 + L_2$	$L_7 + L_5 + L_3$
L_3			$L_6 + L_4 + L_2 + 1$	$L_7 + L_5 + L_3 + L_1$	$L_8 + L_6 + L_4 + L_2$
L_4				$L_8 + L_6 + L_4 + L_2 + 1$	$L_9 + L_7 + L_5 + L_3 + L_1$
L_5					$L_{10} + L_8 + L_6 + L_4 + L_2 + 1$

of characteristic polynomials [51]. We will illustrate this on 2,4-dimethylhexane, the secular determinant of which reduced from 8×8 to 4×4. There are rules for matrix elements of the contracted matrix:

1. The diagonal elements are negative, and zeros are replaced by L_k polynomials, where k indicates the characteristic polynomial of the fragment removed. L_1 stands for vertices without pending bonds.
2. Adjacent off-diagonal elements are the characteristic polynomial of the fragment of L_k obtained by removing the vertex of attachment. In the case of vertices without pending bonds, there remains the entry 1 from the adjacency matrix.

In this way, one arrives at the determinant of Figure 4.9. The L_k polynomials are (Chebyshev) polynomials of half argument $(x/2)$. The initial polynomials are

$$L_1 = x \quad L_2 = x^2 - 1 \quad L_3 = x^3 - 2x \tag{4.28}$$

For higher polynomials, see Table 4.1. The *simultaneous* pruning of all terminal vertices in a graph was a significant improvement in calculating the characteristic polynomials of acyclic graphs. A few year later, Randić and coworkers [49] came up with idea of "ultimate pruning," which is an even more efficient approach for construction of the characteristic polynomial for acyclic graphs. The ultimate pruning approach consists of selecting a single C–C bond (i, j) and constructing a 2×2 determinant:

$$\begin{vmatrix} L_i & L_{ii} \\ L_{jj} & L_j \end{vmatrix} \tag{4.29}$$

Here L_i and L_j are the characteristic polynomials of the fragments of the graph obtained when the bond C_i–C_j has been deleted, and L_{ii} and L_{jj} are characteristic polynomials of the fragments of the graph obtained when the bond C_i–C_j and vertices C_i, C_j have been deleted. Thus, for example, if in the case of 2,4-dimethylhexane,

$$\begin{vmatrix} -L_3 & (L_1)^2 & 0 & 1 \\ 1 & -L_1 & 1 & 0 \\ 0 & L_1 & -L_2 & L_1 \\ 0 & 0 & L_1 & -L_2 \end{vmatrix} = 0 \quad \longleftarrow \quad \begin{vmatrix} 0 & 1 & 0 & 0 \\ 1 & 0 & 1 & 0 \\ 0 & 1 & 0 & 1 \\ 0 & 0 & 1 & 0 \end{vmatrix} = 0$$

FIGURE 4.9 Construction of the reduced characteristic polynomial by elimination of terminal edges following the method of Balasubramanian $L_1 = x$, $L_2 = x^2 - 1$, $L_3 = x^3 - 2x$.

one deletes the central CC bond, one obtains for the determinant $|L_iL_j - L_{ii}L_{jj}| = \{(L_4 - 1)L_4 - L_1L_2L_3\}$. To find the characteristic polynomials of 2,4-dimethylhexane, one has to multiply the small polynomials, and one obtains: $L_8 - 2L_4 - 2L_2 - 1$, or $x^8 - 7x^6 + 13x^4 - 6x^2$.

The even powers of L_k polynomials have no factors (except involving terms of different parity, induced by the center of symmetry), but a linear combination of even power L_k polynomials may have factors. The above linear combination of L_k polynomials has at least factor x^2, which is visible from the expression of the characteristic polynomial expressed in powers of x. But how are we to find factors of characteristic polynomials expressed by L_k polynomials? The answer is simple although it requires some patience. All one has to do is to construct 2×2 matrices, considering the deletion, one by one, of all CC bonds. For example, if one deletes the bond C_1–C_2, the determinant $|L_1L_2 - L_{11}L_{22}| = \{L_1(L_7 - L_3 - L_1) - L_1(L_5 - L_1)\}$, which is $L_1\{(L_7 - L_3 - L_1) - (L_5 - L_1)\}$ or $L_1(L_7 - L_5 - L_3)$.

It is easy to see that the odd power of L_k polynomials have factors. All one is required to do is to construct the determinant $|L_iL_j - L_{ii}L_{jj}|$ for paths $P(3)$, $P(5)$, $P(7)$, and so on. Thus, one obtains

$$L_3 = L_2L_1 - L_1 = L_1(L_2 - 1)$$
$$L_5 = L_3L_2 - L_2L_1 = L_2L_1(L_2 - 2) \qquad (4.30)$$
$$L_7 = L_4L_3 - L_3L_2 = L_1(L_4 - L_2)(L_2 - 1)$$

and so on.

As we see from the above, L_7, L_5, and L_3 have factor L_1, so

$$L_1(L_7 - L_5 - L_3) = (L_1)^2\{(L_4 - L_2)(L_2 - 1) - L_2(L_2 - 2) - (L_2 - 1)\}$$
$$= (L_1)^2(L_6 - 2L_4 + L_2 - 1). \qquad (4.31)$$

A simpler way to find the same is to delete the bond C_2–C_3, in which case the determinant is

$$|L_2L_3 - L_{22}L_{33}| = \{L_3(L_5 - L_1) - L_4(L_1)^2\}, \qquad (4.32)$$

which allows factoring $(L_1)^2$ because L_3 and L_5 have a factor L_1.

In Table 4.3, we show factors for the characteristic polynomials of smaller branched alkanes expressed in terms of characteristic polynomials of linear alkanes. In Table 4.4, we continue with showing factors for 18 isomers of n-octane C_8H_{18} expressed in terms of L_k (the characteristic polynomial of linear alkanes).

Let us look closer at Tables 4.3 and 4.4. The first thing to observe is that in addition to n-octane (and other even power linear alkanes) the only other molecule in Tables 4.3 and 4.4 that has no factors is 3,4-dimethlhexane. Next, one can also observe that almost all graphs shown have factor L_1 except graphs of 3-ethyl-pentane, 3-ethyl-hexane, and 3-ethyl-3-methylpentane, which instead have the factor L_2. As common factors in Tables 4.3 and 4.4, we find L_2, $(L_2 - 1)$, $(L_2 - 3)$, $(L_4 - L_2 - 1)$, the graphs

TABLE 4.3

Characteristic Polynomial Factors for Smaller Branched Alkanes Expressed in Terms of L_k, the Characteristic Polynomial of Linear Alkanes

	Characteristic Polynomial		Characteristic Polynomial
(structure)	$L_1^2(L_2 - 2)$	(structure)	$L_1(L_2 - 1)(L_4 - L_2 - 1)$
(structure)	$L_1(L_4 - L_2)$	(structure)	$L_1(L_6 - L_4 - 1)$
(structure)	$L_1^3(L_2 - 3)$	(structure)	$L_1 L_2^2(L_2 - 3)$
(structure)	$L_1^2(L_4 - 2L_2 + 2)$	(structure)	$L_1(L_6 - L_4 - L_2 - 1)$
(structure)	$L_2(L_4 - L_2 - 1)$	(structure)	$L_1^3(L_2 - 1)(L_2 - 3)$
(structure)	$L_1^2 L_2(L_2 - 3)$	(structure)	$L_1^3(L_4 - 3L_2 + 3)$
(structure)	$L_1^2(L_4 - 2L_2)$	(structure)	$L_1 L_2(L_6 - 2L_4 - 1)$
		(structure)	$L_1^3(L_4 - 3L_2 + 2)$

having these factors are illustrated in Figure 4.10. So the graphs in the first two rows have common eigenvalues ±1, the graphs in the third row have common eigenvalues $\pm\sqrt{2}$, the graphs in the fourth row have common eigenvalues ±2, and the graphs in the last row have common eigenvalues $\pm\sqrt{(2 \pm \sqrt{3})}$, respectively.

An alternative approach to factorization of chemical graphs and their characteristic polynomials, based on a polynomial division approach was described by Kirby [39].

Let us look more closely at the six graphs in the fourth row having common eigenvalues +2. In Figure 4.11, we show the coefficients of the leading eigenvector for graphs having common factor $(L_2 - 3)$, showing that they satisfy the sum rule: Each coefficient is equal to half the sum of coefficients of its adjacent vertices. Thus, if we assume value a for the coefficient of terminal vertices, the coefficients of their

TABLE 4.4

Factorization of Characteristic Polynomial of 18 Octane Isomers in Terms of L_k Polynomials

Structure	Characteristic Polynomial	Structure	Characteristic Polynomial
	L_8		$L_1^2(L_2(L_4-3L_2+3)$
	$L_1^2(L_6-L_4+2L_2-2)$		$L_1^2(L_6-2L_4)$
	$L_8-L_4-L_2$		$L_1^2(L_6-2L_4-2)$
	$L_1^2 L_2(L_2-1)(L_2-3)$		$L_1^2(L_2-1)(L_6-L_4-1)$
	$L_2(L_6-L_4-L_2)$		$L_1^2(L_6-2L_4-L_2-1)$
	$L_1^2(L_2-1)(L_4-2L_2)$		$L_1^2(L_6-2L_4-L_2-2)$
	$L_1^2(L_6-2L_4+L_2-1)$		$L_1^4(L_4-4L_2+6)$
	$L_1^2 L_4(L_2-3)$		$L_2^2(L_4-2L_2-2)$
	$L_8-2L_4-3L_2-1$		$L_1^4(L_4-4L_2+4)$

neighbors have value $2a$. The sum of coefficients of vertex $2a$ must be $4a$, the sum of coefficients of vertex $3a$ must be $6a$, and so on. It is not difficult to see that graphs of Figure 4.11 satisfy the sum rule. Of course, this is what one should expect (according to the results of Longuet-Higgins [36,37]). But what we have not expected is that the coefficient sum rule allows one to construct additional graphs having the same common eigenvalue. Consider the three graphs in the lower part of Figure 4.11. They can be considered to be the initial members of an infinite sequence of graphs having the eigenvalues ± 2, which are obtained by insertion of additional edges in the chain between the two terminal $-CH(CH_3)_2$ groups.

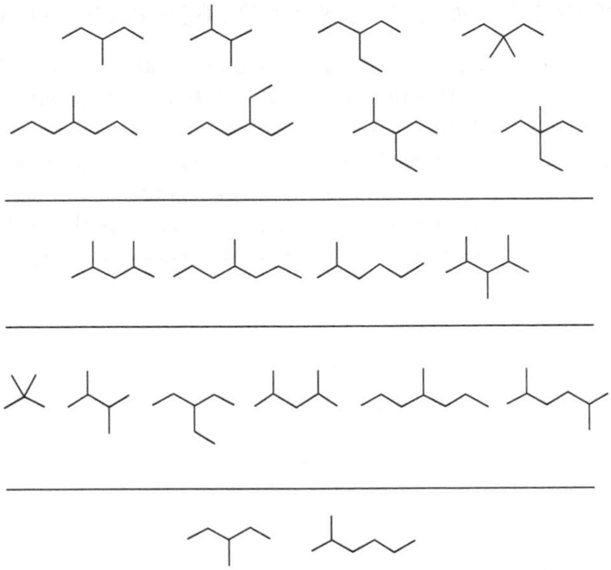

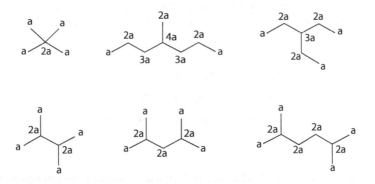

FIGURE 4.10 Smaller alkanes having common factors: L_2 (first two rows), $(L_2 - 1)$ (third row), $(L_2 - 3)$ (fourth row), and $(L_4 - L_2 - 3)$ (bottom row).

FIGURE 4.11 The coefficients of the leading eigenvector for graphs having common factor $(L_2 - 3)$, showing that they satisfy the sum rule for coefficients.

We have not exhausted the topic of spectra of graphs, which includes the characteristic polynomial, isospectral graphs, endospectral vertices, and endospectral graphs, factoring the characteristic polynomial, and searching for graphs having common eigenvalues. We have also considered situations in which a larger graph contains all eigenvalues of a smaller graph, but for example, we have not elaborated why some eigenvalues have high multiplicity, a subject that reflects on the fact that often graphs have higher symmetry than their graphical representation would suggest.

4.8 SYMMETRIC FUNCTION AND NEWTON IDENTITIES

We end with a brief comment on getting the characteristic polynomial of chemical graphs using the symmetric function theory. R. Barakat in his paper has shown that the Frame's method is *"nothing but symmetric functions and Newton identities"* [52]. In the view that a reference is made to Newton, this paper deserves to be included here if even at the very end of this section. Let c_k be the coefficients of the characteristic polynomial. They are the elementary symmetric functions from the eigenvalues of the adjacency matrix (see, for example, Weyl [53] or other books on higher algebra). Thus,

$$c_1 = \Sigma\lambda_1$$
$$c_2 = \Sigma\lambda_1\lambda_2$$
$$c_3 = \Sigma\lambda_1\lambda_2\lambda_3 \tag{4.33}$$
$$\cdots$$
$$c_n = \Sigma\lambda_1\lambda_2\lambda_3\ldots\lambda_n$$

The first coefficient $c_1 = \mathrm{Tr}(A)$, the trace of the adjacency matrix, is the sum of diagonal elements, and the last $c_n = \det A$ is the determinant of the adjacency matrix. For construction of the characteristic polynomial the power sums symmetric functions are required. Newton's identities related the power sums symmetric functions (which can be calculated from known eigenvalues) to coefficients of the characteristic polynomial and allow one to calculate the coefficients. They can be written as

$$s_1 + c_1 = 0$$
$$s_2 + c_1 s_1 + 2c_2 = 0$$
$$s_3 + c_1 s_2 + c_2 s_1 + 3c_3 = 0$$
$$s_4 + c_1 s_3 + c_2 s_2 + c_3 s_1 + 4c_4 = 0$$
$$\cdots$$

The s_k are the traces of A^k. In the time of Newton, matrices were unknown, and the matrix notation adopted here was only subsequently developed by Newton's compatriot A. Cayley. Using these Newton's equations, one can express s_i in terms of the coefficients of the characteristic polynomial. By sequential manipulations, one can obtain explicit formulas. After relatively simple expression of smaller coefficients c_2–c_5 and even c_6 and c_7, the expressions become increasingly cumbersome and complex as reported in the book by Burnside and Panton [54] and shown below:

$$s_1 = -c_1$$
$$s_2 = c_1^2 - 2c_2$$
$$s_3 = -c_1^3 + 3c_1 c_2 - 3c_3$$
$$s_4 = c_1^4 - 4c_1^2 c_2 + 4c_1 c_3 + 2c_2^2 - 4c_4.$$

Observe that subscripts of the coefficients c_i where i can take values between 1 and k, in each term had to add to the subscript of s_k. As one can see, the above equations no longer have a simple pattern seen in Newton's equations, which, in the case of Newton's equations, allows one to write the next line from the lines above.

Barakat [52] has solved Newton's equations for individual coefficients c_k, which we show in Table 4.5 for a selection of coefficients c_2–c_{10}. Barakat arrived at the coefficients c_k by directly solving numerical evaluations using a computer. The computed coefficients (disregarding the sign) show no pattern or regularities and soon start getting very large as is illustrated for c_9 and c_{10}. There is no way one knowing expressions for smaller coefficient can find larger ones, for example, how can one from c_9 find c_{10}. So one gets the numbers, big numbers, but do these numbers have some structure? Recall the statement of Coulson [55]: "*...give us insight, not numbers.*"

Or the same message from Hoffmann [56] in an alternative formulation: "*Even if the...achievements of modern computational chemistry are astounding, one still requires...a simple portable explanation.*"

Of course, we need both the numbers and an explanation! It happens that one of the present authors of this book was asked by the editor of *Theoretica Chimica Acta*, Professor Klaus Ruedenberg, to evaluate the manuscript of Barakat submitted to *Theoretica Chimica Acta*. How editors select reviewers is not often clear, but Barakat in his manuscript included the following comment: "*Since this manuscript was completed, two important papers have appeared: Balasubramaninan [8] and Randić [9]. Each contains extensive references to the literature.*" The references [8] and [9] in the above quote are the references [57] and [58] of this chapter.

So perhaps there was not much mystery in this case as to how the reviewer was selected. In his report to the editor, Randić recommended the manuscript for publication and made a reference to these large numbers occurring for larger coefficients of the characteristic polynomial and pointed out that these numbers relate to selected Young diagrams and have some structure. The editor responded with a comment that the reviewer need not disclose his results to the author, but should publish them after the paper of Barakat had been published. That has been done, and the paper by Randić [58] appeared the next year in which magnitudes of the large coefficients were explained.

TABLE 4.5
Solutions of Newton's Equation for the Coefficients c_k in Terms of Traces s_k

$2!\, c_2 = -s_2$

$3!\, c_3 = 2s_3$

$4!\, c_4 = -6s_4 + 3s_2^2$

$5!\, c_5 = 24s_5 - 20s_3s_2^2$

$6!\, c_6 = -120s_6 + 90s_4s_2 + 40s_3^2 - 15s_2^3$

...

$9!\, c_9 = 40{,}320s_9 - 25{,}920s_7s_2 - 20{,}160s_6s_3 - 18{,}144s_5s_4 + 9072s_5s_2^2 + 15{,}120s_4s_3s_2 + 2240s_3^3 - 2520s_3s_2^3$

$10!\, c_{10} = -362{,}880s_{10} + 226{,}800s_8s_2 + 172{,}800s_7s_3 - 15{,}100s_6s_4 - 75{,}600s_6s_2^2 + 7276s_5^2 - 120{,}960s_5s_3s^2$
$\qquad - 56{,}700s_4^2s_2 - 50{,}400s_4s_3^2 + 18{,}900s_4s_2^3 + 25{,}200s_3^2s_2^3 - 945s_2^5$

There is hidden regularity in these, at first sight, chaotic collections of coefficients. The signs of each term are determined by the parity of the number of rows R of Young diagrams as follows: For even coefficients, the sign of terms are $(-1)^R$, and for odd coefficients, the sign of terms are $(-1)^{R+1}$. The Young diagrams depict partitions of the index of the coefficients of the characteristic polynomial. For example, C_6 (listed in Table 4.5) relates to the following four partitions of number 6:

$$6, \quad 4 + 2, \quad 3 + 3, \quad 2 + 2 + 2,$$

which are all possible partitions not involving a contribution from number 1. The magnitude of the contribution of the partition 6 is 6!/6, which is 120; the magnitude of the contribution of the partition $4 + 2$ is $6!/(4 \times 2)$, which is 90; the magnitude of the contribution of the partition $3 + 3$ is $6!/(3 \times 3 \times 2!)$, which is 40; the magnitude of the contribution of the partition $2 + 2 + 2$ is $6!/(2 \times 2 \times 2 \times 2!)$, which is 15. Observe that when the Young diagram has two, three, or more equal rows, one has to add the factor 2!, 3!, or higher factorials, depending on the number of equal rows. With this introduction to the regularity of the contributions making the individual coefficients of the characteristic polynomial, the reader can verify the correctness of Figures 4.12 and 4.13, in which the contribution to coefficients c_9 and c_{10} are illustrated.

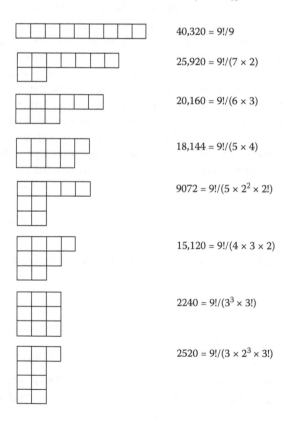

	$40{,}320 = 9!/9$
	$25{,}920 = 9!/(7 \times 2)$
	$20{,}160 = 9!/(6 \times 3)$
	$18{,}144 = 9!/(5 \times 4)$
	$9072 = 9!/(5 \times 2^2 \times 2!)$
	$15{,}120 = 9!/(4 \times 3 \times 2)$
	$2240 = 9!/(3^3 \times 3!)$
	$2520 = 9!/(3 \times 2^3 \times 3!)$

FIGURE 4.12 Young diagrams and the decomposition of the coefficients of c_9.

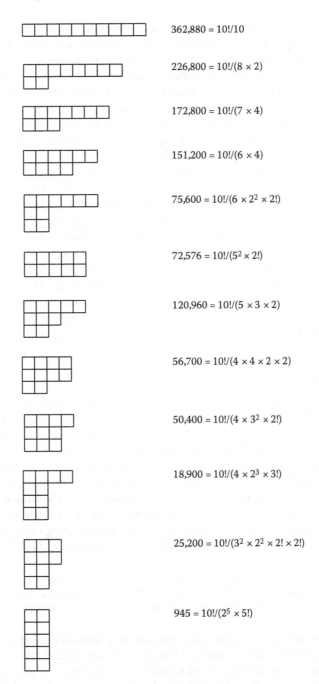

$362,880 = 10!/10$

$226,800 = 10!/(8 \times 2)$

$172,800 = 10!/(7 \times 4)$

$151,200 = 10!/(6 \times 4)$

$75,600 = 10!/(6 \times 2^2 \times 2!)$

$72,576 = 10!/(5^2 \times 2!)$

$120,960 = 10!/(5 \times 3 \times 2)$

$56,700 = 10!/(4 \times 4 \times 2 \times 2)$

$50,400 = 10!/(4 \times 3^2 \times 2!)$

$18,900 = 10!/(4 \times 2^3 \times 3!)$

$25,200 = 10!/(3^2 \times 2^2 \times 2! \times 2!)$

$945 = 10!/(2^5 \times 5!)$

FIGURE 4.13 Young diagrams and the decomposition of the coefficients of c_{10}.

The importance and the significance of the results of Randić is not so much in that he found the Young diagram expressions for the coefficients of Barakat's equations, but that this allows finding the coefficients c_k in terms of traces s_k for solution of higher Newton's equations directly without computer and even without a recursion.

With this discussion of Newton's equations, based on the work of Newton, Weyl, Burnside and Panton, Barakat, and Randić, one would tend to consider the problem of Newton's equation exhausted, nothing new to be added to it, and viewed as "completely solved." By revisiting the Newton equations when writing this book, we came to a new observation. Recollect that it has been mentioned that contributions to individual coefficients (those listed in Figures 4.12 and 4.13) represented by Young diagrams illustrate various partitions of the index of the coefficients but excluding those involving number 1. That has been observed, but why are all the partitions involving number 1 not included? These would be, in the case of number 6, partitions 5 + 1, 4 + 1 + 1, 3 + 2 + 1, 3 + 1 + 1 + 1, and so on. By not including the Young diagrams involving rows with a single cell, we are assured that all the numerical contributions occurring in the solutions for the coefficients will involve different numerical values and will be unique, that is, they cannot occur more than once. This observation is novel; uniqueness of coefficients has not been considered and mentioned earlier. So even when the solution of the problem is "complete," there may be an additional observation that may add novel information to already solved problems!

4.9 SUMMARY

As one can see, we spent considerable time discussing the characteristic polynomials of graphs, their construction, and some of their properties, and in the view that the characteristic polynomials of graphs can today be easily obtained, for example, using MATLAB, why should one need to study their construction? Is this not a waste of time, just as it would be a waste of time today to be concerned about logarithmic tables when calculators can do the same job much faster and more accurately? True, but logarithmic functions are indispensable, so to know about the basic properties of logarithms is essential. In a similar manner, knowing the basic properties of characteristic polynomials is essential for arriving at and understanding other properties of the characteristic polynomials of graphs. Recall the statement of Prelog [59]:

> ...chemists are often satisfied with inspecting and discussing graphs without paying too much attention to their algebraic aspects, but it is evident that some familiarity with the theory of graphs is necessary for deeper understanding of their properties.

We will here no more dwell on the characteristic polynomial except for mentioning additional publications that interested readers can explore and follow. There have been several publications on Le Verrier-Fadeev's method [60,61], which appears to be the most general technique for obtaining the characteristic polynomial. Related to the above is the method of Frame [62], which received some attention in chemical literature [63]. Also techniques for obtaining the characteristic polynomial large graphs have received some attention in chemical literature [45,64–67]. The Chebyshev expansion of characteristic polynomials in terms of L_n polynomials on paths has received some attention [39,68–70]. In the first of these three papers, Kirby considered computer-aided decomposition of the secular determinant, and in the fourth paper, brothers He and He (one being chemist, the other a mathematician) considered properties of "structure factors," which relate to the expansion of characteristic polynomials on terms of L_n. In that paper, is proven two conjectures of Hosoya and Randić concerning the "structure factors," which are of interest for recurrence methods to be used for efficient techniques for expanding and solving characteristic polynomials of complex graphs, such as the partitioning technique [71,72], the pruning technique [50,51], the block-diagonalization method, and so on [2,8,41,51,70–76]. We may add that an early paper on the use of characteristic polynomials of path graphs for calculation of bond lengths in some cyclic compounds was by Goodwin and Vand published in 1955 [77].

There are other topics that we wish to cover, so we will end this section with a small linguistic diversion. However, as we have seen, Chebyshev expansion was fairly central to this section, so we will end with a brief recapitulation on the spelling of the name of Chebyshev, which appears simple in comparison with alternatives but is unfortunately not correct enough. There have been some problems with the spelling of the name of this Russian mathematician Chebyshev or Чебышёв as is his name spelled in the Russian Cyrillic alphabet, which we met in this section. Because Russia has never prescribed an "official" transcription of Чебышёв (or other names), it has become in America known as Chebyshev but has been referred to also as Chebysheff and Chebyshov; Tchebychev and Tchebycheff (French); or Tschebyschev, Tschebyschef, and Tschebyscheff (German). The English Chemical Society recommends for transcription of Russian names to use the Croatian alphabet (a Slavic language just as Russian), which uses č, š, ž for ch, sh, zh, respectively, and which tends to be common in English transcriptions of Russian names. This then also applies to Cyrillic alphabets of Ukrainian, Belorussian, Bulgarian, Macedonian, Serbian, and Montenegrian (which is almost identical to Serbian). If one follows this recommendation, one should write for Чебышёв the transcription Čebišev, at least in the chemical literature (which replaces the Russian ё incorrectly with e). Of the above 10 variations of the name, excluding Чебышёв and Čebišev, the most correct would be the third transcription, Chebyshov, the least used, because the Russian ё is pronounced as o in hot. But, most likely, Чебышёв will remain Chebyshev unless one day Russia adopts (like China) parallel scripts for the Cyrillic and Latin alphabets, which is not to happen soon. If that happens, however, most likely immortal Чебышёв will become Čebišov, a transcript yet to be found in print.

REFERENCES AND NOTES

1. Y. S. Kiang, A. C. Tang, and R. Hoffmann, *Theor. Chim. Acta* 66 (1984) 183.
2. H. Hosoya, Topological index. A newly proposed quantity characterizing topological nature of structural isomers of saturated hydrocarbons, *Bull. Chem. Soc. Jpn.* 44 (1971) 2332–2339.
3. F. H. Clarke, A graph polynomial and its applications, *Discrete Math.* 3 (1971) 305–313.
4. E. Hückel, *Grundzuge der Theorie ungesattiger und aromatischer Verbindugen*, Verlag Chemie, Berlin, 1938.
5. J. N. Murrell, S. F. A. Kettle, and J. M. Tedder, *Valence Theory*, J. Wiley and Sons, London, 1965.

6. R. Hoffmann, S. Shaik, and P. C. Hiberty, A conversation on VB vs MO theory: A never-ending rivalry? *Acc. Chem. Res.* 36 (2003) 750–756.
7. MATLAB (abbreviation for matrix laboratory) developed by MathWorks.
8. E. Heilbronner, Das Komposition Prinzip: Eine anshauliche Methode zur elektron-theoretischen Behandlung nicht oder niedrig symmetrischer Molekeln in Rahmne der MO-Theorie, *Helv. Chim. Acta* 36 (1953) 170–188.
9. V. R. Rosenfeld and I. Gutman, A new recursion relation for the characteristic polynomial of a molecular graph, *J. Chem. Inf. Comput. Sci.* 36 (1996) 527–530.
10. A. J. Schwenk, Computing the characteristic polynomial of a graph, in: R. Bari and F. Harary (Eds.), *Graphs and Combinatorics*, Springer Verlag, Berlin, 1974, pp. 153–172.
11. I. Gutman, Polynomials in graph theory, in: *Chemical Graph Theory, Introduction and Fundamentals*, D. Bonchev and D. H. Rouvray (Eds.), Abacus Press, New York, 1991.
12. H. Sachs, Beziehungen zwischen den in einem Graphen enthaltenen Kreisen und seinem charakteristischen Polynom, *Publ. Math. (Debrecen)* 11 (1964) 119–134.
13. C. A. Coulson, Notes on the secular determinant in molecular orbital theory, *Proc. Camb. Phil. Soc.* 46 (1950) 202.
14. S. Singh, *Fermat Last Theorem*, Fourth Estate Ltd., London, 2002.
15. A. Graovac, D. Plavšić, M. Kaufman, T. Pisanski, and E. C. Kirby, Application of the adjacency matrix eigenvectors method to geometry determination of toroidal carbon molecules, *J. Chem. Phys.* 113 (2000) 1925–1931.
16. T. Pisanski and J. Shawe-Taylor, Characterizing graph drawing with eigenvectors, *J. Chem. Inf. Comp. Sci.* 40 (2000) 567–571.
17. T. Pisanski, M. Razinger, and A. Graovac, Geometry versus topology: Testing self-consistency of the NiceGraph program, *Croat. Chem. Acta* 69 (1996) 827–836.
18. J. R. Dias, Facile determination of molecular graph eigenvalues with symmetrical eigenvectors, *MATCH—Commun. Math. Comput. Chem.* 48 (2003) 55–61.
19. J. R. Dias, Properties and relationships of right-hand mirror-plane fragments and their eigenvectors, *Mol. Phys.* 88 (1996) 407–417.
20. J. R. Dias, An example molecular orbital calculation using Sachs graph method, *J. Chem. Educ.* 69 (1992) 695–700.
21. A. J. Kassman, Generation of eigenvectors of the topological matrix from graph theory, *Theor. Chim. Acta* 67 (1985) 2555.
22. A. K. Mukherjee and K. K. Dutta, Two new graph-theoretical methods for generation of eigenvectors of chemical graphs, *Proc. Indian Acad. Sci.* 101 (1989) 499.
23. M. Randić, M. Vračko, and M. Novič, Eigenvalues as molecular descriptors, chapter 7, in: *QSPR/QSAR Studies by Molecular Descriptors*, M. V. Diudea (Ed.), Nova Science Publishers, Huntington, NY, 2001, pp. 147–211.
24. J. R. Dias, Structural origin of specific eigenvalues in chemical graphs of planar molecules, *Mol. Phys.* 85 (1995) 1043–1060.
25. J. R. Dias, Determining select eigenvalues by embedding smaller structures onto larger ones, *MATCH—Commun. Math. Comput. Chem.* 22 (1987) 257–268.
26. J. R. Dias, Facile calculations of the characteristic polynomials and π-energy levels of molecules using chemical graph theory, *J. Chem. Educ.* 66 (1987) 213–216.
27. J. R. Dias, A facile Hückel molecular orbital solution of buckminsterfullerene using chemical graph theory, *J. Chem. Educ.* 66 (1987) 1012–1015.
28. A. Golbraikh and A. Tropsha, Beware of q^2!, *J. Mol. Graph. Model.* 20 (2002) 269–276.
29. R. Hoffmann, P. v. R. Schleyer, and H. F. Schaefer, III, Predicting molecules—More realism, please!, *Angew. Chem. Int. Ed.* 47 (2008) 7164–7167.
30. Gaussian is computer program initiated by J. A. Pople and his research group W. J. Hehre, W. A. Lathan, R. Ditchfield, and M. D. Newton in 1970 at Carnegie-Mellon University in Pittsburgh, Pennsylvania.

31. Density Functional Theory (DFT) is a continually developing quantum chemical model initiated on two theorems of Pierre Hohenberg and Walter Kohn published in: P. Hohenberg and W. Kohn, Inhomogeneous electron gas, *Phys. Rev.* 136 (1964) B864–B871.
32. S. Fajtlowicz, Written on the wall, a list of conjectures of graffiti. http://www.math.uh.edu/siemion/.
33. S. Fajtlowicz, On conjectures of graffiti III, *Congr. Numerantium* 66 (1988) 23–32.
34. B. Bollobás and P. Erdös, Graphs of extremal weights, *Ars Combin.* 50 (1998) 225–233.
35. C. A. Coulson and A. Streitweiser, *Dictionary of Pi-Electron Calculations*, W. H. Freeman Company, San Francisco, 1965.
36. H. C. Longuet-Higgins, Some studies in molecular orbital theory I. Resonance structures and molecular orbitals in unsaturated hydrocarbons, *J. Chem. Phys.* 18 (1950) 265–274.
37. H. C. Longuet-Higgins, Some studies in molecular orbital theory II. xxx... *J. Chem. Phys.* 18 (1950) 283–291.
38. S. S. D'Amato, B. M. Gimarc, and N. Trinajstić, Isospectral and subspectral molecules, *Croat. Chem. Acta* 54 (1981) 1–52.
39. E. C. Kirby, The factorization of chemical graphs and their polynomials: A polynomial division Approach, *Croat. Chem. Acta* 59 (1986) 635–641.
40. J. R. Dias, Strongly subspectral conjugated molecular systems. From small molecules to infinitely large π-electronic networks, *J. Phys. Chem. A* 101 (1997) 7167–7175.
41. Y.-S. Kiang and A.-C. Tang, A graphical evaluation of characteristic polynomials of Huckel trees, *Int. J. Quantum Chem.* 29 (1986) 229–240.
42. H. Hosoya, Graphical enumerations of the coefficients of the secular polynomials of the Hückel molecular orbitals, *Theor. Chim. Acta* 25 (1972) 215–222.
43. H. Hosoya, What can mathematical chemistry contribute to the development of mathematics? *HYLE—Int. J. Philos. Chem.* 19 (2013) 87–105.
44. I. Gutman, Characteristic and matching polynomials of some compound graphs, *Publ. Inst. Math. (Beograd.)* 27 (1980) 61–66.
45. M. Randić, On alternative form of the characteristic polynomial and the problem of graph recognition, *Theor. Chim. Acta* 62 (1983) 485–498.
46. C. F. Wilcox, Stability of molecules containing (4n)-rings, *Tetrahedron Lett.* 7 (1968) 795.
47. C. F. Wilcox, Stability of molecules containing nonalternant rings, *J. Am. Chem. Soc.* 91 (1969) 2732.
48. M. J. S. Dewar and H. C. Longuet-Higgins, The correspondence between the resonance and molecular orbital theories, *Proc. Roy. Soc. A* 214 (1952) 482.
49. M. Randić, B. Baker, and A. Kleiner, Factoring the characteristic polynomial, *Int. J. Quantum Chem.: Quantum Chem. Symp.* 19 (1986) 107–127.
50. K. Balasubramanian, Spectra of chemical trees, *Int. J. Quantum Chem.* 21 (1982) 581.
51. K. Balasubramanian and M. Randić, The characteristic polynomials of structures with pending bonds, *Theoret. Chim. Acta* 61 (1982) 307–323.
52. R. Barakat, Characteristic polynomials of chemical graphs via symmetric function theory, *Theor. Chim. Acta* 69 (1986) 35–39.
53. H. Weyl, *The Classical Groups, Their Invariants and Representations*, Princeton University Press, Princeton, 1946.
54. W. S. Burnside and A. W. Panton, *Theory of Equations*, Vol. 1, Dover Publications, New York, 1956.
55. C. A. Coulson, d-Electrons in chemical bonding, in: *Proceedings of the Robert A. Welch Foundation Conferences on Chemical Research, vol. XVI: Theoretical Chemistry*, W. O. Milligan (Ed.), Houston, Texas, 1973.
56. R. Hoffmann, Qualitative thinking in the age of modern computational chemistry—Or what Lionel Salem knows, *J. Mol. Struct. (Theochem.)* 424 (1998) 1.

130 Solved and Unsolved Problems of Structural Chemistry

57. M. Randić, On the characteristic equations of the characteristic polynomial, *SIAM J. Algebraic Discrete Methods* 6 (1985) 145–162.
58. M. Randić, On the evaluation of the characteristic polynomial via symmetric function theory, *J. Math. Chem.* 1 (1987) 145–152.
59. V. Prelog, Foreword, in: *Chemical Applications of Graph Theory*, A. T. Balaban (Ed.), Academic Press, London, 1976.
60. U. Le Verrier, Sur les variations séculaires des éléments des orbites pour les sept planètes principales, *J. Math.* 1 (1840) 230.
61. D. K. Faddeev and I. S. Sominskij, *Sbornik Zadatch po Vyshej Algebre*, Nauka, Moscow, 1949; (*Collection of Problems of Higher Algebra*), Moscow-Leningrad, 1949.
62. J. S. Frame, A simple recursion formula for inverting a matrix (abstract). *Bull. Am. Math. Soc.* 55 (1949) 1045.
63. K. Balasubramanian, The use of Frame's method for the characteristic polynomial of chemical graphs, *Theor. Chim. Acta* 65 (1984) 49–58.
64. J. Brocas, Comments on characteristic polynomials of chemical graphs, *Theor. Chim. Acta* 68 (1985) 155–148.
65. P. Křivka, Z. Jeričević, and N. Trinajstić, On the computation of characteristic polynomial of a chemical graph, *Int. J. Quantum Chem.: Quantum Chem. Symp.* 19 (1986) 129–147.
66. M. Randić, On evaluation of the characteristic polynomial for large graphs, *J. Comput. Chem.* 3 (1982) 421–435.
67. S. El-Basil, Characteristic polynomial of large graphs. On alternative form of characteristic polynomial, *Theor. Chim. Acta* 65 (1984) 191–197.
68. E. C. Kirby, The characteristic polynomial: Computer-aided decomposition of the secular determinant as a method of evaluation for homo-conjugated systems, *J. Chem. Res. (S)* 5 (1984) 4–5.
69. E. C. Kirby, The characteristic polynomial: Evaluation as a function of x and as a function of the characteristic polynomials of linear polyenes using small computers, *Comput. Chem.* 9 (1985) 79–83.
70. W. He and W. He, Properties of "structure factor" of characteristic polynomial and the proof of Hosoya-Randić conjectures, *Theor. Chim. Acta* 70 (1986) 35–41.
71. A.-O. Tang and Y.-S. Kiang, Graph theory of molecular orbitals, *Sci. Sin.* 19 (1976) 207.
72. Y.-S. Kiang, Partition technique and molecular graph theory, *Int. J. Quantum Chem.* 20 (1981) 293.
73. A.-O. Tang, Y.-S. Kiang, G.-S. Yan, and S.-S. Dai, *Graph Theoretical Molecular Orbitals* (in Chinese), Publishing House of Science, Bejing, 1980.
74. L. Lovász and J. Pelikán, On the eigenvalues of trees, *Period. Math. Hung.* 3 (1973) 175–182.
75. H. Hosoya and K. Hosoi, Topological index as applied to π-electronic systems. III. Mathematical relations among various bond orders, *J. Chem. Phys.* 64 (1976) 1065–1073.
76. A. J. Schwenk, Computing the characteristic polynomial of a graph, *Lect. Notes Math.* 406 (1973) 153–172.
77. T. H. Goodwin and V. Vand, Calculated bond lengths in some cyclic compounds. Part I. Methods of calculation, *J. Chem. Soc.* (1955) 1683.

5 Structure–Activity

In many interesting areas of chemistry we are approaching predictability, but…I would claim, not understanding.

Roald Hoffmann [1]

5.1 TWO CULTURES

For some time in QSAR coexisted "two cultures" that apparently had no visible interaction. One "culture" cherished the physicochemical modeling of the structure–activity relationship, following the footsteps of Corwin Hansch. The other "culture" cherished and advocated the use of "mathematical descriptors" for modeling of the structure–activity relationship. The latter made QSAR one of the topics of mathematical chemistry. Perhaps in contrast to the case of the "classical" two cultures of C. P. Snow [2], which was about lack of communication between the "arts" and the "sciences," here we have two cultures in the same discipline. The roots of the division of the two cultures of C. P. Snow could be attributed to unfamiliarity of the people professing humanities with the problems relating to science and a lack of understanding of general scientific aims and accomplishments, including some basic ideas and notions of natural science, which, in a broad sense, incorporates medicine and mathematics. This is a half of the story; the other half is that the same appears to be with scientists when relating to humanities: unfamiliarity of most scientists about the issues and problems of humanities, which include languages (modern and ancient), literature, philosophy, religion, painting and sculpture, music, and theater, which then extends to social sciences, including history, anthropology, communication and semiotics, linguistics, cultural and political studies, law, and such. This is not that individual humanists have no interest in natural sciences and that individual scientists have no interest in humanities but that there are not enough bridge builders and even fewer bridge walkers!

That any rule has a few exceptions is well known, but let us mention that there are many exceptions to C. P. Snow's "two cultures," which speaks of two groups of educated people, having interest in either in science and engineering or arts and social sciences. When speaking of individuals, there are too many that excelled in both cultures to be mentioned, so we will mention only two, both chemists, who excelled in both. The first is Carl Djerassi (born 1923), a Bulgarian (by father), Austrian (by mother), Jewish (by father and mother), and American (by choice), organic chemist and novelist, widely known for his contribution to the development of "the pill," the oral contraceptive. The second one is Roald Hoffmann (born 1937), Polish/Ukrainian Jewish (by parents), theoretical organic chemist, author of several books of poetry, survived the Nazi extermination in occupied Europe by hiding for four years in the attic of a school in a small village, kept alive by a

Ukrainian peasant family. Although considerably different by their chemistry and by their artistic directions, Djerassi and Hoffmann found the time to write a joint book, *Oxygen*, on popularizing the meaning of discovery in science [3]. Djerassi, among other activities, has written several novels in a new genre that is neither science nor science fiction, but "science in fiction." In these books, he has described the lives of real scientists and their aspirations and accomplishments as well as frustrations and conflicts. Hoffmann, in the book *Chemistry Imagined: Reflections on Science* [4], describes molecules, which may be as simple as H_2O or that look simple, such as camphor, steroids, and hemoglobin, but are not and that neither look simple nor are simple, such as the immunosuppressant depicted in the book having approximately 50 carbon atoms and a dozen oxygen atoms. This book is also of interest as a rare case of the collaboration of art and science. The book, in addition to numerous figures of molecules, also has 30 full-page colored plates of stimulating images composed by artist Vivian Torrence. In summary, briefly, there is no doubt that there are two cultures, but there is also third culture, a culture of people such as Djerassi, Hoffmann, and Torrence and many others interested in and contributing to both science and art.

We could continue by listing a number of similar cases of scientists involved in arts, whether it is poetry, literature, paintings, music, etc., but a fair listing, just to show that such "bilingual" activities are not so uncommon among scientists and mathematicians, will be too long to consider. Instead we will end with Table 5.1, which gives a short list of very distinguished mathematicians who have been recognized also as writing poetry. Among the 11 mathematicians, we find four whom we have met on pages of this book: Fermat, Leibnitz, Hamilton, and Sylvester. For more on mathematicians and poetry, one should consult a recent paper by Shmukler and Ziskin [5] about "mathematicians in the land of poetry." The paper by Shmukler and Ziskin, published in the *Journal of Mathematics and the Arts*, which is one of several new journals bridging the gap between the "two cultures" and which is a good sign of emerging growth of interest in bridging the gap between the "two cultures."

In the case of QSAR, the number of "bilingual" members in the community in both cultures appears not to be so negligible. Moreover, their number appears to be growing. Nevertheless, a lack of dialog and of communications between the proponents of the two sides of QSAR remain visible [6], which does not serve the common goals and interests of QSAR.

Let us start with a comment on the quantitative and the qualitative character of many structure–property and structure–activity studies. Just because results of a QSAR study are expressed in numbers does not necessarily mean that the result is quantitative if one is interested in the interpretation of the structure–property or structure–activity model considered. The result is a quantitative prediction of numerical values for the *property* considered from the regression equation but, at best, qualitative results relating to the *interpretation* of the model. In order that results qualify as quantitative in that respect, they ought to satisfy certain conditions. One such condition, which is essential for structural interpretation of the results, is that they contain all the information that is necessary to determine, for each descriptor used, how much it *individually* contributed to the resulting regression. In what

TABLE 5.1
Outstanding Mathematicians Writing Poetry

Niccolo Tartaglia (1499–1557)	Solving cubic equations.
Pierre de Fermat (1601–1665)	Among others, famous for his last theorem.
Jacob Bernoulli (1655–1705)	Member of famous Swiss mathematical family. Founder of the calculus of variations, contributed to the development of calculus and probability theory. Studied infinite series.
GottfriedWilhelm Leibniz (1646–1716)	One of the founders of calculus. Besides mathematics, had interests in many fields of the humanities, including philosophy, theology, and linguistics.
Augustine Louis Cauchy (1789–1857)	French mathematician. Contributed to the development of complex analysis, abstract algebra, and differential equations. Wrote 800 papers.
Nikolai Ivanovich Lobachevsky (1793–1856)	Russian mathematician. One of the founders of non-Euclidean geometry.
Sir William Rowan Hamilton (1805–1865)	Among others, the discoverer of the quaternions.
James Joseph Sylvester (1814–1897)	Enriched mathematics by his innovations in algebra, number theory, combinatorics, and graph theory.
Lewis Carroll (1832–1898)	A mathematician, writer, poet and clergyman, the author of *Alice's Adventures in Wonderland*. Had interests in geometry, linear algebra, logic, and recreational mathematics.
Sofia Vasilyevna Kovalevskaya (1850–1891)	The first European woman mathematician to earn a doctoral degree. Worked on partial differential equations, elliptic integrals, mathematical physics, and classical and celestial mechanics.
Felix Hausdorff (1868–1942)	German mathematician, founder of modern topology. Contributed to the theory of measure, the theory of functions, set theory, and functional analysis.

follows, we will see that large majority of published results of multiple regression analysis (MRA) publications, believed to be quantitative, are, at best, "half quantitative." By this, we mean that, in these reports, some important information that would make such works fully quantitative is missing. This does not make the reports *incorrect* but rather *incomplete*.

One of the central oversights and indifferences in use of MRA in QSAR relates not to using orthogonal molecular descriptors. Orthogonal molecular descriptors, which were introduced almost a quarter of a century ago [7], continue to be overlooked by a vast majority of QSAR researchers, who apparently do not realize that use of orthogonal molecular descriptors in multiple regression analysis is not an option but a must if they want to elaborate on the structure–property or structure–activity relationship!

Apparently, it has not been recognized by many users of MRA that without giving *complete* information on the intercorrelation between the descriptors used in MRA, the interpretation of derived regression equations *cannot be considered quantitative*.

It is only when one uses orthogonal descriptors that the coefficients of independent variables give the full information on the contributions of individual variables used in the regression. If one does not use orthogonal descriptors, which may be in well over 90% of publications in QSAR, one cannot *quantitatively* extract information on the relative roles of the descriptors used in MRA. One should therefore consider reporting the regression equations between the descriptors used in MRA, which are necessary for extracting quantitative interpretation from the MRA equation. In the case that one uses MRA merely to predict unknown data, and one is not interested in interpretation of the regression results, a *single* MRA regression equation may suffice. But in such cases, there should be no attempts to speculate on the quantitative importance of different descriptors.

In addition to the problem of interpretation of MRA regression, we would like to point out a widespread misconception concerning the use of *collinear descriptors* in QSAR. It has been known and recognized that highly collinear descriptors are the source of the instability of the coefficients of regression equations, which occur whenever additional descriptors are incorporated into MRA in an effort to reduce the standard error of the regression. So it is not uncommon to read or hear about how one should not use highly collinear descriptors in MRA. Some have even recommended percentage levels of intercorrelations at which collinear descriptors should not be used. This was sound advice some 25 years ago and earlier, but not in the 21st century! As we are approaching the 25th anniversary of the seminal publication on orthogonal MRA, perhaps it is time for awakening the MRA community—and if this book facilitates belated recognition of the necessity of using orthogonal descriptors in MRA when *discussing modeling* structure–property–activity studies, writing about this was worthwhile, regardless of the dozen other topics that the book covers.

We describe in the following sections that it is possible not only to reduce but to completely eliminate collinearity between molecular descriptors, be they mathematical, quantum chemical, or physicochemical. As the result, one obtains stable MRA equations. Continuing ignoring the examination of regressions between interrelated descriptors used in MRA before *discussing* the results of the regression equations is the major unwarranted and needless obstacle for those researchers to derive fully quantitative results from their own labor.

This is not the only oversight of novelties coming from mathematical chemistry that may impact MRA the 21st century. We describe in this chapter that also in 1991, which is close to 25 years ago, there was another novelty deserving wide attention in MRA, which similarly has hardly been noticed. This relates to a novel kind of molecular descriptor, which received little attention and some misunderstanding. The standard MRA uses a set of n descriptors, the so-called "independent" variables, which, as we have pointed out on previous pages, are hardly independent. What is common to all these descriptors, be they mathematical, quantum chemical, or physicochemical, is that their numerical values are determined *before* MRA calculations started. The novelty coming from mathematical chemistry is *variable molecular descriptors*, descriptors the numerical values of which for individual molecules change *during* the search for an optimal regression equation. First, such regressions have used the *variable* connectivity index, and other variable topological indices followed. In such applications, variable indices may involve one or more parameters, the numerical values of which are determined *during* the search for the best MRA equation.

In this respect, the situation (ignoring use of orthogonal descriptors, shunning from collinear descriptors and variable descriptors) is counterproductive, causing a kind of unwarranted stagnation in QSAR. It is a result of a lack of "cross-subcultural communication" between the two seemingly opposing approaches to QSAR, empirical versus nonempirical (mathematical). Let us clarify what "cross-cultural communication" refers to, by citing from [8]:

> Cross-cultural communication endeavors to bring together…relatively unrelated areas…and established areas of communication. Its core is to establish and understand how people from different cultures communicate with each other. Its charge is to also produce some guidelines with which people from different cultures can better communicate with each other.

The above is followed by

> The study of languages other than one's own can serve not only to help one understand what we as humans have in common, but also to assist in the understanding of the diversity which underlines our languages' methods of constructing and organizing knowledge. Such understanding has profound implications with respect to developing a critical awareness of social relationships. Understanding social relationships and the way other cultures work is the groundwork of successful globalization business affairs.

The above says clearly to the two opposing cultures of QSAR what ought to be done, but who is listening? We try to follow this advice and offer an illustration that reflects the benefits of cross-communication.

5.2   QUALITATIVE REGRESSION ANALYSIS

Let us consider a pair of regression equations to illustrate the qualitative nature of them. The selected equation, which represents the Hansch-type QSAR, is

$$\log(1/C) = 1.151\,\pi - 1.464\,\sigma^+ + 7.818$$

This equation relates to the antiadrenergic activity of 3,4-disubstituted N-dimethyl-α-bromophenemethylamines

with X, Y = H, F, Cl, Br, I, Me, where π is the lipophilicity parameter, and σ^+ is Hammett sigma (for benzyl cations). An equation selected to represent the approach to QSAR using mathematical descriptors is [9]

$$\log 1/C = 0.426 \text{ WID} + 0.208 \ ^1\chi + 0.577$$

Here, WID is the identification number based on weighted walks [10], and $^1\chi$ is the connectivity index.

Molecular identification numbers (ID) were introduced in 1984 [10] and have been used as molecular descriptors in QSAR [11] although they were constructed initially as a tool for identification of highly similar molecules. We should add that the problem of identification of highly similar molecules is different and distinct from the problem of graph isomorphism in the sense that for the former occasional occurrence of identical ID numbers for distinct structures is acceptable (as it may point to very similar structures), and in the latter, the occurrence of identical numerical values for different structures points to a failure of the approach. In the following section, we give more information on molecular ID numbers and their properties.

Let us return to the above two QSAR equations to point out their deficiency when one is primarily interested in interpretation of regression equations. The coefficients of these equations do not tell what is the role of the individual descriptors, the role of the lipophilicity parameter (π) and the Hammett sigma (σ^+) in the first equation and the weighted identification number (WID) and the connectivity index ($^1\chi$), because, from these equations, nothing is known about how π and σ^+ are related and similarly how WID and $^1\chi$ are related. If the above equations would have been supplemented with simple regressions of π and σ^+ with 1/C or a regression of π against σ^+, and similarly, if they would have been supplemented with simple regressions of WID and $^1\chi$ with 1/C or a regression of WID against $^1\chi$, one would be able to say more. But the above equations nevertheless contain some useful information, which those familiar with the Hansch approach to QSAR can extract from the first regression equation and those familiar with chemical graph theory can extract from the second regression equation—but, significantly, not vice versa. That is, those not familiar with the Hansch approach to QSAR would have difficulty getting additional information from the first regression equation, and those not familiar with chemical graph theory would have difficulty getting additional information from the second regression equation.

The reason is lack of communication between the two groups, which can be illustrated by the expectations that the Craig plot is unknown to researchers in QSAR using mathematical descriptors, just as variable descriptors are unheard of among physical chemists and medicinal chemists working in QSAR. D. P. Craig was a physical and theoretical chemist from Australia, who, among other studies, was also interested in looking for generalization of the Hückel $4n + 2$ rule to polycyclic compounds in the 1950s. Those familiar with the Hansch approach to QSAR know that, having a regression equation, one can, using a Craig plot, quickly decide which analog to test and possibly synthesize. The Craig plot depicts one parameter against another, like σ against π for various molecular functional groups [12]. Returning to the Hansch equation, one can now deduce from the Craig plot that, to obtain high value for the activity (1/C), it is necessary to select among substituents having positive values of π and negative values of σ, that is, substituents such as CH_3, C_2H_5, $n\text{-}C_4H_9$, SCH_3, and $N(CH_3)_2$ or similar functional groups.

Coming to the second regression equation involving WID, a way to improve the regression is to consider variable molecular descriptors. Thus, instead of $^1\chi$, one could use the variable connectivity index, and instead of WID, the variable weighted walks. The variable connectivity index can be viewed as a generalization of the connectivity index. The bond contributions of the connectivity index: $1/\sqrt{(m\,n)}$, where m and n are the valences of atoms forming a bond, are now viewed as variables. The values of weights are determined during the optimization of the regression, minimizing the standard error of the regression.

In summary, it seems to us that all *researchers in QSAR*, those following in the steps of Hansch as well as those using topological indices, should learn about the Craig plot and variable molecular descriptors! We wish to be constructive and will try to continue to build bridges although we may occasionally appear provocative. In particular, we would like to encourage the researchers of QSAR who might be in between to be better informed on the similarities and the differences between the physicochemical modeling and mathematical modeling of structure–activity.

In the next section, we point out the difference between *QSAR* and *qsar* or "*QSAR*," the term introduced for the first time in this book, a provocation of a kind. We will show that many, if not most of the structure–activity contributions are neither *QSAR* nor *qsar*, but in between, by which we mean neither fully quantitative nor merely qualitative.

5.3   QSAR VERSUS "QSAR"

As previously mentioned, the acronym QSAR stands for the quantitative structure-activity relationship. However, there may be some ambiguity associated with the attribute "quantitative." It does not necessarily follows that results expressed or having numerical representation are necessarily quantitative. Qualitative results can equally be numerically represented. Strictly speaking, we define and view QSAR models as *quantitative* only when the numerically expressed models allow *meaningful* interpretation of the numerical results obtained for the structure–activity relationship within the basic concepts of the particular model. This means that the physicochemical models should allow quantitative interpretation of the numerical physicochemical descriptors used and that the structure–mathematical models should allow quantitative interpretation of the numerical structure–mathematical descriptors used. We will use the symbol *qsar* and "*QSAR*" as the abbreviation for *qualitative structure–activity relationship*. Such are the relationships that are non-numerical and the relationships that may be numerical but the variables used are interrelated and thus do not allow unique interpretation of the MRA equations. Because all molecular descriptors hitherto used in QSAR, whether they are based on physicochemical properties, quantum mechanical calculations, or molecular graphs, are all interrelated, it follows that all such hitherto reported results, without further elaboration, remain essentially qualitative, being *qsar* rather than QSAR.

From the above, it also follows that *simple* structure–property or structure–activity regressions (that is, regressions using a *single* descriptor) represent quantitative QSAR, provided that the descriptors used have meaningful physicochemical, meaningful quantum chemical, or meaningful graph theoretical structural interpretation. However, the situation is not so simple in the case of MRA. We elaborate later on the inherent ambiguities of MRA, which are twofold: (i) The instabilities of the coefficients of the regression equations when an independent variable (descriptor) is added, removed, or changed, and (ii) the interpretation of the individual contributions of the independent variables used from the resulting regression equation. The problem is that the so-called "independent" variables are, as a rule, not independent.

Although most researchers in QSAR are aware of the first problem, the second problem, the rigorous interpretation of the individual contributions of the independent variables, seems to be largely overlooked or ignored based on recognition that "weakly" correlated variables will not cause large instabilities for coefficients of a regression equation. That is correct, but this still means they will cause small instabilities and hence not represent the exact solution by not offering precise information on the relative role of the descriptors used! *Approximate* solutions are those that we would classify as *qsar* or "*QSAR*", not *QSAR*.

In order to illustrate the situation, we have selected two cases. The first case illustrates the use of partial order as an approach to arrive at quantitative characterization of proteome maps. The second case illustrates the use of partial order to illustrate the systematic search for the pharmacophore.

5.4 ON QUANTITATIVE CHARACTERIZATION OF MOLECULAR STRUCTURE

We may mention that Wiener [13], in the same year that Platt introduced paths as potential molecular descriptors (1947), reported correlation between the boiling points of smaller alkanes and a selected pair of mathematical descriptors for octanes. Wiener used two mathematical descriptors: W, which represents the sum of all distances between carbon atoms in a molecule, measured by the number of bonds separating them (to which Platt referred as the Wiener number), and P, the count of paths of length three. Wiener found the two descriptors W and P to produce very good correlations not only for the boiling points of alkanes but also several other physicochemical properties of alkanes. Although the descriptor W increases with molecular size, the descriptor P was found important to differentiate between structures having the same size but having a different number of close contacts between hydrogen at neighboring carbon atoms. Thus, P, to some degree, reflects molecular shape.

In order to get *quantitative* relationships between structures and their properties, one needs to find suitable numerical characterizations of structures. Numerical characterizations of the structure can then generate numerical descriptors of structures to be used for construction of structure–property correlations. *Suitable* structure descriptors could be tested to see if they correlate with molecular bioactivity in modeling effects and responses of such substances on living tissue or organisms. In this way, one can identify bioactive natural products or synthesized compounds, which can show novel therapeutic activities and be able to protect against diseases or combat diseases or harmful chemicals that have toxic or mutagenic effects, etc.

The almost hidden but key word here is *suitable*. Chemical structures can be described in very many different ways using common chemical vocabulary and terminology. One can start by focusing on atoms and bonds: the hybridization of atoms; atomic charges; the types of bonding, localized or delocalized; and other quantum chemical parameters, describing local groups, stereochemistry and conformation, chirality, and so on. Eventually, one may also include selective properties of molecules, such as thermodynamic information on reactivity and even the spectral data. But the basic question remains: Are experimental molecular properties *suitable* for quantitative (numerical) characterization of structure–property or structure–activity relationships? If one uses such, then, strictly speaking, one is dealing with property–property relationships and not structure–property relationships.

In our view, molecular properties are not optimal molecular descriptors if one is interested in structure–property relationships because physicochemical properties are the output of molecular structure, not the input! Although selection of molecular properties may offer insights into mechanistic modeling, at best, they relate to *parallelisms* between such properties and activity but say nothing about the structure–activity relationship *directly*. This appears to have been overlooked for too long a time. Having said that, this does not mean that the property–property correlations, in which correlations are found between some parallelisms between selected *physical* and *chemical* properties and bioactivities, are not useful and are of no interest in QSAR. Of course they are of interest and have already offered numerous useful insights in QSAR. However, such correlations are simply not *structure*–activity studies, but *property*–activity studies.

We should add that descriptors of molecules that only record the presence or absence of some atoms or functional groups should be viewed as *qualitative* and considered to be an index unless they *count* the number of such atoms or functional groups. A list of such can be viewed as a *molecular code*, the same as one can view empirical formulas of molecules, but at the same time, such codes can be viewed as molecular invariants. In general, there is an important difference between *codes* and *invariants*. Codes are essential for representing structures in some *useful* practical fashion that allows computer manipulations with structures, starting from structure enumerations (e.g., isomer enumeration), structure construction, building combinatorial libraries, searching combinatorial libraries, and possibly seeking similarities or dissimilarities among molecules. Good codes should allow reconstruction of a structure from the code.

For structure–property and structure–activity studies, one should use solely structure *invariants*, and when structures are represented by molecular graphs, graph invariants could serve as structure descriptors. The mathematical term *invariant* means any quantity that is *independent of* graphical representation of a molecule or adopted numbering of atoms. In other words, *invariants* represent intrinsic *mathematical properties* of molecular structure. They differ from the physicochemical molecular properties in that they are either obtained by counting or can be expressed with mathematical formulas and are always therefore numerically *exact*. The same can be said for quantum chemical molecular descriptors. In contrast, physicochemical molecular descriptors that are molecular properties are obtained by measuring, which may be sufficiently precise and thus numerically satisfactory; nevertheless, with time, they may undergo some, even if minor, revisions.

The structure–property correlations also represent property–property correlations with an important distinction: they represent the correlation between structural (mathematical) property–physicochemical property or structural (mathematical) property–activity property, with which *structural property* encrypts the structure.

5.5 THE CRITICISM THAT IS NOT

A few years ago, Dearden and collaborators [14] published a paper with the very provocative and pretentious title "How Not to Develop a Quantitative Structure–Activity or Structure–Property Relationship (QSAR/QSPR)." In this publication, in the section "Errors in QSARs/QSPRs," the authors listed, in total, 21 *"types of error in QSAR/QSPR development and use."* They ended their list with *"Lack of mechanistic interpretation,"* as an error, which itself is not an error but a matter of politics!

QSAR, as the acronym implies, is about quantitative structure–activity relationships and not about quantitative mechanistic–activity relationships! We will comment on few of the "errors" on the list of 21 types of error, which, in our view, are not errors and on a few questions for which have been offered answers for those interested in considering them. It is not our aim here to critically review the paper "How Not to Develop a Quantitative Structure-Activity or Structure-Property Relationship" by Dearden et al., but to make our contribution to "How to Develop a Quantitative Structure–Activity or Structure–Property Relationship." One particular matter that we will address is, in our view, a misrepresentation in the article by Dearden et al. of the connectivity index and the variable connectivity index.

Let us start with the problem of collinearity, which is one of the central problems and greatest discomforts associated with MRA. It used to be referred to as an MRA "nightmare." Dearden et al. start their section "Use of collinear descriptors" with *"The use of collinear descriptors is undesirable..."* But one may immediately ask why? No specific reasons are given explicitly, but it is implied that this is because of the well-known instabilities of the coefficients of the regression equations. Hearing that *"the use of collinear descriptors is undesirable..."* one is entitled to ask a question: Is this because we do not know how to find an answer to the problem or is this some inherent fundamental substantial contradiction involving collinear descriptors that make them undesirable? Observe, however, that the latter cannot be the case because all molecular descriptors are intercorrelated, and these correlations only differ in *degree*, not in *substance*. Therefore, the recommendation that *"the use of collinear descriptors is undesirable..."* is tantamount to *"we don't understand it, so it is undesirable"*! This is the language of defeatists; there may be people who understand the problem and could solve it. If solved, this would make such a request obsolete and irrelevant.

As mentioned, the problem has been solved, and it was solved almost 25 years ago [7]! But to know about it, one should read QSAR publications coming from mathematical chemistry circles. Obviously, lack of cross-communication between the Hansch QSAR school and the circle around chemical graph theory is the culprit with equal guilt on both sides.

The issue of collinearity should not be raised to the level of doctrine, dogmatism, and politics. It is a matter of mathematics. That one so important result can, for so long, be overlooked by a sizable group of researchers in QSAR, if not the majority, is hardly just a matter of unfamiliarity with the literature on the topic by so many people.

As to their reasons for undesirable use of collinear descriptors, Dearden et al. state that (i) two highly collinear descriptors are effectively contributing the same information twice, and (ii) that the use of collinear descriptors can have adverse effects on the statistical analysis, for example, by causing instability in the regression coefficients.

We will see that it is irrelevant whether two descriptors duplicate the same information, but what is important is whether the parts in which they differ make *significant contributions* to the regression. As we show, the parts in which two descriptors overlap can be eliminated; thus, all that is important for collinear descriptors is to find if the *residuals* of their intercorrelation (parts in which they differ) are relevant for structure–property or structure–activity regression or not. To illustrate their point, Dearden et al. consider two simple regressions (that is, based on a single descriptor) from a paper by Randić and Basak on the toxicity of a series of alkyl ethers to mice [9]:

$$\log 1/C = 0.634 \ (\pm 0.066) \ ^1\chi + 0.789 \ (\pm 0.197)$$

$$\log 1/C = 0.325 \ (\pm 0.027) \ \mathrm{WID} + 0.602 \ (\pm 0.169)$$

that have the regression coefficients $r = 0.9099$ and $r = 0.9418$, respectively. The corresponding standard errors are $s = 0.176$ and $s = 0.143$, and the probability (p) values for all coefficients is $p < 0.001$. Dearden et al. continue and combine both descriptors $^1\chi$ and WID into a single regression:

$$\log 1/C = 0.426 \ (\pm 0.133) \ \mathrm{WID} - 0.208 \ (\pm 0.268) \ ^1\chi + 0.577 \ (\pm 0.174)$$

$$r = 0.9434; \ s = 0.144; \ p(^1\chi) = 0.447.$$

Observe that the above regression was not considered by Randić and Basak at all, but was introduced by Dearden et al. By pointing out that $^1\chi$ and WID are highly collinear ($r = 0.9793$), recognizing that both WID and $^1\chi$ give good correlations with log 1/C, they combine them together and conclude that $^1\chi$ *"appears not to be a good descriptor, because both the standard error of its coefficient and its p-value are very high."* However, they have overlooked the fact that the regression using both $^1\chi$ and WID does not decrease the standard error of the regression, but *increases* it. This ought to be a sufficiently strong signal to disregard such a regression. What the regression involving both $^1\chi$ and WID signifies is that the difference between the two collinear descriptors in this case is not improving the construction of the structure–property space. This means that parts in which WID and $^1\chi$ differ are not relevant for this case. In other words, the approach is adding a dimension to the structure–property space that is irrelevant for the considered structure–property relationship.

Looking back at the three regression equations listed earlier, observe that the equation of Dearden et al. has the highest coefficient of regression ($r = 0.9434$) compared with the two simple regressions having $r = 0.9418$ and $r = 0.9099$ for WID and $^1\chi$, respectively. Be aware that one can judge the relative quality of different regressions by considering the coefficient of regression but only for regressions having *the same number* of descriptors. Thus, from r values, one can deduce that WID is a better descriptor than $^1\chi$ for log 1/C (which we know from the corresponding standard errors). However, it would be invalid to use r to judge the relative quality of regressions that use a different number of descriptors. In such cases, one should always use the standard error s as a measure of quality.

5.6 HIGHLY COLLINEAR DESCRIPTORS: THE PROBLEM THAT IS NOT

The use of highly collinear descriptors was viewed to be improper, and in some circles, this view is still maintained. Only few years ago, in 2009, in the earlier mentioned article of Dearden and coworkers [14] one can read:

> The use of collinear descriptors is undesirable, for two reasons. First, two highly collinear descriptors are effectively contributing the same information twice, and thus add nothing to mechanistic interpretation of a QSAR or QSPR; in fact, they can detract from it. Second, it is known that the use of collinear descriptors can have adverse effects on the statistical analysis, for example by causing instability in the regression coefficients.

There are several misconceptions and misdirections in the above statement:

1. Two highly collinear descriptors may contain the same information, but there is no harm in having duplicate information. Harm is not having information in which two descriptors *differ*, which *may be* essential for regression analysis. It is easy to find if highly collinear descriptors carry different *relevant* information for regression analysis. If they do, then the standard error when both descriptors are used will visibly decrease and the regression coefficient r will visibly increase in comparison with when only one of those descriptors is used. If that is the case, both such descriptors, regardless of their collinearity, will be useful and should be used.
2. It is true that using collinear descriptors has adverse effects on the coefficients of the regression equation. It is not true that they have adverse effects on the statistical analysis because when they are replaced by a pair of orthogonalized descriptors (which can be constructed for each pair of highly collinear descriptors) this neither changes the standard errors or the correlation coefficient r of the regression nor the coefficients of the regression equation when the additional descriptor is included in the regression.
3. The instability of the regression coefficients has nothing to do with the regression *statistics* and everything to do with *interpretation* of the results of the regression. To elaborate, even the "respectable regressions," based on weakly correlated descriptors, have an inherent problem with the *quantitative* interpretation. However, such regression equations, when it is known that the descriptors used are not highly collinear, may allow *qualitative* interpretations (which often may suffice).

Finally, a word on that "*collinear descriptors...add nothing to mechanistic interpretation of QSAR or QSPR, in fact they detract from it.*" This is a political statement, an opinion of few or many (it does not matter how many). Recall a statement of L. Margulis [15]: *"So I don't see how people can have strong opinions... Let me put it this way: opinions aren't science. There is no scientific basis! It is just opinion!"*

Why should QSAR models have mechanistic interpretation? The above may apply and hold for models based on concepts of physical chemistry. They claim to represent a structure–property model but do not use *structural* descriptors. Strictly speaking, the central issue here is not whether highly collinear descriptors or slightly collinear descriptors contribute a large amount or a little amount of the same information into the regression equation, which adds nothing new to the structural interpretation of QSAR or QSPR, but whether highly collinear descriptors contribute *different* structural information and, if they do, whether that *different* information is relevant for QSAR or QSPR. Let us illustrate a regression of two highly correlated descriptors, which apparently *"are effectively contributing the same information twice"* and show that they can complement each other.

Consider the following illustration: Let us start with a *simple* regression (that is, a regression using one independent variable) for molar refraction or refractivity of octanes. Molar refractivity, MR, is a measure of the total polarizability of a mole of a substance and depends on the temperature, the index of refraction, and the pressure. It is defined as

$$MR = (4\pi/3) \, N_A \, \alpha$$

where $N_A = 6.02214 \times 10^{23}$ is Avogadro's constant (the number of molecules in a mole), and α is the mean polarizability (the ability of molecules to be polarized). In terms of density (ρ) and molecular weight (M), it can be shown that

$$MR = (n^2 - 1) \, M/(n^2 + 2) \, \rho.$$

We will first consider regressions based on single descriptors, using the connectivity indices $^1\chi$ and $^2\chi$ [16]. One easily finds that the corresponding regressions are very poor, giving for the coefficients of regressions r_1 and r_2 the following $r_1 = 0.087$ and $r_2 = 0.187$, respectively.

Most people would immediately discard these two descriptors as utterly useless in regressions for MR. For example, the QSRP/QSAR statistical software **CO**mprehensive **DE**scriptors for **S**tructural and **S**tatistical **A**nalysis (CODESSA) by Alan R. Katritzky, Mati Karelson, and Ruslan Petrukhin [17,18] and many other practicing chemists in searching for optimal descriptors for a regression would automatically eliminate from consideration descriptors having regressions with regression coefficients $r < 0.200$. Thus, in the above case, both connectivity indices would be rejected. For sure, both should not be considered if one considers simple (one-variable) regressions in a search for optimal regression descriptors.

But here is a great surprise: If one decides to use both "useless" descriptors, one obtains a very respectable regression equation:

$$MR = 4.695 \, ^1\chi + 1.372 \, ^2\chi + 17.548,$$

which is characterized by the following statistical information: $n = 18$, $r = 0.971$, $s = 0.047$, and $F = 115$. Here n is the number of octane isomers used, r is the regression coefficient, s is the standard error, and F is the Fisher ratio. How is this possible? How is one to understand this? Observe that the regression coefficient is very acceptable.

In Figure 5.1, we have illustrated schematically a hypothetical structure–property space. We represented the two connectivity indices as two almost parallel vectors oriented close to the axis x, and the property is represented by a point some distance away but close to the y axis (that is, away from the directions of the two connectivity indices). It is clear that connectivity vectors are not directed toward the direction of the property considered. However, the two vectors define a plane and, when combined, can reach any point in the plane, including the property "X." It is well known that two vectors span a plane, so one can form their linear combination directed to any point in the same plane. Because the property and the vectors are in the same plane, one can obtain a very respectable regression. Had the two connectivity vectors been in a plane perpendicular to the structure–property plane (as was the case with WID and $^1\chi$), they would not be able to offer such a good correlation with the property (MR).

It is interesting to mention that Xu and Zhang [19] constructed a *single* molecular descriptor to replace the linear combination of two connectivity descriptors. They have shown that when $^2\chi$ is made orthogonal to $^1\chi$, one can construct a *single* descriptor producing very good regression for *MR*. Because $^1\chi$ comes with a very low regression coefficient, and $^1\chi$ and $^2\chi$ give very good regression, one knows that $^1\chi$ points to the wrong "direction." It should then not be surprising that the direction that is orthogonal to $^1\chi$ may lead to a satisfactory correlation.

We repeat, for the third time, that the orthogonalization of molecular descriptors was introduced into structure–property–activity studies about 25 years ago [7,20–28] and has only occasionally been used since then [29–41]. It is still today not widely known, not recognized as *indispensible* in QSAR and QSPR modeling if people want to interpret their models in terms of descriptors used in the construction of the regression equation. Repeating the same message three times deliberately is

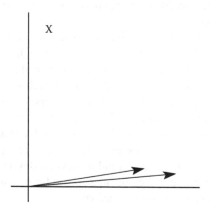

FIGURE 5.1 Schematic representation of the structure space in the case of two vectors spanning the space but individually not pointing to the property considered.

likely to make it known to all readers. Often a message spoken once is overlooked by most listeners; if spoken twice, a few in the audience will notice it, but if spoken three times, everybody will recognize its importance!

5.7 THE SINGLE EQUATION FALLACY

A single equation determines structure space but offers no information on the relative importance of the descriptors that define the space, whether they are mathematical, quantum chemical, physicochemical, or molecular properties.

It is interesting, in a way, that in their lengthy list of 21 "errors" in QSAR that Dearden, Cronin, and Kaiser [14] constructed, a few of which, in our view, should not have been listed as errors, the most serious and most widespread error in QSAR and QSPR is missing! This missing error we will refer to as the "*Single Equation Error*" or, to be more precise, the "*Single Regression Equation Fallacy.*" As an illustration, let us consider

$$\log 1/C = 0.426 \ WID - 0.208 \ ^1\chi + 0.577.$$

Here we are not interested in the fact that introduction of the second descriptor did not improve the standard error. Let's pretend that it did—something one can know only a posteriori. Why do we consider this above regression equation to illustrate an *erroneous* approach to QSAR? The equation as such is not in error, and one may say that it is formally true, but it represents, at best, *half a truth!* The error is in trying to interpret the contributions of the individual descriptors, WID and $^1\chi$. This cannot be done from a single equation because the descriptors WID and $^1\chi$ are not independent despite the traditional reference to "*independent variables.*" To refer to descriptors used in QSAR as *independent variables* can be viewed as an illustration of a historical blunder. For more on historical blunders in science, interested readers may browse through several publications of Calabrese on hormesis [42]. The so-called independent variables correlate with one another as any pair of molecular descriptors mutually correlate, whether they belong to physical chemistry, quantum chemistry, or chemical graph theory. As already said, there are no *independent variables* unless one undergoes the orthogonalization process (outlined in the previous section) and makes them *independent.* If this is not done, one has to report not only the "final" regression equation but also the simple regressions:

$$\log 1/C = 0.634 \ ^1\chi + 0.789$$

$$\log 1/C = 0.325 \ WID + 0.602.$$

Alternatively, one can report a pair of equations:

$$\log 1/C = 0.634 \ ^1\chi + 0.789$$

$$\log 1/C = -0.208 \ ^1\chi + 0.426 \ WID + 0.577$$

or the pair of equations

$$\log 1/C = 0.325 \text{ WID} + 0.602$$

$$\log 1/C = 0.426 \text{ WID} - 0.208 \, ^1\chi + 0.577.$$

Observe that the second equation is the same one. These two cases differ only in the order in which the variables have been selected, the order being WID, $^1\chi$ in the first case and $^1\chi$, WID in the second. Now one can construct a single equation that will have *complete* information on the relative contributions of the two descriptors, as outlined in the section on orthogonal descriptors, by combining coefficients of the two equations. So if one orders the variables first WID, second $^1\chi$, one obtains the regression

$$\log 1/C = 0.325 \text{ WID} - 0.208 \, ^1\chi^* + 0.602.$$

If one orders the variables first $^1\chi$, second WID, one obtains the regression

$$\log 1/C = 0.634 \, ^1\chi + 0.426 \text{ WID}^* + 0.789.$$

Hence, it is possible to express the results of QSAR multivariate regressions by a *single* equation, but only if the descriptors in such an equation are truly independent (read: orthogonal, which means that the regression coefficient of their mutual regression is zero). In the cases of regressions involving two variables, there are two equations to choose from. In the above case, one can choose between WID, $^1\chi^*$ and $^1\chi$, WID*, where asterisks indicate that the second descriptors $^1\chi$ and WID are modified so that the part in which they duplicate the first descriptor is eliminated.

Let us compare the equation reported by Dearden, Cronin, and Kaiser:

$$\log 1/C = 0.426 \, (\pm 0.133) \text{ WID} - 0.208 \, (\pm 0.268) \, ^1\chi + 0.577 \, (\pm 0.174)$$

or in the simple form

$$\log 1/C = 0.426 \text{ WID} - 0.208 \, ^1\chi + 0.577$$

with the above two alternative regression equations. Observe that the two coefficients, 0.426 and −0.208, both appear in both orthogonal equations, but they are not the coefficients of WID and $^1\chi$, respectively, but they are the coefficients of the residuals of regression of $^1\chi$ against WID and WID against $^1\chi$, which we, for brevity, have shown as $^1\chi^*$ and WID*. In short, a single regression equation, except in the case of truly independent variables (orthogonal variables), per se is not an error, but any discussion and conclusion on the relative importance of individual descriptors is misleading and represents an error, turning half a truth into a false statement.

It is needless to say that in the case of three variables there will be six alternative equations to choose from (that is three factorial, or 3!, which is 6); in the case of four variables, there will be 24 alternative equations to choose from (that is four factorial, or 4!, which is 24); and so on. Thus, in QSAR modeling, in addition to the "nightmare" of having to choose a few descriptors from a large pool of descriptors, there is an additional apparent "nightmare" of choosing a single ordering of n variables from $n!$ alternatives. The problem has been present in physics all the time when one considers the Gram-Schmidt orthogonalization process, so the problem is as old as the orthogonalization of vectors. One can now even better appreciate the use of sets of descriptors that can be ordered in some "natural" way, such as paths of increasing lengths or connectivity indices of increasing order, etc., as such sets of descriptors avoid the $n!$ problem in the construction of truly independent descriptors.

The situation with the problem of $n!$ choices of ordering descriptors is not so alarming as it may appear at first sight. One practical approach could be to order the descriptors with respect to their standard errors obtained in their single variable regressions. In the above case, the two criteria coincide with suggesting the ordering of WID as the first descriptor and $^1\chi$ as the second descriptor.

REFERENCES AND NOTES

1. R. Hoffmann, Qualitative thinking in the age of modern computational chemistry—or what Lionel Salem knows, *J. Mol. Struct. (Theochem.)* 424 (1998) 1–6.
2. C. P. Snow, *The Two Cultures*, Cambridge University Press, London, 2001.
3. C. Djerassi and R. Hoffmann, *Oxygen*, Wiley-VCH, Weinheim, 2001.
4. R. Hoffmann and V. Torrence, *Chemistry Imagined. Reflections on Science*, Smithsonian Institution Press, Washington and London, 1993.
5. A. Shmukler and C. Ziskin, Through the looking glass of history: Mathematicians in the land of poetry, *J. Math. Arts* 8 (2014) 78–86.
6. H. Kubinyi, Home page: Chemoinformatics in Drug Discovery—Quo Vadis? 2nd Strasbourg Summer School on Chemoinformatics, June 20–24, 2010, slide 4 and 5 of PowerPoint presentation.
7. M. Randić, Resolution of ambiguities in structure-property studies by use of orthogonal descriptors. *Chem. Inf. Comput. Sci.* 37 (1991) 311–320.
8. http://en.wikipedia.org/wiki/cross-cultural_communication, August 2014.
9. M. Randić and S. C. Basak, Multiple regression analysis with optimal molecular descriptors, *SAR QSAR Environ. Res.* 11 (2000) 1–23.
10. M. Randić, On molecular identification numbers, *J. Chem. Inf. Comput. Sci.* 24 (1984) 164–175.
11. S. Carter, N. Trinajstić, and S. Nikolić, On the use of ID numbers in drug research: A QSAR of neuroleptic pharmacophores, *Med. Sci. Res.* 16 (1988) 185–186.
12. A. Kar, *Medicinal Chemistry*, 3rd ed., Section 2.11, New Age Int.- Publ., New Delhi, 2005.
13. H. Wiener, Structural determination of paraffin boiling points, *J. Am. Chem. Soc.* 69 (1947) 17–20.
14. J. C. Dearden, M. T. D. Cronin, and K. L. E. Kaiser, How not to develop a quantitative structure-activity or structure-property relationship (QSAR/QSPR), *SAR QSAR Environ. Res.* 20 (2009) 241–266.
15. L. Margulis, as cited in J. Horgan: *The End of Science* (Croatian translation p. 131), *The End of Science: Facing the Limits of Knowledge in the Twilight of the Scientific Age*, Broadway Books, New York, 1996.
16. M. Randić, On characterization of chemical structure, *J. Chem. Inf. Comput. Sci.* 37 (1997) 672–687.
17. CODESSA (Comprehensive Descriptors for Structures and Statistical Analysis) reports most of the descriptors designed by A. R. Katritzky, V. Lobanov, and M. Karelson (Eds.), University of Florida, Gainesville, FL, 1995.

18. COODESSA PRO, QSPR/QSAR SOFTWARE (*COmprehensive DEscriptors for Structural and Statistical Analysis*) by A. R. Katritzky, M. Karelson, and R. Petrukhin, University of Florida, Gainesville, FL, 2001–2005.

19. L. Xu and W.-J. Zhang, A comparison of different methods for variable selection, *Anal. Chim. Acta* 446 (2001) 475–481.

20. M. Randić, Orthogonal molecular descriptors, *New J. Chem.* 15 (1991) 517–525.

21. M. Randić, Search for optimal molecular descriptors, *Croat. Chem. Acta* 64 (1991) 43–54.

22. M. Randić, Correlation of enthalpy of octanes with orthogonal connectivity indices, *J. Mol. Struct. (Theochem.)* 233 (1991) 45–49.

23. M. Randić, Fitting of non-linear regressions by orthogonalized power series, *J. Comput. Chem.* 14 (1993) 363–370.

24. M. Randić, Curve-fitting paradox, *Int. J. Quantum Chem.: Quantum Biol. Symp.* 21 (1994) 215–225.

25. M. Randić, Orthosimilarity, *J. Chem. Inf. Comput. Sci.* 361 (1996) 1092–1097.

26. M. Randić, Retro-regression—Another important multivariate regression improvement, *J. Chem. Inf. Comput. Sci.* 41 (2001) 602–606.

27. M. Randić and M. Pompe, Retro-regression—A way to resolve multivariate regression ambiguities, *Acta Chim. Slov.* 52 (2005) 408–416.

28. D. J. Klein, M. Randić, D. Babić, B. Lučić, S. Nikolić, and N. Trinajstić, Hierarchical orthogonalization of descriptors, *Int. J. Quantum Chem.* 63 (1997) 215–222.

29. M. Šoškić, D. Plavšić, and N. Trinajstić, Link between orthogonal and standard multiple linear regression models. *J. Chem. Inf. Comput. Sci.* 36 (1996) 829–832.

30. O. Araujo and D. A. Morales, An alternative approach to orthogonal graph theoretical invariants, *Chem. Phys. Lett.* 257 (1996) 393–396.

31. O. Araujo and D. A. Morales, A theorem about the algebraic structure underlying orthogonal graph invariants. *J. Chem. Inf. Comput. Sci.* 36 (1996) 1051–1053.

32. O. Araujo and D. A. Morales, Properties of new orthogonal graph theoretical invariants in structure–property correlations, *J. Chem. Inf. Comput. Sci.* 38 (1998) 1031–1037.

33. B. Lučić, S. Nikolić, N. Trinajstić, and D. Juretić, The structure-property models can be improved using the orthogonal descriptors, *J. Chem. Inf. Comput. Sci.* 35 (1995) 532–538.

34. B. Lučić and N. Trinajstić, New developments in QSPR/QSAR modeling based n topological indices, *SAR QSAR Environ. Res.* 7 (1997) 45–62.

35. B. Lučić and N. Trinajstić, Multivariate regression outperforms several robust architectures of neural networks in QSAR modeling, *J. Chem. Inf. Comput. Sci.* 39 (1999) 121–132.

36. B. Lučić, N. Trinajstić, S. Sild, M. Karelson, and A. R. Katritzky, A new efficient approach for variable selection based on multiregression: Prediction of gas chromatographic retention times and response factors, *J. Chem. Inf. Comput. Sci.* 39 (1999) 610–621.

37. N. R. Draper and H. Smith, *Applied Regression Analysis*, 2nd ed., John Wiley & Sons, New York, 1981.

38. D. Amić and D. Davidović-Amić, Correlation of flavones derivatives for inhibition of cAMP phosphodiestgerase, *J. Chem. Inf. Comput. Sci.* 35 (1995) 1034–1038.

39. D. Amić, D. Davidović-Amić, D. Bešlo, B. Lučić, and N. Trinajstić, The use of the ordered orthogonalized multivariate linear regression in a structure-activity study of coumarin and flavonoid derivatives as inhibitors of Aldose Reductase, *J. Chem. Inf. Comput. Sci.* 37 (1997) 581–586.

40. M. Šoškić, D. Plavšić, and N. Trinajstić, Inhibition of the Hill reaction by 2-methylthio-4,6-bis(monoalkylamino)-1,3,5-triazines, *J. Mol. Struct. (Theochem.)* 394 (1997) 57–65.

41. F. M. Fernández, P. R. Duchowitz, and E. A. Castro, About orthogonal descriptors in QSPR/QSAR theories, *MATCH—Commun. Math. Comput. Chem.* 51 (2004) 39–57.

42. E. J. Calabrese, Historical blunders: How toxicology got the dose-response relationship half right, *Cell. Mol. Biol.* 51 (2005) 643–654.

6 Molecular Descriptors

6.1 STRUCTURE DESCRIPTORS

Let us comment on structure descriptors, most of which have been derived from studying mathematical properties of molecular structure and molecular graphs. The first nontrivial mathematical descriptors of chemical structure emerged in 1947, independently introduced by J. R. Platt, who suggested paths of different lengths as molecular descriptors [1], and by H. Wiener, who introduced his "mysterious" number, referred to as the Wiener number W, along with using paths of length three (P) [2]. Wiener has shown that regressions using W, P led to a number of good structure–property correlations. This is how Wiener described calculation of his descriptor W, which he called the path number:

> The path number W is defined as the sum of the distances between any two carbon atoms in the molecule, in terms of carbon–carbon bonds. Brief method of calculation: Multiply the number of carbon atoms on one side of any bond by those on the other side; W is the sum of these two values for all bonds.

The Wiener number W can be obtained by summing the elements of the distance matrix above the main diagonal [3]. The article by Wiener may be viewed as an illustration of a "Sleeping Beauty," which is defined as important papers having significant novelty that, however, have been overlooked for many years as reflected in the low count of citations, until it was finally rediscovered [4–7]. For Wiener's number W, this happened in mid-1970, when, with the revival of chemical graph theory, his paper was rediscovered. Currently W is one among very few widely used and widely studied topological indices in QSAR.

The next mathematical structural invariant emerged almost 25 years later, in 1971, when H. Hosoya introduced his "topological index" Z, which is given by summing the count of the number of sets of disjointed k edges in a molecular graph, z_k [3]. Hosoya has shown that the Z index gives a good correlation for the boiling points of alkanes. The numbers z_k, which are components contributing to Z, arise in mathematics as the absolute values of the coefficients of the characteristic polynomial of acyclic graphs. Thus, the characteristic polynomials became a source of an important molecular descriptor, and Z illustrates an unexpected and, by now, well-established connection between mathematics and chemistry.

6.2 THE CONNECTIVITY INDEX

In the mid-1970s, the connectivity index χ (chi) was introduced in physical chemistry—if one views chemical graph theory and mathematical chemistry in a broad sense to be areas overlapping with physical chemistry. The connectivity index represents another important early molecular descriptor, which continues to be of considerable interest not only in chemistry but also mathematics. This index is the first molecular descriptor *designed* to yield good structure–property correlations for a selection of physicochemical properties. Since its introduction in the mid-1970s, the paper in which it was described as the connectivity index χ has been cited well over 2500 times, which apparently is confirming interest in χ and its utility.

Despite this respected visibility, details of its design, which are instructive, appear not to be well known, showing that many users have not seen the original publication. Some people even expressed a view that the index was found accidentally and that it may serve as an illustration of "good luck." In the original paper, the connectivity index χ or $^1\chi$ index (in the view that later "higher order" connectivity indices will be introduced) was called the branching index, but it was renamed the connectivity index and given symbol χ by L. B. Kier soon after. Because the index can be calculated for linear molecules and cycles, which have no branching atoms, the suggestion by L. B. Kier to refer to χ as the connectivity index is a much better name and a quite adequate name. The index also has been referred to as the Randić index, for example, in chemistry by Katritzky, Karelson, and coworkers [8,9] and is known in mathematical literature only as the Randić index. Observe that this is not only the first index that is given as real numbers, not as integers, as was the case with the indices of Wiener, Platt, and Hosoya, and it is also the first index constructed by a design to *parallel* at least some physicochemical properties of molecules (the boiling points in alkanes).

Kier and Hall demonstrated very shortly after the connectivity index was proposed that this index, in combination with structurally related "higher order" connectivity indices, turned out to correlate well with many physical properties of molecules, in particular, alkanes, and a few molecular properties of molecules having heteroatoms, such as amines, alcohols, etc. In the next section, we outline the construction of the connectivity index.

In the introductory part of the article "On Characterization of Molecular Branching," [10] after the opening sentences, *"Some molecular properties depend upon molecular shape and vary regularly within a series of homologous compounds. The degree of branching of the molecular skeleton is the critical factor involved,"* the following can be found:

> This paper is concerned with the problem of characterizing the degree of molecular branching. The problem is approached in two stages. First, a procedure is outlined which allows ordering members of homologous series in a sequence which parallels the increase in ramification of molecular skeletons, and at least gives satisfying answers in cases beyond dispute. Next, we consider associating each individual structure with an index (parameter) which will form the same sequential ordering of the members of the series.

By ordering isomers of smaller alkanes C_4H_{10}, C_5H_{12}, and C_6H_{14} by decreasing boiling points, one finds that the boiling points of linear isomers are the largest, and the boiling points of the most branched isomers are the smallest. This was known, but, by associating with each structure a number that forms the same sequential

ordering of isomers as obtained if alkanes are ordered in decreasing order relative
to their boiling points, it was possible to arrive at *numerical* representations of isomers that parallel the relative magnitudes of the physical property—here, the boiling
points.

Now, we need an index, a number, to represent the molecule (isomer), and the
number should have some transparent structural meaning. The elementary structural
concepts of molecules are atoms and bonds, and the basic discrimination between
atoms is their valence, which allows one to classify the bonds into distinct bond
types. In graph theory, the vertex valence is given by the number of adjacent vertices in the graph. In the case of hydrogen-suppressed graphs, the vertex valence
discriminates between primary, secondary, tertiary, and quaternary carbon atoms,
the valences that take values 1, 2, 3, and 4, respectively. This leads to the following
10 bond types (edge types):

$$(1, 1), (1, 2), (1, 3), (1, 4), (2, 2), (2, 3), (2, 4), (3, 3), (3, 4), (4, 4)$$

Briefly, (m, n) signifies the bond types that take any of the available values of m,
$n = 1, 2, 3, 4$. In the case of alkanes, the bond type $(1, 1)$ appears only in ethane and is
therefore of no interest. In Table 6.1, we show 10 isomers of smaller alkanes, two of
butane, three of pentane, and five of hexane, which have seven out of 10 bond types.
Then we have decomposed each molecule into its bond types arriving thus at seven
inequalities shown at the right-hand side of the table. If one could find a solution to
these seven inequalities, that is, find some numerical values for the seven (m, n) bond
types of Table 6.1, one would arrive at a numerical characterization of the smaller
alkane isomers "by design." So constructed molecular descriptors ought to give good
regressions because they uphold the same relative magnitudes for neighboring isomers that hold for their boiling points.

As is known, inequalities can have numerous solutions, so one should look for a
simple solution, if there is such. One way to find solutions for a set of inequalities
is to test a selection of simple expressions to see if they satisfy the inequalities. In
our case, for example, one could consider (m, n) to be $m = 1/m$ and $n = 1/n$, which
is incidentally the choice of (m, n) values used in the construction of one of the
Zagreb indices [11–15]. However, this simple choice is not a solution as one can
verify, because it does not satisfy all the seven inequalities. After testing a few
similar expressions, it was found that by taking $m = 1/\sqrt{m}$ and $n = 1/\sqrt{n}$ all the
seven inequalities of Table 6.1 are satisfied. This is the way the new *bond additive*
molecular descriptor, the connectivity index $^1\chi = \Sigma 1/\sqrt{m\,n}$, was born. The numerical values for the seven inequalities using the weights $1/\sqrt{(m\,n)}$ are shown in Table
6.1. In Table 6.2, we show additional solutions of the nine inequalities that offer an
alternative form of the connectivity index $^1\chi$, which was more recently found but
has not yet been tested.

We should make two brief comments: (i) One should keep in mind that
$m = 1/\sqrt{m}$ and $n = 1/\sqrt{n}$ is one of the possible solutions of the set of considered
inequalities, and in addition to those of Table 6.2, there may be additional solutions of interest in different applications. (ii) If some expressions, such as the
mentioned Zagreb indices, do not satisfy the inequalities of Table 6.1 (based on

TABLE 6.1

The Inequalities for Butane, Pentane, and Hexane Isomers Based on Their Boiling Points

Isomers	Bond Decompositions

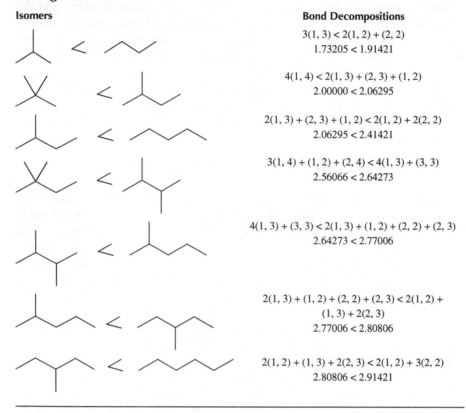

$3(1, 3) < 2(1, 2) + (2, 2)$
$1.73205 < 1.91421$

$4(1, 4) < 2(1, 3) + (2, 3) + (1, 2)$
$2.00000 < 2.06295$

$2(1, 3) + (2, 3) + (1, 2) < 2(1, 2) + 2(2, 2)$
$2.06295 < 2.41421$

$3(1, 4) + (1, 2) + (2, 4) < 4(1, 3) + (3, 3)$
$2.56066 < 2.64273$

$4(1, 3) + (3, 3) < 2(1, 3) + (1, 2) + (2, 2) + (2, 3)$
$2.64273 < 2.77006$

$2(1, 3) + (1, 2) + (2, 2) + (2, 3) < 2(1, 2) + (1, 3) + 2(2, 3)$
$2.77006 < 2.80806$

$2(1, 2) + (1, 3) + 2(2, 3) < 2(1, 2) + 3(2, 2)$
$2.80806 < 2.91421$

the boiling points of smaller alkanes), this does not mean that they should not be of interest in QSAR. All that this may indicate is that such indices are not likely to produce as good regressions for the boiling points of smaller alkanes as the connectivity index.

6.2.1 ON THE SUCCESS OF THE CONNECTIVITY INDEX

Over the past almost four decades, the connectivity index has received considerable attention. For most of its early recognition, clear responsibility goes to Kier and Hall [16,17], who were very much interested in both structure–property and structure–activity studies. Besides hydrocarbons, they paid attention to compounds involving oxygen, nitrogen, fluorine, chlorine, bromine, sulfur, and more. Kier and Hall immediately realized a need to differentiate among atoms of different kinds, which brought them to formulate a modification of the connectivity index known as the "valence connectivity indices" [18], which, from the start in 1976, have continued to play an important role in QSAR even today.

TABLE 6.2
The Same Bond Inequalities for Butane, Pentane, and Hexane Isomers "Increased" Valence and Valence "Squared"

Modified Bond Composition Inequalities $m \rightarrow m + 1$	Modified Bond Composition Inequalities $m \rightarrow m^2$
$3(2, 4) < 2(2, 3) + (3, 3)$	$3(1, 3) < 2(1, 2) + (2, 2)$
$1.06066 < 1.14983$	$1.00000 < 1.25000$
$4(2, 5) < 2(2, 4) + (3, 4) + (2, 3)$	$4(1, 4) < 2(1, 3) + (2, 3) + (1, 2)$
$1.26492 < 1.40403$	$1.00000 < 1.33333$
$2(2, 4) + (3, 4) + (2, 3) < 2(2, 3) + 2(3, 3)$	$2(1, 3) + (2, 3) + (1, 2) < 2(1, 2) + 2(2, 2)$
$1.40403 < 1.48316$	$1.33333 < 1.50000$
$3(1, 4) + (1, 2) + (2, 4) < 4(1, 3) + (3, 3)$	$3(1, 4) + (1, 2) + (2, 4) < 4(1, 3) + (3, 3)$
$1.48316 < 1.66421$	$1.37500 < 1.44444$
$4(1, 3) + (3, 3) < 2(1, 3) + (1, 2) + (2, 2) + (2, 3)$	$4(1, 3) + (3, 3) < 2(1, 3) + (1, 2) + (2, 2) + (2, 3)$
$1.66421 < 1.73736$	$1.44444 < 1.58333$
$2(1, 3) + (1, 2) + (2, 2) + (2, 3) < 2(1, 2) + (1, 3) + 2(2, 3)$	$2(1, 3) + (1, 2) + (2, 2) + (2, 3) < 2(1, 2) + (1, 3) + 2(2, 3)$
$1.73736 < 1.74740$	$1.58333 < 1.66667$
$2(1, 2) + (1, 3) + 2(2, 3) < 2(1, 2) + 3(2, 2)$	$2(1, 2) + (1, 3) + 2(2, 3) < 2(1, 2) + 3(2, 2)$
$1.74740 < 1.81650$	$1.66667 < 1.75000$

The arrival of a novelty—here we are speaking of the connectivity index—is sometimes independently recognized only by a few until the novelty gets the momentum of being recognized and accepted by a wide circle of users. We may mention a group of scholars from Auburn University in Alabama, J. M. Wells, C. Randall Clark, and R. M. Patterson, as early users of the connectivity indices who appreciated the novelty. Let us quote the opening paragraph of their paper "Structure–Retention Relationship Analysis for Some Mono and Polycyclic Aromatic Hydrocarbons in Reverse-Phase Liquid Chromatography Using Molecular Connectivity" [19]:

> In a treatise "On Characterization of Molecular Branching" [1*] a topological branching index was proposed for the study of molecular graphs. In benchmark articles [2*–4*] the concept of molecular connectivity as an indexing system was expanded and refined. Now, some 10 years later, the predictions from the original paper [1*] describing the utility of molecular connectivity as (1) "providing an accurate and precise scheme for predicting a particular property" and (2) "revealing novel relationships among unsuspected quantities and discerning the topological nature in others" have been realized.

References 1*–4* are to early work of Randić, Kier, and Hall and their collaborators.

The connectivity index received significant unexpected public recognition when, in 2000, L. L. Hall and L. B. Kier organized, within the Division of Computers in Chemistry of the American Chemical Society, "Molecular Connectivity: The First

Quarter Century Symposium" to celebrate 25 years of the publication of the article on the connectivity index. The symposium was attended by some 25 scientists for a two-day meeting offering an opportunity to participants to present some more recent results for the connectivity index [20]. Equally, as a significant development, one may mention various generalizations and extensions of the connectivity index, one of which is the variable connectivity index, which is discussed later in this chapter.

Last but not least, we may mention considerable activity in some mathematical circles interested in mathematical properties of molecular descriptors. *Mathematical Aspects of Randić-Type Molecular Structure Descriptors* is the title of one of two books on the mathematical properties of the connectivity index χ, in which selected papers on these topics were presented [21,22]. Clearly, the connectivity index represents one successfully "solved problem" in the search for useful mathematical molecular descriptors. But a question can be raised: Is there anything "unsolved" relating to the topic of the connectivity index? Is there something that can still be improved? We will come up with some answers to these questions later.

6.3 TOPOLOGICAL INDICES

The Wiener number W, Hosoya's topological index Z, and the Randić connectivity index χ were followed during the next two decades by continuing efforts of a sizable group of researchers in constructing additional mathematical molecular descriptors, primarily for use in structure–property and structure–activity studies. This "invasion" of mathematical descriptors to QSAR was neither appreciated nor understood by many bystanders. Some opponents came with simplistic arguments and tried to characterize the emergence of mathematical descriptors as a flood, overabundance, profusion, as being an extravagant, over-sufficient, plenitude—briefly, a plethora. If you consult several dictionaries, you will find that a plethora refers not only to a very large amount or number, but to an amount that is much greater than what is necessary. Is this so? One should observe that all the "complaints" have been coming from people who have not been users of mathematical descriptors, physical chemists and medicinal chemists who were followers of the traditional Hansch QSAR.

There have been a number of books and reviews on topological indices, and we will mention only three books: *Topological Indices and Related Descriptors in QSAR and QSPR*, edited by Devillers and Balaban [23]; the already mentioned *Comprehensive Descriptors for Structures and Statistical Analysis (CODESSA)* by Katritsky et al.; [8,9] and *Molecular Descriptors for Chemoinformatics* by Todeschini and Consonni [24,25]. For introductory information on mathematical molecular descriptors, one may consult a chapter on topological indices in *The Encyclopedia of Computational Chemistry* [26]. Instead of elaborating on specific molecular descriptors, we have summarized in Table 6.3 the numerous areas of physical chemistry and chemistry in general in which topological indices have found useful applications.

Of the hundreds and hundreds of proposed topological indices, we selected to mention here only two. One is an index of Estrada [27], proposed in 2000, which is one of very few indices that have *several* applications beyond the customary

TABLE 6.3
Areas of Physical Chemistry and Chemistry in General
in Which Topological Indices Have Found Applications

Physicochemical properties of molecules
Biological activity of drugs, including toxicity
Search for pharmacophore
Screening of combinatorial libraries
Search of large databases
Molecular similarity
Molecular diversity
Drug design
Characterization of folded proteins
Characterization of DNA sequences
Characterization of t-RNA
Characterization of proteins
Characterization of molecular shape
Characterization of chirality
Characterization of graph and network centrality
Characterization of network irregularities
Recognition of molecular docking
Numerical characterization of proteome maps
DNA and protein alignment

structure–property–activity studies. The other is the pair of shape indices of Randić [28], proposed in 2001, which is one of very few indices that have shown very good regressions with *several* physicochemical properties of octane isomers. There are not two but at least two dozen that would be worth mentioning here, but this is a book about solved and unsolved problems in structural chemistry and not a book about topological indices. As one can see, both abovementioned indices came after a good 25 years of intensive search for *simple* novel mathematical descriptors for QSAR, which resulted in several hundred reports on new indices of various complexities. In contrast, the Estrada index and the proposed shape indices of Randić are conceptual and computationally unusually simple and elegant.

The Estrada index is based on the eigenvalues of the adjacency matrix of a graph. Let $\lambda_1, \lambda_2, \lambda_3, \lambda_4, \lambda_5, \ldots \lambda_n$ be the eigenvalues of the adjacency matrix; then the Estrada index is defined as

$$E = \Sigma\, e^{\lambda_i}$$

where the summation goes from $i = 1, \ldots n$.

A number of applications of the Estrada index were reported, and they cover such diverse fields of science as the folding of proteins in bioinformatics [27,29,30], global properties of complex networks [31–33], quantum chemistry [34], and statistical thermodynamics [33]. Motivated by these applications, several studies were undertaken, aimed at elucidating the dependence of the Estrada index on the structure of the considered molecular graph [35–39]. For some mathematical properties of the Estrada index, which are related to Chebyshev polynomials, see a publication by Ginosar et al. [40].

As already mentioned, Chebyshev polynomials (Table 6.4) were named after P. L. Čebišev (1821–1894) who introduced them in 1854 [41] as a tool important for approximation theory. Since then, they have found applications in other areas of mathematics, including algebra, combinatorics, and number theory [42]. Chebyshev polynomials have also been used in chemical graph theory as a tool for factorization of the characteristic polynomials [43,44]. In Table 6.4 we have illustrated the initial Chebyshev polynomials of the first and of the second kind. For a brief history of orthography of transcription of the name of Chebyshev, Чебышёв (Russian) see the book by Davis, *The Thread* [45]. We have already listed 11 alternative transcriptions of Chebyshev, including the Croatian Čebišev (recommended by the English Chemical Society), but if one insists on only using English characters, then the most correct would be the 12th version: Chebyshoev. This would be in the style of writing the German Hückel, Austrian Schrödinger, and Danish Ørsted as Hueckel, Schroedinger, and Oersted, respectively.

TABLE 6.4

Chebyshev Polynomials

The first few Chebyshev polynomials of the first kind:

$T_0(x) = 1$

$T_1(x) = x$

$T_2(x) = 2x^2 - 1$

$T_3(x) = 4x^3 - 3x$

$T_4(x) = 8x^4 - 8x^2 + 1$

$T_5(x) = 16x^5 - 20x^3 + 5x$

$T_6(x) = 32x^6 - 48x^4 + 18x^2 - 1$

$T_7(x) = 64x^7 - 112x^5 + 56x^3 - 7x$

$T_8(x) = 128x^8 - 256x^6 + 160x^4 - 32x^2 + 1$

$T_9(x) = 256x^9 - 576x^7 + 432x^5 - 120x^3 + 9x$

The first few Chebyshev polynomials of the second kind:

$U_0(x) = 1$

$U_1(x) = 2x$

$U_2(x) = 4x^2 - 1$

$U_3(x) = 8x^3 - 4x$

$U_4(x) = 16x^4 - 12x^2 + 1$

$U_5(x) = 32x^5 - 32x^3 + 6x$

$U_6(x) = 64x^6 - 80x^4 + 24x^2 - 1$

$U_7(x) = 128x^7 - 192x^5 + 80x^3 - 8x$

$U_8(x) = 256x^8 - 448x^6 + 240x^4 - 40x^2 + 1$

$U_9(x) = 512x^9 - 1024x^7 + 672x^5 - 160x^3 + 10x$

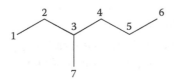

FIGURE 6.1   Numbering of atoms in 3-methylhexane.

TABLE 6.5
Paths and Walks of Lengths Two and Three
for 3-Methylhexane

	p_2	w_2	p_2/w_2		p_3	w_3	p_3/w_3
1	1	2	1/2	1	2	4	2/4
2	2	4	2/4	2	1	7	1/7
3	2	5	2/5	3	1	12	1/12
4	3	5	3/5	4	1	8	1/8
5	1	3	1/3	5	2	7	2/7
6	1	2	1/2	6	1	3	1/3
7	2	3	2/3	7	2	5	2/5
		3.50000				1.87024	

Let us return briefly to introduce the shape indices of Randić [28], which are calculated for individual vertices as the ratio of the number of paths of length k and the number of walks of length k. Consider 3-methylhexane (Figure 6.1) for illustration of paths of length $k = 2$ and $k = 3$. It is not difficult to count the paths in acyclic graphs. In Table 6.5, we show the counts of paths and walks of lengths two and three for 3-methylhexane. As the result for 3-methylhexane C_7H_{16} one obtains $(p_2/w_2) = 3.5000$ and $(p_3/w_3) = 1.87024$.

Because isomers are of the same size, clearly major variations in these indices reflect on their shapes. They do slightly increase as the size of the molecule increases, just as was the case with connectivity indices, because path and walk quotients for additional atoms have to be added. For example, for 3-methylheptane C_8H_{18}, we have $(p_2/w_2) = 4.0000$ and $(p_3/w_3) = 2.09841$. In Table 6.6, we have listed (p_2/w_2) and (p_3/w_3) for the 18 isomers of octane, and in Table 6.7, we have listed a selection of regressions for different properties of isomers of octane using the (p_2/w_2) and (p_3/w_3) descriptors.

6.4   COMPLEX BIOMOLECULAR NETWORKS AND THE CONNECTIVITY INDEX $^1\chi$

Recently, the connectivity index has found application in an unexpected area: as a tool for characterization of irregularities of complex biomolecular networks. Complex biological networks arise in studies of interactions between large numbers of proteins (counted in thousands) occurring in individual cells. The number

TABLE 6.6

The Path/Walks Descriptors p_2/w_2 and p_3/w_3 for Isomers of Octane

Molecule	p_2/w_2	p_3/w_3
n-octane	0.45833	0.22321
2-M	0.50208	0.21151
3-M	0.50000	0.26230
4-M	0.49167	0.25524
3-E	0.49167	0.30227
2, 2-MM	0.55625	0.20000
2, 3-MM	0.54167	0.28769
2, 4-MM	0.53958	0.24681
2, 5-MM	0.54583	0.20328
3, 3-MM	0.55417	0.28297
3, 4-MM	0.54167	0.33173
3-E, 2-M	0.53810	0.32756
3-E, 3-M	0.55982	0.36191
2, 2, 3-MMM	0.60268	0.30717
2, 2, 4-MMM	0.59345	0.17950
2, 3, 3-MMM	0.60774	0.34454
2, 3, 4-MMM	0.58810	0.30989
2, 2, 3, 3-MMMM	0.66964	0.32143

of interactions can be two or more times larger. For example, this is how, in a paper titled "Randić Index, Irregularity and Complex Biomolecular Networks" [46], Estrada points to a novelty of this field:

> It goes beyond any imagination to ask how quantum chemistry or even molecular mechanics can deal with more than 6000 interactions between more than 2700 proteins in the main connected component of the human protein—protein interaction network.

This clearly points that these problems are outside quantum chemistry and belong to new disciplines that are emerging in Bioinformatics. Estrada continues the above quote to point out the following:

> In contrast, methods based on network theory, which rest basically on the study of topological properties of these interaction networks, has become the standard analysis in systems biology and chemistry.

We may mention also that large networks occasionally occur in chemistry. For example, in a study of degenerate isomerization of P_7^{3-}, one comes across the "baby" monster: a graph having $7!/3 = 1680$ vertices [47]. Still larger, the "monster" graph, which has $10!/3 = 1,209,600$ vertices [48,49], describes the degenerate isomerization of bullvalene $C_{10}H_{10}$ [50]. These are regular graphs; their properties are more regular and are therefore susceptible to analysis. It has been known from the beginning

TABLE 6.7

The Regressions of Selected Properties of Octanes Using the Path/Walk Shape Descriptors p_2/w_2 and p_3/w_3

Property	Correlation Coefficient (r)	Standard Error (s)
MR	0.9948	0.0201
r	0.9908	0.0017
P_c	0.9868	0.208
Steric	0.9781	0.394
DH	0.9542	0.123
H_v	0.9520	0.660
S	0.9503	1.50
DH_f	0.9405	0.449
H_c	0.9395	0.460
H_f	0.9387	0.459
Acentric	0.9197	0.0127
BP	0.9141	2.64
C^{13}	0.9115	817
T_c	0.9098	3.88
R^2	0.8538	0.100
CT	0.8508	5.45
DF	0.8177	0.519
Octane #	0.8050	0.114
CP	0.7400	1.03
V_c	0.7031	0.0112

of the studies of graph spectra as reported by Collatz [51] that for regular graphs the difference between the leading eigenvalue (the largest positive eigenvalue designated as λ_1) and the average eigenvalues is zero. The regular graphs are graphs in which all vertices have the same degree (valence). Such are the graphs of Platonic and Archimedean polyhedra, graphs of degenerate rearrangements, and regular infinite lattices such as the graphite lattice. Most graphs are not regular, including the protein–protein interaction networks. Collatz and Sinogovitz (CS) thus proposed, as a measure of network irregularity, the difference

$$CS = \lambda_1 - d_{av}$$

where d_{av} is the average degree.

Interest in irregular networks revived since the 1980s when an additional two measures for irregularity of networks were proposed based on the variance of vertex degrees [52] and on the sum of absolute values of differences in degrees of adjacent vertices [53]. We should mention that the index that Bell used as a measure of network irregularities was introduced about 10 years before by Snijders [54] as a measure of graph centrality. The concept of graph centers goes to the very beginning of chemical graph theory and is traced to Sylvester, who, with Cayley, can be viewed as a founder of chemical graph theory, just as Euler is viewed as a father of graph theory.

The concept of centrality is more general in that it tries to assign to all vertices in a graph some measure of how close or how far they are from the molecular center, which is in all graphs well defined and is a single vertex or a single edge (i.e., two vertices). For more on the development of the notion of centrality, consult the recent publication [55]. Another quantity that is of interest in large networks is the degree of network irregularity. Here one tries to measure overall variations in the valence of network vertices by considering an index that would be zero or minimum for regular graphs and have the maximal value for star graphs. A star graph of order k is a tree on k vertices with one vertex having degree $k - 1$ and the other $k - 1$ vertices having degree 1. It was Collatz and Sinogowitz in 1957 who conjectured that their index, CS, satisfies this requirement, but Cvetković and Rowlinson in 1988 found counterexamples [56]. Hence, there are graphs that have even larger irregularities than star graphs. Similarly, in the case of Bell's index of network irregularities, Hansen and Mélot [57] reported that the Albertson index also is not always maximized for star graphs. In view of this, in 2005, Gutman, Hansen, and Mélot [58] urged a search for a new index that would have minimal irregularity measures for regular graphs and maximal measures of irregularity for the star graphs.

This was about 10 years ago, but five years later, Estrada brought a surprise: He found that the Randić connectivity index plays an important role in the definition of the new network irregularity measure [46]. Estrada recognized a good feature of the Albertson index, which is defined as the difference of valences of adjacent vertices. Estrada also noticed its deficiency, which is due to the fact that with an increase in the valence of the central index in star graphs, the limit of the maximal difference of the valence of the central vertex and its neighbors approaches 1. From this, follows that for some higher valences of the central vertex the star graphs will not have maximal irregularity. Accordingly, Estrada proposed to minimize, instead of Albertson's sum of absolute differences of vertex degrees, the related quantity

$$\sum (1/\sqrt{k_i} - 1/\sqrt{k_j})^2$$

where k_i denotes the valence of vertex i and summation goes over all edges. This quantity can be viewed as a generalization of the approach by Albertson, particularly as one can also consider a variable exponent instead of $-1/2$ (reciprocal square root). Another distinction is that instead of using the modulus (the absolute values of the difference of vertex valence), Estrada assumed the difference squared. This is reminiscent of the early development of the calculus of statistics.

Joseph Boscovic (1711–1787; Figure 6.2), living mostly in Italy, was invited by the Royal Society to take part in the measurement of the arch of the globe in Brazil, but on request of the Pope that it be done in Italy, between Rome and Rimini, he stayed in Italy. He added his result to the earlier three measurements and decided to minimize the residuals of the four measurements using the modulus. This was the

FIGURE 6.2 Joseph Boscovic (Ruggiero or Rudjer Bošković).

first time that this was proposed and represents an important novelty in statistics, so rightly one can view Boscovic as the father of modern statistics. What some 50 years later Legendre and Gauss did was to replace use of modules by using quadratic function. Approaches based on modulus are today referred to as L1-norm, and the least square approaches are known today as L2-norm.

Let us return to the Estrada approach to characterization of network irregularities. He generalized the difference between the reciprocal square roots of the vertex degrees to

$$\sum \left\{ k_i^\alpha - k_j^\alpha \right\}^2$$

summed over all adjacent vertices, and then, in order to eliminate increasing the limit for $[k_i - k_j]^2$, which was the problem, Estrada introduced the Laplacian matrix **L**. The Laplacian marix is defined as: $\mathbf{L} = \mathbf{K} - \mathbf{A}$, where **K** is the diagonal matrix of vertex degrees, and **A** is the adjacency matrix. With this, Estrada arrived at a very elegant formula for the irregularity measure of a graph or a network:

$$(-\tfrac{1}{2}) < k_i^\alpha \, | \, L \, | \, k_j^\alpha >,$$

which breaks down in two sums:

$$\sum \left(k_i^\alpha - k_j^\alpha \right)^2 = \sum (k_i)^{2\alpha+1} - 2 \sum (k_i k_j)^\alpha$$

The first sum on right hand side of the equation is over all the vertices, and the second sum is over all the edges. The second sum represents the generalized Randić index R_α [59,60], which can be written as

$$(-\tfrac{1}{2}) < k_i^\alpha \, | \, A \, | \, k_j^\alpha >.$$

A is the adjacency matrix and $|k^\alpha\rangle = \left(k_1^\alpha, k_2^\alpha, \cdots, k_n^\alpha\right)$ represents a column vector where k_i^α is the valence of the vertex i. This is a novel and elegant formula for the Randić index R_α, which in the special case of $\alpha = -1/2$ gives the Randić connectivity index $^1\chi$. Rather than paraphrasing Estrada, we will reproduce two short paragraphs at the end of his paper from the section on ruminations:

> There is a vast literature in mathematics and in chemistry about the Randić index. Eminent mathematicians like Erdös and Bollobás already paid attention to this index at the end of the 20th century. Gutman and others [58,61,62], have found many extreme properties and bounds for this index. Many chemists and biologist have found an immeasurable amount of applications for this index.

> The simplicity of the Randić index is astonishing:

$$R_\alpha = \frac{1}{2}\left\langle k_i^\alpha | A | k_j^\alpha \right\rangle.$$

Graham Farmelo has edited a selection of "Elegant Formulae" [63], which includes Einstein $E = mc^2$, Schrödinger's $H\Psi = E\Psi$, Shannon's $I = -p\, log_2\, pl$ among others. No doubt about the elegance, simplicity, and importance of all theses formulae. Randić equation is elegant, simple and fundamental for understanding many properties of interaction networks. Only time will tell whether the "Randić Equation" deserves a place among Farmelo's "elegant formulae." In the meantime, we continue exploring it to discover new facts that support its inclusion.

6.4.1 An Illustration

We will illustrate use of the Randić formula $(-1/2) < k_i^{-1/2} | A | k_j^{-1/2} >$ as proposed by Estrada on 3-methyhexane. In Table 6.8 at the top, we show multiplication of the

---

TABLE 6.8
Calculation of the Randić Index Using the Estrada Formula

$$\left(-\tfrac{1}{2}\right) < k_i^{-1/2} | A | k_j^{-1/2} >$$

	1	2	3	4	5	6	7	$\lvert k_j^{-1/2} >$		$\lvert A \lvert k_j^{-1/2} >$
1	0	1	0	0	0	0	0	1		$1/\sqrt{2}$
2	1	0	1	0	0	0	0	$1/\sqrt{2}$		$1 + 1/\sqrt{3}$
3	0	1	0	1	0	0	1	$1/\sqrt{3}$		$1/\sqrt{2} + 1/\sqrt{3} + 1$
4	0	0	1	0	1	0	0	$1/\sqrt{2}$	=	$1/\sqrt{3} + 1/\sqrt{2}$
5	0	0	0	1	0	1	0	$1/\sqrt{2}$		$1/\sqrt{2} + 1$
6	0	0	0	0	1	0	0	1		$1/\sqrt{2}$
7	0	0	1	0	0	0	0	1		$1/\sqrt{3}$

Note: The scalar product gives $1/\sqrt{2} + 1/\sqrt{2} + 1/\sqrt{6} + 1/\sqrt{6} + 1/\sqrt{6} + 1/\sqrt{3} + 1/\sqrt{6} + 1/2 + 1/\sqrt{2} + 1/\sqrt{2} + 1/\sqrt{2} + 1/\sqrt{3}$, which is twice the connectivity index: $1/\sqrt{2} + 1/\sqrt{6} + 1/\sqrt{6} + 1/2 + 1/\sqrt{2} + 1/\sqrt{6}$.

adjacency matrix with vector $|\mathbf{A}|\mathbf{k}_j^{-1/2}>$, and below, we show the scalar product $<\mathbf{k}_i^{-1/2}|$ and $|\mathbf{A}|\mathbf{k}_j^{-1/2}>$. Scalar product of two vectors is a number. As one can see, the reason for the factor $-1/2$ is because multiplication gives twice the connectivity index.

6.5 JOSEPH BOSCOVIC

We will briefly introduce here Ruggiero Boscovich or, as is he known in his native Croatia, Rudjer Bošković, who was an outstanding scientific and cultural figure in 18th century Europe characterized as a physicist, astronomer, mathematician, diplomat, philosopher, theologian, and Jesuit priest, all of which he was, and he excelled in all these. We already mentioned his founding steps in statistics. As an astronomer, he was first to detect the equator of rotating planets, the absence of atmosphere on the Moon, and evaluated orbits of planets and comets by making *three* measurements. In 1758, he published a very influential book: *Theoria Phylosophicae Naturalic Redacta ad Unicam Lagem in Natura Existentium* (Theory of Natural Philosophy Derived to the Single Law of Forces which Exist in Nature). This book was published again in Venice (1763), Vienna (1764), London (1922), the United States (1966), and Zagreb, Croatia (1974).

In 1760, Boscovic became a member of the Royal Society (London) and, in 1761, a member of The Imperial Academy of Sciences and Arts (St. Petersburg). He traveled to different countries. He visited Ljubljana three times and was in contact with Jurij Vega (1754–1802), the Slovene mathematician, physicist, surveyor, meteorologist, nobility, and military officer for guns, known for his logarithmic tables and for the design of better mortars. He advised Austria on the use of the metric system, but it took 100 years to be adopted in Austro-Hungary. We may add that Vega's improved mortars were used for the first time during the battle for liberation of Belgrade, the capitol of Serbia, from Ottoman Empire after 350 years of occupation.

Boscovic is one of the founders of the Italian Academy (limited to 40 members), and he built the observatory in Brera (part of Milan). Apparently, he had many enemies, who prevented him from visiting California to observe the transit of Venus in 1769 on invitation of the Royal Society. After suppression of the Jesuit order in 1773 in Italy on invitation of the King of France, he became director of Optics for the French Navy, a position created for him, where he stayed 10 years. He returned to Brera Observatory, and near the end of his life was measuring the orbit of comets (making three observations for each comet in order to use the medians).

We include this information on Rudjer Bošković with a single purpose: to illustrate the great breadth of scientists of the 18th century (not of Boscovic alone, as is also seen from the fragmentary information on Jurij Vega). Today, many scientists are so deeply involved in their own research areas, and our time is different, so that few can afford to follow work in other areas, much less to consider getting involved in related sciences and even to follow areas close to their own possible interest. For example, how many quantum chemists know or are aware of what else is going on in theoretical chemistry, including chemical graph theory and graphical bioinformatics? This is not likely to improve as the fragmentation of science will continue as science continues to widen and its topics continue to grow!

It is true that our progress was made possible by advances in technology and novel tools available, which are good and which our forefathers in science did not have. But we also see a lack of understanding and lack of communications between tool makers and tool users. One should realize that current unsolved problems in different disciplines may be unsolved to a significant degree due to lack of appropriate tools. Problems are not unsolved because there were no clever people around but because the appropriate tool may be missing. So tool making should be encouraged rather than restrained by requesting tool makers to solve the problems of tool users as has been heard from time to time.

6.6 DESIDERATA FOR MOLECULAR DESCRIPTORS

We have recently found in Wikipedia [64] the basic requirements for optimal molecular descriptors, shown in Table 6.9. There is a list of 12 desirable properties for molecular descriptors. After examining them, we fully agreed with this list of the desired attributes of mathematical molecular descriptors. Although there was no reference given, the list looked familiar, and because we fully agreed with it, we searched the literature and found that these "rules" had been proposed about 15 years ago in a 50-page article: "In Search of Structural Invariants" by one of the present authors [65]. This explains why the list "Desiderata for molecular descriptors" looked so familiar, and we agreed with it. Unfortunately, since that time, there have been reported numerous topological indices that do not satisfy a number of the recommendations of "Desiderata" listed, so it was not a bad idea that somebody reproduce them in Wikipedia. Those "rules" have been composed by appropriate modification of similar rules for construction of molecular *codes* proposed by R. C. Read [66]. It may be a worthy exercise, when using available molecular descriptors or constructing new descriptors, to see the degree to which they overlap with the "desired" qualities, such as having structural information, having a good correlation

---

TABLE 6.9
Basic Requirements for Molecular Descriptors

1. Should have structural interpretation
2. Should have good correlation with at least one property
3. Should preferably discriminate among isomers
4. Should be possible to apply to local structure
5. Should be possible to generalize to "higher" descriptors
6. Should be simple
7. Should not be based on experimental properties
8. Should not be trivially related to other descriptors
9. Should be possible to construct efficiently
10. Should use familiar structural concepts
11. Should change gradually with gradual change in structures
12. Should have the correct size dependence, if related to the molecule size

---

with at least one property, being simple, using familiar structural concepts, changing gradually with small changes in structures and being size consistent, etc.

Let us comment on several less obvious desirable attributes of molecular descriptors. From what we have already said, rules 1 and 7 should be considered essential for structural descriptors of molecules. Rule 2 relates to the search for structure–property space, at least for the property for which the particular descriptor is a good descriptor. Rule 3 is not a necessary requirement and is one that may not be easy to satisfy for larger molecules using substructure counts as a basis for the construction of components of the descriptor considered but may be more important for topological indices for searching combinatorial libraries. Rule 4 is important when one considers molecular fragments in local similarity searches, such as searching for a pharmacophore. Rule 5 allows one not only to have a new topological index, but also to construct a novel "vector" basis for structure–property space, which is not an ad hoc combination of existing topological indices, but a collection of indices that have some common structural features. As mentioned earlier, having set of structurally related "higher order" indices makes ordering of descriptors easy when constructing regressions using orthogonal descriptors. An illustration of such indices is the connectivity index and higher order connectivity indices and several similar constructions [67]. Rule 6 embodies the Occam's razor rule, which, according to the *Merriam-Webster Dictionary* [8], has been described as

...a scientific and philosophic rule that entities should not be multiplied unnecessarily which is interpreted as requiring that the simplest of competing theories be preferred to the more complex or that explanations of unknown phenomena be sought first in terms of known quantities.

It should be mentioned here that the Occam's razor rule has been mentioned from time to time by those who continue to object to the apparent "explosion" of topological indices, which has occurred in more recent years in QSAR, but *"the simplest of competing theories"* being preferred does not extend to replacing "theories" with "descriptors." Therefore, *"the simplest of competing descriptors"* should not necessarily *"be preferred."* What is important for molecular descriptors is not how similar they are and how much they may duplicate one another, but do they *differ* one from another in some details that are important (for the structure–property–activity relationship).

Rule 7 excludes the use of experimental properties for construction of molecular descriptors, which characterizes the Hansch QSAR as an empirical approach rather than structural. This in no way diminishes the significant accomplishments, success, and triumph of the Hansch QSAR. It more relates to a linguistic issue of interpretation of the word "structure." Looking in different dictionaries, one finds a variety of definitions, including the following: (i) something arranged in a definite pattern; (ii) the aggregate of elements of an entity to their relationship to each other; (iii) anything composed of parts arranged together in some way, an organization; (iv) a complex system considered from the point of view of the whole rather than of any single part.

Clearly cases (i–iii) are not broad enough to include molecular properties as constituents of a structure, but with some imagination (and good will) perhaps case (iv) can even if properties are not a synonym for structures.

In simple language, when speaking of a building as a structure, does this mean an empty building or filled? In the analogy, does molecular structure mean the skeletal frame or include properties of such frame? In this respect, mathematical properties of molecular graphs are legitimate structure descriptors. Why should physicochemical properties not be legitimate structure descriptors? Well, the problem is in the fact that the so-called molecular physicochemical properties are not molecular properties (that is properties of a *single* molecule), but bulk properties (that is, properties of an *assembly* of molecules). So here we have an apparent paradox in both, the quantitative structure–property relationship (QSPR) and the QSAR studies. In QSPR, when exploring the quantitative structure–property relationship, one uses mathematical *structural* descriptors (the Wiener numbers W and P, the Hosoya topological index Z, the connectivity index χ, and so on) based on a *single* molecule to construct the regressions for the molecular bulk properties (physicochemcal properties). In contrast, in QSAR, when exploring the quantitative structure–activity relationship, one uses physicochemical descriptors (such as molecular weight, molar refraction, octanol/water partition, Hammett sigma constant, lipophilicity, molecular density, and so on) to construct the regressions for the molecular properties (molecular bioactivity), but descriptors relate to molecular *bulk* properties.

Resolving this paradox appears desirable for both QSPR and QSAR, but the problem apparently has hardly received attention. Understanding this paradox remains an unresolved theoretical problem of structural chemistry, but apparently the paradox is working, both in QSPR and QSAR, and this may be the main reason why the problem, at best, received limited attention.

Returning back to the issue of potential linguistic ambiguities in science, broad and somewhat questionable use of the term "structure" for both the structure as something built by arranging the parts of the whole and something including external properties of such is not an isolated case. Another illustration is the term "topological indices" widely used for mathematical descriptors characterizing molecular connectivity. Molecular topology and molecular connectivity are not the same. In topology, one can only speak of a neighborhood, but molecular connectivity (and graph theory) has metric. Topological indices should have been referred to as graph theoretical indices instead, but it may be too late now for a change, just as it is equally too late to consider renaming the historically named *atom*, which means indivisible, from the time when this was believed to be so, to something else, such as "tom." All this only proves that languages and linguistics are not mathematics.

Rule 8 is evident without proof or reasoning; the emphasis here is on "trivially." Some descriptors eventually may be found to be related, and finding such relationships in the case of descriptors that have wider use is of interest. Recall the statement of Richard Feynman (who shared the Nobel Prize in Physics in 1965 for his contributions to quantum electrodynamics) [68]: *"The formulation is mathematically equivalent to the more usual formulation. There are therefore, no fundamentally new results. However, there is a pleasure in recognizing old things from a new point of view."*

From what we have seen before, if two descriptors are highly intercorrelated, this does not mean that they are trivially related. Trivially related are two descriptors whose coefficient of regression is 0.9999999...or 1 exactly!

Rule 9 should not be overlooked, and it may not be important in the case of smaller molecules and small numbers of molecules, but when one considers combinatorial libraries, which may have several thousand compounds and up to several hundred thousands of virtual compounds, one needs descriptors that can be constructed quickly and efficiently. Use of molecular descriptors that involve higher complexities will be prohibitive unless their computation is very fast.

Rule 10 perhaps, in part, overlaps to a degree with rule 1. Formally speaking, any construction that does not depend on vertex labels or particular ways a graph is represented qualifies as a structural descriptor. A construction, in fact, need not even be very complex to result in a descriptor that need not have a simple direct structural interpretation. An illustration is the Wiener number W, for which Platt tried to find some simple structural interpretation in terms of molecular volume [69]. The emphasis here is on clearly interpretable terms of familiar structural concepts; otherwise, they will have limited use in our understanding of the structure–property relationship that they are supposed to explain. On rules 11 and 12, we have previously commented.

The mathematical "beauty" of descriptors may be an asset, but it is not an attribute that guarantees such descriptors are necessarily suitable in structure–property–activity studies. Consider, for instance, the case of the Szeged index (Sz index) proposed by Gutman 20 years ago [70], which is a generalization of the Wiener index. The Wiener index was defined only for acyclic graphs, and the Szeged index is its generalization to cyclic graphs. The "beauty" of this index is in using essentially the same algorithm by Wiener for construction of the Wiener index but modified so that it holds for cyclic structures. The modification consists of replacing "*multiply the number of atoms on one side of a bond by the number of atoms on the other side of a bond*" by "*multiply the number of atoms closer to one side of a bond by the number of atoms closer to the other side of a bond.*" Observe that the so-modified definition of the Wiener index to the generalized Wiener index still continues to hold for acyclic graphs but now applies also to cyclic graphs.

However, if one applies this algorithm to cyclic molecules having odd cycles, one finds that bonds making odd cycles have atoms at *equal* distances from both ends of such bonds. According to the above generalization, the atoms of odd rings do not contribute equally to the Szeged index. Because of this, in different molecules, there will be different numbers of "ignored atoms," and therefore, the Szeged index for molecules having even and odd rings of similar size may be quite different [71,72]. Hence, the Szeged index is not size-consistent, which could be fatal for applications of the Szeged index in correlations involving cyclic molecules with odd rings. However, instead of ignoring contributions from atoms at the same distances from both ends of bonds, one can divide the count of atoms at the *same* distance from each end of a bond equally between the atoms making such bonds, and one obtains a revised Szeged index, which maintains the size consistency and can thus be used as a descriptor in structure–property correlations [72].

6.7 HIGHER-ORDER CONNECTIVITY INDICES

There is no doubt that the success of the connectivity index $^1\chi$ is closely tied to the possibility of its extension to structurally related higher-order connectivity indices $^2\chi$, $^3\chi$, $^4\chi$ ... $^k\chi$ [73]. In the spirit of Platt, the suggestion of use of paths of different lengths for characterization of molecules, the higher order connectivity indices do just that in going beyond the bond additivity model and considering the contribution of paths of length two, paths of length three, and so on. Thus, $^2\chi$ is defined as

$$^2\chi = \sum 1/\sqrt{m_i m_j m_k}$$

where m_i, m_j, and m_k are the degrees of vertices forming a path of length two, and the summation is carried over all paths of length two in a molecule. Similarly, $^3\chi$ is defined as

$$^3\chi = \sum 1/\sqrt{m_i m_j m_k m_l}$$

where m_i, m_j, m_k, and m_l are the degrees of vertices forming a path of length three, and the summation is carried over all paths of length three in a molecule.

One can extend the above approach and construct the higher-order connectivity indices up to the longest paths in a molecule although applications have shown that usually only the higher-order indices based on shorter paths make significant contributions in regressions. Kier and Hall also considered the higher-order descriptors based on smaller molecular fragments, such as clusters of three bonds with a common atom. The significance of the higher-order connectivity indices is that they define higher-order structure–property space and extend the use of the connectivity index $^1\chi$ beyond the simple regressions.

The connectivity indices $^1\chi$, $^2\chi$, $^3\chi$, $^4\chi$ ... $^k\chi$ form a basis, which allows use of the same set of indices for different molecular properties and different sets of molecules. This makes the comparative study of different properties and different molecules simpler because all the MRA results relate to the same structure–property space. This is not the case when, in MRA, one combines an ad hoc collection of molecular descriptors. It appears that the problem of construction of alternative simple structural bases has hardly been considered. The problem is, can we find alternative bases of structurally related descriptors to be used in MRA analogous to the set of the connectivity index and the higher-order connectivity indices? In other words, can one generalize an index into higher-order indices?

An early contribution to this has been considered, and a few very simple illustrations were presented 20 years ago [74], but no systematic search for potential other bases for use in structure–property–activity studies by MRA, particularly when compounds studied involve a half dozen heteroatoms appear to have been undertaken. We may consider this issue as an open problem, an unsolved problem, of structural chemistry.

Let us outline an algorithm for construction of different higher-order connectivity indices. Consider modified higher-order connectivity indices defined as

$$^{2}\chi = \sum 1/\sqrt{(m_i m_j)(m_j m_k)}$$

Here m_i, m_j, and m_k are the degrees of vertices forming a path of length two, and the summation is carried over all paths of length two in a molecule. Similarly, $^{3}\chi$ is defined by

$$^{3}\chi = \sum 1/\sqrt{(m_i m_j)(m_j m_k)(m_k m_l)}$$

where m_i, m_j, m_k, and m_l are the degrees of vertices making up a path of length three, and the summation is carried over all paths of length three in a molecule and so on for longer paths. In this modification, instead of considering a product of vertex degrees for vertices forming a path, one considers the product of contributions of edges forming the paths.

6.8 MOLECULAR ID NUMBERS

The modified connectivity indices have been used for construction of the molecular identification (ID) numbers, a new kind of molecular descriptor, which have been found to have exceptional discrimination power [75]. In Table 6.10, we have illustrated the construction of the molecular ID number for a small molecule containing

---

TABLE 6.10
The Atomic Contributions to Molecular ID Number

	Path 1	Path 2	Path 3	Atomic ID
1	$1/\sqrt{3}$	$(2/\sqrt{3})(1/\sqrt{6})$	$(2/\sqrt{3})(1/\sqrt{6})(1/\sqrt{4})$	2.284457
2	$1/\sqrt{3} + 2/\sqrt{6}$	$(2/\sqrt{6})(1/\sqrt{4})$		2.802095
3	$1/\sqrt{6} + 1/\sqrt{4}$	$(1/\sqrt{6})(1/\sqrt{3}) + (1/\sqrt{4})(1/\sqrt{6}) +$ $(1/\sqrt{6})(1/\sqrt{6})$	$(1/\sqrt{4})(1/\sqrt{6})(1/\sqrt{3})$	2.632592
4	$1/\sqrt{6} + 1/\sqrt{4}$	$(1/\sqrt{6})(1/\sqrt{3}) + (1/\sqrt{4})(1/\sqrt{6}) +$ $(1/\sqrt{6})(1/\sqrt{6})$	$(1/\sqrt{4})(1/\sqrt{6})(1/\sqrt{3})$	2.632592
	Numerical			**Sum**
1	0.577350	0.471405	0.235702	1.284457
2	1.393847	0.408248	0	1.802095
3	0.908248	0.606493	0.117851	1.632592
4	0.908248	0.606493	0.117851	1.632592
Σ	3.787693	2.092639	0.471404	6.351736

Note: Atomic ID = 1 + sum; molecular ID = sum of atomic ID = 4 + 6.351736 = 10.351736.

---

four carbon atoms, a methylcyclopropane represented by the labeled hydrogen depleted molecular graph:

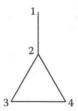

Each row of Table 6.10 corresponds to one atom and lists the modified higher-order connectivity indices. The sum of each row augmented by one, which represents the count of "paths of length zero," gives the atomic ID number for that atom. When all paths of all atoms in a molecule are added, one obtains the molecular ID number. These numbers have been found useful in the search for a pharmacophore as has been outlined elsewhere [76]. Molecular IDs have shown considerable power of discrimination among similar molecules. Thus, the smallest pair of alkanes having the same molecular ID have been found by Knop et al. [77], have 15 vertices and are illustrated in Figure 6.3. This is one pair of graphs among 4347 [78].

Because bonds (1, 4) and (2, 2) make the same contributions to the connectivity index, it is possible that molecules of different bond composition have the same ID number. If one constructs connectivity indices by using different weight contributions for bonds of different types, molecular ID numbers of even higher discriminatory power would result. So by selecting the set of smallest prime numbers as weights, one obtains the prime number connectivity indices [79]. When tested by Knop et al., it was found that all alkane graphs having 19 and fewer carbon atoms have been fully discriminated [80]. There was one pair of graphs on 20 vertices ID with an identical prime number. These are two graphs among 366,319 alkane graphs on 20 vertices or two graphs among 618,050 alkane graphs having 20 or fewer vertices.

The search for pairs of graphs having the same molecular ID number or molecular prime number ID can be viewed as an illustration of "experimental" mathematics or, if you wish, "experimental" chemical graph theory. The concept of molecular ID numbers has been extended later to cover rings [81] and alternative ID numbers were considered [82]. The examination of the discrimination power of molecular identification numbers has been very recently revisited [83–86], clearly indicating that this topic has not yet been completely explored. New indices have outperformed earlier ones and show limits of classical measures, such as the Balaban index J.

FIGURE 6.3　Pair of graphs on 15 vertices having the same molecular ID number.

6.9 RECENT MODIFICATION OF THE HIGHER-ORDER CONNECTIVITY INDICES

There is an additional modification of the higher-order connectivity indices, which we consider important or at least interesting and which has never been implemented until very recently [87,88]. Observe that higher-order connectivity indices have an increasingly greater number of factors under the square root sign. Suppose that those numbers are not based on properties of the molecular graph but rather on some properties of molecules as indeed some modifications of the connectivity index $^1\chi$ have been. Then different higher-order connectivity indices would have different dimensions. A way to correct this is to take the kth root, not the square root, in construction of higher-order indices. Thus, although the definition of the $^1\chi$ remains unaffected, the higher-order connectivity indices have to be modified as shown below. For $^2\chi$, we thus have

$$^2\chi = \sum 1/\sqrt[3]{m_i m_j m_k}$$

Here, m_i, m_j, and m_k are the degrees of vertices forming a path of length two, and the summation is carried over all paths of length two in a molecule. Similarly, $^3\chi$ is defined as

$$^3\chi = \sum 1/\sqrt[4]{m_i m_j m_k m_l}$$

where m_i, m_j, m_k, and m_l are the degrees of vertices making up the path of length three, and the summation is carried over all paths of length three in a molecule and so on. An effect of this modification is that the connectivity index and the higher-order connectivity indices maintain similar magnitudes instead of gradually decreasing in magnitude as has been the case with the standard higher-order connectivity indices.

In Table 6.11, we have illustrated the difference between the calculated contributions to the second-order connectivity index on a graph having 20 vertices. This is one of the graphs by Ivanciuc et al. [89], illustrated in Figure 6.4, which has the same values for several "advanced" topological indices. The modified connectivity index is based on the weighted adjacency matrix, in which the bond contributions are based on the partition of the Wiener index [2]. As one can see from Table 6.11, the modified second-order connectivity index, based on the generalized connectivity index $\chi(W)$, which is 0.354314 and which is by an order of magnitude greater than the "traditional" second-order connectivity index 0.043597, based on use of the square roots of path contributions.

6.10 VARIATIONS OF THE CONNECTIVITY INDEX

In several studies, the connectivity index has been based on alternative characterizations of heteroatoms as will be briefly outlined in the next section. Here we will describe a modification of the connectivity index that is of a mathematical nature

TABLE 6.11

Calculation of the Second Order Connectivity Index for the Graph of Figure 6.4

Path	$(p_{ijk})^{-1/2}$	$(p_{ijk})^{-1/3}$	Path	$(p_{ijk})^{-1/2}$	$(p_{ijk})^{-1/3}$
1,2,3	0.002551	0.018672	5,16,17	0.001554	0.013414
2,3,4	0.000708	0.007941	5,16,18	0.001554	0.013414
2,3,9	0.001500	0.013101	6,7,8	0.002020	0.015977
3,4,5	0.000335	0.004826	6,7,20	0.002020	0.015977
3,4,11	0.000708	0.007941	6,5,15	0.001217	0.011400
3,4,13	0.000708	0.007941	6,5,16	0.000562	0.006813
3,9,10	0.002551	0.018672	7,6,19	0.002020	0.015977
4,5,6	0.000338	0.004848	8,7,20	0.005579	0.031456
4,5,15	0.000933	0.009545	4,3,9	0.000708	0.007941
4,5,16	0.000933	0.009545	5,4,11	0.000548	0.006698
4,11,12	0.001968	0.015706	5,4,13	0.000548	0.006698
4,13,14	0.001968	0.015706	15,5,16	0.001554	0.013414
5,6,7	0.000562	0.006813	17,16,18	0.005579	0.031456
5,6,19	0.001217	0.011400	11,4,13	0.001157	0.011020
			Sum	**0.043597**	**0.354314**

Note: The two numerical columns list the contributions of paths of length two to $^2\chi$ when square root and cub root algorithms were used.

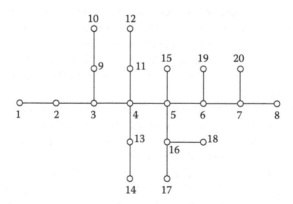

FIGURE 6.4 Graph on 20 vertices for which Table 6.11 has listed all paths of lengths two and their contributions to the second-order generalized connectivity index.

rather than empirical, which follows in the next section. Recall that construction of the connectivity index started with the set of inequalities presented in Table 6.1 and finding that the algorithm $1/\sqrt{(m\ n)}$ is a solution without indicating that this is but one of the possible solutions. In Table 6.2, we illustrated two additional solutions of the same set of inequalities that offer alternative sets of connectivity indices. At left in Table 6.2 are shown calculated connectivity indices based on an increase of the valence of the carbon atom by +1, and at right in Table 6.2 are shown calculated connectivity indices based on using squared valence values for the carbon atom. Both choices have no a priori rationale, but they are on the same footing as the original solution based on atom valence values (m, n). Clearly, there are additional possible variations, including assuming m, n as variables and seeking the optimal values of the so-modified m and n, which produce the smallest standard error for the considered regression. This is precisely what the variable connectivity index represents, which will be discussed in the next section.

Here, we would like to comment on mathematical modification, not of the connectivity index, but the higher-order connectivity indices.

6.11 ON CHARACTERIZATION OF HETEROATOMS

Standard graph theory usually considers systems in which all vertices belong to the same class. In some mathematical problems, however, one discriminates between vertices of different kinds by viewing them to be of a different "color." Chemical graph theory has a more demanding request in that one wants not only *qualitatively* ("different colors") but also *quantitatively* to discriminate vertices representing different elements. Thus, chemical graph theory goes beyond mathematical graph theory when modeling structure–property–activity relationships for heteroatomic systems.

Soon after the connectivity index was introduced, Kier and Hall realized that in expanding its application to chemistry of heteroatomic molecules one needs to differentiate vertices belonging to heteroatoms from those belonging to carbon atoms. They proposed a modification of the graph valences of vertices belonging to heteroatoms and thus arrived at the valence connectivity indices [18]. Later, following Kier and Hall, several other topological indices were similarly modified to take into account the presence of heteroatoms in molecules (e.g., ref. [88–96]). All these approaches have assumed the existence of *universal* heteroatom descriptors, which will *adequately* discriminate between carbon and other atoms. However, with the construction and applications of the variable connectivity index, it has been realized that such an assumption is invalid. There are no universal heteroatom descriptors that can satisfactorily characterize heteroatoms in all regression equations. Thus considerations of different molecular properties for the same set of heteroatom compounds may require a different set of heteroatom descriptors, each set serving well only in correlations involving a limited set of properties.

In this respect, it is interesting to observe that in several early applications of the valence connectivity indices Kier and Hall reported regressions using both the valence connectivity indices and the "ordinary" connectivity indices. A number of regressions involving both ordinary and valence connectivity indices produced better regressions with both indices used simultaneously. Why would use of both indices give, in some cases, better regressions than use of valence connectivity alone?

In addition, in their book on application of connectivity indices in drug research [16], Kier and Hall have listed a dozen illustrations of regressions involving compounds having heteroatoms in which a better regression was obtained using the "ordinary" connectivity index alone instead of the valence connectivity. Thus, for example, although $^1\chi$ gives as good a correlation for the molar refraction R_m of ethers as that obtained with $^1\chi^v$ [97], it gives a better regression for van der Waals constants of mixed compounds [98]. Other properties with which $^1\chi$ gives a better regression than $^1\chi^v$ include the molar magnetic susceptibility of alcohols [99], the heat of vaporization of aliphatic alcohols [100], the boiling points of aliphatic alcohols [101] and amines [102], the solubility of aliphatic alcohols [103], and the partition coefficient ($\log P$) of aliphatic ethers [104]. Why? Why should the "simple" connectivity index in these particular cases produce better correlations than the valence connectivity index that differentiates nitrogen and oxygen in these compounds? This question could have been asked 40 years ago, but was either not asked or, if asked, not answered.

Clearly, if a regression using the selected weighting factors specific for a different heteroatom does not give a better result than a regression based on not using weights at all, it means that the selected modification of the connectivity index is *inadequate* for the particular property considered, and some other *weights* (which include the weight "zero" associated with the simple connectivity index) will produce better regression. Hence, in general, *different* properties may require *different* weights for the same heteroatoms. Of course, it is needless to say that, in many other regressions, the valence connectivity indices produced significantly better regressions than those that would be obtained by using the "simple" connectivity indices, suggesting that for those properties the weights selected for heteroatoms by Kier and Hall have been adequate. As will be seen, occasionally for some properties, the contribution of the same heteroatom may be visibly different than it is for other properties. In other words, it can be that the heteroatom weighting appropriate for a particular property is different (possibly even opposite) to what was assumed to hold for most other instances. As will be outlined in the next section on the variable connectivity index, in some cases, a heteroatom can make zero contribution and even *negative* contribution to the magnitude of molecular descriptor. This would mean that such heteroatom is responsible for decreasing the magnitude of the property considered relative to contributions of carbon atoms, which has led to the concept of anti-connectivity [105,106].

In summary, one may say that in regression analyses heteroatoms need to be differentiated from carbon atoms as has been shown by Kier and Hall 40 years ago, but instead of prescribing a single set of modified indices for any individual heteroatoms, one should consider *two or more* such sets of descriptors. Two heteroatom descriptors, when combined, will cover a wider structure space "territory" than a single index having preselected weight to represent heteroatoms covers even if "helped" by the "plain" connectivity index χ. That indeed this happened we have seen from a few regressions reported in the book by Kier and Hall.

Very recent and quite independent support for this comes from a study of Pogliani and Julian-Ortiz on QSPR with descriptors based on averages of vertex invariants [107]. They introduced a new type of indices, the mean molecular connectivity indices, based on a number of different concepts of mean. They have been interested in comparative study of multiple regression analysis (MRA) and artificial neural network (AAN) approaches. For illustration, they examined a selection of physicochemical properties of organic solvents, using the mean molecular connectivity indices together with molecular connectivity indices, experimental parameters, and random variables, as descriptors. Of interest here is to mention that in their report on MRA, which used five descriptors (out of a pool of 121) on 10 properties considered, that in the case of the refraction index and the dipole moments they reported among the five best descriptors $^1\chi$, not the valence $^1\chi^v$. We should add that among the 64 solutions considered there were more than half a dozen different heteroatoms. This result by Pogliani and Julian-Ortiz thus illustrates that there are properties for which the valence connectivity indices are not optimal.

6.12 THE VARIABLE CONNECTIVITY INDEX

Rather than prescribing fixed bond contributions $1/\sqrt{(m\,n)}$ to the connectivity index, where m, n are the valences of vertices in a molecular graph, and similarly, the fixed contributions of various heteroatoms as proposed by Kier and Hall for construction of valence connectivity indices, the variable connectivity indices assume now variable contributions $1/\sqrt{[(m + x)\,(n + y)]}$. Here x, y are the variables yet to be determined for each regression considered independently. Because the connectivity index can be calculated from the row sums of the adjacency matrix, the introduction of variables x, y corresponds formally to augmenting the graph adjacency matrix by introducing variables along the main diagonal, where variables x, y belong to different kinds of atoms, forming the corresponding molecular graph.

In Table 6.12, we illustrate variations of the connectivity indices $^1\chi(x, y)$, $^2\chi(x, y)$, and $^3\chi(x, y)$ for 1-aminohexane as a function of the weights for nitrogen atoms while assuming $x = 0$ for carbons atoms. In this example, we assumed $x = 0$ for carbon atoms and varied only the weights y for nitrogen in the interval [0.50, −0.95]. As one can see, in this way, one obtains a set of values for the connectivity indices to choose from.

TABLE 6.12
Variation of the Connectivity Indices for 1-Aminohexane with Weights for Nitrogen (Assuming $x = 0$ for Carbons)

y	$^1\chi(0, y)$	$^2\chi(0, y)$	$^3\chi(0, y)$
+0.5	3.2845	1.3922	0.5711
0	3.4142	1.4571	0.6036
−0.25	3.5336	1.5118	0.6309
−0.50	3.7071	1.6036	0.6768
−0.75	4.1213	1.8107	0.7803
−0.80	4.2883	1.8941	0.8221
−0.85	4.5326	2.0164	0.8832
−0.90	4.9432	2.2216	0.9858
−0.95	5.8694	2.6847	1.2175

This approach has been applied to a set of 16 amino-alkanes, from 1-aminopropane to 1-aminononane, including several branched alkanes with nitrogen attached in addition to terminal carbon and to carbons at positions 2 and 3. In Table 6.13, we have illustrated variations of the carbon atoms valence parameter, shown as x, and variations of the nitrogen valence parameter shown as y in a regression of the boiling points of smaller amino-alkanes. As one can see, when $x = 0$ and $y = 0$, this becomes the regression using the connectivity index $^1\chi$ for all atoms (not differentiating between carbon and nitrogen atoms). In this case, one obtains for the standard error $s = 3.488$ (right top of the last column in Table 6.13). If one varies x (the weight of carbon atoms) but keeps $y = 0$ for the nitrogen atom, one obtains for the optimal $x = 1.25$ some improvement for the standard error of the regression, which becomes $s = 3.008$. If one varies y for the nitrogen atom but keeps $x = 0$ for carbon atoms, one obtains for the optimal $y = -0.65$, which results in visible improvement of the standard error of the regression, which is now $s = 2.081$. Finally, when the weights for both carbon and nitrogen are simultaneously varied, the standard error further decreases for optimal $x = 1.25$, $y = -0.65$ to $s = 1.907$. After changing both x and y, which has been illustrated in Table 6.13, the optimal values for x and y, which give the smallest standard error in a regression of the boiling points of the 16 amino alkanes, are found to be 1.25 and -0.65, respectively, for carbon and nitrogen atoms. Observe also the efficiency of the search for the optimal value. It took only 20 steps to find the optimal numerical values for the variable parameters. The approach does not require one to calculate the full six-by-six table of the regressions for all 36 possible values. This was possible because in the search for optimal parameters one need not to be concerned with local minima, which typify such problems for general functions $f(x, y)$.

In Table 6.14, we show the variable connectivity index $^1\chi(x, y)$, the experimental boiling points (bp), the calculated bp, and their difference for the regression of the boiling points of amino-alkanes. The final regression equation,

$$bp = 83.456 \ ^1\chi(1.25, -0.65) - 83.992 \text{ with } s = 1.91 \text{ and } r = 0.9990$$

is illustrated in Figure 6.5. Here s is the standard error, and r is the regression coefficient. Observe also from Table 6.14 that all differences between of bp_{exp} and bp_{calc}

TABLE 6.13

Variation of the Standard Error for Different Weights for Carbon (x) and Nitrogen (y) in Smaller Amino-Alkanes

$x \downarrow y \rightarrow$	−0.80	−0.70	−0.65	−0.60	−0.50	0
0			2.081			3.488
1.00	2.603	1.963	1.913	1.947	2.121	
1.25	2.537	1.946	1.907	1.951	2.134	3.008
1.50	2.470	1.929	1.914	1.967	2.162	
1.75		1.925	1.927			
2.00			1.951			

TABLE 6.14
The Optimal Variable Connectivity Index, Experimental and Calculated Boiling Points and Their Difference

Molecule	$^1\chi(x, y)$	bp_{exp}	bp_{calc}	Difference
1-Aminononane	3.46126	201	204.9	−3.9
1-Aminooctane	3.15357	180	179.2	+0.8
1-Aminoheptane	2.84588	155	153.5	+1.5
1-Aminohexane	2.53818	130	127.8	+2.2
1-Amino-4-methylpentane	2.46883	125	122.0	+2.9
2-Aminohexane	2.39755	114.5	116.1	−1.6
1-Aminopentane	2.23049	104	102.2	+1.8
1-Amino-2-methylbutane	2.16893	96	97.0	−1.0
1-Amino-3-methylbutane	2.16114	96	96.4	−0.4
2-Aminopentane	2.08986	92	90.4	+1.6
3-Aminopentane	2.09766	91	91.1	−0.1
2-Amino-2-methylbutane	1.93152	78	77.2	+0.8
1-Aminobutane	1.92280	77	76.5	+0.5
1-Amino-2-methylpropane	1.85344	69	70.7	−1.7
2-Aminobutane	1.78217	63	64.7	−1.7
1-Aminopropane	1.61511	49	50.8	−1.8

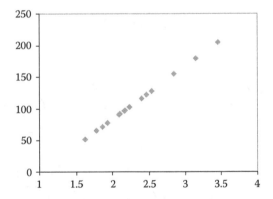

FIGURE 6.5 Correlation of the experimental boiling points of 16 primary amines versus the calculated boiling points using the variable connectivity index.

are well below two standard errors. This justifies referring to such regression results as "the high quality structure–property regressions" and "high quality structure–activity regressions" [108].

The illustration of the correlation of the calculated boiling points against the experimental boiling points for the amino-alkanes, although impressive, does not necessarily reflect fully on the potential of variable connectivity indices for structure–property–activity regressions because the regularity in this case could have been expected to parallel, to a great extent, the already known regularities for the boiling points of alkanes.

A better test for the potential of the variable connectivity descriptors in QSAR is to choose a more "difficult" case. So we will report on a regression of a set of imidazolidines with respect to their hypotensive activity. Let us examine the hypotensive activities of 18 methyl and chloro derivatives of 2-(arylimino)imidazolidine. The compounds are illustrated schematically in Figure 6.6.

They are ordered from left to right and from top to bottom relative to their hypotensive activity (experimental ED_{50} values in µg/kg) obtained from dose-response curves following intravenous administration to anaesthetized, nonmotentive rats, that is, in vivo effective doses which produce anesthesia in 50% of the population. These compounds were used to illustrate construction of the variable connectivity indices in the seminal paper on the variable connectivity index [109]. All the data has been taken from particularly detailed research by Timmermans and Van Zweiten [110], who studied a set of 27 hypotensive imidazolidines, which in addition to the 17 choro derivatives considered here, include nine additional compounds having other heteroatoms: fluorine, bromine, nitrogen, and oxygen. In the report that we selected, these compounds were excluded because they represent too small a sample to allow one to determine with acceptable certainty their relative weights for the construction of the variable connectivity indices.

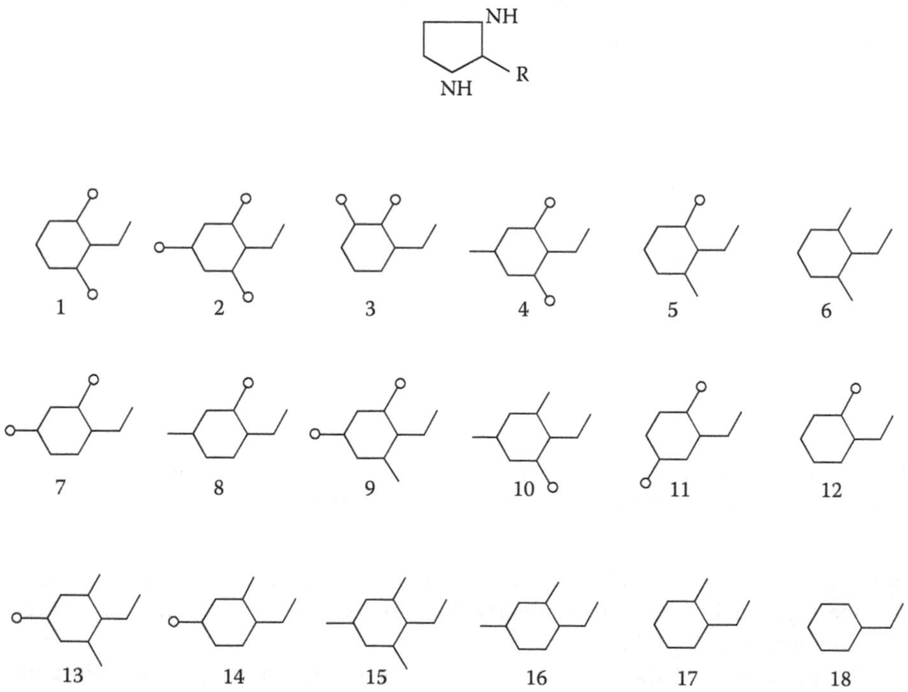

FIGURE 6.6 The diagrams of molecular graphs of 18 variable fragments R of 2-(arylimino)-imidazolidines considered. The locations of the chlorine atoms are indicated by small circles.

In Table 6.15, we have listed the variable connectivity indices $^1\chi(x, y)$, $^2\chi(x, y)$, and $^3\chi(x, y)$ for the 18 clonidine-like compounds, assuming $x = 0$ for carbon atoms and $y = -0.20$ for chlorine. Observe the rather small domains of the three variable connectivity indices. In the case of $^1\chi(0, -0.20)$ and $^2\chi(0, -0.20)$, the variations appear to be within 10% of their magnitudes, and in the case of $^3\chi(0, -0.20)$, the variations are even smaller. The magnitudes of $^3\chi(0, -0.20)$ vary around 5% of their values. A close look at Table 6.15 shows that the changes are much smaller and even zero between compounds having the same number of chlorine and methyl groups. The presence of methyl groups does lower the magnitude of the variable connectivity indices, or in other words, the presence of chlorine increases the magnitudes of the corresponding variable connectivity indices involving chlorine. This is the consequence of having y negative ($y < 0$), which decreases the denominator in $1/\sqrt{[(m + x)(n + y)]}$, and thus assigns to C–Cl bonds a larger contribution to the variable connectivity indices.

In Table 6.16, we have listed the calculated antihypertensive activities, the predicted antihypertensive activities, and the experimental antihypertensive values based on the regression using the variable connectivity indices $^1\chi(x, y)$, $^2\chi(x, y)$, and $^3\chi(x, y)$ with $x = 0$ and $y = -0.20$. The standard error and the regression coefficient for the calculated antihypertensive activities are $s = 0.222$ and $r = 0.9773$, and for

TABLE 6.15
The Variable Connectivity Indices for the 18 Clonidine-Like Imidazolidines (Carbon Is Assumed $x = 0$; Chlorine Is Computed $y = -0.20$)

No.	Derivative	$^1\chi(x, y)$	$^2\chi(x, y)$	$^3\chi(x, y)$
1	2,6-Cl$_2$	4.278	2.015	0.978
2	2,4,6-Cl$_3$	4.418	2.120	0.969
3	2,3-Cl$_2$	4.278	2.015	0.963
4	2,6-Cl$_2$-4-Me	4.384	2.092	0.957
5	2-Cl-6-Me	4.244	1.989	0.964
6	2,6-Me$_2$	4.210	1.964	0.949
7	2,4-Cl$_2$	4.262	2.036	0.982
8	2-Cl-4-Me	4.228	2.008	0.970
9	2,4-Cl$_2$-6-Me	4.384	2.095	0.955
10	2,6-Me$_2$-Cl	4.350	2.067	0.944
11	2,5-Cl$_2$	4.262	2.036	0.982
12	2-Cl	4.122	1.939	0.982
13	2,6-Me$_2$-4-Cl	4.350	2.069	0.942
14	2-Me-4-Cl	4.228	2.011	0.968
15	2,4,6-Me$_3$	4.316	2.042	0.931
16	2,4-Me$_2$	4.194	1.983	0.956
17	2-Me	4.088	1.914	0.966
18	Unsubstituted	3.966	1.869	0.979

TABLE 6.16

Calculated and Predicted (Cross-Validation Values) for Antihypertensive Activities Based on Variable Connectivity Indices

No.	Compound	Regression	Cross-Validation	Experiment
1	$2,6\text{-}Cl_2$	2.034	1.977	2.14
2	$2,4,6\text{-}Cl_3$	1.460	1.478	1.41
3	$2,3\text{-}Cl_2$	1.298	1.286	1.37
4	$2,6\text{-}Cl_2\text{-}4\text{-}Me$	1.061	1.035	1.22
5	$2\text{-}Cl\text{-}6\text{-}Me$	1.372	1.451	1.12
6	$2,6\text{-}Me_2$	0.697	0.627	0.85
7	$2,4\text{-}Cl_2$	0.566	0.536	0.68
8	$2\text{-}Cl\text{-}4\text{-}Me$	0.111	0.061	0.68
9	$2,4\text{-}Cl_2\text{-}6\text{-}Me$	0.850	0.901	0.57
10	$2,6\text{-}Me_2\text{-}Cl$	0.459	0.448	0.52
11	$2,5\text{-}Cl_2$	0	0.607	0.32
12	$2\text{-}Cl$	0.259	0.285	0.15
13	$2,6\text{-}Me_2\text{-}4\text{-}Cl$	0.249	0.300	0.04
14	$2\text{-}Me\text{-}4\text{-}Cl$	−0.080	−0.084	−0.05
15	$2,4,6\text{-}Me_3$	−0.150	−0.193	−0.07
16	$2,4\text{-}Me_2$	−0.532	−0.527	−0.56
17	$2\text{-}Me$	−0.448	−0.406	−0.61
18	Unsubstituted	−2.076	−2.047	−2.10
r		0.9773	0.9676	
s		0.222	0.248	

the corresponding values for leave-one-out cross-validation, which can be viewed as predicted values, the standard error and the regression coefficients are $s = 0.248$ and $r = 0.9676$, respectively.

It is instructive to compare the standard error and the regression coefficients of the regressions using $^1\chi(x, y)$ only, pairs of descriptors $^1\chi(x, y)$ and $^2\chi(x, y)$, and triplet $^1\chi(x, y)$, $^2\chi(x, y)$, $^3\chi(x, y)$. One finds then the following:

$$^1\chi(x, y) \qquad\qquad\qquad s = 0.712 \qquad r = 0.690$$
$$^1\chi(x, y), ^2\chi(x, y) \qquad\qquad s = 0.634 \qquad r = 0.781$$
$$^1\chi(x, y), ^2\chi(x, y), ^3\chi(x, y) \qquad s = 0.223 \qquad r = 0.977$$

As one can see, even though variations in $^3\chi(0, -0.20)$ between 18 different compounds are not very large, the improvement in the regression is dramatic. Although the selected variables $x = 0$ and $y = -0.20$ are not necessarily optimal, the above regression illustrates the ability of the variable connectivity indices to cover the structure space better than the descriptors of traditional QSAR, which involved five descriptors in a nonlinear regression equation, obtaining $s = 0.301$ and $r = 0.964$.

In other applications of the variable connectivity indices, similar and even more impressive results have been reported. Despite the successes of the variable connectivity indices, which is in part due to the novelty of the approach, optimization of independent variables *during* the construction of regression equations, apparently they received at best limited attention. Recently, the regression of a collection of clonidine-like compounds having hypotensive action has been revisited [111]. Recall that, initially, in 1977, the hypotensive action of clonidine-like compounds was studied using the traditional Hansch-type QSAR approach [110]. In 1991, the same was reexamined using modified connectivity indices adjusted for heteroatoms as representative mathematical descriptors [109]. At that time, chlorine atoms were differentiated from carbon atoms by modifying their effective valence. Most recently, the nontraditional variable connectivity indices were used, but the role of carbon atoms and chlorine were optimized. The best Hansch-type regression used *five* descriptors having accompanying standard deviation $s = 0.350$. The best correlation based on *three* connectivity indices was accompanied with standard deviation $s = 0.233$. The best correlation based on *a single* optimized connectivity index, reported in the abovementioned paper, has standard deviation $s = 0.186$. The best optimized single connectivity index is $^3\chi(x, y)$, not $^1\chi(x, y)$ or $^2\chi(x, y)$. This, however, should not be surprising because it has been already observed earlier [112] that $^3\chi(x, y)$ is the best single variable descriptor for several volume-dependent properties, such as density and molar refraction. Hence, with considerable confidence, one can therefore conclude that hypotensive activity of clonidine-like compounds also belongs to the same class of volume-sensitive molecular properties.

When one looks more closely into different connectivity indices, there is a significant difference between the connectivity indices $^1\chi$ or $^2\chi$, and $^3\chi$. In numerous correlations with physicochemical properties in the case of 18 octanes using different molecular descriptors, the connectivity indices $^1\chi$ and $^2\chi$ are found to give good correlations for the properties that depend on the molecular surface while $^3\chi$ is found to give good correlation for the properties that depend on molecular volume. This is illustrated in Table 6.17, which in addition to the abovementioned properties, molar refraction and density, include also molecular volume, [112]. Most experimental properties examined in Ref. [112] have been reported as published in Ref. [113]. Because the best single connectivity index regression was found to be obtained with $^3\chi$, it follows that experimental ED_{50} for hypotensive activity of clonidine-like compounds is volume-dependent molecular property. We wished to give to this conclusion some visibility in order to illustrate how important conclusions may be extracted from regression equations by analyzing descriptors used without quantitative information from the regression equation.

The variable connectivity indices may have been viewed with suspicion in some circles, not as being illegitimate, but as being "composite" and representing more than a single parameter. For example, Zefirov and Palyulin [114] attribute the success of the regressions using variable connectivity indices due to some "hidden" variables (which were never defined), implying that this represents an increase in the count of variables used in a regression. But the notion of "hidden" variables has no merits.

TABLE 6.17
The Best Regression Coefficients (*r*)
and Corresponding Descriptors
for Selection of Physicochemical
Properties of 18 Octane Isomers

Property	*r*	Descriptor
Heat of vaporization	0.942	$^1\chi$
Heat of formation	0.931	$1/^2\chi$
Heat of atomization	0.931	$1/^2\chi$
Surface tension	0.964	$^2\chi - ^3\chi$
Pitzer acentric factor	0.992	$^2\chi$
Molecular volume	0.978	$^3\chi$
Molar refraction	0.970	$^3\chi$

The number of independent variables used in a regression equation is given by *the count of variables* used in a regression equation, not on how are variables selected or calculated. A simple regression has a *single* descriptor, so regression using the connectivity index or the variable connectivity index is simple regression. After all, use of the variable connectivity index, in fact, means computing two dozen or more simple (single variable) regressions. So how can selecting *one* simple regression out of two dozen simple regressions, make it not simple? The situation is different if one considers, for example, if the same descriptor is used in a quadratic equation. This counts as a regression with two variables because the regression equation has, in addition to the constant, two terms and indicates a 2-D structure–property space. Regressions using three variable descriptors signify 3-D structure space and so on.

A totally different question is how are the variables selected and how are they calculated, which has nothing to do with how many variables are used. To see that variation of descriptors during the search for the optimal regression equation has nothing to do with implication of undefined "hidden" variables, let us consider a case of using two variables (x, y) for which, prior to the search for optimization, one prepares tables of 20 values of each variable. This amounts to calculating 400 $^1\chi(x, y)$ descriptors. In parallel, look at the computer software such as CODESSA [8,9], which offers about 400 molecular descriptors. Now consider a regression using a single variable connectivity index, such as the regression on the boiling points of nitro-alkanes outlined earlier. One possibility is to search through the prepared table of 400 descriptors for the best pair of variables (x, y). If that was done, one would find $x = 1.25$ and $y = -0.20$ as optimal. Alternatively, one can use CODESSA and find a few combinations of probably two to three descriptors that will have close statistical parameters (the standard error and the regression coefficient). Clearly there are no "hidden" variables in sight whether one used a prepared set of 400 descriptors or one used CODESSA, which would calculate 400 descriptors for each molecule before searching for optimal regression. The difference between the search for optimal descriptors (using 400 descriptors prepared in advance) or searching for optimal (x, y) by computing the individual regression at each step separately and selecting the direction in changing new values for variables based on observed trends in x, y toward a minimal overall standard error is that the latter is more efficient because involves fewer computations. As we have seen in the case of searching for (x, y) to get the best regression for nitro-alkanes, we found the best regression in 20 steps rather than 400.

There are no "hidden" variables in QSAR, but there are "hidden" virtues of the novel approach in QSAR based on variable descriptors. We took the liberty of para-phrasing E. B. Wilson [115], who said,

> ...every once in a while some new theory or new experimental method or appara-tus makes it possible to enter a new domain. Sometimes it is obvious to all that this opportunity has arisen, but in other cases recognition of the opportunity requires more imagination.

Instead we may say

> ...every once in a while some new idea or new theoretical approach makes it possible to introduce a new approach to old problems. Sometimes it is obvious to all that this opportunity has arisen, but in other cases, recognition of the opportunity requires more imagination.

6.13   SURPRISE, SURPRISE

Of course, science thrives on novelties. Novelty is something not expected, not anticipated, unrelated to known things. Novelty is also sudden recognition of a rela-tionship between known phenomena considered hitherto totally unrelated. A clas-sic illustration is that of the discovery of electromagnetism. Hans Christian Ørsted (Oersted), (1777–1851), a Danish physicist and chemist, during a lecture in 1820, noticed that when turning off current a magnetic needle was affected. According to some, he was trying to show students that electricity and magnetism are not related but during the demonstration found that they appear to be related. Ørsted found the mathematical law that governs the strength of the field. Ørsted's discovery was the first connection found between electricity and magnetism. The second, possibly even more important, discovery was the law of induction of Michal Faraday (1791–1867), who, 10 years later, found that electric current is induced in a wire moving in a magnetic field. These are great discoveries, but whenever two apparently unrelated phenomena are found to be connected, this is an important novelty contributing to unifying the relevant parts of science. As we see in this section, one such unknown and surprising discovery relates to the variable connectivity indices, which have unexpectedly shown that the calculated optimal variable weights may have physico-chemical interpretation!

Consider a regression for a set of 66 halogenides and their boiling points [116]. The molecules considered include branched and cyclic halo-alkanes and several halo-alkenes involved as heteroatoms chlorine, bromine, and iodide. All the mole-cules and their boiling points, which cover the range from −24°C (chloromethane) to

+238°C (1-bromodecane) are listed in Table A12.1 of Appendix 12. The best regression equations, using as descriptors various topological indices, including also the relative count of heteroatoms, and involving from one to five descriptors have been reported in the literature. The regression equation with five descriptors is shown in Table 6.18, and the best regression equation having fewer descriptors can be seen in Table A12.2 in Appendix 12. The resulting regression equation based on five descriptors is accompanied with respectable statistics: coefficient $r^2 = 0.9848$ and standard error $s = 7.58°C$. In Figure 6.7, we show the plot of calculated boiling points against the experimental.

The optimal variables found were the following: for carbon, the optimal weight was 1.27; for chlorine, the optimal weight −0.235; for bromine, the optimal weight −0.653; and for iodine, the optimal weight was −0.8005. It is known from the early calculations with the variable connectivity index that small positive values for optimal weights have little influence on affecting bond contributions to the overall

TABLE 6.18

The Coefficients of the Best Regression Equation Using Five Descriptors for the Boiling Points of 66 Halogenides [114]

	Equation Coefficients	Descriptors	r^2	s
	−122.68	Constant		
1	50.832	Kier and Hall (order 1)		
2	−1.3185	Information content (order 0)		
3	36.613	Randić index (order 1)		
4	13.252	Balaban index		
5	76.491	Relative number of iodine atoms		
			0.9848	7.58

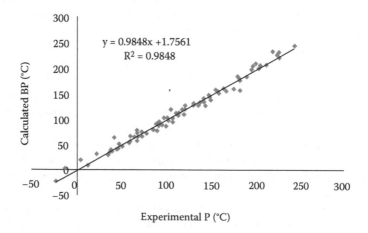

FIGURE 6.7 Regression for a set of 66 halogenides and their boiling points. The molecules considered include branched and cyclic halo-alkanes involving chlorine, bromine, and iodine.

connectivity index, and negative values, even if small, can make a dramatic increase in the role of the corresponding bonds. Thus, the above results show that halogen atoms, going from chlorine to bromine and iodide will have increasing contributions to the boiling point of halogen compounds.

But now comes a surprise: The four optimal weights of C, Cl, Br, and I, the numbers 1.27, 0.235, 0.653, and 0.8005 correlate inversely very well with atomic masses of carbon (12), chlorine (35.5), bromine (79.9), and iodine (126.9) as can be see from Figure 6.8. We show the plot of reciprocal mass ($1/M$) against the optimal weights for C, Cl, Br, and I. Observe that the plot is a linear regression with the regression coefficient $r = 0.99995$. The novelty is that statistical parameters (weights) for variable connectivity indices have a direct structural interpretation! When the variable connectivity index was introduced, the focus was on the search for the best regression, and some insight on the relative roles of hetero bonds C-X could be obtained.

It was only more recently, a good 20 years later, that Pompe and Randić realized optimal weights may correlate with some *atomic* properties [116]. After considering regressions of some alcohols, ethers, esters, and ketones, it was noticed that the calculated atomic charges correlate with charge of oxygens in different local environments. As the saying (due to Aristotle) goes, *"One swallow does not make a spring."* Pompe and Randić waited for another illustration, which is of a more recent time, and, as outlined above, fully and dramatically support the novelty that regressions based on variable connectivity indices may lead to unknown correlations of optimal weights with selected *atomic* properties. This may also be the case with other variable molecular descriptors—time will tell.

The correlations of optimal weights with some *atomic* properties introduced some novel information for QSAR and, in a way, are reminiscent of Craig plots, which we have described in Chapter 5 and which similarly introduced novelty to QSAR, allowing researchers to decide on the direction of improving their analyses. In view of this relationship, we suggest a temporary reference to this *additional* regression that *follows* optimal variable connectivity index regressions be referred to as Variable Weights diagram. The Variable Weights plot may not always be easy to

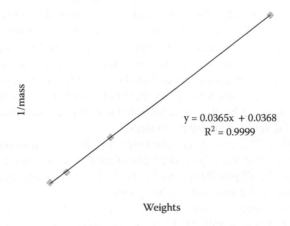

FIGURE 6.8   Plot of reciprocal atomic mass against the optimal weights for C, Cl, Br, and I.

find because, for some molecular properties, the weights may depend on more than a single atomic property. Be it as it may, it seems that the Variable Weights plot opens a new direction in QSAR and may facilitate finding even better molecular descriptors with simpler interpretation than the best that we currently have. In addition, the Variable Weights plot raise the unexpected possibility of interpretation of some mathematical molecular descriptors in terms of some physicochemical properties— potentially a novel bridge between two "subcultures" of QSAR.

In 1991, Randić has generalized the connectivity index $^1\chi$ by allowing the parameters (m, n) that it contains, which signify vertex (atom) valence in bonds and allowed them to vary. The optimal values were selected so that the standard error of a regression using $^1\chi$ is the smallest. Despite that this index has been around 25 years and has shown *dramatic* improvements in structure-property regressions, the variable connectivity index, at best, received limited attention. In a number of illustrations, it was shown how a single variable connectivity index gives better regression than the use of three to four standard molecular descriptors. Suspicion of many chemists to mathematical descriptors that have no apparent physicochemical interpretation may be a hurdle to consider. However, the above-mentioned linear regression of the reciprocal mass of atoms against the optimal weights involving chlorine, bromine, and iodine as heteroatoms, found using simplex algorithm, which gave correlation with the coefficient of regression $r = 0.999995$, ought to suffice to convince skeptics that objections to mathematical descriptors have no merit.

6.14   THE STRUCTURE–STRUCTURE SPACE

In the next section, we consider the partial ordering of octane isomers based on the count of p_2 and p_3. One can look at such a diagram as an illustration of two-dimensional structure space. In general, the structure space can be three-dimensional and higher dimensional space, in which structure descriptors are used as the space coordinates. The distributions of structures in the structure space will depend on selected structure descriptors, and if they are well selected, meaningful trends in molecular properties may be observed as will be seen later.

In addition to the use of paths as molecular descriptors, the connectivity indices $^1\chi$, $^2\chi$, $^3\chi$, $^4\chi$ ... $^k\chi$ offer an alternative basis. As mentioned before, in the case of 18 octanes, it was found that despite that the connectivity indices $^1\chi$ and $^2\chi$ parallel one another to a considerable degree (i.e., are highly intercorrelated) and give good correlations for the properties that depend on the molecular surface, when combined, they also give good correlation for the properties that depend on molecular volume. In view of this, it would be of considerable interest to examine structure–structure space based on the set of connectivity indices.

Kier and Hall in 2001 considered structure space based on the connectivity indices $^1\chi$ (as the y coordinate) and $^3\chi$ (as the x coordinate) [117]. In Figure 6.9, we show a simplified Kier and Hall plot ($^1\chi$ against $^3\chi$) for 10 hydrocarbons, acyclic and cyclic, having eight atoms and a similar branching pattern.

The above plot considers only 10 structures while Kier and Hall considered 18 octane isomers and an additional 18 cyclic and polycyclic structures having eight atoms [118–120]. In analyzing $^1\chi$ against $^3\chi$ plot, they summarized their finding that

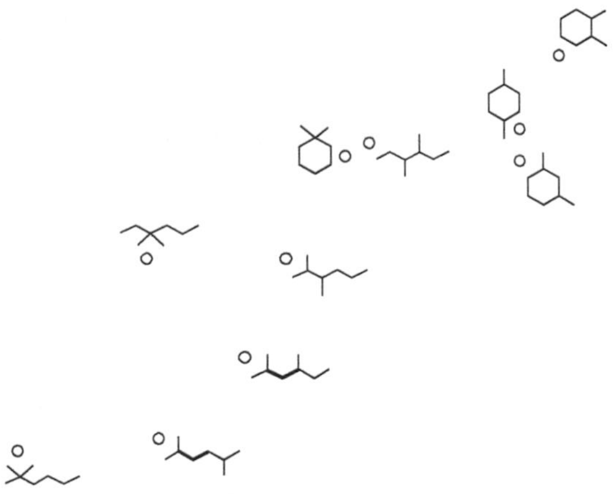

FIGURE 6.9 Kier-Hall plot ($^1\chi$ against $^3\chi$) for 10 hydrocarbons acyclic and cyclic having eight atoms and similar branching pattern. This figure was kindly supplied by L. H. Hall.

a decrease in $^1\chi$ indicates increased branching, and an increase in $^3\chi$ goes with an increased number of branching adjacencies, and along the diagonals is a constant number of branch points. Let's quote their summary:

> Based on this brief analysis, it is clear that these two chi indices provide a chemically meaningful organization of octane skeletons, suggesting that simple chi indices provide the basis for a structure space in which molecular skeletons are arrayed in a useful manner. Much more useful information is encoded than in the simple path counts. Simple chi indices provide the basis for database searching when molecular skeletons are the important issues.

In Figure 6.10, we show partial ordering for 27 cyclic and polycyclic hydrocarbons having eight carbons atoms based on the count of p_2 and p_3. As one can see from Figure 6.10, at five locations, there is a pair of structures having the same count of p_2 and p_3, and at a single location, there are three structures having the same count of p_2 and p_3. However, none of the structures have the same count when one considers paths of all lengths, which have been listed in Table 6.19. These compounds, together with the 18 isomers of n-octane, except for structures 18 and 22 in Table 6.19, is the pool of compounds from which Kier and Hall constructed their $^1\chi$, $^3\chi$ structure space, illustrated in Figure 6.11.

A close comparison between the Kier-Hall plot, based on ($^1\chi$, $^3\chi$), and the corresponding partial ordering, based on (p_2, p_3), considered in the next chapter, are instructive, showing some similarities in grouping the same molecules together. This is true for acyclic isomers of n-octane independently, cyclic and polycyclic compounds of Table 6.19. This book is about solved and unsolved problems in structural chemistry and the topic of structure–structure spaces, including comparative study of different structure spaces, such as, for example, the space based on ($^1\chi$, $^3\chi$) and the space based on (p_2, p_3) can be viewed as an unsolved problem. The (p_2, p_3) structure space for octane has been around for 35 years, which is very instructive, although mostly *qualitative*, allowing one to discern structure–property *trends*

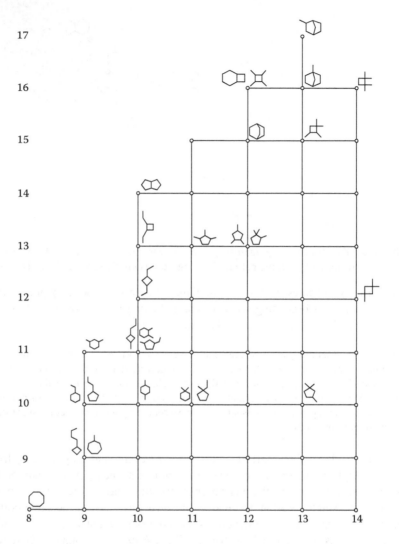

FIGURE 6.10 Partial ordering of C_8H_{16} isomers based on the count of p_2 and p_3 (along the coordinate x and y axes, respectively).

because (p_2, p_3) is confined to a grid based on integers. Nevertheless, it can occasionally predict an unknown molecular property by averaging data on the nearest neighbors in a similar fashion how the Mendeleev periodic table led to a number of predictions for unknown elements from known data of neighbors. The $(^1\chi, {}^3\chi)$ structure space for octane has been around for almost 15 years, and the overall local similarity with the (p_2, p_3) structure space suggests that this space is likely to be *more quantitative* because its geometry uses real numbers; molecules that are neighbors can be at close distance or not. For example, in a Kier-Hall plot, 2,5-dimethylhexane and 2,2-dimethylhexane are relatively close, and so also are their boiling points (differing by 2.3°C) despite that the molecules are not very similar. In contrast, the

TABLE 6.19

The Path Counts for Cyclic Structures Having Eight Carbons Shown in Figure 6.11

	(x, y)	Path Count		(x, y)	Path Count
1	8, 8	8, 8, 8, 8, 8, 8, 8	15	14, 12	8, 14, 12, 16
2	9, 9	8, 9, 9, 7, 6, 4, 2	16	10, 13	8, 10, 13, 10, 6, 2, 1
3	9, 9	8, 9, 9, 8, 9, 9, 2	17	11, 13	8, 11, 13, 12, 14, 2
4	9, 10	8, 9, 10, 10, 10, 4	18	12, 13	8, 12, 13, 11, 3
5	9, 10	8, 9, 10, 11, 6, 4, 2	19	12, 13	8, 12, 13, 11, 4, 2
6	10, 10	8, 10, 10, 10, 12, 4	20	10, 14	8, 10, 14, 16, 14, 12, 8
7	11, 10	8, 11, 10, 10, 10, 2	21	12, 15	8, 12, 15, 12, 12, 6, 6
8	11, 10	8, 11, 10, 11, 6, 2	22	13, 15	8, 13, 15, 12, 3
9	13, 10	8, 13, 10, 13, 8	23	12, 16	8, 12, 16, 17, 16, 11, 10
10	9, 11	8, 9, 11, 10, 10, 4, 2	24	12, 16	8, 12, 16, 12, 6
11	10, 11	8, 10, 11, 10, 6, 4	25	13, 16	8, 13, 16, 17, 16, 8
12	10, 11	8, 10, 11, 10, 10, 8	26	14, 16	8, 14, 16, 8, 4
13	10, 11	8, 10, 11, 12, 8, 3	27	13, 17	8, 13, 17, 22, 19, 11, 2
14	10, 12	8, 10, 12, 10, 8, 2			

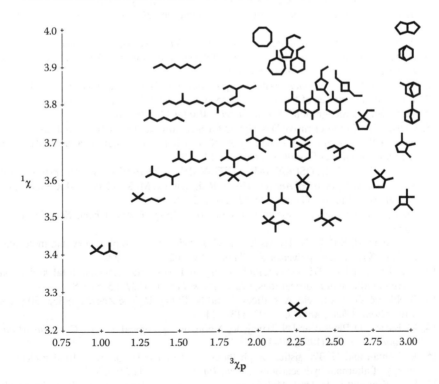

FIGURE 6.11 Kier-Hall plot ($^1\chi$ against $^3\chi$) for C_8 cyclic and polycyclic hydrocarbons. The figure was kindly supplied by L. L. Hall.

boiling points of 2,5-dimethylhexane and 2-methylheptane, which have very close $^3\chi$ values, are not so close in their $^1\chi$ values (differing by 8.5°C), which suggests that it depends strongly on $^1\chi$. To be more quantitative, one would have to represent structures by points (as we did in Figure 6.9) so that the distance between structures can be measured more accurately.

Although the $(^1\chi, {}^3\chi)$ structure space and the (p_2, p_3) structure space have been around for a long time, apparently they caught, at best, limited attention, and that is likely an overstatement. It is therefore too early to speculate whether they will remain an interesting curiosity or develop into an auxiliary tool of multivariate regression analysis (MRA). We would like to suggest that MRA researchers, in addition to reporting regression equations, also consider plotting for selected pairs of descriptors (d_i, d_j) the corresponding (d_i, d_j) structure space to see if the selected descriptors can offer useful additional insights on the regressions considered.

REFERENCES AND NOTES

1. J. R. Platt, Influence of neighbor bonds on additive bond properties in paraffins, *J. Chem. Phys.* 15 (1947) 419–420.
2. H. Wiener, Structural determination of paraffin boiling points, *J. Am. Chem. Soc.* 69 (1947) 17–20.
3. H. Hosoya, Topological index. A newly proposed quantity characterizing topological nature of structural isomers of saturated hydrocarbons, *Bull. Chem. Soc. Jpn.* 44 (1971) 2332–2339.
4. A. F. J. van Raan, Sleeping Beauties in science, *Scientometrics* 59 (2004) 467–472.
5. E. Garfield, Premature discovery or delayed recognition—Why? *Essays of an Information Scientist* 4 (1980) 488–493.
6. M. Kozak, *Current Science* has its "Sleeping Beauties," *Curr. Sci.* 104 (2013) 1129–1130.
7. J. Li, Citation curves of "all-elements-sleeping-beauties": "Flash in the pan" first and then "delayed recognition," *Scientometrics* 100 (2014) 595–601.
8. CODESSA (Comprehensive Descriptors for Structures and Statistical Analysis) reports most of the descriptors designed by A. R. Katritzky, V. Lobanov, and M. Karelson (Eds.), University of Florida, Gainesville, FL, 1995.
9. COODESSA PRO, QSPR/QSAR SOFTWARE (*COmprehensive DEscriptors for Structural and Statistical Analysis*) by A. R. Katritzky, M. Karelson, and R. Petrukhin, University of Florida, Gainesville, FL, 2001–2005.
10. M. Randić, Characterization of molecular branching, *J. Am. Chem. Soc.* 97 (1975) 6609–6615.
11. I. Gutman, B. Ruščić, N. Trinajstić, and C. F. Wilcox Jr., Graph theory and molecular orbitals. XII. Acyclic polyenes, *J. Chem. Phys.* 62 (1975) 3399.
12. I. Gutman and N. Trinajstić, Graph theory and molecular orbitals. Total π-electron energy of alternant hydrocarbons, *Chem. Phys. Lett.* 17 (1972) 535–538.
13. S. Nikolić, G. Kovačević, A. Miličević, and N. Trinajstić, The Zagreb indices 30 years after, *Croat. Chem. Acta* 76 (2003) 113–124.
14. M. Barysz, D. Plavšić, and N. Trinajstić, A note on topological indices, *Commun. Math. Comput. Chem.* 19 (1986) 89–116.
15. I. Gutman and N. Trinajstić, Graph theory and molecular orbitals. Total π-electron energy of alternant hydrocarbons, *Chem. Phys. Lett.* 17 (1972) 535–538.
16. L. B. Kier and L. H. Hall, *Molecular Connectivity in Chemistry and Drug Research*, Academic Press, New York, 1976.

17. L. B. Kier, Indexes of molecular shape from chemical graphs, in: D. H. Rouvray (Ed.), *Computational Chemical Graph Theory*, Nova Science Publishers, New York, 1990, pp. 152–174.
18. L. B. Kier and H. L. Hall, Molecular connectivity. VII. Specific treatment of heteroatoms. *J. Pharm. Sci.* 65 (1976) 1806–1809.
19. J. M. Wells, C. Randall Clark, and M. Paterson, Structure-retention relationship analysis for some mono and polycyclic aromatic hydrocarbons in reverse-phase liquid chromatography using molecular connectivity, *Anal. Chem.* 58 (1986) 1625–1633.
20. M. Randić, The connectivity index 25 years after, *J. Mol. Graph. Model.* 20 (2001) 19–35.
21. X. Li and I. Gutman, *Mathematical Aspects of Randić-type Molecular Structure Descriptors*, Mathematical Chemistry Monographs No. 1, Kragujevac, Serbia, 2006.
22. I. Gutman and B. Furtula (Eds.), *Recent Results in the Theory of Randić Index*, Mathematical Chemistry Monographs, Kragujevac No. 6, Serbia, 2008.
23. J. Devillers and A. T. Balaban (Eds.), *Topological Indices and Related Descriptors in QSAR and QSPR*, Gordon and Breach, Amsterdam, 1999.
24. R. Todeschini and V. Consonni, *Handbook of Molecular Descriptors*, Wiley-VCH, Weinheim, 2008.
25. R. Todeschini and V. Consonni, *Molecular Descriptors for Chemoinformatics* (the revised and enlarged edition of the Handbook of Molecular Descriptors), Wiley-VCH, Weinheim, 2009.
26. M. Randić, Topological indices, in: P. v. R. Schleyer, N. L. Allinger, T. Clark, J. Gasteiger, P. A. Kollman, H. F. Schaefer III, and P. R. Schreiner (Eds.), *The Encyclopedia of Computational, Chemistry*, Wiley, Chichester, 1999, pp. 3018–3032.
27. E. Estrada, Characterization of 3D molecular structure, *Chem. Phys. Lett.* 319 (2000) 713–718.
28. M. Randić, Novel shape descriptors for molecular graphs, *J. Chem. Inf. Comput. Sci.* 41 (2001) 607–613.
29. E. Estrada, Characterization of the folding degree of proteins, *Bioinformatics* 18 (2002) 697–704.
30. E. Estrada, Characterization of the amino acid contribution to the folding degree of proteins, *Proteins* 54 (2004) 727–737.
31. E. Estrada and J. A. Rodríguez-Velázquez, Subgraph centrality in complex networks, *Phys. Rev. E* 71 (2005) 056103.
32. E. Estrada and J. A. Rodríguez-Velázquez, Spectral measures of bipartivity in complex networks, *Phys. Rev. E* 72 (2005) 046105.
33. E. Estrada, Topological structural classes of complex networks, *Phys. Rev. E* 75 (2007) 016103.
34. E. Estrada, J. A. Rodríguez-Velázquez, and M. Randić, Atomic branching in molecules, *Int. J. Quantum. Chem.* 106 (2008) 823–832.
35. E. Estrada and N. Hatano, Statistical-mechanical approach to subgraph centrality in complex networks, *Chem. Phys. Lett.* 439 (2007) 247–251.
36. I. Gutman, E. Estrada, and J. A. Rodríguez-Velázquez, On a graph-spectrum-based structure descriptor, *Croat. Chem. Acta* 80 (2007) 151–154.
37. I. Gutman and A. Graovac, Estrada index of cycles and paths, *Chem. Phys. Lett.* 436 (2007) 294–296.
38. I. Gutman, S. Radenković, A. Graovac, and D. Plavšić, Monte Carlo approach to Estrada index, *Chem. Phys. Lett.* 446 (2007) 233–236.
39. I. Gutman, S. Radenković, B. Furtula, T. Mansour, and M. Schork, Relating Estrada index with spectral radius, *J. Serb. Chem. Soc.* 72 (2007) 1321–1327.
40. Y. Ginosar, I. Gutman, T. Mansour, and M. Schork, Estrada index and Chebyshev polynomials, *Chem. Phys. Lett.* 454 (2008) 145–147.

41. P. L. Chebyshev, Théorie des mécanismes connus sous le nom de parallélogrammes, *Mém. Acad. Sci. Pétersb.* 7 (1854) 539–586.
42. T. Rivlin, *Chebyshev Polynomials. From Approximation Theory to Algebra and Number Theory*, Wiley, New York, 1990.
43. M. Randić, On alternative form of the characteristic polynomial and the problem of graph recognition, *Theor. Chim. Acta* 62 (1983) 485–498.
44. H. Hosoya and M. Randić, Analysis of the topological dependency of the characteristic polynomial in its Tschebyschev expansion, *Theor. Chim. Acta* 63 (1983) 473–495.
45. P. J. Davis, *The Thread: A Mathematical Yarn*, Birkhausser Boston, Inc., Cambridge, MA, 1983.
46. E. Estrada, Randić index, irregularity and complex biomolecular networks, *Acta Chim. Slov.* 57 (2010) 597–603.
47. M. Randić, D. O. Oakland, and D. J. Klein, Symmetry properties of chemical graphs. IX. The valence tautomerism in the P_7^{3-} skeleton, *J. Comput. Chem.* 7 (1986) 35–54.
48. J. Brocas, The reaction graph of the Cope rearrangement in bulvallene, *J. Math. Chem.* 15 (1994) 389–395.
49. T. Živković, Bullvalene reaction graph, *Croat. Chem. Acta* 69 (1996) 215–222.
50. W. von E. Doering and W. R. Roth, A rapidly reversible degenerate Cope rearrangement: Bicyclo[5.1.0]octa-2,5-diene, *Tetrahedron* 19 (1963) 715–737.
51. L. Collatz and U. Sinogowitz, Spektren endlicher Grafen. *Abh. Math. Sem. Univ. Hamburg* 21 (1957) 63–77.
52. F. K. Bell, A note on the irregularity of graphs, *Linear Algebra Appl.* 161 (1992) 45–54.
53. M. O. Albertson, The irregularity of a graph, *Ars Combin.* 46 (1997) 219–225.
54. T. A. B. Snijders, The degree variance: An index of graph heterogeneity, *Social Networks* 3 (1981) 163–174.
55. M. Randić, M. Novič, M. Vračko, and D. Plavšić, On the centrality of vertices of molecular graphs, *J. Comput. Chem.* 34 (2013) 1409–1419.
56. D. Cvetković and P. Rowlinson, On connected graphs with maximal index, *Publ. Inst. Math (Beograd.)* 44 (1988) 29–34.
57. P. Hansen and H. Mélot, Graphs and discovery, in: S. Fajtlovicz (Ed.), *DIMACS Series in Discrete Mathematics and Theoretical Computer Science*, Vol. 69, American Mathematical Society, Providence, RI, 2005, pp. 253–264.
58. I. Gutman, P. Hansen, and H. Mélot, Variable neighborhood search for extremal graphs. 10. Comparison of irregularity indices for chemical trees. *J. Chem. Inf. Model.* 45 (2005) 222–230.
59. B. Bollobás and P. Erdös, Graphs of extremal weights, *Ars Combin.* 50 (1998) 225–233.
60. X. Li and Y. Shi, A survey on the Randic index, *MATCH—Commun. Math. Comput. Chem.* 59 (2008) 127–156.
61. X. Li and I. Gutman, *Mathematical Aspects of Randić-type Molecular Structure Descriptors*, Mathematical Chemistry Monographs No. 1, Kragujevac, Serbia, 2006.
62. X. Li and I. Gutman, *Mathematical Aspects of Randić-type Molecular Structure Descriptors*, Mathematical Chemistry Monographs No. 1, Kragujevac, Serbia, 2006.
63. G. Farmelo (Ed.), *It Must Be Beautiful: Great Equations of Modern Science*, Granta, UK, 2003.
64. en.wikipedia.org/wiki/Molecular_descriptor, August 2014.
65. M. Randić, In search of structural invariants, *J. Math. Chem.* 9 (1992) 97–146.
66. R. C. Read, A new system for the designation of chemical compounds, I: Theoretical preliminaries and the coding of acyclic compounds, *J. Chem. Inf. Comput. Sci.* 23 (1983) 135–149.
67. M. Randić and N. Trinajstić, Viewpoint 4—Comparative structure-property studies: The connectivity basis, *THEOCHEM.—J. Mol. Struct.* 284 (1993) 209–221.

68. R. Feynman, Space-time approach to non-relativistic quantum mechanics, *Rev. Modern Phys.* 20 (1948) 367–387.

69. J. R. Platt, Prediction of isomeric differences in paraffin properties, *J. Phys. Chem.* 56 (1952) 328–336.

70. I. Gutman, A formula for the Wiener number of trees and its extension to graphs containing cycles, *Graph Theory Notes N. Y.* 27 (1994), pp. 9–15.

71. T. Pisanski and M. Randić, Use of the Szeged index and the revised Szeged index for measuring network bipartivity, *Discrete Appl. Math.* 158 (2010) 1936–1944.

72. M. Randić, On generalization of Wiener index for cyclic structures, *Acta Chim. Slov.* 49 (2002) 483–496.

73. L. B. Kier, W. J. Murray, M. Randić, and L. H. Hall, Molecular connectivity. 5. Connectivity series concept applied to density, *J. Pharm. Sci.* 65 (1976) 1226–1230.

74. M. Randić, Characterization of atoms, molecules, and classes of molecules based on path enumerations, *MATCH—Commun. Math. Comput. Chem.* 5 (1979) 3–60.

75. M. Randić, On molecular identification numbers, *J. Chem. Inf. Comput. Sci.* 24 (1984) 164–175.

76. M. Randić, Design of molecules with desired properties. A molecular similarity approach to property optimization, in: *Concepts and Applications of Molecular Similarity*, M. A. Johnson and G. Maggiora (Eds.), John Wiley & Sons, New York, 1990, pp. 77–145.

77. K. Szymanski, W. R. Müller, J. V. Knop, and N. Trinajstić, On the identification numbers for chemical structures, *Int. J. Quantum Chem.* 30 (1986) 173–183.

78. J. V. Knop, W. R. Müller, K. Szymanski, and N. Trinajstić, *Computer Generation of Certain Classes of Molecules*, SKTH/Kemija u industriji, Zagreb, 1985.

79. M. Randić, Molecular ID numbers: By design, *J. Chem. Inf. Comput. Sci.* 26 (1986) 134–136.

80. K. Szymanski, W. Müller, J. Knop, and N. Trinajstić, On Randić's molecular identification numbers, *Int. J. Quantum Chem.* 25 (1985) 413–415.

81. M. Randić, Ring ID numbers, *J. Chem. Inf. Comput. Sci.* 28 (1988) 142–147.

82. O. Ivanciuc, A. Balaban, Design of topological indices. Part 3. New identification numbers for chemical structures: MINID and MINSID, *Croat. Chem. Acta* 69 (1996) 9–16.

83. M. Dehmer and M. Grabner, The discrimination power of molecular Identification Numbers revisited, *MATCH Commun. Math. Comput. Chem.* 69 (2013) 785–794.

84. M. Dehmer and M. Grabner, K. Varmuza, Information indices with high discriminative power for graphs, *PLoS One* 7 (2012) e31214.

85. M. Dehmer, L. Sivakumar, and K. Varmuza: Uniquely discriminating molecular structures using novel eigenvalue-based descriptors, *MATCH Commun. Math. Comput. Chem.* 67 (2012) 147–172.

86. M. Dehmer, M. Emmert-Streib, and M. Grabner, A computational approach to construct a multivariate complete graph invariant, *Inf. Sci.* 260 (2014) 200–208.

87. M. Randić, D. Bhattacharya, and D. J. Klein, On molecular descriptors for screening combinatorial libraries, *J. Comput. Chem.* (submitted).

88. M. Randić, Reported at the 10th Annual Meeting of the Int. Acad. Math. Chem. held in Split, Croatia June 7–10, 2014.

89. O. Ivanciuc, T. S. Balaban, and A. T. Balaban, Chemical graphs with degenerate topological indices based on information on distances, *J. Math. Chem.* 14 (1993) 21–33.

90. A. T. Balaban, Chemical graphs. 48. Topological index *J* for heteroatom-containing molecules taking into account periodicities of element properties, *MATCH Commun. Math. Comput. Chem.* 21 (1986) 115–122.

91. E. J. Kupchik, General treatment of heteroatoms with the Randić molecular connectivity index, *Quant. Struct.-Act. Relat.* 8 (1989) 98–103.

92. I. S. Antipin, N. A. Arslanov, V. A. Palyulin, A. I. Konovalov, and N. S. Zefirov, Solvation topological index. Topological description of dispersion interactions (Russ.), *Dokl. Akad. Nauk, SSSR* 316 (1991) 925–927 (*Chem. Abstr.* 115, 91390).

93. J. Gálvez, R. Garcia, M. T. Salabert, and R. Soler, Charge indexes. New topological descriptors, *J. Chem. Inf. Comput. Sci.* 34 (1994) 520–525.

94. O. Ivanciuc, T. Ivanciuc, and A. T. Balaban, Design of topological indices. Part 10. Parameters based on molecular graph descriptors for heteroatom-containing molecules, *J. Chem. Inf. Comput. Sci.* 38 (1998) 395–401.

95. L. Pogliani, Modeling with molecular pseudoconnectivity descriptors. A useful extension of the intrinsic I-state concept, *J. Phys. Chem. A* 104 (2000) 9029–9045.

96. Č. Podlipnik, T. Šolmajer, and J. Koller, Lipophilic connectivity indices, *MATCH Commun. Math. Comput. Chem.* 49 (2003) 7–14.

97. L. B. Kier and L. H. Hall, *Molecular Connectivity in Chemistry and Drug Research*, Academic Press, New York, 1976, p. 102.

98. L. B. Kier and L. H. Hall, *Molecular Connectivity in Chemistry and Drug Research*, Academic Press, New York, 1976, p. 120.

99. L. B. Kier and L. H. Hall, *Molecular Connectivity in Chemistry and Drug Research*, Academic Press, New York, 1976, p. 131.

100. L. B. Kier and L. H. Hall, *Molecular Connectivity in Chemistry and Drug Research*, Academic Press, New York, 1976, p. 135.

101. L. B. Kier and L. H. Hall, *Molecular Connectivity in Chemistry and Drug Research*, Academic Press, New York, 1976, p. 136.

102. L. B. Kier and L. H. Hall, *Molecular Connectivity in Chemistry and Drug Research*, Academic Press, New York, 1976, p. 137.

103. L. B. Kier and L. H. Hall, *Molecular Connectivity in Chemistry and Drug Research*, Academic Press, New York, 1976, p. 151.

104. L. B. Kier and L. H. Hall, *Molecular Connectivity in Chemistry and Drug Research*, Academic Press, New York, 1976, p. 165.

105. M. Pompe, Variable connectivity index as a tool for solving "anticonnectivity" problem, *Chem. Phys. Lett.* 404 (2005) 296–299.

106. M. Randić and M. Pompe, "Anticonnectivity": A challenge for structure-property-activity studies, *J. Chem. Inf. Model.* 46 (2006) 2–8.

107. L. Pogliani and J. V. de Julian-Ortiz, *QSPR with Descriptors Based on Averages of Vertex Invariants. An ANN Study*, Royal Society of Chemistry Advances, 4 (2014) 44733–44740.

108. M. Randić and S. C. Basak, Construction of high-quality structure-property-activity regressions: The boiling points of sulfides, *J. Chem. Inf. Comput. Sci.* 40 (2000) 899–905.

109. M. Randić, Novel graph theoretical approach to heteroatom in quantitative structure-activity relationship, *Chemom. Intell. Lab. Syst.* 10 (1991) 213–227.

110. P. B. Timmermans and P. A. van Zweiten, Quantitative structure-activity relationship in centrally acting imidazolidines structurally related to clonidine, *J. Med. Chem.* 20 (1977) 1636–1644.

111. M. Randić, J. Markelj, and M. Pompe, Traditional and non traditional QSAR correlations revisited, *SAR QSAR Environ. Chem.* (to be submitted).

112. M. Randić, Comparative regression analysis. Regressions based on a single descriptor *Croat. Chem. Acta* 66 (1993) 289–312.

113. D. E. Needham, I.-C. Wei, and P. G. Seybold, Molecular modeling of the physical properties of alkanes, *J. Am. Chem. Soc.* 110 (1988) 4186–4194.

114. N. S. Zefirov and V. A. Palyulin, QSAR for boiling points of "small" sulfides. Are the "high quality structure–property–activity regressions" the real high quality QSAR models? *J. Chem. Inf. Comput. Sci.* 41 (2001) 1022–1027.

115. E. B. Wilson, *Introduction to Scientific Research*, McGraw-Hill, New York, 1952.
116. M. Pompe and M. Randić (work in progress).
117. L. H. Hall and L. B. Kier, Issues in representation of molecular structure. The development of molecular connectivity, *J. Mol. Graph. Model.* 20 (2001) 4–18.
118. L. H. Hall and L. B. Kier, Molecular connectivity chi indices for data analysis and structure-property modeling, in: J. Devillers and A. T. Balaban (Eds.), *Topological Indices and Related Descriptors in QSAR and QSPR*, Amsterdam, Netherlands, Gordon and Breach, 1999.
119. L. H. Hall, L. B. Kier, and L. M. Hall, "Topological QSAR applications: Structure information representation in drug discovery," applications to drug discovery—Ligand-based lead optimization, in: *Comprehensive Medicinal Chemistry II*, Vol. 4, D. J. Triggle and J. B. Taylor (Eds.), Elsevier, Oxford, UK, 2007, pp. 537–574.
120. L. H. Hall, Development of structure information from molecular topology for modeling biological properties: A tribute to the creativity of Lemont Burwell Kier on his 80th birthday, *Curr. Comp.-Aided Drug Design* 8 (2012) 93–106.

7 Partial Ordering

7.1 SEARCHING FOR REGULARITIES IN STRUCTURE–PROPERTY DATA

The basic hypothesis of chemistry, which extends to physics, biology, medicine, and other natural and even social sciences, is that *similar* structures (or *similar* systems) have *similar* properties. Thus, searching for regularities in structure–property data becomes one of the central topics of the natural sciences and quantification of the structure–property data one of the major challenges. Consider, for example, a set of experimental data, such as the boiling points of alkanes [1,2], the chromatographic retention data [3,4], the C^{13} chemical shifts in alkanes [5], the diamagnetic susceptibilities of alkanes and alkenes [6], the determination of pK_a values of selected organic acids [7], the reaction rate constants of volatile unsaturated hydrocarbons with OH radicals [8], the resonance energies and aromaticities of conjugated hydrocarbons [9–33], the ring currents in non-benzenoid hydrocarbons [34–38], the activity of nonspecific anesthesia [39], the mutagenicity of nitroarenes [40], the taste of sulfamates [41], the similarity in benzomorphans [42], the quantitative structure–activity relationship of flavonoid p56lck protein tyrosine kinase inhibitors by neural network approach [43], or the characterization of trans-membrane proteins [44]. There is no end to the list; we just mention a selection of topics with which the present authors are more familiar.

When considering a set of experimental data, for example, the boiling points of alkanes, the first question to ask is, "Is there any regularity between the structures and their boiling points?" One way to find out is to order molecules relative to the magnitude of the property found, hoping that molecules having similar physicochemical property will also have similar mathematical characterizations. A straight ordering of structures by a selected physicochemical property may nevertheless obscure observing structural factors important for the property considered. This is because unknown regularity may involve several different structural features, such as the degree of branching, the sites of branching, the lengths of branches, and, of course, beyond molecular connectivity, the spatial distributions of bonds, etc. As we will demonstrate here, the partial ordering can be a suitable tool in the search for regularities in the experimental data of structurally related compounds. This is, because the partial ordering relates to sequences, not individual numbers, and different sequences may capture different structural features of molecules.

7.2 ON CHARACTERIZATION OF MOLECULAR STRUCTURES

Most physical and chemical properties of molecules are expressed as *numbers*, but structures are not numbers! Sequences are ordered *collections* of numbers, not numbers, so characterization of structures and characterization of sequences may have something in common! In order to find out the degree to which similar structures or similar molecules will have similar properties (here the boiling points), one needs to develop a suitable *tool*. In other words, one needs a mathematical *language* to characterize the degree of similarity between the objects considered. The tool or the language to be used or to be developed will depend on the problem considered. Only if the language is mathematical will one be able to represent structures numerically and thus possibly arrive at a quantitative characterization of the degree of similarity and quantitative characterization of the structure–property relationship. It may be of interest here to recall a statement of Galileo Galilei (1564–1642) [45], Italian physicist, mathematician, astronomer, and philosopher, who is considered a founder of modern science:

> Philosophy is written in this grand book the universe, which stands continually open to our gaze. But the book cannot be understood unless one first learns to comprehend the language and to read the alphabet in which it is composed. It is written in the language of mathematics, and its characters are triangles, circles and other geometrical figures, without which it is humanly impossible to understand a single word of it; without these, one wonders about in a dark labyrinth.

The first step, however, is to observe a regularity if there is one! After that comes the second step, a quantitative characterization of the observation. By developing an adequate structure-based numerical language for characterizing similarity between molecules and by observing some regularity, one obtains some *clue* as to which structural elements may be of importance for a quantitative representation of the structure–property relationship. Hence, the search for structure–property–activity studies need not be "fishing in the dark" (the black box approach) as are many current MRA studies, which are based on computer software packages, which, as output, give, using optimal molecular descriptors associated with the MRA, statistical parameters and numerical characterization, if available.

In Figure 7.1 are depicted carbon skeletons of 18 octane isomers, and below them are shown their boiling points. Looking at Figure 7.1 a little bit longer, one may see that several "long" molecules have high bp, but there are also a few "long" molecules (e.g., 2,2-dimethylhexane, 2,4-dimethylhexane, and 2,5-dimethylhexane) that have small bp. Similarly, one may see that several "short" molecules have low bp, but there are also "short" molecules (e.g., 3,3-diethylpentane) that have large bp. Clearly the attributes "long" or "short" are not adequate descriptors for quantitative characterization of molecules.

More than 65 years ago, J. R. Platt [46] suggested the count of *paths of different lengths* in molecules as a *tool* for numerical characterization of molecules. The suggestion by Platt is that the *sequences* of path numbers are the language of interest in comparative study of chemical structures. For some, such a proposition may appear to be a chemically futile mathematical exercise that may entertain some. In the central column of Table 7.1, we show the result of such a mathematical "exercise," which offers a numerical representation of octane isomers. For example, 2,3-dimethyhexane

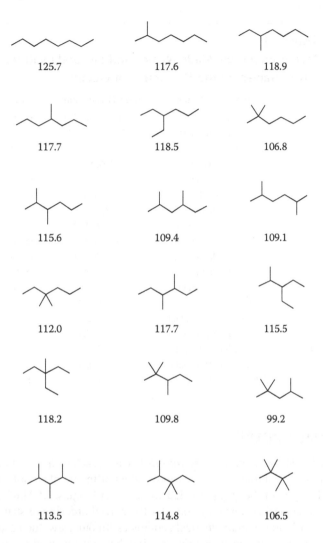

FIGURE 7.1   Skeletal forms of octane isomers and their boiling points in °C.

becomes 7, 8, 7, 4, 2 because it has seven bonds (or paths of length one), $p_1 = 7$; eight
pairs of adjacent bonds (or paths of length two), $p_2 = 8$; seven segments of three consecutive CC bonds (so $p_3 = 7$); four segments of four consecutive CC bonds ($p_4 = 4$);
and two segments of five consecutive CC bonds ($p_5 = 2$). Clearly, all octane isomers
have seven bonds, thus paths of length one are of no interest for discussing *variations*
of octane boiling points.

     As one can see from Table 7.1, octane isomers differ in the number of paths p_2,
p_3, p_4, and so on. But Table 7.1 does not answer the central question that we have
asked: Is there any regularity in the data of Figure 7.1 or do similar isomers have
similar bp?

TABLE 7.1

Men (Committee) Made Names and the God (Nature) Given Names for the 18 Isomers of Octane

	Men Made Names	God Given Names	(x, z)
1	n-octane	7, 6, 5, 4, 3, 2, 1	(6, 5)
2	2-methylheptane	7, 7, 5, 4, 3, 2	(7, 5)
3	3-methylheptane	7, 7, 6, 4, 3, 1	(7, 6)
4	4-methylheptane	7, 7, 6, 5, 2, 1	(7, 6)
5	3-ethylhexane	7, 7, 7, 5, 2	(7, 7)
6	2,2-dimethylhexane	7, 9, 5, 4, 3	(9, 5)
7	2,3-dimethylhexane	7, 8, 7, 4, 2	(8, 7)
8	2,4-dimethylhexane	7, 8, 6, 5, 2	(8, 6)
9	2,5-dimethylhexane	7, 8, 5, 4, 4	(8, 5)
10	3,3-dimethylhexane	7, 9, 7, 4, 1	(9, 7)
11	3,4-dimethylhexane	7, 8, 8, 4, 1	(8, 8)
12	3-ethyl-2methylpentane	7, 8, 8, 5	(8, 8)
13	3-ethyl-3methylpentane	7, 9, 9, 3	(9, 9)
14	2,2,3-trimethylpentane	7, 10, 8, 3	(10, 8)
15	2,2,4-trimethylpentane	7, 10, 5, 6	(10, 5)
16	2,3,3-trimethylpentane	7, 10, 9, 2	(10, 9)
17	2,3,4-trimethylpentane	7, 9, 9, 3	(9, 9)
18	2,2,3,3-tetramethylbutane	7, 12, 9	(12, 9)

7.3 PARTIAL ORDERING

As this book is about *quantitative* and *qualitative* approaches in structure–property–activity studies, we continue the discussion of the boiling points of alkanes by seeking, for regularity, in the boiling point data of octanes of Figure 7.1. Having structures represented by sequences, one may consider the partial ordering of structures based on the partial ordering of mathematical sequences. In our view, the partial ordering as a tool in structure–property–activity studies has not received its deserved visibility, and it should get a wider attention. The concept of partial order, an important concept of discrete mathematics, apparently has been mostly overlooked in the QSAR community except for scientists in chemometrics, chemical graph theory, and mathematical chemistry [47–67].

Partial ordering is an important tool of discrete mathematics, which has found visible application in chemistry [68–75]. The beginning started in 1903 with Scottish mathematician Muirhead [76], who considered ordering of sequences of the same length, such as A $(a_1, a_2, a_3, \ldots a_n)$ and B $(b_1, b_2, b_3, \ldots b_n)$. To do this, he first constructed the corresponding sequences of partial sums, the elements of which involve an increasing number of sequence terms:

$$a_1, a_1 + a_2, a_1 + a_2 + a_3, \ldots, a_1 + a_2 + a_3 + \ldots + a_n \text{ and } b_1, b_1 + b_2, b_1 + b_2 + b_3, \ldots b_1 + b_2 + b_3 + \ldots + b_n$$

Then he constructed the set of inequalities for the corresponding partial sums:

$$a_1 \geq b_1$$
$$a_1 + a_2 \geq b_1 + b_2$$
$$a_1 + a_2 + a_3 \geq b_1 + b_2 + b_3$$
$$\ldots \geq \ldots$$
$$a_1 + a_2 + a_3 + \ldots + a_n = b_1 + b_2 + b_3 + \ldots + b_n$$

If, for two sequences, the above set of inequalities are satisfied, which means that all the inequalities are satisfied, according to Muirhead, then one may conclude that the sequence A dominates the sequence B, shortly $A \geq B$. If at least one of the inequalities is not satisfied, then the sequences A and B are viewed as not comparable. Observe that both sequences must have the same number of elements and that the last expression in the list of inequalities is the equality (equation), not inequality.

Let us outline the construction of a partial ordering on the set of octane isomers of Figure 7.1. We start by representing each isomer of octane by a pair of path counts (p_2, p_3) [77–79], shown in the last columns of Table 7.1. Observe that there are two pairs of isomers having the same (p_2, p_3). Let us immediately mention that having a duplicate count of paths for two or more structures is not necessarily a disadvantage if one is interested in the structure–property relationship because two or more molecules may have identical or very similar magnitudes for selected molecular properties. By representing isomers by pairs (p_2, p_3) one can make a diagram, viewing p_2 and p_3 as the (x, y) coordinates, in which isomers will be depicted as points in the structure space. For the set of 18 octane isomers, the result is shown in Figure 7.2,

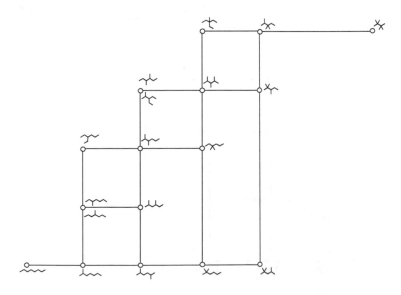

FIGURE 7.2   Molecular carbon skeletons of octane isomers superimposed on the (p_2, p_3) grid.

which may be viewed as a graphical representation of a partially ordered set of (x, y) coordinates, which represents one particular partially ordered diagram of C_8H_{18} isomers.

In Figure 7.2, which illustrates the partial ordering of 18 octane isomers graphically, individual structures are located at the sites of their (p_2, p_3) coordinates. The structures on the left are dominated by those at right, and those below are dominated by those above. Here "dominated" means that the structure has a larger count of both paths, p_2 and p_3. Well, that sounds nice, but what has this to do with chemistry? To see the connection, just replace each structure in Figure 7.2 by its boiling points as shown in Figure 7.3. As one does this, immediately one can see regularity in the boiling points of octanes. The magnitudes of the boiling points are not scattered at random over the partial ordering diagram but show trends and gradual variations between the neighboring points in the diagram. Because the ordering diagram is based on structural elements (the paths p_2 and p_3), Figure 7.3 demonstrates that the boiling points of octanes manifest structure–property regularity. One can see from Figure 7.3 that the relative magnitudes of the boiling points decrease with an increase in the p_2 path count and increase with the count of the p_3 paths.

Hence, we have found a quite simple, although *qualitative*, regularity between octane structures and octane boiling points. This finding, the first of its kind for physicochemical properties, is very significant—not because it is the first, but because it is not the only. Subsequently, similar regularities have been found to hold not only for a number of other physicochemical properties of octanes, but as well for other alkanes. Observe, that despite being a qualitative relationship, it allows one, by knowing the boiling points of the neighboring isomers in the grid of the diagram in Figure 7.3, to approximately estimate the missing property for the isomer

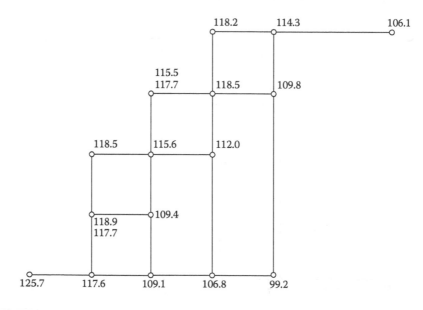

FIGURE 7.3 The boiling points of octane isomers superimposed on the (p_2, p_3) grid.

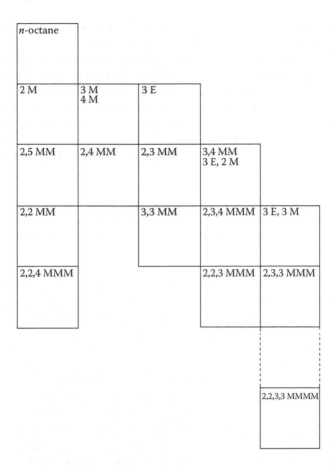

FIGURE 7.4   The periodic table for octane isomers.

surrounded by neighbors. This is reminiscent of the properties of the periodic table of elements. There is thus here some conceptual similarity of the (p_2, p_3) diagram with the periodic table of elements, which justifies us to redraw Figure 7.2 as a table illustrated in Figure 7.4, to which one can refer as the "periodic table of octane isomers" [80,81].

7.4 ^{13}C CHEMICAL SHIFTS SUMS IN ALKANES

Since about mid-1950, NMR chemical shifts have been found to be highly powerful spectroscopic tools in assisting determination of chemical structures. It is particularly important relating to determining hydrogen locations inside a molecule, which are generally not detectable by X-ray studies. Atomic chemical shifts arise when molecules are placed in outside magnetic fields. The external magnetic field interacts with the magnetic moment of electrons, which causes slight shielding of the magnetic field to which are exposed the magnetic moment of atomic nuclei. As a result, spectral frequencies of different nuclei in NMR are shifted by different amounts, which are measured in parts per million (ppm; in view that the shielding factor is about 10^{-5} for protons and less than 10^{-3} for other nuclei).

It is needless to say how useful have been NMR spectra in chemistry. A simple proof of this is tables and tables of chemical shifts for various types of atomic nuclei, particularly for proton chemical shifts and ^{13}C chemical shifts. Clearly, chemical shifts are atomic properties, but if one would add all carbon chemical shifts in a hydrocarbon, one would obtain a numerical quantity that can be viewed as a molecular *entity*. Let us elaborate on this. Paraphrasing what one finds in Wikipedia and the *Merriam-Webster Dictionary*, one can write the following explanations for "entity":

> An entity is something that exists actually or hypothetically and need not be of material nature. It is something that exists by itself: something that is separate from other things, something having conceptual reality. Among others, an entity is often used to refer to ghosts and other spirits.

From the above, it follows that we may view the chemical shift sum, obtained when one adds all ^{13}C carbon chemical shifts in a hydrocarbon, an entity, as a molecular property. Entities have properties, including ghosts and spirits, even if they do not exist! It is therefore very surprising that almost no reference was to be found in chemical literature to ^{13}C carbon chemical shift sums although it has been almost 35 years since the publication introducing the ^{13}C carbon chemical shift sum as a novel molecular property appeared. Ghosts may be entities, but ^{13}C carbon chemical shifts are real and are not ghosts; they could be useful in estimating missing chemical shifts as can be verified by averaging the chemical shifts of isomers in adjacent sites of the periodic table of isomers in the case of octanes. So where are the periodic tables of alkene isomers, the periodic tables of isomers of alcohols, the periodic tables of isomers of amines, and so on? Currently they exist as ghosts!

Figure 7.5 shows the periodic table of octane isomers with the value of ^{13}C chemical shift sum for each isomer. As one can see, chemical shift sums increase with increase of p_2 and decrease with increase of p_3. Now this result is impressive, not because we found regularity in the ^{13}C chemical shift sums of octanes—there have been several additional similar results for collections of physicochemical properties of octanes in addition to the boiling points shown in Figure 7.3—but the ^{13}C chemical shift sums of octanes are not one of the *experimental* molecular physicochemical properties. It is a *virtual* molecular property, *constructed* by combining sets of *experimental* atomic chemical shifts.

To the best of our knowledge, this may well be the first virtual physicochemical molecular property so far reported. The opposite of this would be atomic or bond virtual properties obtained by partitioning of (experimental) molecular properties, and these have been known for quite a while. One illustration is the bond dipole moments, which are not experimentally measured, except for diatomic molecules, but are constructed by decomposition of experimentally measured molecular dipole moments along individual bonds in a molecule. A similar illustration is the calculated π electron ring currents for cyclic conjugated hydrocarbons, which are obtained by decomposition of the electron density currents, that could be assigned

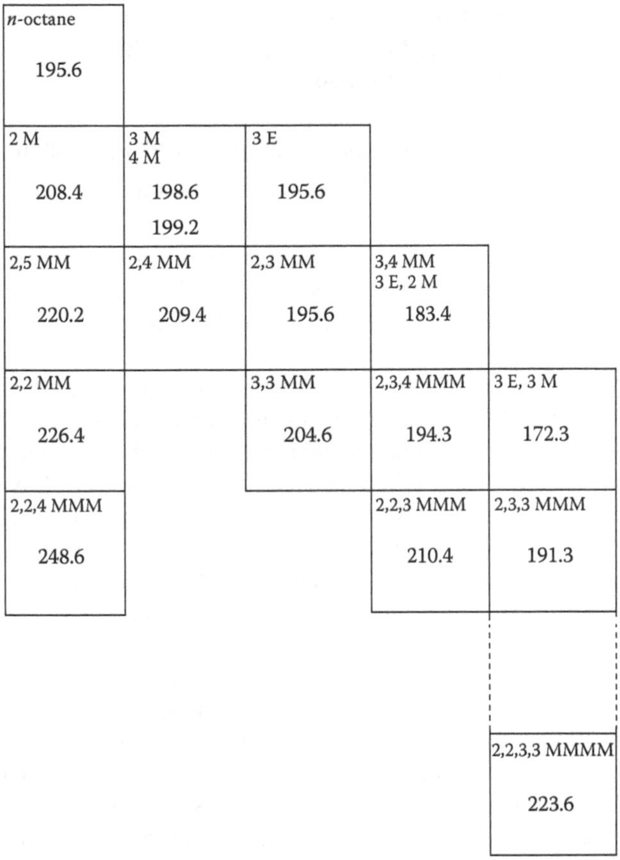

n-octane 195.6				
2 M 208.4	3 M 4 M 198.6 199.2	3 E 195.6		
2,5 MM 220.2	2,4 MM 209.4	2,3 MM 195.6	3,4 MM 3 E, 2 M 183.4	
2,2 MM 226.4		3,3 MM 204.6	2,3,4 MMM 194.3	3 E, 3 M 172.3
2,2,4 MMM 248.6			2,2,3 MMM 210.4	2,3,3 MMM 191.3
				2,2,3,3 MMMM 223.6

FIGURE 7.5 The ^{13}C molecular chemical shifts of octane isomers.

to individual CC bonds, into ring contributions. We will later discuss ring currents at greater length. Finally, the same could be said about molecular orbitals obtained by solving the Schrödinger equation, which are nonobservables, but when combined to generate the square of the total molecular wave function, become observable (as molecular electron density).

The seminal paper on ^{13}C chemical shift sums [82], thus, has opened a novel avenue in the studies of molecules, but apparently the new road has hardly been traveled. There have been few publications worth noticing, beyond the extension of the paper on ^{13}C chemical shifts of octanes to ^{13}C chemical shifts of nonane isomers [83]. Thus Randić and Trinajstić [84] found good correlation of ^{13}C chemical shift sums in octanes with the difference of path numbers ($p_2 - p_3$), and Miyashita and coworkers [85] found a graph theoretical descriptor, a combination of path numbers ($p_0 + p_1 + p_2 - p_3$) to be an excellent descriptor for correlation of ^{13}C chemical shift sums in alkanes. Rücker and Rücker [86] followed by using as descriptors the count of walks of length two (W_2) and of walks of length three (W_3) in a molecule, which can be easily obtained from the second and the third power of the adjacency matrix, respectively, to correlate ^{13}C chemical shift sums in alkanes. Apparently the two descriptors W_2 and W_3 produced better correlations for highly branched alkanes. Finally, Ivanciuc [87] came across the equation:

$$^{13}C \text{ Chemical Shift Sum} = 27.85 \, p_1 + 12.29 \, p_2 - 11.92 \, p_3 + 2.08 \, p_4 - 22.55$$

$$n = 66 \qquad r = 0.996 \qquad s = 4.62$$

where n is the number of alkane molecules considered, r is the coefficient of regression, and s is the standard deviation. Observe how close are the coefficients of p_2 and p_3, which explains why the graph theoretical descriptor $(p_2 - p_3)$ of Randić and Trinajstić was found to be so good.

Using artificial neural networks (ANN) Ivanciuc et al. compared those results with multi-linear regression (MLR or MRA) results and found that, in this case, the neural network approach "offers only slightly better results, both for calibration and cross validation." Apparently the same conclusion concerning prediction of ^{13}C NMR chemical shifts of alkanes was reached by Svozil et al. [88] indicating, according to Ivanciuc, *"that the relationship between the path counts and CSS has a small non-linear character and the ANN model could not provide much better result than the MLR model."*

The chemical shift sum publication, which, after almost 35 years, has only 20 citations, is good illustration of a "Sleeping Beauty" that eventually will be awaked by a prince kissing her (if he comes), a work worthy of greater attention but apparently overlooked by most researchers in the NMR area. It also illustrates that there are Sleeping Beauties that need to be kissed more than once to be awakened! Skeptics may raise the question of whether this Beauty is worth kissing, whether it is worth awakening? After all, the ^{13}C NMR chemical shift sum is an "artificial" construction, perhaps just a curiosity. So why should it be of interest to other scientists?

Instead us trying to answer the constructed "criticism" let us quote a fragment from the paper by Ivanciuc et al., which shows how chemical shift sums (CSS) can help to detect systematic errors in real experimental data:

> The CSS can be used to detect systematic deviations in ^{13}C NMR spectra determined in different laboratories. In the case of systematic errors, the deviations for a single chemical shift may be too small to be noticeable; the chemical shift sum accumulates such systematic deviations, making more efficient and easy the detection of significant errors in the ^{13}C NMR spectra.

Ivanciuc, Rabine, and Cabrol-Bass [82] continue to illustrate that *"there is a systematic difference between the chemical shifts reported by Grant and Paul [89] and those reported by Lindeman and Adams [90], which constantly have a greater values. This difference is hardly detected when comparing the individual ^{13}C chemical shift values."*

Should this be enough to get the attention of the princes at large? Clearly, it is not that only the seminal paper on chemical shifts sums did not receive due attention, but the same holds for the above cited follow-up papers by Randić and Trinajstić [84], by Miyashita and coworkers [85], by Rücker and Rücker [86], and the paper by Ivanciuc and coworkers [87]. Another four Sleeping Beauties!

Finally, when speaking of "due attention," we do not mean solely to these five publications but *the topic* that they are trying to bring to the attention of the NMR community, the topic of partial ordering and open problems of constructing diagrams for other classes of organic compounds beyond alkanes, which could, in addition to use in detection of possible systematic errors in reported NMR data, also be used directly in QSAR studies as has been already illustrated by Sakamoto and Watanabe in 1986 [91] and Aoki et al. 10 years later [92]. They have used the average ^{13}C NMR chemical shifts for aromatic carbon for modeling carcinogenicity of polycyclic and chlorinated conjugated hydrocarbons. The average ^{13}C NMR chemical shifts are trivially related to the ^{13}C NMR chemical shifts sums.

Let us end with one more quotation from the paper by Ivanciuc et al., this time how they have summarized the work of Aoki et al. [92]:

This model suggests a more fundamental relationship between the molecular structure (expressed as path counts) and carcinogenicity with the main advantage that it would be possible to predict the carcinogenicity of a compound without the need to synthesize it and to determine its ^{13}C NMR spectrum.

Before leaving the topic of partial ordering based on path counts, let us mention that the seminal publication by Platt on the use of paths as molecular descriptors was the Sleeping Beauty that was to die sleeping were it not for arrival of chemical graph theory in early 1970. Interestingly, there was a single paper on paths in smaller alkanes published in 1961 by K. Altenburg in East Germany in which path numbers were used for calculating the radius of branched molecules [93]. Altenburg does not cite the paper by Platt, so apparently he did not see it and discovered path numbers independently. The paper by Altenburg has historical value even if he would have known of the paper by Platt, which he did not, in being the first paper demonstrating use of path numbers for characterization of molecular properties (in alkanes). The paper by Altenburg would also remain a Sleeping Beauty expecting to die were it not for chemical graph theory, which stimulated interest in path numbers even if it took some time.

A review of diverse uses of partial order in chemistry can be found in a special issue of *MATCH Communication in Mathematical and in Computer Chemistry* [52], in the books of P. R. Duchowicz and E. A. Castro, *The Order Theory in QSPR—QSAR Studies* [94], and Brüggemann and Larsen, *Partial Order in Environmental Sciences and Chemistry* [95].

7.5 STUDY OF PROTEOME MAPS USING PARTIAL ORDERING

Proteome maps (or proteomics maps) are arrived at by an experimental approach, which simultaneously combines electrophoresis and chromatography, and separates thousands of different proteins from cellular material by charge and by mass. The results are obtained as collection of spots on 2-D films, from which one can extract the (x, y) coordinates of the protein spots and, by measuring the density of the spot, their abundance. Such results one can use to construct a "bubble" diagram, 2-D map in which the size of the "bubbles" represents the abundance of proteins, and the location (the x, y coordinates) indicate the corresponding change and mass of individual proteins. Typically, of the few thousands of proteins in a cell, around 200 proteins can be currently detected in sufficient abundance for comparative studies of proteome maps. Of considerable interest is to investigate how abundances of proteins change under various experimental conditions, but without some numerical characterization of the proteome maps, it is difficult to initiate a *quantitative* study of such maps. Visual comparisons that have been used, and continue to be used, have offered some qualitative insights, but this is not enough if one is interested in studying the degree of similarity of different proteome maps quantitatively.

With the year 2000, the situation has been changed, when for the first time, it was described how one can make a quantitative comparison of different proteome maps. One starts with associating a matrix to a proteome map, the elements of which depend on the charge, the mass and the abundance of proteins. The basic idea behind this breakthrough, which upgraded the *qualitative* studies of proteome maps to a *quantitative* level, was to associate with a map a suitable geometric object, such as a graph or other geometrical object, the vertices of which belong to selected protein spots of a map. Simply stated, the resulting matrices are based on adjacencies of vertices of geometrical objects *superimposed* over a proteome maps, and distances between vertices are used for construction of matrix elements [96–102].

A partial order diagram is one such geometrical object that has been suggested for quantitative characterization of proteome maps [68,96], which we will outline here. For illustration, we have selected 20 of the most abundant protein spots of a map, each being characterized by charge, mass, and abundance. First, one orders the 20 proteins with respect to their charge and with respect to their mass and then seeks the domination relationships, which will show proteins that have both the charge and the mass greater than other selected protein spots in the map. The 20 most abundant proteins selected for this illustration have been taken from experimental data of Anderson et al. [103]. In Figure 7.6, we have illustrated schematically the map for the selected 20 most abundant proteins of Anderson et al. In the map, the x-coordinates represent the magnitude of the charge, and the y-coordinates represent the protein mass. Here we do not (yet) need information on the abundances. Labeling of vertices is based on the relative abundance of vertices, by assigning the label 1 to the most abundant protein spot.

A close look at Table 7.2 shows that spot 1 and spot 2 cannot be ordered or, in the terminology of partial ordering, are not comparable because although spot 1 has bigger mass (coordinate y), spot 2 has greater charge (coordinate x). Neither are spots 2 and 3 comparable because $x_2 > x_3$ while $y_2 < y_3$. However, if one compares the x, y coordinates of spot 1 and spot 3, one find that $x_1 > x_3$ and $y_1 > y_3$; thus, spot 1 dominates spot 3.

Partial ordering consists of finding all relationships $x_i \geq x_j$ and $y_i \geq y_j$ ($i, j = 1$–20). Observe that the condition $x_i > x_j$ and $y_i > y_j$ geometrically corresponds, if the points (x_i, x_j) and (y_i, y_j) are connected by a line, to lines with a positive slope. To be more precise, one should allow for the condition $x_i \geq x_j$ and $y_i \geq y_j$, but the equality cannot be simultaneously satisfied because that corresponds to the same spot. Thus, equalities of the more general condition correspond to vertical and horizontal connecting lines, respectively. By knowing these, and also knowing that the relationships $x_i \geq x_j$

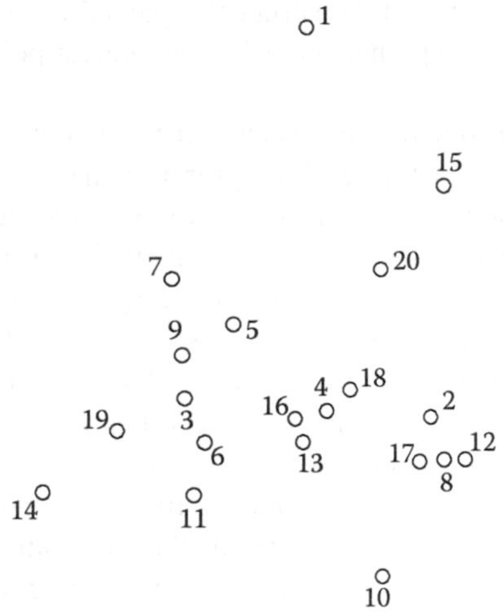

FIGURE 7.6 Schematic representation of the proteome map of Table 7.2 as a set of points in 2-D determined by their position as separated by electrophoresis (charge) and chromatography (mass).

TABLE 7.2
The Charge and the Mass of 20 Most Abundant Spots of the Proteome Map of Anderson et al. [103] Schematically Shown in Figure 7.6

	Charge	Mass	Abundance
1	2112	2279	1444
2	2804	904	1436
3	1418	960	1367
4	2182	929	1272
5	1686	1196	1186
6	1528	826	1149
7	1346	1353	1123
8	2869	778	1089
9	1406	1118	982
10	2540	409	936
11	1474	665	900
12	2975	773	867
13	2068	823	848
14	642	670	825
15	2861	1650	820
16	2033	903	800
17	2753	766	798
18	2334	982	728
19	1054	864	722
20	2520	1366	696

and $y_i \geq y_j$ are transitive, one can construct the partial order diagram directly over Figure 7.6 by connecting the points above by the nearest points below and at left as is illustrated in Figure 7.7.

Once the partial order diagram has been constructed, before one starts to search for useful graph invariants of the resulting graph of the partial order (to be used as proteome map descriptors), one should try to incorporate the information on spots abundance into the analysis. A straightforward way to do this is to replace the zeros on the main diagonal of the adjacency matrix by the magnitudes of the spots abundance.

One should recognize that this procedure is not as simple as it appears because the scales used to measure the experimental abundances are unrelated to non-dimensional matrix elements. Namely, the experimentally reported abundances have to be adjusted so that spot abundances neither take the dominant role, simply by being expressed as large numbers, nor play a marginal role if expressed as small numbers obtained by using different scales in their measurements. One has to find a balance, which is not a trivial problem. Clearly a selection of the units used to measure abundance should not take preference, and some kind of normalization is essential. Fortunately, this problem has been already addressed. As outlined by Randić et al. [101], a way to arrive at a matrix in which both the off-diagonal elements and the diagonal elements will have balanced roles is to normalize the

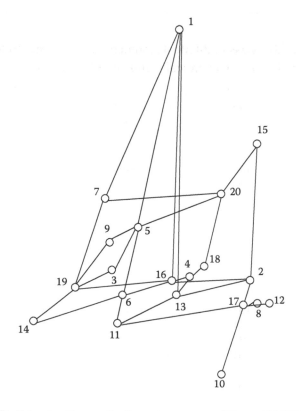

FIGURE 7.7   Partial order diagram for the proteome map of Figure 7.6 (Table 7.2).

diagonal elements so that the sum of the non-diagonal matrix elements are equal to the sum of the diagonal elements.

Once a partial order diagram has been constructed, one should select some suitable graph invariants to be used as map descriptors or consider construction of novel graph invariants to serve as map descriptors. Some molecular descriptors may be of interest, but others need not, in the view that bond-additivity, which is of considerable importance for calculating molecular properties is here of no particular relevance. In contrast, distance-related topological indices may be of interest, in the view that they tend to have better discriminatory properties in comparison with similar structures. At the same time, some distance-based topological indices have a disadvantage because the distance matrix is dense; that is, most or all of its off-diagonal matrix elements are not zero. This considerably increases computations involving such matrices in comparison with sparse matrices (the most of its off-diagonal matrix elements are zero).

In view that calculation with dense matrices are more time-consuming, one can combine advantages of having information on distances with advantages of sparse matrices by construction of the distance-adjacency matrix (D-A matrix). In this matrix, only the information on distances of adjacent vertices is used. In this case, the partial order diagram means considering only the distance between adjacent vertices, that is, the lengths of edges. Another type of sparse distance matrices is the so-called D_{MAX} matrix [104–106], which is obtained from the distance matrix by retaining only the largest entries in each row and each column of the matrix, and setting all the other entries equal to zero.

In addition to the diagram of partial ordering, one can construct other graphs embedded over proteome maps. One of the advantages of the diagram of partial ordering is that the resulting graph is not very sensitive to minor variations in the (x, y) coordinates of the protein spots. This is very important because experimental proteome maps coming from different laboratories for the same experiments may vary considerably. It is well known that currently very few laboratories have elaborate enough protocols to ensure high reproducibility of their results, so comparison of proteome maps from different sources could be very problematic. This points to a need for *robust* theoretical models, and the partial ordering diagrams have the advantage that they can be applied to proteome maps obtained by skipping protein spots, whether of medium or lower abundance, that are close in distances from one another. Such spots could produce partial order diagrams that are different for the same experiment made in different laboratories. A robust approach in quantitative characterization of proteome maps is an essential component, and the approach based on partial ordering satisfies such expectations, being insensitive enough to well-known experimental variability of proteome maps.

REFERENCES AND NOTES

1. M. Randić and N. Trinajstić, Isomeric variations in alkanes: Boiling points of nonane, *New J. Chem.* 18 (1994) 179–189.
2. M. Randić, High quality structure-property regressions. Boiling points of smaller alkanes, *New J. Chem.* 24 (2000) 165–171.

3. M. Randić, On structural origin of chromatographic retention data, *J. Chromatogr.* 161 (1978) 1–14.
4. M. Randić, S. C. Basak, M. Pompe, and M. Novič, Prediction of gas chromatographic retention indices using variable connectivity index, *Acta Chim. Slov.* 48 (2001) 169–180.
5. M. Randić, Survey of structural regularities in molecular properties. I. Carbon-13 chemical shifts in alkanes, *Int. J. Quantum Chem.* 23 (1983) 1707–1722.
6. M. Randić, Graph theoretical analysis of diamagnetic susceptibilities of alkanes and alkenes, *Chem. Phys. Lett.* 53 (1978) 602–605.
7. M. Pompe and M. Randić, Variable connectivity model for determination of pK(a) values of selected organic acids, *Acta Chim. Slov.* 54 (2007) 605–610.
8. M. Pompe, M. Veber, M. Randić, and A. T. Balaban, Using variable and fixed topological indices for the prediction of reaction rate constants of volatile unsaturated hydrocarbons with OH radicals, *Molecules* 9 (2004) 1160–1176.
9. M. Randić, A. T. Balaban, and D. Plavšić, Applying the conjugated circuits method to Clar structures of [n]phenylenes for determining resonance energies, *Phys. Chem. Chem. Phys.* 13 (2011) 20644–20648.
10. M. Randić and X. F. Guo, Giant benzenoid hydrocarbons. Superphenalene resonance energy, *Polycycl. Aromat. Comp.* 18 (2000) 49–69.
11. M. Randić and X. Guo, Resonance energy of giant benzenoid hydrocarbon $C_{78}H_{26}$, *J. Quantum Chem.* 74 (1999) 697–708.
12. M. Randić and X. Guo, Giant benzenoid hydrocarbons. Superphenalene resonance energy, *New J. Chem.* 23 (1999) 251–260.
13. M. Randić, Resonance in catacondensed benzenoid hydrocarbons, *Int. J. Quantum Chem.* 63 (1997) 585–600.
14. M. Randić, Giant benzenoid hydrocarbons, Supernaphthalene resonance energy, *Acta Chim. Slov.* 44 (1997) 361–374.
15. M. Randić, D. J. Klein, S. El-Basil, and P. Calkins, Resonance in large benzenoid hydrocarbons, *Croat. Chem. Acta* 69 (1996) 1639–1660.
16. M. Randić and N. Trinajstić, Critical test for resonance energies, *J. Am. Chem. Soc.* 109 (1987) 6923–6926.
17. M. Randić, V. Solomon, S. C. Grossman, D. J. Klein, and N. Trinajstić, Resonance energies of large conjugated hydrocarbons by a statistical method, *Int. J. Quantum Chem.* 32 (1987) 35–59.
18. M. Randić, A statistical approach to resonance energies of large molecules, *Chem. Phys. Lett.* 128 (1986) 193–197.
19. M. Randić, B. Ruščić, and N. Trinajstić, Herndon's structure-resonance theory—On the valence structure count for conjugated radicals, *Croat. Chem. Acta* 54 (1981) 295–308.
20. M. Randić, Resonance energies of very large benzenoid hydrocarbons, *Int. J. Quantum Chem.* 17 (1980) 549–586.
21. M. Randić, A graph theoretical approach to conjugation and resonance energies of hydrocarbons, *Tetrahedron* 33 (1977) 1905–1920.
22. M. Randić, Conjugated circuits and resonance energies of benzenoid hydrocarbons, *Chem. Phys. Lett.* 38 (1976) 68–70.
23. M. Randić, π-Electron currents in polycyclic conjugated hydrocarbons of decreasing aromatic character and novel structural definition of aromaticity, *Open Org. Chem. J.* 5 (Suppl. 1-M2) (2011) 11–26.
24. A. T. Balaban and M. Randić, Structural approach to aromaticity and local aromaticity in conjugated polycyclic systems, Chapter 8, in: *Carbon Bonding and Structures, Advances in Physics and Chemistry*, M. H. Putz (Ed.); *Carbon Materials: Chemistry and Physics 5*, (Series Editors: F. Cataldo and P. Milani), Springer Science+Business Media B. V., 2011, pp. 159–204.

25. A. T. Balaban and M. Randić, Partitioning of π-electrons in rings of polycyclic conjugated hydrocarbons: Part 6. Comparison with other methods for estimating the local aromaticity of rings in polycyclic benzenoids, *J. Math. Chem.* 37 (2005) 443–453.
26. M. Randić, Aromaticity of polycyclic conjugated hydrocarbons, *Chem. Rev.* 103 (2003) 3449–3605.
27. M. Randić, D. Plavšić, and N. Trinajstić, Aromaticity in polycyclic conjugated hydrocarbon dianions, *THEOCHEM.—J. Mol. Struct.* 54 (1989) 249–274.
28. M. Randić, B. M. Gimarc, S. Nikolić, and N. Trinajstić, On the aromatic stability of helicenic systems, *Gazz. Chim. Ital.* 119 (1989) 1–11.
29. M. Randić, L. L. Henderson, R. Stout, and N. Trinajstić, Conjugation and aromaticity of macrocyclic systems, *Int. J. Quantum Chem.: Quantum Chem. Symp.* 22 (1988) 127–141.
30. M. Randić, S. Nikolić, and N. Trinajstić, Aromaticity in heterocyclic molecules containing divalent sulfur, *Coll. Czechoslovak Chem. Commun.* 53 (1988) 2023–2054.
31. M. Randić and N. Trinajstić, Conjugation and aromaticity of corannulenes, *J. Am. Chem. Soc.* 106 (1984) 4428–4434.
32. M. Randić, Aromaticity and conjugation, *J. Am. Chem. Soc.* 99 (1977) 444–450.
33. M. Randić, Graph theoretical approach to local and overall aromaticity of benzenoid hydrocarbons, *Tetrahedron* 31 (1975) 1477–1481.
34. M. Randić, M. Novič, and D. Plavšić, π-Electron currents in fixed π-sextet aromatic benzenoids, *J. Math. Chem.* 50 (2012) 2755–2774.
35. M. Randić, D. Vukičević, M. Novič, and D. Plavšić, π-Electron currents in larger fully aromatic benzenoids, *Int. J. Quantum Chem.* 112 (2012) 2456–2462.
36. M. Randić, M. Novič, M. Vračko, D. Vukičević, and D. Plavšić, π-Electron currents in polycyclic conjugated hydrocarbons: Coronene and its isomers having five and seven member rings, *Int. J. Quantum Chem.* 112 (2012) 972–985.
37. M. Randić, D. Vukičević, A. T. Balaban, M. Vračko, and D. Plavšić, Conjugated circuits currents in hexabenzocoronene and its derivatives formed by joining proximal carbons, *J. Comput. Chem.* 33 (2012) 1111–1122.
38. M. Randić, Graph theoretical approach to π-electron currents in polycyclic conjugated hydrocarbons, *Chem. Phys. Lett.* 500 (2010) 123–127.
39. L. B. Kier, L. H. Hall, W. J. Murray, and M. Randić, Molecular connectivity. I. Relationship to nonspecific anesthesia, *J. Pharm. Sci.* 64 (1975) 1971–1974.
40. M. Randić, S. C. Grossman, B. Jerman-Blažič, D. H. Rouvray, and S. El-Basil, An approach to modeling the mutagenicity of nitroarenes, *Math. Comput. Model.* 11 (1988) 837–842.
41. T. Okuyama, Y. Miyashita, S. Kanaya, H. Katsumi, S.-I. Sasaki, and M. Randić, Computer-assisted structure taste studies of sulfamates by pattern-recognition method using graph theoretical invariants, *J. Comput. Chem.* 6 (1988) 636–646.
42. M. Randić and C. L. Wilkins, Graph theoretical study of structural similarity in benzomorphans, *Int. J. Quantum Chem.: Quantum Biol. Symp.* 6 (1979) 55–71.
43. M. Novič, Z. Nikolovska-Coleska, and T. Šolmajer, Quantitative structure–activity relationship of flavonoid p56^{lck} protein tyrosine kinase inhibitors. A neural network approach, *J. Chem. Inf. Comput. Sci.* 37 (1997) 990–998.
44. M. Randić, M. Novič, A. Roy Choudhury, and D. Plavšić, On graphical representation of trans-proteins, *SAR QSAR Environ. Res.* 23 (2012) 327–343.
45. Galileo Galilei.
46. J. R. Platt, Influence of neighbor bonds on additive bond properties in paraffins, *J. Chem. Phys.* 15 (1947) 419–420.
47. R. Brüggemann and G. Welz, in: *Order Theoretical Tools in Environmental Sciences, Order Theory (Hasse DiagramTechnique) Meets Multivariate Statistics*, K. Vogt and G. Welz (Eds.), Shaker-Verlag, Aachen, 2002.

48. D. J. Klein and D. Babić, Partial orderings in chemistry, *J. Chem. Inf. Comput. Sci.* 37 (1997) 656–671.
49. D. J. Klein and L. Bytautas, Directed reaction graphs as posets, *MATCH Commun. Math. Comput. Chem.* 42 (2000) 261–289.
50. D. J. Klein, Similarity and dissimilarity in posets, *J. Math. Chem.* 18 (1995) 321–348.
51. A. B. Shtarev, E. Pinkhassik, M. D. Levin, I. Stibor, and J. Michl, Partially fluorinated dimethyl bicycle (1.1.1) pentane-1,3-dicarboxylates: Preparation and NMR spectra, *J. Am. Chem. Soc.* 123 (2001) 3484–3492.
52. D. J. Klein and J. Brickmann (Eds.), Partial orderings in chemistry, *MATCH Commun. Math. Comput. Chem.* 42 (2000) 1–290.
53. E. A. Castro, F. M. Fernandez, and P. R. Duchowicz, QSPR modeling of the enthalpy of formation based on partial order ranking, *J. Math. Chem.* 37 (2005) 433–441.
54. P. R. Duchowicz, M. G. Vitale, and E. A. Castro, Partial order ranking for the aqueous toxicity of aromatic mixtures, *J. Math. Chem.* 44 (2008) 541–549.
55. R. Brüggemann and K. Voigt, Basic principles of Hasse diagram technique in chemistry, *Comb. Chem. High Throughput Screen.* 11 (2008) 756–769.
56. R. Brüggemann and L. Carlsen, An improved estimation of averaged ranks of partial orders, *MATCH Commun. Math. Comput. Chem.* 65 (2011) 383–414.
57. T. Ivanciuc and D. J. Klein, Parameter-free structure-property correlation via progressive reaction posets for substituted benzenes, *J. Chem. Inf. Comp. Sci.* 44 (2004) 610–617.
58. M. Pavan and R. Todeschini, New indices for analyzing partial ranking diagrams, *Anal. Chim. Acta* 515 (2004) 167–181.
59. D. J. Klein, T. Ivanciuc, A. Ryzhov, and O. Ivanciuc, Combinatorics of reaction-network posets, *Comb. Chem. High Throughput Screen.* 11 (2008) 723–733.
60. D. J. Klein, Prolegomenon on partial orderings in Chemistry, *MATCH Commun. Math. Comput. Chem.* 42 (2000) 7–21.
61. R. Brüggemann, A. Kerber, and G. Restrepo, Ranking objects using fuzzy orders with an application to refrigerants, *MATCH Commun. Math. Comput. Chem.* 66 (2011) 581–603.
62. T. Ivanciuc, O. Ivanciuc, and D. J. Klein, Posetic quantitative superstructure/activity relationships (QSSARs) for chlorobenzenes, *J. Chem. Inf. Model.* 45 (2005) 870–879.
63. T. Došlić and D. J. Klein, Splinoid interpolation on finite posets, *J. Comput. Appl. Math.* 177 (2005) 175–185.
64. D. Lerche and P. B Sørensen, Evaluation of the ranking probabilities for partial orders based on random linear extensions, *Chemosphere* 53 (2003) 981–992.
65. G. Restrepo and D. J. Klein, Predicting densities of nitrocubanes using partial orders, *J. Math. Chem.* 49 (2011) 1311–1321.
66. G. Restrepo, R. Brüggemann, and D. J. Klein, Partially ordered sets: Ranking and prediction of substances' properties, *Curr. Comput. Aided Drug Design* 7 (2011) 133–145.
67. D. Babić and N. Trinajstić, Möbius inversion on a poset of a graph and its acyclic subgraphs, *Discrete Appl. Math.* 67 (1996) 5–11.
68. M. Randić, M. Novič, M. Vračko, and D. Plavšić, Study of proteome maps using partial ordering, *J. Theor. Biol.* 266 (2010) 21–28.
69. M. Randić, G. A. Kraus, and B. Jerman-Blažič, Ordering graphs as an approach to structure-activity studies, in: *Chemical Applications of Topology and Graph Theory*, R. B. King (Ed.), *Studies Phys. Theor. Chem.* 28 (1983) 192–205.
70. M. Randić, B. Jerman-Blažič, D. H. Rouvray, P. G. Seybold, and S. C. Grossman, The search for active substructures in structure-activity studies, *Int. J. Quantum Chem.: Quantum Biol. Symp.* 14 (1987) 245–260.
71. E. Rincón-Villamizar and G. Restrepo, Rules relating hepatotoxicity with structural attributes of drugs, *Toxicol. Environ. Chem.* 96 (2014) 594–613.

72. G. Restrepo and A. Bernal, Ordering molecular energies by moving boxes, *Chem. Phys. Lett.* 612 (2014) 51–55.

73. N. Quintero, I. M. Cohen, and G. Restrepo, Relating β+ radionuclides' properties by order theory, *J. Radioanal. Nucl. Chem.* 298 (2013) 1937–1946.

74. R. Brüggemann, G. Restrepo, K. Voigt, and P. Annoni, Weighting intervals and ranking, exemplified by leaching potential of pesticides, *MATCH Commun. Math. Comput. Chem.* 69 (2013) 413–432.

75. G. Restrepo, S. C. Basak, and D. Mills. Comparison of SAR and QSAR approaches to mutagenicity of aromatic and heteroaromatic amines, *Curr. Comput.-Aided Drug Design* 7 (2011) 109–121.

76. R. F. Muirhead, Some methods applicable to identities and inequalities of symmetric algebraic functions of n letters, *Edinburgh Math. Soc.* 21 (1903) 144.

77. M. Randić, G. M. Brissley, R. B. Spencer, and C. L. Wilkins, Use of self-avoiding paths for characterization of molecular graphs with multiple bonds, *Comput. Chem.* 4 (1980) 27–43.

78. M. Randić and C. L. Wilkins, Graph theoretical analysis of molecular properties—Isomeric variations in nonanes, *Int. J. Quantum Chem.* 18 (1980) 1005–1027.

79. M. Randić and C. L. Wilkins, On a graph theoretical basis for ordering of structures, *Chem. Phys. Lett.* 63 (1979) 332–336.

80. M. Randić, Chemical structure—What is she?, *J. Chem. Educ.* 69 (1992) 713–718.

81. M. Randić and C. L. Wilkins, Graph theoretical ordering of structures as basis for systematic searches for regularities in molecular data, *J. Phys. Chem.* 83 (1979) 1525–1540.

82. M. Randić, Chemical shift sums, *J. Magn. Res.* 39 (1980) 431.

83. M. Randić, M. Novič, D. Vikić-Topić, and D. Plavšić, On intramolecular average ^{13}C chemical shift in nonanes, *Int. J. Quantum Chem.* 109 (2009) 3094–3102.

84. M. Randić and N. Trinajstić, Composition as a method for data reduction: Application to carbon-13 MNR chemical shifts, *Theor. Chim. Acta* 73 (1988) 233–246.

85. Y. Miyashita, H. Ohsako, T. Okuyama, S.-I. Sasaki, and M. Randić, Computer-assisted studies of structure–property relationship using graph invariants, *Magn. Reson. Chem.* 29 (1991) 362–365.

86. G. Rücker and C. Rücker, Counts of all walks as atomic and molecular descriptors, *J. Chem. Inf. Comput. Sci.* 33 (1993) 683–695.

87. O. Ivanciuc, J.-P. Rabine, and D. Cabrol-Bass, ^{13}C NMR chemical shift sum prediction for alkanes using neural networks, *Comput. Chem.* 21 (1997) 437–443.

88. D. Svozil, J. Pospíchal, and V. Kvasnička, Neural-network prediction of carbon-13 NMR chemical shifts of alkanes, *J. Chem. Inf. Comput. Sci.* 35 (1995) 924–928.

89. D. M. Grant and E. G. Paul, Carbon-13 magnetic resonance. II. Chemical shift data for the alkanes, *J. Am. Chem. Soc.* 86 (1964) 2984–2990.

90. L. P. Lindeman and J. Q. Adams, Carbon-13 nuclear magnetic resonance spectrometry. Chemical shifts for the paraffins through C9, *Anal. Chem.* 43 (1971) 1245–1252.

91. Y. Sakamoto and S. Watanabe, On the relationship between the chemical structure and the carcinogenicity of polycyclic and chlorinated monocyclic aromatic compounds as studied by means of ^{13}C NMR, *Bull. Chem. Soc. Jpn.* 59 (1986) 3033.

92. T. Aoki, S. Ohshima, and Y. Sakamoto, On the relationship between the chemical structure and the carcinogenicity of polycyclic and chlorinated monocyclic aromatic compounds as studied by means of ^{13}C NMR, *Polycycl. Aromat. Compd.* 11 (1996) 245–252.

93. K. Aktenburg, Zur Berechnung des Raius verzweigter Moleküle, *Kolloid Zeit.* 178 (1961) 112–117.

94. P. R. Duchowicz and E. A. Castro, *The Order Theory in QSPR—QSAR Studies*, Mathematical Chemistry Monographs No. 7, Kragujevac, Serbia, 2006.

95. R. Brüggemann and L. Carlsen, (Eds.), *Partial Order in Environmental Sciences and Chemistry*, Springer, Heidelberg, 2006.

96. M. Randić, A graph theoretical characterization of proteomics maps, *Int. J. Quantum Chem.* 90 (2002) 848–858.

97. M. Randić, On graphical and numerical characterization of proteomics maps, *J. Chem. Inf. Comput. Sci.* 41 (2001) 1330–1338.

98. M. Randić, J. Zupan, and M. Novič, On 3-D graphical representation on proteomics maps and their numerical characterization, *J. Chem. Inf. Comput. Sci.* 41 (2001) 1339–1344.

99. M. Randić, F. Witzmann, M. Vračko, and S. C. Basak, On characterization of proteomics maps and chemically induced changes in proteomes using matrix invariants: Application to peroxisome proliferators, *Med. Chem. Res.* 10 (2001) 456–479.

100. M. Randić and S. C. Basak, A comparative study of proteomics maps using graph theoretical biodescriptors, *J. Chem. Inf. Comput. Sci.* 42 (2002) 983–992.

101. M. Randić, M. Novič, and M. Vračko, On characterization of dose variations of 2-D proteomics maps by matrix invariants, *J. Proteome Res.* 1 (2002) 217–226.

102. M. Randić, Quantitative characterization of proteomics maps by matrix invariants, in: *Handbook of Proteomics Methods*, P. M. Conn (Ed.), Humana Press, Inc. Totowa, NJ, 2003, pp. 429–450.

103. N. L. Anderson, R. Esquer-Blasco, F. Richardson, P. Foxworthy, and P. Eacho, The effect of peroxisome proliferation on protein abundances in mouse liver, *Toxicol. Appl. Pharmacol.* 137 (1996) 75–89.

104. M. Randić, D_{MAX}—Matrix of dominant distances in a graph, *MATCH—Commun. Math. Comput. Chem.* 70 (2013) 221–238.

105. M. Randić, R. Orel, and A. T. Balaban, D_{MAX} matrix invariants as graph descriptors. Graphs having the same Balaban index *J*, *MATCH—Commun. Math. Comput. Chem.* 70 (2013) 239–258.

106. M. Dehmer and Y. Shi, The uniqueness of matrix graph invariants, *PLoS One* 9 (2014) e83868.

8 Novel Molecular Matrices

8.1 NOVEL MOLECULAR DESCRIPTORS

Previously discussed misconceptions in QSAR are not the only misconceptions floating around in several research circles of QSAR. From time to time one hears cries, particularly from those who are in QSAR using exclusively physicochemical properties as molecular descriptors, that there are enough topological indices. First, they should be the least concerned with this issue, because (i) they are neither using topological indices, (ii) nor are they familiar with topological indices. They apparently lament that there are enough studies of indices and that researchers in mathematical chemistry should awake and start solving "real world" problems! Well, it is amusing that those who are, with respect to research in mathematical chemistry, "sleeping" should call others to awaken! Such "criticism" reminds us of an old anonymous saying: *"People with the least expertise have the most opinions."* These criticisms reflect more on those who raise objections than on those being criticized.

It is true that continuous growth of the pool of molecular descriptors, widely called topological indices, created the problem of how to choose the best descriptors in various applications. This topic has received some attention. The more one knows about the properties of considered topological indices, the more likely it is that one will make a better selection of the topological indices to be tested. One can also view such indices as the content of a "black box," and search systematically for optimal descriptors, running a risk of missing useful alternatives when applying some routine statistical procedures that eliminate highly correlated descriptors without examining whether there are parts in which highly collinear descriptors differ, and which are relevant for the characterization of the considered property.

Do we still need additional molecular descriptors? Well, there will always be room for new and novel molecular descriptors that will incorporate some structural information that has not been well captured by current molecular descriptors. But these days, novel molecular descriptors are hard to find, particularly if they are to be conceptually and computationally simple and have straightforward structural interpretation. Finding such will continue to be not an everyday event. However, recall a statement of E. Bright Wilson [1]: *"... every once in a while some new theory or new experimental method or apparatus makes it possible to enter a new domain."*

One such new theory has been the common vertex matrix [2], which has led to a novel approach to graph eccentricities and network eccentricities, based on the count of pairs of vertices at equal distance from each vertex in a graph [3]. A preliminary study has shown that the connectivity-type index based on the vertex centrality values displays considerable discriminatory power in differentiating between similar structures, including very similar structures among isomers. The discriminatory power of a centrality-based connectivity index exceeds visibly the discriminatory power of Balaban's J index [4–6], which is a distance matrix analog of the connectivity index.

It does appear that novel matrices of graphs, such as, for example, the recently constructed n-dimensional (pseudo) distance matrix [7], the D_{MAX} matrix containing only the maximal distances in each row and each column of the distance matrix [8,9], and the already mentioned common vertex matrix [2], may lead to a number of novel graph descriptors, some of which may be of interest in the search for structurally highly similar molecules.

There are a number of novel descriptors that have been introduced during the past few years that deserve attention. Let us mention several novel descriptors all that appeared in the same issue of *Current Computer-Aided Drug Design*: Ivanciuc [10] outlined how the molecular descriptors and topological indices can be developed with a more general approach based on molecular graph operators, which define a family of graph indices related by a common formula. Graph descriptors and topological indices for molecules containing heteroatoms and multiple bonds are computed with weighting schemes based on atomic properties, such as the atomic number, covalent radius, or electronegativity. Marrero-Ponce et al. [11] defined strategies that generalize the definition of global or local invariants from atomic contributions (local vertex invariants, LOVIs), introducing related metrics (norms), means, and statistical invariants. Lučić, Trinajstić, and collaborators introduced "The Sum-Connectivity Index" [12], an additive variant of the Randić connectivity index. L. Jäntschi and collaborators [13] presented the methodology of the molecular descriptors family (MDF) as an integrative tool in molecular modeling and its abilities as a multivariate QSAR/QSPR modeling tool. Toropov et al. [14] elaborated on the logic and evolution of the optimal descriptors OCWLGI. These descriptors were based on molecular graphs and demonstrated their ability as a tool for modeling biological and physicochemical parameters of chemical compounds. Ivanciuc et al. [15] defined a class of network-QSAR models based on molecular networks induced by a sequence of substitution reactions on a chemical structure that generates a partially ordered set (or *poset*)-oriented graph. They may be used to predict various molecular properties with quantitative superstructure–activity relationships (QSSAR).

Although some may view the continuing interest of a number of scientists in construction of novel indices, when there are thousands that have been already introduced, as a violation of the Occam's razor doctrine, one should not overlook another doctrine that characterizes the importance of scientific results, which says the importance of novelty can be measured by the number of existing "items" it makes irrelevant. This comes from David Hilbert (1862–1943), a very distinguished German mathematician at the turn of the century, whose exact words are, *"One can measure the importance of a scientific work by the number of earlier publications rendered superfluous by it."*

From that point of view, new mathematical descriptors of recent times are more likely than not to make many ad hoc descriptors of the past less relevant for QSAR than they appear today.

Before continuing this "never-ending" topic on the plethora of topological indices and elaborate on a novel class of connectivity indices designed not so much for use in MRA applications in QSAR, but for screening of combinatorial libraries and searching for highly similar molecules, let us briefly comment on the name OCWLGI. The awkward name OCWLGI for new molecular invariants that deserve a wider attention is in our view an unnecessary drawback. It is an acronym for "optimal correlation weights of local and global invariants" of molecular graphs, obtained using the Monte Carlo method. Generally people have difficulties remembering acronyms that cannot be easily pronounced. However, by some rearrangement of the words forming the acronym, one can construct the pronounceable COWGIL (for correlation optimal weights global invariants and local), which is close enough in content to the original wording. This is easy to pronounce and easy to remember even though the meaning is biologically absurd because cows have lungs and do not have gills; fishes do. Maybe in chemistry, in addition to tool makers, we need acronym makers as OCWLGI is not the first awkward acronym in chemical literature.

8.2   NOVEL MATRICES IN CHEMISTRY

Matrices have a long history and were known for a time merely as rectangle arrays associated with coefficients of systems of linear equations used for construction of determinants that would be combined to arrive at the solutions of such equations. An illustration is the well-known Cramer rule of Algebra for solving linear equations, which dates to 1750. The word "matrix" was coined by Sylvester 100 years later. It was based on the Latin word *Mater* for mother, as such matrices (arrays) were giving birth to many minors (determinants of smaller matrices obtained by deleting one row and one column of the original matrix). It was Arthur Cayley who viewed matrices independently from equations as these are used today [16]. Cayley introduced mathematical operation on matrices, such as addition, subtraction, multiplication, and division of matrices. He published the first book on matrices, *Memoir on the Theory of Matrices*, in 1858, but it was Sylvester who significantly further developed the early theory of matrices [17].

In his *Memoir on the Theory of Matrices*, Cayley mentioned the important theorem for matrices, known as the Cayley-Hamilton theorem, which states that a square matrix satisfies its own characteristic polynomial. The significance of the Cayley-Hamilton theorem is that for a matrix of size n × n all information is in the first A^n matrices, n = 1, ... n. Thus, there is no new information to be obtained by calculating higher powers of matrices.

We are interested here in matrices associated with graphs and similar chemical objects. For convenience, we grouped matrices with respect to the time they emerged, into "old," "not so old," "recent," and "novel," as shown in Table 8.1. Here we selected as "old" graph matrices that were introduced before the mid-1970s, when chemical graph theory took roots and started to grow. As "recent" we consider matrices published in the year 2000 or later. The matrices given the label "novel," have this for two reasons: (i) They are new in time; not yet published; or, if published, they appeared after the writing of this book started, and (ii) they have some novelty in their content and application, not being, in that sense, related to most other chemical matrices designed for construction of QSAR molecular descriptors. Matrices can also be classified with respect to their entries, which could be (i) integers, (ii) algebraic numbers, (iii) real numbers, (iv) complex numbers, (v) sets, or (vi) non-numerical. There is no need to list rational numbers as a separate class because matrices with rational numbers can always be scaled to matrices having integers as elements.

TABLE 8.1
Selection of Matrices of Interest in Chemistry

Old	Not So Old	Recent	Novel
Adjacency	Wiener	Revised Szeged	χ (Wiener sparse)
Distance	Distance/distance	n-dim. distance	χ (n-dim. distance)
Kirchhoff	Szeged	D_{MAX}	χ (valence sum)
Laplace	Resistance distance	Amino acid adjacency	χ (augmented valence)
Terminal	12×12 Fullerene	Sequential adjacency	χ (common vertex)
	Graphical	Common vertex	χ (centrality)
	Ulam subgraphs		Square root distance
	Path		n-root distance
			Distance squared
			Ring closure

Matrices with integer elements often arise from counting. Such are the Wiener matrix, the Szeged matrix, the amino acid adjacency matrix, the common vertex matrix, and the n-dimensional pseudo-distance matrix. In Table 8.2 are illustrated, for the molecular graph of 3-methylhexane, the adjacency matrix, the distance matrix (counting edges between two vertices), the Wiener matrix (counting vertices on each side of an edge), the common vertex matrix (counting pairs of vertices at the same distance from an edge), and the pseudo-distance matrix (counting different elements between rows [columns] of the adjacency matrix). For each of the matrices shown, we have also shown their corresponding row sums. The matrix row sums have recently become of considerable interest for construction of a new class of molecular descriptors of high very discriminatory power. The algorithm for construction of new molecular descriptors was used in 1982 by Balaban for the construction of his distance-topological index J. So here the same algorithm has been resurrected once again.

The elements of the n-dimensional distance matrix are obtained by viewing the columns of the adjacency matrix as vectors in n-dimensional space and considering the Hamming distances between all columns (or rows). The Hamming distance between two sequences of the same length is given by the number of sites in which the two sequences have different entries. Also matrices based on graph theoretical distances, such as the distance matrix; the terminal distances; the D_{MAX} matrix, which is obtained from the distance matrix by retaining only the largest distances in each row and column of the distance matrix; and the most recent matrix, the ring closure matrix, which has been initiated with the writing of this book (see the second half of this chapter), are matrices having integers as their elements.

Graphical matrices, which as matrix elements do not have numbers but graphs, are converted to numerical matrices by replacing graphical elements by some invariant of the graph or subgraphs involved. The Ulam matrix appears to be the only matrix that uses subgraphs directly as its elements. The Ulam matrix, so far, can be solely used to

TABLE 8.2
The Adjacency Matrix, the Distance Matrix, the Wiener Matrix, the Common Vertex Matrix, and the Pseudo-Distance Matrix of 3-Methylhexane and Their Row Sums

Adjacency Matrix							RS
0	1	0	0	0	0	0	1
1	0	1	0	0	0	0	2
0	1	0	1	0	0	1	3
0	0	1	0	1	0	0	2
0	0	0	1	0	1	0	2
0	0	0	0	1	0	0	1
0	0	1	0	0	0	0	1

Distance Matrix							RS
0	1	2	3	4	5	3	18
1	0	1	2	3	4	2	13
2	1	0	1	2	3	1	10
3	2	1	0	1	2	2	11
4	3	2	1	0	1	3	14
5	4	3	2	1	0	4	19
3	2	1	2	3	4	0	15

Wiener Matrix							RS
0	6	0	0	0	0	0	6
6	0	10	0	0	0	0	16
0	10	0	12	0	0	6	28
0	0	12	0	10	0	0	22
0	0	0	10	0	6	0	16
0	0	0	0	6	0	0	6
0	0	6	0	0	0	0	6

Common Vertex Matrix							RS
0	0	1	0	2	0	0	3
0	0	0	2	0	1	4	7
1	0	0	0	1	0	0	2
0	2	0	0	0	1	3	6
2	0	1	0	0	0	0	3
0	1	0	1	0	0	1	3
0	4	0	3	0	1	0	8

Pseudo-Distance Matrix							RS
0	3	0	0	0	0	0	3
3	0	5	0	0	0	0	8
0	5	0	5	0	0	4	14
0	0	5	0	4	0	0	9
0	0	0	4	0	3	0	7
1	0	0	0	3	0	0	4
0	0	4	0	0	0	0	4

Full Pseudo-Distance Matrix							RS
0	3	2	3	3	2	1	14
3	0	5	2	4	3	1	18
2	5	0	5	3	4	4	23
3	2	5	0	4	1	1	16
3	4	3	4	0	3	3	20
2	3	4	1	3	0	2	15
1	1	4	1	3	2	0	12

check if a given "deck of cards" in the graph reconstruction problem is legitimate, that is, if any of the subgraphs included in the deck of cards has been altered.

In the continuation of this chapter, we will focus on the "novel" matrices of the last column of Table 8.1. As one can see, there are only three groups of new matrices. In the first group are matrices the row sums of which are used to construct indices, resembling the connectivity index. In contrast to the adjacency matrix, the row sums of which give the graph valences (degrees) that cover very a limited range of values, and the distance matrix, the row sums of which cover an intermediate range of values, the row sums of the new matrices cover a wide range of values in comparison with other matrices. Because of a greater range of values of the row sums, the generalized connectivity indices of these matrices will have very high discriminatory power. We have symbolized these matrices with the connectivity index symbol χ adding in brackets for the matrix symbol. In this notation, $^1\chi$ is $\chi(A)$ and the Balaban index J is $\chi(D)$, where A is the adjacency matrix and D is the distance matrix.

In the second group are two distance matrices; the first has been known in physics [18] and the second in mathematics [19–21] and mathematical chemistry [22–28]. The last matrix is a newly modified terminal matrix, to be explained shortly, which offers highly compact matrix representation of polycyclic molecules and may be of potential interest in chemical documentation.

We start with the first group. As already mentioned, the connectivity index χ can be formally constructed combining the row sums of the adjacency matrix using the algorithm represented by $\Sigma_{(i,j)}$ $1/\sqrt{(R_i R_j)}$, where R_i and R_j are the row sums of a pair of adjacent vertices summed over all bonds.

Balaban was the first to introduce this algorithm in construction of his J index, applying it to the row sums of the distance matrix. Because the row sums of the distance span a greater range of values, the resulting bond contributions to the calculated J values varied widely, and in J indices for small alkanes having less than $n = 12$ carbon atoms all J values differed. In Figure 8.1, we show the six pairs of dodecane isomers that have the same Balaban's index J. In the case of the connectivity index, there are graphs having the same index already among 18 octanes.

Nevertheless, the connectivity index continues to be one of the most widely used molecular descriptors, and the degeneracy is not so important for structure–property correlations because there are also molecules having some of the same or very similar values for molecular properties. The degeneracy, however, is the critical factor if one wishes to discriminate among molecules as much as possible, which is our interest in this section.

There are two aspects of the Balaban J index that are worth emphasizing: (i) a relatively high discriminating power and (ii) a relatively simple construction algorithm. That indices of higher discriminatory power are desirable has been recognized for some time. Thus, in 1993, Ivanciuc et al. [29] proposed several indices of higher discriminatory power, which have been based on local molecular invariants derived from distance degree sequences (DDS) of individual vertices. New indices were different for all alkane graphs having less than $n = 18$, but being based on information theory, calculation of these indices required the use of information of DDS and use of the binary logarithms in their construction. Computationally, this was not necessarily a handicap, but the question remains: Can one find a conceptually and computationally simpler alternative?

Our answer is yes. And the approach to be outlined in this section is based on revisiting Balaban's index J. Observe that J has increased its discriminatory power in comparison with the connectivity index by using the *same* row sum algorithm but a *different* matrix! We speculated therefore that a fruitful approach to new highly discriminatory indices may be found by doing the same: Keep the row sum algorithm but search for matrices that would have a greater range of their row sums than has the distance matrix. Such matrices having a greater range for their row sums are expected to produce molecular descriptors of greater discriminatory power. However, in order to keep the conceptual and computational concept surrounding new molecular descriptors as simple as possible, it is important that (i) the matrices considered have transparent structural interpretation; (ii) if possible, be sparse (matrices having many zero matrix elements); and (iii) if possible, be based on integers for easy of construction and calculation of such matrices.

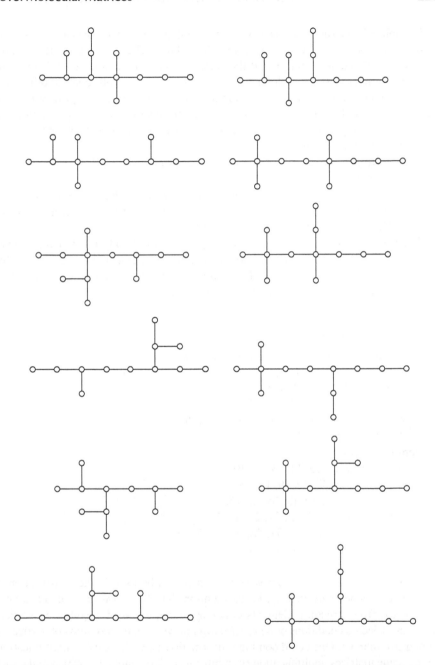

FIGURE 8.1   The six pairs of dodecane graphs that have identical Balaban's index *J*.

In Table 8.2, we show the row sums for the adjacency matrix, the (sparse) Wiener matrix, the common vertex matrix, and the n-Dim. The elements of the Wiener matrix are defined as $a_{i,j} = a_i \times a_j$ if the vertices i and j are connected, where a_i and a_j are the number of vertices closer to vertex i and j, respectively, and = 0 otherwise.

The elements of the common vertex matrix are defined as $a_{i,i} = p_{ii}$ where p_{ii} is the number of pairs of vertices at the same distance from vertex a_{ii}, $a_{i,j} = p_{ij}$ if i and j are connected where p_{ij} is the number of pairs of vertices at the same distance from both a_i and a_j, and = 0 otherwise.

In the original publication on the common vertex matrix [3], the diagonal elements were assumed to be zero.

The elements of the pseudo distance matrix, n-Dim, are defined as follows: $a_{i,j} = c_i H c_j$ if i and j are connected, where c_i and c_j are the columns of adjacency matrix, and H is Hamming distance, and = 0 otherwise.

The Hamming distance between two sequences, as already mentioned, is given by the number of sites at which the two corresponding elements in sequences differ.

Below we have listed the ordered row sums for numerous matrices, illustrated on the graph of 3-methylhexane, having integer entries:

Sparse matrices:
    Wiener 28, 22, 16, 16, 6, 6, 6
    Augmented valence 41, 27, 23, 30, 12, 9, 8
    Connectivity 15, 10, 6, 6, 3, 2, 2
    Centrality 20, 18, 8, 8, 8, 2, 2
    n-D distance 14, 9, 8, 7, 4, 3, 3
    Walk of length two 50, 40, 28, 21, 15, 8, 5
    DDS 69, 58, 54, 46, 35, 33, 31
Dense matrices:
    Distance 19, 18, 15, 14, 13, 11, 10
    Wiener 39, 35, 32, 30, 19, 18, 16
    n-D distance 23, 20, 18, 16, 15, 15, 13
    Valence sum 35, 27, 22, 22, 22, 17, 17
    Augmented valence 76, 71, 66, 61, 56, 52, 51
    Common vertex 8, 7, 6, 4, 3, 3, 3

Before discussing the above row sums, which can be used in the construction of generalized connectivity indices, let us mention that the above matrices have been constructed from structural elements either by *enumeration* of neighboring vertices satisfying selected conditions or by *combining information* on row sums of vertices in an additive manner. One could combine information on vertices using multiplication, but for some matrices, multiplication soon introduces large numbers, which make such approaches less desirable. What is desirable when constructing molecular descriptors of high discriminatory power is that (i) the row sum entries cover a relatively large range, and (ii) the row sums of symmetry nonequivalent vertices are different.

The difference between the largest and the smallest row sum gives the range that the row sum covers. Because in 3-methylhexane all carbon atoms are symmetry non-equivalents, it is desirable that the row sum sequences have no equal entries.

From a look at the sequences listed above, one see that in the case of 3-methylhexane only the following matrices satisfy both conditions:

Sparse matrices: The augmented valence, the walks of length two, and the distance degree sequence matrix

Dense matrices: The distance, the Wiener matrix and the augmented valence matrix

That there are matrices with all row sums different need not be surprising because all the matrices mentioned involve information beyond the nearest neighbor connectivity. It is possible that for these matrices the row sums of symmetry nonequivalent vertices are the same for some vertices of other graphs and that some matrices that "fail" the test on 3-methylhexane may not fail on other graphs—hence, the above requirements should be viewed as a guidance and not as necessary or sufficient conditions that the novel descriptors based on such matrices will always discriminate between similar graphs. It is equally possible that all symmetry nonequivalent vertices have different row sums but that the generalized connectivity indices constructed from using such sums are the same for non-isomorphic graphs. The ordered row sums listed should therefore be viewed as a list that may be useful, not to suggest which matrices satisfying the conditions will produce and which will not produce highly discriminatory indices.

As one can see, there is an excessive occurrence of the same row sums for symmetry unrelated vertices for the following sparse matrices: the Wiener, the connectivity, and the centrality matrices, and for the following dense matrices: valence sum and common vertex matrix. Observe that row sums in some of these matrices cover a respectable range, but that is not enough to ensure diversity among their row sums. Finally, we should add that one should expect there will be pairs of graphs that will have the same generalized connectivity indices for various different matrices. This, in itself, need not be viewed as a serious disadvantage of these descriptors as long as the so-identified structures are also *very* similar. Recall that our prime aim is detection of *similar* structures and is not an alternative approach for testing graph isomorphism.

8.3 GENERALIZED DISTANCE MATRICES

In Table 8.1, we have listed as "novel" three distance-based matrices: the square root distance matrix, the n-root distance matrix, and the distance squared matrix, but not because they are new in content and not resembling something already known or that they are original in conception. They are certainly not new because they have been already used in physics and mathematics. However, because they appear to be of potential interest for chemistry and are unknown in chemical literature, one may consider them as a novelty for chemistry. In Table 8.3, we have illustrated the square root distance matrix and the n-root distance matrix, again for the case of 3-methylhexane and norbornane, a bicyclic also seven carbon atom structure. We start with the square root and n-root distance matrix and will discuss the distance squared matrix later.

TABLE 8.3
Root Distance Matrices

Square Root Distance Matrix

3-Methylhexane

$$\begin{bmatrix} 0 & 1 & \sqrt{2} & \sqrt{3} & \sqrt{4} & \sqrt{5} & \sqrt{3} \\ 1 & 0 & 1 & \sqrt{2} & \sqrt{3} & \sqrt{4} & \sqrt{2} \\ \sqrt{2} & 1 & 0 & 1 & \sqrt{2} & \sqrt{3} & 1 \\ \sqrt{3} & \sqrt{2} & 1 & 0 & 1 & \sqrt{2} & \sqrt{2} \\ \sqrt{4} & \sqrt{3} & \sqrt{2} & 1 & 0 & 1 & \sqrt{3} \\ \sqrt{5} & \sqrt{4} & \sqrt{3} & \sqrt{2} & 1 & 0 & \sqrt{4} \\ \sqrt{3} & \sqrt{2} & 1 & \sqrt{2} & \sqrt{3} & \sqrt{4} & 0 \end{bmatrix}$$

Norbornane

$$\begin{bmatrix} 0 & 1 & \sqrt{2} & \sqrt{3} & \sqrt{2} & 1 & \sqrt{2} \\ 1 & 0 & 1 & \sqrt{2} & \sqrt{3} & \sqrt{2} & \sqrt{2} \\ \sqrt{2} & 1 & 0 & 1 & \sqrt{2} & \sqrt{2} & 1 \\ \sqrt{3} & \sqrt{2} & 1 & 0 & 1 & \sqrt{2} & \sqrt{2} \\ \sqrt{2} & \sqrt{3} & \sqrt{2} & 1 & 0 & 1 & \sqrt{2} \\ 1 & \sqrt{2} & \sqrt{2} & \sqrt{2} & 1 & 0 & 1 \\ \sqrt{2} & \sqrt{2} & 1 & \sqrt{2} & \sqrt{2} & 1 & 0 \end{bmatrix}$$

n-Root Distance Matrix

3-Methylhexane

$$\begin{bmatrix} 0 & 1 & \sqrt{2} & \sqrt[3]{3} & \sqrt[4]{4} & \sqrt[5]{5} & \sqrt[3]{3} \\ 1 & 0 & 1 & \sqrt{2} & \sqrt[3]{3} & \sqrt[4]{4} & \sqrt{2} \\ \sqrt{2} & 1 & 0 & 1 & \sqrt{2} & \sqrt[3]{3} & 1 \\ \sqrt[3]{3} & \sqrt{2} & 1 & 0 & 1 & \sqrt{2} & \sqrt{2} \\ \sqrt[4]{4} & \sqrt[3]{3} & \sqrt{2} & 1 & 0 & 1 & \sqrt[3]{3} \\ \sqrt[5]{5} & \sqrt[4]{4} & \sqrt[3]{3} & \sqrt{2} & 1 & 0 & \sqrt[4]{4} \\ \sqrt[3]{3} & \sqrt{2} & 1 & \sqrt{2} & \sqrt[3]{3} & \sqrt[4]{4} & 0 \end{bmatrix}$$

Norbornane

$$\begin{bmatrix} 0 & 1 & \sqrt{2} & \sqrt[3]{3} & \sqrt{2} & 1 & \sqrt{2} \\ 1 & 0 & 1 & \sqrt{2} & \sqrt[3]{3} & \sqrt{2} & \sqrt{2} \\ \sqrt{2} & 1 & 0 & 1 & \sqrt{2} & \sqrt{2} & 1 \\ \sqrt[3]{3} & \sqrt{2} & 1 & 0 & 1 & \sqrt{2} & \sqrt{2} \\ \sqrt{2} & \sqrt[3]{3} & \sqrt{2} & 1 & 0 & 1 & \sqrt{2} \\ 1 & \sqrt{2} & \sqrt{2} & \sqrt{2} & 1 & 0 & 1 \\ \sqrt{2} & \sqrt{2} & 1 & \sqrt{2} & \sqrt{2} & 1 & 0 \end{bmatrix}$$

8.3.1 SQUARE ROOT AND N-ROOT DISTANCE MATRICES

One may ask why should the square root matrix be of interest in chemistry? Two research groups have independently introduced a topological index based on the reciprocal values of distances [30,31]. This index has been referred to as the Harary index, to honor the late Frank Harary (1921–2005), not only because Harary made visible contributions in graph theory, including writing one of the earlier textbooks on graph theory, and contributed to popularizing graph theory in mathematics and other sciences, but he also collaborated with chemists.

The Harary index received attention in both chemical and mathematical literature. In chemistry, one finds several references on the Harary index [32–34] and several chapters in the book on topological indices and related descriptors in QSAR and QSPR [35–39]. According to Trinajstić and coworkers [32], the Harary index "*is based on the chemists' intuitive expectation that distant sites in a structure should influence each other less than the near sites.*"

However, for a molecular descriptor to be of interest in structure–property–activity studies, it is not enough that distant sites in a structure should play a lesser role than neighboring sites, but also that "external" atoms, those contributing mostly to the molecular surface, play a greater role than the "interior" atoms, which should make smaller such contributions [40]. In this respect, the Harary index and other distance-related indices, also including the famous Wiener index, are at some disadvantage as has been recognized by some advocates of such indices. According to Trinajstić [32],

> A problem of the Harary index, as well as of many other topological indices, such as the Wiener index [41] and the reverse Wiener indices, [33] is that they give greater weights to the inner (interior) edges and smaller weights to the outer (terminal) edges of an alkane tree [40]. In contrast the connectivity index, which has given many very good correlations for several alkane physico-chemical properties, gives greater weights to the outer edges and smaller weights to the inner edges of alkanes.

This observations on the Harary index and Wiener index oppose intuitive reasoning that the outer bonds, the more exposed bonds, should have greater weights than the inner bonds because the outer bonds are associated with the larger part of the molecular surface and are consequently expected to make a greater contribution to the physical and chemical properties.

The square root distance matrix does reduce automatically the influence of more distant neighbors in a natural way, so it is of interest. In that respect, even better is the n-root matrix, illustrated at bottom in Table 8.3 on 3-methylhexane and norbornane because now with increases in distance the root exponent also increases, which more drastically decreases the role of vertices at larger distances. We are not proposing the square root matrix and the related n-root matrix as an answer that will cure the ill features of the distance matrix as a source for construction of molecular descriptors to be used in structure–property–activity studies, but more to illustrate an alternative modification of the distance matrix for construction of topological indices.

8.3.2 DISTANCE SQUARED MATRIX AND HIGHER POWERS

The Hadamard product of two matrices A and B of the same size is defined as a multiplication of their corresponding elements:

$$[AB]_{ij} = [A]_{ij}[B]_{ij}$$

Such matrices have found applications for compression of data. Of considerable interest to chemistry is the Euclidean squared distance matrix of a graph [42], which has only four eigenvalues different from zero regardless of its size. This then allows the use of eigenvalues of the squared distance matrix for comparative study of matrices for graphs of different sizes. For larger matrices, four eigenvalues may appear a very drastic reduction in data to be used for comparative studies, but one can consider the higher powers of the Hadamard product of the matrix with itself, A^3, A^4, A^5, etc., which have 9, 16, and 25 nonzero eigenvalues, respectively.

8.4 RING CLOSURE MATRIX

Of the additional novel matrices, we would like to mention the modified terminal matrix listed in Table 8.1 under the name "ring closure matrix." The terminal matrix is an important novelty in graph theory, introduced by Zaretsky in 1965 in his paper on the reconstruction of a tree from the distances between its terminal vertices [43–45]. According to the theorem by Zaretsky, regardless of how big an acyclic graph (a tree) is, a small matrix that includes only the distance between the terminal vertices has all the necessary information on the graph and allows one to reconstruct the graph from its terminal matrix. This thus allows very condensed representations of acyclic molecules. Besides the potential use of a terminal matrix in chemical documentation, this matrix has also been used in graphical bioinformatics for numerical characterization of proteins [46]. It would be of great interest in chemistry if the terminal matrix could be generalized to *polycyclic* molecules. Because matrices of cyclic compounds have no terminal vertices, this translates to the question: *Can one, in a cyclic compound, select a smaller number of vertices, such that information on their distance suffices for reconstruction of the whole graph?*

Our past interest in problems of chemical nomenclature [47–53] never lets us forget this remarkable result by Zaretsky, but as is well known, characterization of properties of polycyclic structures is much more intricate than are those of acyclic systems. This has been so well illustrated, for example, not only with difficulties in finding a generalization of the famous Hückel $4n + 2$ rule, which holds for monocyclic systems, to polycyclic systems, but once a generalization is found, there may be additional difficulties in that many may fail to recognize that the solution was found. Some problems, after being elusive for too long a time, even after solutions have been found, have been considered by many as still unsolved! A good illustration of this was the solution of the inverse X-ray diffraction patterns by Karle and Hauptman [54,55], which, after being solved, took some time to be accepted [56].

Another illustration relates to generalizations of the Wiener number W, to cyclic molecules. Wiener defined W only for acyclic molecules. This time, after a solution was found [57], named the Szeged index, it took some time to realize that the solution was deficient, being size inconsistent. The size consistency means that invariants of molecules having the same number of atoms should have numerical values of the same order of magnitude. In the case of the Szeged index, the size inconsistency was later corrected [58,59] by suggesting that in the revised Szeged index vertices at the same distance from terminal vertices of any bond are not ignored but are equally shared by the terminal vertices of bonds. There are still a number of unsolved problems that require our attention. In our view, one of these is the problem of construction of highly condensed matrices for polycyclic molecules.

The ring closure matrix [60], which we will briefly outline here, is a solution to the construction of condensed matrices for cyclic systems. However, in contrast to the case of acyclic graphs, for which the terminal matrix is unique, except for the assignment of labels to terminal vertices, in the case of polycyclic graphs, there are many ways in which one can select vertices to be used for construction of the "terminal" matrix to be associated with cyclic graphs. The basic strategy is to find a rule for how to select among numerous spanning trees of a cyclic graph one spanning tree to be used and how to augment the list of terminal vertices of the selected spanning tree by including vertices of the deleted edges forming the spanning tree. In application, one has first to decide how to select a particular spanning tree out of many. One way is to construct canonical labeling for the graph and delete ring closure edges having the smallest canonical labels.

8.5   SPANNING TREES

We introduce spanning trees and the terminal matrix of Zaretsky by considering azulene, a bicyclic graph represented by a fused pentagon and heptagon. A spanning tree is an acyclic graph obtained from a cyclic graph by retaining all its vertices but eliminating the smallest number of edges, which, when erased, results in an acyclic graph. In bicyclic graphs, such as azulene, this means erasing two edges, but there are numerous spanning graphs, depending on the selection of edges to be erased. The number of spanning trees, even in small polycyclic graphs, can be very large. Thus there are 34 different spanning trees in azulene; the 17 symmetry nonequivalent spanning trees have been illustrated in Figure 8.2.

The remaining 17 spanning trees can be obtained by reflecting these spanning trees through the horizontal axis. Of the 17 nonequivalent spanning trees of azulene, there are six spanning trees with two terminal vertices, there are seven spanning trees with three terminal vertices, and there are four spanning trees with four terminal vertices. Enumeration of spanning trees can be teduous. John and Sachs

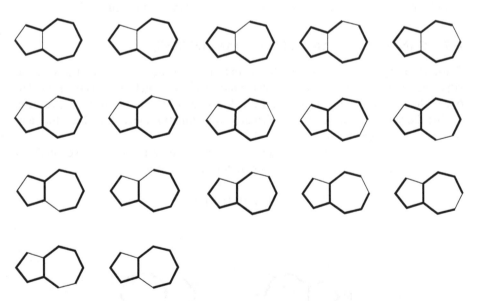

FIGURE 8.2   The symmetry nonrelated spanning trees of azulene. There are 34 spanning trees for azulene molecular graphs. Above are 17 symmetry nonequivalent spanning trees of the azulene graph. The remaining 17 spanning trees can be obtained by reflecting through the horizontal axis.

constructed an algorithm that is efficient in enumeration of spanning trees in benzenoid systems [61].

8.6 CANONICAL LABELS FOR GRAPHS

In this section, we consider construction of a canonical spanning tree. Instead of considering rules for how to select one of the spanning trees as canonical, we will use canonical labeling of a graph and select as ring closure bonds the bonds containing the vertex with the smallest canonical labels. The following three steps define the canonical spanning tree and allow construction of the ring closure matrix for the polycyclic graph considered:

1. For a graph G considered one has to select canonical labels for vertices. Recall that there are several alternative canonical labelings.
2. We have selected canonical labels that will produce, for the adjacency matrix, the smallest binary number when its rows are read from left to right and from top to bottom.
3. To obtain the canonical spanning tree, one deletes ring opening edges involving a vertex having the smallest canonical label in each ring.

In Figure 8.3a, we show azulene canonical labels. The smallest labels for the pentagon has the edge (1, 9) involving as the smallest the label 1, and for the heptagon, the edge (3, 7) involves as the smallest the label 3. When these two edges are erased, one obtains the canonical spanning tree of azulene (shown in Figure 8.3b).

8.7 CANONICAL LABELS FOR ADAMANTANE

Starts by assigning the smallest label 1 first, and in order that the first row of the adjacency matrix is the smallest possible binary number, label 1 has to be assigned to one of the vertices having the smallest valence. In adamantane, there are six vertices having the smallest valence two, which are all symmetry equivalent. Therefore any of those can be selected and given label 1.

The two (symmetry equivalent) adjacent vertices to vertex 1 will have labels 10 and 9, which are the largest labels that give for the first row:

$$0\,0\,0\,0\,0\,0\,0\,0\,1\,1,$$

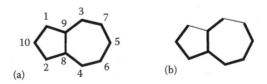

(a)   (b)

FIGURE 8.3   (a) Azulene and its canonical labels. (b) The canonical spanning tree of azulene.

which is the smallest binary row number for a 10×10 matrix, having vertices with smallest valence 2 (Figure 8.4a). The next smallest label 2 should be placed on one of the two symmetry equivalent sites next to vertex 10, and its nearest vertex will have label 8, the largest available label, giving for the second row:

$$0\ 0\ 0\ 0\ 0\ 0\ 0\ 1\ 0\ 1.$$

This is the smallest possible binary number for the second row of the 10×10 adjacency matrix of adamantane. Observe that assignment of vertex 8 created a vertex between vertices 8 and 9 that has no label (Figure 8.4b), to which one should assign label 3, which is the smallest number possible for the third row of the adjacency matrix:

$$0\ 0\ 0\ 0\ 0\ 0\ 0\ 1\ 1\ 0.$$

In continuation, the best location for the next smallest label 4 is next to vertex 10 (Figure 8.4d), giving for the fourth row of the adjacency matrix:

$$0\ 0\ 0\ 0\ 0\ 0\ 1\ 0\ 0\ 1.$$

The other vertex next to vertex number 4 has received the largest available label 7. At this stage, eight vertices have received their canonical labels. The remaining two vertices that have not yet been assigned have as their neighbors vertices (7, 9) and (7, 8). Clearly, the remaining label 5 takes as neighbors (7, 9) and label 6 takes as its neighbors vertices (7, 8) to make the corresponding rows the smallest binary

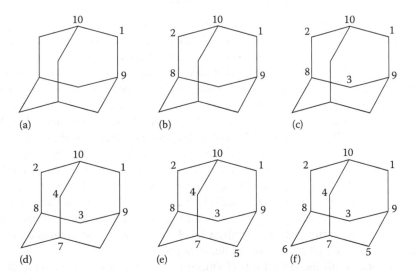

FIGURE 8.4 Construction of the canonical labels for carbon skeleton of adamantane. (a) Placing label 1 on one of six equivalent sites, with the largest labels as neighbors. (b) Placing the next smallest label 2 next to largest label 10. (c) Placing the next smallest label 3 between label 8 and 9. (d) Placing the next smallest label 4 next to largest available label 10. (e) Placing the next smallest label 5 between 7 and 9. (f) Completing labeling of vertices by placing the label 6 between 7 and 8.

TABLE 8.4

The Canonical Adjacency Matrix of Adamantane

$$
\begin{bmatrix}
0 & 0 & 0 & 0 & 0 & 0 & 0 & 0 & 1 & 1 \\
0 & 0 & 0 & 0 & 0 & 0 & 0 & 1 & 0 & 1 \\
0 & 0 & 0 & 0 & 0 & 0 & 0 & 1 & 1 & 0 \\
0 & 0 & 0 & 0 & 0 & 0 & 1 & 0 & 0 & 1 \\
0 & 0 & 0 & 0 & 0 & 0 & 1 & 0 & 1 & 0 \\
0 & 0 & 0 & 0 & 0 & 0 & 1 & 1 & 0 & 0 \\
0 & 0 & 0 & 1 & 1 & 1 & 0 & 0 & 0 & 0 \\
0 & 1 & 1 & 0 & 0 & 1 & 0 & 0 & 0 & 0 \\
1 & 0 & 1 & 0 & 1 & 0 & 0 & 0 & 0 & 0 \\
1 & 1 & 0 & 1 & 0 & 0 & 0 & 0 & 0 & 0
\end{bmatrix}
$$

numbers. The complete canonical adjacency matrix is shown in Table 8.4. Having found canonical labeling, illustrated at bottom right in Figure 8.4, we can start construction of the canonical spanning tree.

8.8   WALK ABOVE CODE FOR POLYCYCLIC GRAPHS

Before continuing, let us outline the construction of unique binary code for acyclic and cyclic graphs. The recently introduced Walk Above code [59], represents a modification and important upgrading of the Walk Around code proposed by mathematician R. C. Read [62]. The Walk Around binary code was based on circling around an acyclic graph with unlabeled vertices drawn in a plane. The code records the history of the walk around a graph by writing 0 when one observes an edge for the first time and by writing 1 as one comes around for the second time to the same edge. The deficiency of the Walk Around codes is that they are not unique as they depend on the selection of the starting vertex and the sense of circling around the graph. The Walk Above codes differ from the Walk Around codes in two respects: (i) they consider labeled graphs only, and (ii) they start the walk at vertex 1 and move toward the neighboring vertex having the *smallest* label, continuing the walk and, always, if is there is a choice, (iii) giving preference to vertices not previously visited, and (iv) selecting the vertex with the smallest label. In this way, one obtains a *unique* binary code for a graph cyclic, polycyclic, or acyclic.

Let us first recall the rules for performing the Walk Above and arriving at the Walk Above code:

1. Start with the vertex having the smallest label and move to the vertex having the next smallest label.
2. Continue the same until you come to the terminal vertex. For each passed edge, add 0 in the code.

3. On returning to each passed edge, add 1 in the code until you come to a branching vertex. Continue moving toward the vertex with the smallest label, giving priority to vertices not visited before. Again add 0 to the code for bonds not visited before and adding 1 for bonds visited before.
4. End when all bonds have been crossed twice and you have returned to the starting vertex.
5. If you come to a vertex that you have previously crossed, treat it as a terminal vertex and start returning. In this way, one will never cross any edge more than twice.

We will consider now adamantane, for which we have already described the assignment of the canonical labels. We have selected adamantane for the following reasons: (i) It requires about half a dozen trials to find canonical labels one labeling among $10! = 3,628,800$; (ii) it is a beautiful molecule, and (iii) it was, for the first time, synthesized by Prelog and Seiwerth [63] in our neighborhood, Zagreb, Croatia, about 80 years ago. Admittedly, at the time, some of us were not yet around.

8.8.1 WALK ABOVE CODE FOR ADAMANTANE

Having canonical labels for vertices of adamantane, we can initiate construction of the Walk Above code. To obtain the Walk Above code, one starts with vertex label 1, and following the rules listed before, arrive at the walk:

$$1 - 9 - 3 - 8 - 2 - 10 - 4 - 7 - 5 - 9 - 5 - 7 - 6 - 8 -$$
$$6 - 7 - 4 - 10 - 1 - 10 - 2 - 8 - 3 - 9 - 1.$$

After one has completed the walk, one has to select the ring closure bonds. One starts with the smallest label (1) and erases the edge involving vertex 1 and the adjacent vertex having the smaller label, which is vertex 9. Next, one looks for the next smallest label that belongs to a closed ring, which, in this case, is vertex 2, a member of a ring $(2 - 8 - 6 - 7 - 4 - 10 - 2)$. One therefore erases the edge $(2 - 8)$. The next vertex with the smallest label is vertex 3, which belongs to the ring $(3 - 8 - 6 - 7 - 5 - 9 - 3)$, which is broken by erasing edge $3 - 8$. In this way, one obtains a canonical spanning tree with ring closure edges $1 - 9$, $2 - 8$, and $3 - 8$, which is shown in Figure 8.5. The terminal spanning tree matrix is shown in Table 8.5 at left.

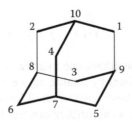

FIGURE 8.5 The canonical spanning tree of adamantane.

TABLE 8.5
The Terminal Matrix of the Canonical Spanning Tree and the Ring Closure Matrix for Adamantane

	1	2	3	8
1	0	2	6	5
2	2	0	6	5
3	6	6	0	5
8	5	5	5	0

	1	2	3	8	9
1	2	2	6	5	5
2	2	2	6	5	5
3	6	6	2	5	1
8	5	5	5	3	4
9	5	5	1	4	3

8.8.2 WALK ABOVE BINARY CODE

Let us return to the list of vertices in the Walk Above the graph of adamantane. In order to arrive at the binary code for adamantane as one is moving above the graph, one assigns 0 for each edge crossed for the first time and label 1 when crossing the same edge for the second time. The code starts with five zeros as one starts from vertex 1 and reaches vertex 10, passing vertices 9, 3, 8, and 2. To continue, we have to choose between going to vertex 1 or vertex 4. According to our rules, we go to vertex 4 despite it having larger magnitude because we have been at vertex 1 before (this was our starting vertex). Arriving at vertex 7, we have a choice between two vertices, 5 and 6, not visited before, so we choose the smaller label 5 and arrive at vertex 9. So far, all nine edges that we crossed were crossed for the first time, so the binary codes starts with nine initial zeros. From vertex 9, we have to return back to vertex 7, which gives for the code so far 0 0 0 0 0 0 0 0 0 1 1. In continuation, we go to vertex 6 and 8 and return back to 7, which gives for continuation of the code 0 0 1 1. From vertex 7, we return to vertex 10, which adds 1 1 to the code. Before returning to vertices 2, 8, 3, 9, and 1, we have to visit vertex 1 because the edge (10, 1) has not been so far crossed, which gives for the end part of the code 0 1 for edge (10,1) and 1 1 1 1 1 for bonds 10 – 2 – 8 – 3 – 9 – 1. So the Walk Above binary code of adamantane is 24 binary digits:

$$0 0 0 0 0 0 0 0 0 1 1 0 0 1 1 1 1 0 1 1 1 1 1 1.$$

Following Rouse-Ball [64], one can replace 0 and 1 in the code with left and right brackets which give

$$(((((((( )) (( )))) ( ))))).$$

This trivial sign substitution allows one to arrive at a geometrical representation of the binary code. One continues by joining adjacent left and right brackets into a circle, which transforms the above code to set of inscribed circles (ellipses) shown in Figure 8.6. Interested readers can further explore the geometrical representation of the binary code to find out that dual of the set of inscribed circles allows one to construct the spanning tree and identify the ring closure bonds [65].

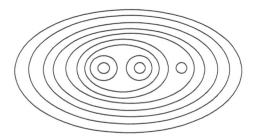

FIGURE 8.6 The geometrical representation of the canonical binary Walk Above code of adamantane. Dual of this graph will give the spanning trees of adamantane with double labels for ring closure bonds.

8.9 RING CLOSURE MATRIX OF ADAMANTANE

The ring closure matrix is obtained by augmenting the terminal matrix of the canonical spanning tree by including the ring closure vertices that are not terminal and using diagonal entries to indicate for each vertex its degree. The so augmented matrix for adamantane is shown in Table 8.5 at right. Using diagonal entries to indicate the valence of vertices may facilitate reconstruction of a cyclic graphs, but may be viewed as optional, when canonical labels are known and used for reconstruction.

Reconstruction of a graph from the ring closure matrix is fairly straightforward after the augmented canonical matrix has been constructed. Construction of the spanning tree may call for a few trial-and-error steps, but one is assured by the theorem by Zaretsky that the reconstruction is guaranteed. One starts with the reconstruction of the canonical spanning tree. In Figure 8.7a and c, we show reconstruction of the adamantane graph from the ring closure matrix, in three steps. At Figure 8.7a is shown the beginning of the reconstruction: One starts by combining the two largest ring closure matrix elements $(1, 3) = (2, 3) = 6$ with the information on the distance $(1, 2) = 2$. This allows construction of a tree with tree terminal ring closure vertices, the vertices 1, 2, and 3. Next, we need to locate two additional ring closure vertices, 8 and 9. In the next step, one should use the information on the distance $(3, 9) = 1$, which located the vertex 9 indicated in Figure 8.7a. It remains to find the position of vertex 8, which is at equal distance from the vertices 1, 2, and 3. For this, we have first to locate the vertex that is at equal distance from the vertices 1, 2, and 3, which is the central vertex of the so far reconstructed graph. Next we have to add two more vertices in order to obtain the required distance of 5 between the vertices 1, 2, 3, and vertex 8. The reconstructed resulting spanning tree of adamantane is shown at Figure 8.7c.

After reconstruction of the canonical spanning tree, we have to introduce the ring closure edges $(1, 9)$, $(2, 8)$, and $(3, 8)$, which is easy because we already know the locations of vertices 1, 2, 3, 8, 9. In Figure 8.7b is shown the reconstructed graph of adamantane, and in Figure 8.7d is shown the same graph depicted in a more symmetrical form, which allows it to be recognized as a molecular graph of adamantane.

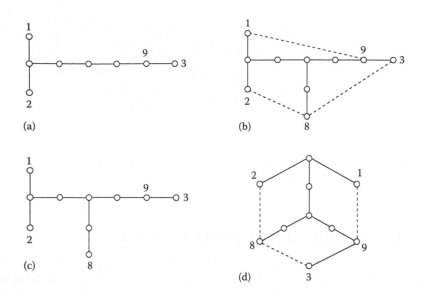

FIGURE 8.7 Reconstruction of adamantane graph from the canonical ring closure matrix. (a) The beginning and (c) full reconstruction of the canonical spanning tree; (b) reconstruction of adamantine graph; (d) adamantine graph having more symmetrical form.

8.9.1 ADDITIONAL ILLUSTRATION: FIBONACENES

For illustration of construction of canonical spanning trees based on canonical labels and subsequent construction of the ring closure matrix, we selected fibonacenes [66–70]. These are cata-condensed benzenoid hydrocarbons in which every benzene ring except the two terminal rings is a "kink" ring (in the terminology of Gordon and Davison [71]). That is, each following ring along the chain of fused benzene rings is in a *cis* rather than *trans* position. As a consequence, the algorithm of Gordon and Davison for the count of Kekulé valence structures (K) in cata-condensed benzenoid hydrocarbons gives 2, 3, 5, 8, 13, 21… and corresponds to benzene (K = 2), naphthalene (K = 2 + 1), phenanthrene (K = 2 + 3), chrysene (K = 3 + 5), picene (K = 5 + 8), fulminene (K = 8 + 13), and so on. The numbers, 2, 3, 8, 13, 21, and so on are the well-known Fibonacci numbers.

In Figure 8.8, we show molecular structures of the leading members of Fibonacenes phenanthrene, chrysene, picene, and fulminene, having 14, 18, 22, and 26 carbon atoms. For more on the properties of fibonacenes see [66,68–72] and a review on Fibonacci numbers in the topological theory of benzenoid hydrocarbons and related graphs by El-Basil and Klein [73]. In the first row of Figure 8.8, we also show the canonical labels for the four fibonacenes, which are not difficult to find. Clearly, the small labels 1, 2, 3, etc., have to be placed on carbon atoms of degree 2, but the terminal benzene rings are optimal locations as there are four carbon atoms adjacent, which makes possible for carbon atoms 1, 2, and 3 each to have only two neighbors.

The canonical labeling of fibonacenes appears chaotic, but in fact, there is a hidden regularity, which can be seen in the second row of Figure 8.8, which illustrates the canonical spanning trees for the four fibonacenes, which have simple appearance. They

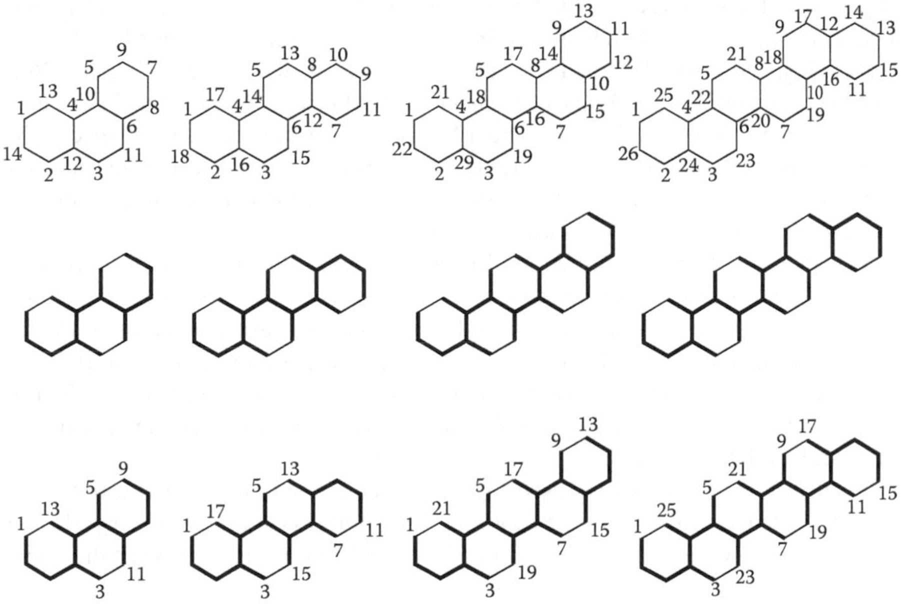

FIGURE 8.8 In the first row for the leading members of fibonacenes: phenanthrene, chrysene, picene, and fulminene (first row) we show their canonical labels. In the second row we show their canonical spanning trees. In third row we show canonical labels of the spanning trees, which belong to the ring closure bonds.

are constructed by erasing in each benzene ring the CC bond that has the smallest canonical label. Thus, in phenanthrene, the smallest label 1 is in the terminal benzene ring, and the adjacent carbon with the smaller label is carbon 13. Hence, C_1–C_{13} is the ring closure bond to be erased. In the next benzene ring (at right), the smallest label is 3, and the adjacent carbon with the smaller label is carbon 11. Hence, the C_3–C_{11} bond is the ring closure bond to be erased. Finally, in the last benzene ring of phenenthrene, the smallest label is 5, and the adjacent carbon with the smaller label is carbon 9. Hence, the C_5–C_9 bond is the ring closure bond to be erased. In the case of chrysene, the ring closure bonds are C_1–C_{17}, C_3–C_{15}, C_5–C_{13}, and C_7–C_{11}. The same pattern for the labels of the ring closure continues in higher fibonacenes: The smallest labels 1, 3, 5, 7, 9 increase regularly, and the labels of the adjacent ring closure carbon atoms for the terminal ring jump by +4. Similar regularity in canonical labels is reflected also on carbon atoms of degree 3, which form the branching graph of fibonacenes. Branching graphs for cata-condensed benzenoid hydrocarbons were introduced Kirby and Gutman [74], and in the case of the family of fibonacenes shown in Figure 8.8, they are regular chain graphs superimposed on graphite lattice, in which three consecutive CC bonds alternate in a *cis* and *trans* CC pattern.

In the last row of Figure 8.8, we again show the canonical spanning trees of the four fibonacenes with the labels of the terminal vertices (which in the case of chrysene are 1, 3, 5, 7, 11, 13, 15, 17). This is how the canonical tree would be reconstructed according to the Zaretsky theorem. But because we know the pattern of paring of the smallest and the largest canonical labels, it is easy to identify and recover the ring closure bonds (which in the case of chrysene are bonds C_1–C_{17}, C_3–C_{15}, C_5–C_{13}, and C_7–C_{11}). Thus instead of using 14 × 14, 18 × 18, 22 × 22, and 26 × 26 adjacency matrices to represent phenanthrene, chrysene, picene, and fulminene, we can use 6 × 6, 8 × 8, 10 × 10, and 12 × 12 ring closure matrices, respectively, without any loss of information. In Table 8.6, we show the ring closure matrices for phenanthrene and crysene, which are the same as their canonical terminal matrices, because all ring closure vertices are terminal.

In conclusion, let us recall that until now the canonical labels have been used on *individual* molecules. This is the first time that the canonical labels have been constructed for a *family* of molecules, if one disregards trivial cases: path graphs, monocyclic graphs, and star graphs. Hence, in this chapter, not only have we introduced the novelty of ring closure matrices, but we have opened a new topic in the field of canonical labels concerned with regularities of the canonical labels for families of structurally related molecules.

Finally, in the view that we used highly aromatic fibonacenes for illustration of canonical labels, construction of canonical spanning trees, and their ring closure matrices, it is in order to remind readers of a few articles on aromaticity of these and

TABLE 8.6

The Terminal Matrices of the Canonical Spanning Trees and the Ring Closure Matrices for Phenanthrene and Chrysene Ring Closure Matrices for Phenanthrene and Chrysene

Phenanthrene

	1	3	5	9	11	13
1	0	4	6	8	7	5
3	4	0	4	6	5	3
5	6	4	0	5	3	3
9	8	6	5	0	4	6
11	7	5	3	4	0	4
13	5	3	3	6	4	0

Chrysene

	1	3	5	7	11	13	15	17
1	0	4	6	8	11	9	7	5
3	4	0	4	6	9	7	5	3
5	6	4	0	4	7	5	3	3
7	8	6	4	0	5	3	3	5
11	11	9	7	5	0	4	6	8
13	9	7	5	3	4	0	4	6
15	7	5	3	3	6	4	0	4
17	5	4	3	5	8	6	4	0

Phenanthrene

	1	3	5	9	11	13
1	2	4	6	8	7	5
3	4	2	4	6	5	3
5	6	4	2	5	3	3
9	8	6	5	2	4	6
11	7	5	3	4	2	4
13	5	3	3	6	4	2

Chrysene

	1	3	5	7	11	13	15	17
1	2	4	6	8	11	9	7	5
3	4	2	4	6	9	7	5	3
5	6	4	2	4	7	5	3	3
7	8	6	4	2	5	3	3	5
11	11	9	7	5	2	4	6	8
13	9	7	5	3	4	2	4	6
15	7	5	3	3	6	4	2	4
17	5	4	3	5	8	6	4	2

related benzenoid and non-benzenoid compounds because these compounds may be next in line to be similarly studied. First, let us mention several recent publications that relate to characterization of aromaticity in terms of conjugated circuits and ring currents [75–80]. Conjugated circuits have led to quantitative representation of the qualitative Clar Aromatic Sextet Theory [81,82]. These recent articles supplement and update an earlier comprehensive review on aromaticity [83], which covered the topic of aromaticity extensively. As one can see, the topic of aromaticity continues to expand in its width and depth.

In conclusion of this chapter, which involved canonical labels based on the smallest binary number for graphs when the rows of the adjacency matrix are read from left to right and from top to bottom, it seems appropriate to comment on the novelty and lack of its wider use despite early recognition of merits, at least as mentioned in "News and Views" in *Nature* in 1974 [84], the year the paper on canonical labeling of vertices [85] was published. Use of the canonical labels was illustrated in a number of publications for determining the symmetry of chemical rearrangement graphs [86–97] during the following dozen years. Despite its demonstrated utility, canonical labeling, at best, was receiving limited attention.

This may be not surprising and even be exceptional as it is not uncommon for novelty, while being recognized by a few, to be overlooked by the majority. Rather than arguing in favor of assisting those who do not know much about canonical labels, we have in Appendix 13 reproduced a paragraph from the report titled, "Toward Chemical Topology," from *Nature* in 1974, written by their correspondent, which speaks for itself.

REFERENCES AND NOTES

1. E. B. Wilson, *Introduction to Scientific Research*, McGraw-Hill, New York, 1952.
2. M. Randić, M. Novič, and D. Plavšić, Common vertex matrix: A novel characterization of molecular graphs by counting, *J. Comput. Chem.* 34 (2013) 1409–1419.
3. M. Randić, M. Novič, M. Vračko, and D. Plavšić, On the centrality of vertices of molecular graphs, *J. Comput. Chem.* 34 (2013) 1409–1419.
4. A. T. Balaban, Topological indices based on topological distances in molecular graphs, *Pure Appl. Chem.* 55 (1983) 199–206.
5. A. T. Balaban, Highly discriminating distance–based topological index, *Chem. Phys. Lett.* 89 (1982) 399–404.
6. A. T. Balaban and O. Ivanciuc, Historical development of topological indices, in: *Topological Indices and Related Descriptors in QSAR and QSPR*, J. Devillers and A. T. Balaban (Eds.), Gordon and Breach, Amsterdam, 1999, pp. 21–57.
7. M. Randić, T. Pisanski, M. Novič, and D. Plavšić, Novel graph distance matrix, *J. Comput. Chem.* 31 (2010) 1832–1841.
8. M. Randić, D_{MAX}—Matrix of dominant distances in a graph, *MATCH—Commun. Math. Comput. Chem.* 70 (2013) 221–238.
9. M. Randić, R. Orel, and A. T. Balaban, D_{MAX} matrix invariants as graph descriptors. Graphs having the same Balaban index *J*, *MATCH—Commun. Math. Comput. Chem.* 70 (2013) 239–258.
10. O. Ivanciuc, Chemical graphs, molecular matrices and topological indices in chemoinformatics and quantitative structure-activity relationships, *Curr. Comput. Aided Drug Design* 9 (2013) 153–163.

11. S. J. Barigye, Y. Marrero-Ponce, O. Martínez Santiago, Y. Martínez López, F. Pérez-Giménez, and F. Torrens, Shannon's, mutual, conditional and joint entropy information indices: Generalization of global indices defined from local vertex invariants, *Curr. Comput. Aided Drug Design* 9 (2013) 164–183.

12. B. Lučić, I. Sović, J. Batista, K. Skala, D. Plavšić, D. Vikić-Topić, D. Beslo, S. Nikolić, and N. Trinajstić, The sum-connectivity index—An additive variant of the Randić connectivity index, *Curr. Comput. Aided Drug Design* 9 (2013) 184–194.

13. S. D. Bolboacă, L. Jäntschi, and M. V. Diudea, Molecular design and QSARs/QSPRs with molecular descriptors family, *Curr. Comput. Aided Drug Design* 9 (2013) 195–205.

14. A. A. Toropov, A. P. Toropova, E. Benfenati, and G. Gini, OCWLGI descriptors: Theory and praxis, *Curr. Comput. Aided Drug Design* 9 (2013) 226–232.

15. O. Ivanciuc, T. Ivanciuc, and D. J. Klein, Flow network QSAR for the prediction of physicochemical properties by mapping an electrical resistance network onto a chemical reaction poset, *Curr. Comput. Aided Drug Design* 9 (2013) 233–240.

16. A. Cayley, A memoir on the theory of matrices. *Philos. Trans. Roy. Soc. London* 148 (1858) 17–37.

17. See: N. J. Higham, *Cayley, Sylvester, and Early Matrix Theory*, Manchester Institute for Mathematical Sciences, School of Mathematics, The University of Manchester, Manchester, MIMS EPrint: 2007.119.

18. R. Reams, Hadamard inverses, square roots and products of almost semidefinite matrices, *Linear Algebra and it Applications*, 288 (1999) 35–43.

19. A. Y. Alfakih, On the nullspace, the rangespace and the characteristic polynomial of Euclidean distance matrices, *Linear Algebra Appl.* 416 (2006) 348–354.

20. J. Dattorro, *Convex Optimization and Euclidean Distance Geometry*, Meboo Publishing, Stanford, 2005.

21. I. J. Schoenberg, Remarks to Maurice Frechet's article: Sur la definition axiomatique d'une classe d'espace distancies vectoriellement applicable sur l'espace de Hilbert, *Ann. Math.* 36 (1935) 724–732.

22. M. Kunz, About eigenvalues of quadratic distance matrices of alkanes, Firefox HTLM Document.

23. M. Kunz, Inverses of perturbed Laplace-Kirchhoff and some distance matrices, *MATCH, Commun. Math. Comput. Chem.* 32 (1995) 221–234.

24. M. Kunz, On equivalence relation between distance and coordinate matrices, *MATCH, Commun. Math. Comput. Chem.* 32 (1995) 193–203.

25. M. Kunz, A Möbius inversion of the Ulam subgraphs conjecture, *J. Math. Chem.* 9 (1992) 297–305.

26. M. Kunz, Distance matrices yielding angles between arcs of the graphs, *J. Chem. Inf. Comput. Chem.* 34 (1994) 957–959.

27. M. Kunz, Path and walk matrices of trees, *Coll. Czech. Chem. Commun.* 54 (1989) 2148–2155.

28. M. Kunz, On topological and geometrical matrices, *J. Math. Chem.* 13 (1993) 145–151.

29. O. Ivanciuc, T. S. Balaban, and A. T. Balaban, Chemical graphs with degenerate topological indices based on information on distances, *J. Math. Chem.* 14 (1993) 21–33.

30. D. Plavšić, S. Nikolić, N. Trinajstić, and Z. Mihalić, On the Harary index for the characterization of chemical graphs, *J. Math. Chem.* 12 (1993) 235–250.

31. O. Ivanciuc, T.-S. Balaban, and A. T. Balaban, Design of topological indices. Part 4. Reciprocal distance matrix, related local vertex invariants and topological indices, *J. Math. Chem.* 12 (1993) 309–318.

32. B. Lučić, A. Miličević, S. Nikolić, and N. Trinajstić, Harary index—Twelve years later, *Croat. Chem. Acta* 75 (2002) 847–868.

33. A. T. Balaban, D. Mills, O. Ivanciuc, and S. C. Basak, Reverse Wiener indices, *Croat. Chem. Acta* 73 (2000) 923–941.

34. M. Diudea, T. Ivanciuc, S. Nikolić, and N. Trinajstić, Matrices of reciprocal distance, polynomials and derived numbers, *MATCH Commun. Math. Comput. Chem.* 35 (1997) 41–64.

35. J. Devillers and A. T. Balaban (Eds.), *Topological Indices and Related Descriptors in QSAR and QSPR*, Amsterdam, Netherlands, Gordon and Breach, 1999.

36. J. Devillers and A. T. Balaban (Eds.), *Topological Indices and Related Descriptors in QSAR and QSPR*, Amsterdam, Netherlands, Gordon and Breach, 1999, p. 40.

37. J. Devillers and A. T. Balaban (Eds.), *Topological Indices and Related Descriptors in QSAR and QSPR*, Amsterdam, Netherlands, Gordon and Breach, 1999, p. 111.

38. J. Devillers and A. T. Balaban (Eds.), *Topological Indices and Related Descriptors in QSAR and QSPR*, Amsterdam, Netherlands, Gordon and Breach, 1999, p. 202.

39. J. Devillers and A. T. Balaban (Eds.), *Topological Indices and Related Descriptors in QSAR and QSPR*, Amsterdam, Netherlands, Gordon and Breach, 1999, p. 227.

40. M. Randić and J. Zupan, On the interpretation of the well-known topological indices, *J. Chem. Inf. Comput. Sci.* 41 (2001) 550–560.

41. H. Wiener, Structural determination of paraffin boiling points, *J. Am. Chem. Soc.* 69 (1947) 17–20.

42. B. Horvat, G. Jaklič, I. Kavkler, and M. Randić, Rank of Hadamard powers of Euclidean distance matrices, *J. Math. Chem.* 52 (2014) 729–740.

43. K. A. Zaretsky, Reconstruction of a tree from the distance between its pendant vertices, *Uspekhi Math. Nauk (Russian Math. Surveys)* 20 (1965) 90–92.

44. Y. A. Smolenski, A method for linear recording of graphs, *USSR Comput. Math. Phys.* 2 (1969) 396–397.

45. V. Levenshtein, E. Konstatinova, E. Konstatinov, and S. Molodstov, Reconstruction of a graph from 2-vicinities of its vertices, *Discrete Appl. Math.* 156 (200) 1399–1406.

46. B. Horvat, T. Pisanski, and M. Randić, Terminal polynomials and star-like graphs, *MATCH—Commun. Math. Comput. Chem.* 60 (2008) 493–512.

47. M. Randić, Citations versus limitations of citations: Beyond Hirsch index, *Scientometrics* 80 (2009) 809–818.

48. A. T. Balaban and M. Randić, Proposal for using an untapped source of citations characteristic scientific areas, *Scientometrics* 49 (2000) 517–521.

49. T. Pisanski, D. Plavšić, and M. Randić, On numerical characterization of cyclicity, *J. Chem. Inf. Comput. Sci.* 40 (2000) 520–523.

50. D. Plavšić, D. Vukičević, and M. Randić, On canonical numbering of carbon atoms in fullerenes: C_{60} buckminsterfullerene, *Croat. Chem. Acta* 78 (2005) 493–502.

51. M. Randić, S. Nikolić, and N. Trinajstić, Compact codes: On nomenclature of acyclic chemical compounds, *J. Chem. Inf. Comput. Sci.* 35 (1995) 357–365.

52. M. Randić, Compact molecular codes, *J. Chem. Inf. Comput. Sci.* 26 (1986) 136–148.

53. M. Randić, On unique numbering of atoms and unique codes for molecular graphs, *J. Chem. Inf. Comput. Sci.* 15 (1975) 105–108.

54. H. Hauptman and J. Karle, Phases of Fourier coefficients directly from crystal diffraction data, *Acta Cryst.* 1 (1948) 70–75.

55. H. Hauptman and J. Karle, The probability distribution of the magnitude of a structure factor. I. The centrosymmetric crystal, *Acta Cryst.* 6 (1953) 131–135.

56. H. A. Hauptman (as told to and edited by D. J. Grothe), *On the Beauty of Science (A Nobel Laureate Reflects on the Universe, God, and the Nature of Discovery)*, Prometheus Books, Amherst, NY.

57. I. Gutman, A formula for the Wiener number of trees and its extension to graphs containing cycles, *Graph Theory Notes N. Y.* 27 (1994) 9–15.

58. M. Randić, On generalization of Wiener index for cyclic structures, *Acta Chim. Slov.* 49 (2002) 483–496.

59. T. Pisanski and M. Randić, Use of the Szeged index and the revised Szeged index for measuring network bipartivity, *Discrete Appl. Math.* 158 (2010) 1936–1944.

60. M. Randić, M. Novič, and D. Plavšić, Ring closure matrix, in: *Current Organic Chemistry*, M. V. Putz (Ed.), Special Issue Hot Topic: Graph Theory and Molecular Topology in Organic Chemistry, in press.

61. P. John and H. Sachs, Calculating the characteristic polynomial, eigenvectors and number of spanning trees of a hexagonal system, *J. Chem. Soc. Faraday Trans.* 86 (1990) 1033–1039.

62. R. C. Read, The coding of various kinds of unlabeled trees, in: *Graph Theory and Computing*, R. C. Read (Ed.), Academic Press, New York, 1972, pp. 153–182.

63. V. Prelog and R. Seiwerth, Über die Synthese des Adamantans, *Berichte der Deutschen Chemischen Gesellschaft (A and B Series)*, 74 (1941) 1644–1648. doi:10.1002 /cber.19410741004.

64. W. W. Rouse Ball, *Mathematical Recreations and Essays*, Macmillan, New York, 1960.

65. M. Randić, M. Novič, and D. Plavšić (work in progress).

66. A. T. Balaban, Chemical graphs. 50. Symmetry and enumeration of fibonacenes (unbranched catacondensed benzenoids isoarithmic with helicenes and zigzag catafusenes), *MATCH Commun. Math. Comput. Chem.* 24 (1989) 29–38.

67. A. T. Balaban and F. Harary, Chemical graphs V. Enumeration and proposed nomenclature of benzenoid cata-condensed polycyclic aromatic hydrocarbons, *Tetrahedron* 24 (1968) 2505–2516.

68. I. Gutman and S. J. Cyvin, *Introduction to the Theory of Benzenoid Hydrocarbons*, Springer–Verlag, Berlin, 1989.

69. I. Gutman and S. Klavžar, Chemical graph theory of fibonacenes, *MATCH Commun. Math. Comput. Chem.* 55 (2006) 39–54.

70. I. Gutman, A. T. Balaban, M. Randić, and C. Kiss-Tóth, Partitioning of π-electrons in rings of fibonacenes, *Z. Naturforsch.* 60a (2005) 171–176.

71. M. Gordon and W. H. T. Davison, Theory of resonance topology of fully aromatic hydrocarbons, *J. Chem. Phys.* 20 (1952) 428–435.

72. I. Gutman and S. Klavžar, Chemical graph theory of fibonacenes, *MATCH Commun. Math. Comput. Chem.* 55 (2006) 39–54.

73. S. El-Basil and D. J. Klein, Fibonacci numbers in the topological theory of benzenoid hydrocarbons and related graphs, *J. Math. Chem.* 3 (1989) 1–23.

74. E. C. Kirby and I. Gutman, The branching graphs of polyhexes: Some elementary theorems, conjectures and open questions, *J. Math. Chem.* 13 (1993) 359–374.

75. M. Randić, π-Electron currents in polycyclic conjugated hydrocarbons of decreasing aromatic character and a novel structural definition of aromaticity, *Open Org. Chem. J.* 5 (Suppl. 1-M2) (2011) 11–26.

76. M. Randić, M. Novič, and D. Plavšić, π-Electron currents in fixed π-sextet aromatic benzenoids, *J. Math. Chem.* 50 (2012) 2755–2774.

77. M. Randić, D. Vukičević, M. Novič, and D. Plavšić, π-Electron currents in larger fully aromatic benzenoids, *Int. J. Quantum Chem.* 112 (2012) 2456–2462.

78. M. Randić, M. Novič, M. Vračko, D. Vukičević, and D. Plavšić, π-Electron currents in polycyclic conjugated hydrocarbons: Coronene and its isomers having five and seven member rings, *Int. J. Quantum Chem.* 112 (2012) 972–985.

79. M. Randić, D. Vukičević, A. T. Balaban, M. Vračko, and D. Plavšić, Conjugated circuits currents in hexabenzocoronene and its derivatives formed by joining proximal carbons, *J. Comput. Chem.* 33 (2012) 1111–1122.

80. M. Randić, Graph theoretical approach to π-electron currents in polycyclic conjugated hydrocarbons, *Chem. Phys. Lett.* 500 (2010) 123–127.

81. M. Randić, Novel insight into Clar's aromatic π-sextets, *Chem. Phys. Lett.* 601 (2014) 1–5.

82. M. Randić and D. Plavšić, Algebraic Clar formulas—Numerical representation of Clar Structural formulas, *Acta Chim. Slov.* 58 (2011) 448–457.
83. M. Randić, Aromaticity of polycyclic conjugated hydrocarbons, *Chem. Rev.* 103 (2003) 3449–3605.
84. From our molecular physics correspondent: Toward chemical topology, news and views, *Nature* 252 (1974) 188–189.
85. M. Randić, On the recognition of identical graphs representing molecular topology, *J. Chem. Phys.* 60 (1974) 3920–3928.
86. M. Randić, D. J. Klein, V. Katović, D. O. Oakland, W. A. Seitz, and A. T. Balaban, Symmetry properties of chemical graphs. X. Rearrangement of axially distorted octahedral, *Stud. Phys. Theor. Chem.* 51 (1987) 266–284.
87. M. Randić, D. O. Oakland, and D. J. Klein, Symmetry properties of chemical graphs. IX. The valence tautomerism in the P_7^{3-} skeleton, *J. Comput. Chem.* 7 (1986) 35–54.
88. M. Randić, Symmetry properties of chemical graphs. VIII. On complementarity of isomerization models, *Theor. Chim. Acta* 67 (1985) 137–155.
89. M. Randić, V. Katović, and N. Trinajstić, Symmetry properties of chemical graphs. VII: Enatiomers of tetragonal-pyramidal rearrangements, *Stud. Phys. Theor. Chem.* 23 (1983) 399–408.
90. M. Randić and M. I. Davis, Symmetry properties of chemical graphs. VI. Isomerization of octahedral complexes, *Int. J. Quantum Chem.* 26 (1984) 69–89.
91. M. Randić, Symmetry properties of chemical graphs. V. Internal rotation in $XY_3XY_2XY_3$, *J. Comput. Chem.* 4 (1983) 73–83.
92. M. Randić and V. Katović, Symmetry properties of chemical graphs. IV. Rearrangement of tetragonal pyramidal complexes, *Int. J. Quantum Chem.* 21 (1982) 647–663.
93. M. Randić, G. M. Brissley, and C. L. Wilkins, Computer perception of topological symmetry via canonical numbering of atoms, *J. Chem. Inf. Comput. Sci.* 21 (1981) 52–59.
94. M. Randić, Symmetry properties of graphs of interest in chemistry. III. Homotetrahedryl rearrangement, *Int. J. Quantum Chem.: Quantum Chem. Symp.* 14 (1980) 557–577.
95. M. Randić, Symmetry properties of graphs of interest in chemistry. II. Desargues-Levi graph, *Int. J. Quantum Chem.* 15 (1979) 663–682.
96. M. Randić, A systematic study of symmetry properties of graphs. I. Petersen graph, *Croat. Chem. Acta* 49 (1977) 643–655.
97. M. Randić, On discerning symmetry properties of graphs, *Chem. Phys. Lett.* 42 (1976) 283–287.

9 On Highly Similar Molecules

9.1 SEARCH FOR STRUCTURALLY SIMILAR MOLECULES

Searching for structurally *very similar* molecules is a relatively new direction in structure–property–activity studies, which is gaining momentum in view of considerable interest in screening huge combinatorial libraries for a few structures of great similarity to selected standards. Such explorations will undoubtedly receive wide attention in view of the astounding success of Professor Lahana and coworkers, who introduced a computer-assisted method for identifying molecules having a desired pharmacological activity by screening huge combinatorial libraries for the target structure [1]. Starting with a single biologically active molecule and selecting a set of molecular descriptors as a tool in searching a combinatorial bank, they were able to identify about two dozen potentially interesting molecules out of the pool of 100,000 virtual structures. By virtual structures, we mean structures that exist in the computer, that is, in computer-created combinatorial libraries a possible minor number of which may have also been earlier synthesized.

In our view, this (now more than 15 years old) approach represents a *significant* advance in structure–activity studies that ought to have greater visibility in the SAR and QSAR community. Instead, using the words of Lahana and coworkers in outlining their work, we prefer to cite the complete abstract of their publication as this will best describes the novelty of their approach [1]:

> We describe the rational design of immunosuppressive peptides without relying on information regarding their receptors or mechanisms of action. The design strategy uses a variety of topological and shape descriptors in combination with an analysis of molecular dynamics trajectories for the identification of potential drug candidates. This strategy was applied to the development of immunosuppressive peptides with enhanced potency. The lead compounds were peptides, derived from the heavy chain of HLA class I, that modulate immune responses in vitro and in vivo. In particular, a peptide derived from HLA-B2702, amino acids 75-84 (2702.75-84) prolonged skin and heart allograft survival in mice. The biological activity of the rationally designed peptides was tested in a heterotopic mouse heart allograft model. The molecule predicted to be most potent displayed an immunosuppressive activity approximately 100 times higher than the lead compound.

With this, we would like to end the introduction into justification for searching for novel molecular descriptors specifically designed for searching combinatorial libraries and highly similar structures. We hope that this may alert researchers looking for novel drugs to discuss this problem and, despite a few additional applications, construction of molecular descriptors of very high discrimination power is essentially virgin territory worth further exploration.

Hence, one may consider construction of new molecular descriptors of very high discriminatory power specifically designed for screening combinatorial libraries one unsolved problem in this area. Observe that Lahana and coworkers used *available* molecular descriptors, most of which have limited discrimination powers and have been intended for use in MRA. There have been few molecular descriptors of very high discrimination power available, such as the molecular ID numbers [2,3–5], but they may not be suitable for screening combinatorial libraries because they are computationally intensive.

The topic of construction of molecular descriptors of very high discrimination power for QSAR studies has been very recently revisited [6–9]. Common to several of these novel molecular descriptors is use of the row sums of sparse matrices for construction of the generalized connectivity indices. These are indices closely related to the Balaban J index, the first molecular descriptor of higher power of discrimination, in that they used matrix row sums and the algorithm of the connectivity index involving $1/\sqrt{}$ but applied it to other than the adjacency matrix. Such indices have shown not only high discrimination power, but appear also to well characterize molecules so that they can be used to measure the degree of molecular similarity. We will illustrate construction of one such index for the map of Figure 7.6. These indices can be easily constructed and are not computationally intensive. The intended use of novel descriptors is for screening combinatorial libraries. Here, we will outline their use for characterization of proteome maps, which open an area of their additional applications.

When the row sums as entries in construction of the connectivity-type indices are applied to the adjacency matrix of molecular graphs, one obtains the connectivity index χ of Randić [10,11]. When they are applied to the graph distance matrix, one obtains Balaban's index J [12]. Both indices, χ and J, have been very useful in QSAR, which is an incentive to explore properties of additional structural indices of this type. At our disposal are several novel sparse matrices that have been recently introduced in chemical graph theory, which, in the view that their row sums cover a wider range of values, can be expected to produce novel highly discriminatory topological indices.

Consider the sparse matrix ^nD if possible for the proteome map of the adjacency matrix of the partial order diagram of Figure 7.7. The matrix elements of ^nD are given as the Hamming distance between the columns (rows) of the adjacency matrix A [13], shown in Matrix 9.1, but we consider *only* the matrix elements for adjacent vertices in the diagram of Figure 7.7.

For example, the vertices 1 and 7 are adjacent, and the corresponding rows are

| 0 | 0 | 0 | 0 | 1 | 0 | 1 | 0 | 0 | 0 | 0 | 0 | 1 | 0 | 0 | 1 | 0 | 0 | 0 | 0 |
| 1 | 0 | 0 | 0 | 0 | 0 | 0 | 0 | 0 | 0 | 0 | 0 | 0 | 0 | 0 | 0 | 0 | 0 | 1 | 1 |

The Hamming distance between these two binary sequences is 7, which is the number of columns in which the two sequences (of equal length) have different entries.

Incidentally, observe that, in this case, the result is the same as the sum of the valence (degrees) of the pair of adjacent vertices. Thus $v_1 + v_7 = 7$, where v_1 and v_7 are the degrees of the two vertices forming the edge $(1, 7)$. This is to be true for all triangle-free graphs as is the case with the partial order diagram of Figure 7.7.

Adjacency matrix for the partial order diagram for proteome map of Figure 7.7

$$
\begin{bmatrix}
0 & 0 & 0 & 0 & 1 & 0 & 1 & 0 & 0 & 0 & 0 & 0 & 1 & 0 & 0 & 1 & 0 & 0 & 0 & 0 \\
0 & 0 & 0 & 0 & 0 & 0 & 0 & 0 & 0 & 0 & 0 & 0 & 1 & 0 & 1 & 1 & 1 & 0 & 0 & 0 \\
0 & 0 & 0 & 0 & 1 & 0 & 0 & 0 & 0 & 0 & 0 & 0 & 0 & 0 & 0 & 0 & 0 & 0 & 1 & 0 \\
0 & 0 & 0 & 0 & 0 & 0 & 0 & 0 & 0 & 0 & 0 & 0 & 1 & 0 & 0 & 1 & 0 & 1 & 0 & 0 \\
1 & 0 & 1 & 0 & 0 & 1 & 0 & 0 & 1 & 0 & 0 & 0 & 0 & 0 & 0 & 0 & 0 & 0 & 0 & 1 \\
0 & 0 & 0 & 0 & 1 & 0 & 0 & 0 & 0 & 0 & 1 & 0 & 0 & 1 & 0 & 1 & 0 & 0 & 0 & 0 \\
1 & 0 & 0 & 0 & 0 & 0 & 0 & 0 & 0 & 0 & 0 & 0 & 0 & 0 & 0 & 0 & 0 & 0 & 1 & 1 \\
0 & 0 & 0 & 0 & 0 & 0 & 0 & 0 & 0 & 0 & 0 & 0 & 0 & 0 & 0 & 0 & 1 & 0 & 0 & 0 \\
0 & 0 & 0 & 0 & 1 & 0 & 0 & 0 & 0 & 0 & 0 & 0 & 0 & 0 & 0 & 0 & 0 & 0 & 1 & 0 \\
0 & 0 & 0 & 0 & 0 & 0 & 0 & 0 & 0 & 0 & 0 & 0 & 0 & 0 & 0 & 0 & 1 & 0 & 0 & 0 \\
0 & 0 & 0 & 0 & 0 & 1 & 0 & 0 & 0 & 0 & 0 & 0 & 1 & 0 & 0 & 0 & 1 & 0 & 0 & 0 \\
0 & 0 & 0 & 0 & 0 & 0 & 0 & 0 & 0 & 0 & 0 & 0 & 0 & 0 & 0 & 0 & 1 & 0 & 0 & 0 \\
1 & 1 & 0 & 1 & 0 & 0 & 0 & 0 & 0 & 0 & 1 & 0 & 0 & 0 & 0 & 0 & 0 & 0 & 0 & 0 \\
0 & 0 & 0 & 0 & 0 & 1 & 0 & 0 & 0 & 0 & 0 & 0 & 0 & 0 & 0 & 0 & 0 & 0 & 1 & 0 \\
0 & 1 & 0 & 0 & 0 & 0 & 0 & 0 & 0 & 0 & 0 & 0 & 0 & 0 & 0 & 0 & 0 & 0 & 0 & 1 \\
1 & 1 & 0 & 1 & 0 & 1 & 0 & 0 & 0 & 0 & 0 & 0 & 0 & 0 & 0 & 0 & 0 & 0 & 1 & 0 \\
0 & 1 & 0 & 0 & 0 & 0 & 0 & 1 & 0 & 1 & 1 & 1 & 0 & 0 & 0 & 0 & 0 & 0 & 0 & 0 \\
0 & 0 & 0 & 1 & 0 & 0 & 0 & 0 & 0 & 0 & 0 & 0 & 0 & 0 & 0 & 0 & 0 & 0 & 0 & 1 \\
0 & 0 & 1 & 0 & 0 & 0 & 1 & 0 & 1 & 0 & 0 & 0 & 0 & 1 & 0 & 1 & 0 & 0 & 0 & 0 \\
0 & 0 & 0 & 0 & 1 & 0 & 1 & 0 & 0 & 0 & 0 & 0 & 0 & 0 & 1 & 0 & 0 & 1 & 0 & 0 \\
\end{bmatrix}
$$

$$(9.1)$$

From the ^nD matrix (Matrix 9.2), it is easy to calculate the row sums, which, when ordered, give the sequence

$$45, 41, 39, 35, 33, 32, 31, 30, 28, 22, 22, 20, 14, 14, 13, 12, 12, 6, 6, 6.$$

The ordered row sums can be used as a proteome map code. Clearly, all maps of the same cell type, which are solely based on the partial order of the control group, will have the same map code. But that is desirable. If we assign to the proteome map of the control group having 20 spots numbers 1–20 as the sequence code of the proteome map with number 1 belonging to the most abundant protein spot and number 20 to the least abundant protein spot, then one can differentiate between different proteome maps having the same 20 proteins but having perturbed abundances. Maps within each group can be differentiated by ordering the spot labels, which are given by the relative initial abundances, relative to the row sums of the Hamming distance matrix for the partial order diagram of proteome map. The abundance code for the above map is then

$$16, 5, 19, 17, 1, 6, 2, 13, 20, 7, 11, 4, 3, 9, 14, 15, 18, 8, 10, 12.$$

Hamming distance matrix for the partial order diagram of proteome map of Figure 7.7

$$
\begin{bmatrix}
0 & 0 & 0 & 0 & 9 & 0 & 7 & 0 & 0 & 0 & 0 & 0 & 8 & 0 & 0 & 9 & 0 & 0 & 0 & 0 \\
0 & 0 & 0 & 0 & 0 & 0 & 0 & 0 & 0 & 0 & 0 & 0 & 8 & 0 & 6 & 9 & 9 & 0 & 0 & 0 \\
0 & 0 & 0 & 0 & 7 & 0 & 0 & 0 & 0 & 0 & 0 & 0 & 0 & 0 & 0 & 0 & 0 & 0 & 7 & 0 \\
0 & 0 & 0 & 0 & 0 & 0 & 0 & 0 & 0 & 0 & 0 & 0 & 7 & 0 & 0 & 8 & 0 & 5 & 0 & 0 \\
9 & 0 & 7 & 0 & 0 & 7 & 0 & 0 & 7 & 0 & 0 & 0 & 0 & 0 & 0 & 0 & 0 & 0 & 0 & 9 \\
0 & 0 & 0 & 0 & 9 & 0 & 0 & 0 & 0 & 0 & 7 & 0 & 0 & 6 & 0 & 9 & 0 & 0 & 0 & 0 \\
7 & 0 & 0 & 0 & 0 & 0 & 0 & 0 & 0 & 0 & 0 & 0 & 0 & 0 & 0 & 0 & 0 & 0 & 8 & 7 \\
0 & 0 & 0 & 0 & 0 & 0 & 0 & 0 & 0 & 0 & 0 & 0 & 0 & 0 & 0 & 0 & 6 & 0 & 0 & 0 \\
0 & 0 & 0 & 0 & 7 & 0 & 0 & 0 & 0 & 0 & 0 & 0 & 0 & 0 & 0 & 0 & 0 & 0 & 7 & 0 \\
0 & 0 & 0 & 0 & 0 & 0 & 0 & 0 & 0 & 0 & 0 & 0 & 0 & 0 & 0 & 0 & 6 & 0 & 0 & 0 \\
0 & 0 & 0 & 0 & 0 & 7 & 0 & 0 & 0 & 0 & 0 & 0 & 7 & 0 & 0 & 0 & 8 & 0 & 0 & 0 \\
0 & 0 & 0 & 0 & 0 & 0 & 0 & 0 & 0 & 0 & 0 & 0 & 0 & 0 & 0 & 0 & 6 & 0 & 0 & 0 \\
8 & 8 & 0 & 7 & 0 & 0 & 0 & 0 & 0 & 0 & 7 & 0 & 0 & 0 & 0 & 0 & 0 & 0 & 0 & 0 \\
0 & 0 & 0 & 0 & 0 & 6 & 0 & 0 & 0 & 0 & 0 & 0 & 0 & 0 & 0 & 0 & 0 & 0 & 7 & 0 \\
0 & 6 & 0 & 0 & 0 & 0 & 0 & 0 & 0 & 0 & 0 & 0 & 0 & 0 & 0 & 0 & 0 & 0 & 0 & 6 \\
9 & 9 & 0 & 8 & 0 & 9 & 0 & 0 & 0 & 0 & 0 & 0 & 0 & 0 & 0 & 0 & 0 & 0 & 10 & 0 \\
0 & 9 & 0 & 0 & 0 & 0 & 0 & 6 & 0 & 6 & 8 & 6 & 0 & 0 & 0 & 0 & 0 & 0 & 0 & 0 \\
0 & 0 & 0 & 5 & 0 & 0 & 0 & 0 & 0 & 0 & 0 & 0 & 0 & 0 & 0 & 0 & 0 & 0 & 0 & 5 \\
0 & 0 & 7 & 0 & 0 & 0 & 8 & 0 & 7 & 0 & 0 & 0 & 0 & 7 & 0 & 10 & 0 & 0 & 0 & 0 \\
0 & 0 & 0 & 0 & 9 & 0 & 7 & 0 & 0 & 0 & 0 & 0 & 0 & 0 & 6 & 0 & 0 & 5 & 0 & 0 \\
\end{bmatrix}
$$

$$(9.2)$$

If necessary in smaller map libraries, these codes can be shortened, and in large libraries made longer by including additional protein spots. The row sums of ^nD matrix can be used also for construction of the generalized connectivity index of the map, a single number to be used as a map ID number. We will illustrate appearance of the proposed proteome map codes on proteome maps of liver cells of mice treated with six different concentrations of peroxisome proliferator Ly171883. We have selected 10 of the most abundant protein spots, listed as 1–10 in the first column of Table 9.1, followed by their abundances. In the next column are listed abundances of the same 10 spots, which belong to the group of animals that received a dose with a concentration of 0.003 peroxisome proliferator Ly171883, preceded by numbers 1–10, assigned to spots in decreasing relative abundance. Observe that although spots 1, 2, and 3 are still the same three most abundant spots, now the fourth most abundant spot is protein 6 of the control group, followed by proteins 4 and 5 as the next most abundant. In the case of higher concentrations of peroxisome proliferator Ly171883, proteome maps show increased variations in relative abundances of the protein spots, which result in different codes for different maps with one exception. The exception is the proteome maps belonging to concentrations 0.01 and 0.03, which, although not the same, do not differ enough to change the magnitudes of abundances of their corresponding spots. Thus, the same code belongs to maps that differ little. This is interesting as this points to

TABLE 9.1

Abundance of 10 Most Abundant Proteins in the Proteome Maps of Liver Cells of Mice in the Control (Column 0) and Abundances of the Same Protein Spots for Six Different Concentrations of Peroxisome Proliferator Ly171883

0		0.003		0.01		0.03		0.1		0.3		0.6	
1	54.1	1	54.1	1	59.1	1	59.5	1	59.5	1	70.3	1	75.7
2	40.1	2	45.0	3	28.6	3	28.6	4	36.8	5	28.6	6	20.5
3	26.4	3	26.4	4	26.4	4	26.4	5	29.0	4	29.0	4	31.7
4	24.6	5	24.6	2	32.0	2	34.4	2	41.8	3	46.7	2	56.6
5	17.6	6	21.1	6	21.1	6	19.4	6	19.4	6	19.4	5	21.1
6	17.2	4	25.8	5	22.4	5	24.1	3	37.8	2	49.9	3	55.0
7	15.3	7	18.4	7	18.4	7	16.8	7	16.8	7	15.3	9	16.8
8	13.9	8	16.7	8	16.7	8	15.3	8	15.3	8	15.3	8	18.0
9	10.7	9	8.6	9	11.8	9	12.8	9	12.8	9	15.0	7	20.3
10	9.8	10	6.9	10	6.9	10	5.9	10	5.9	10	5.9	10	5.9

Note: The columns with integers show the relative abundances for different concentrations.

the concentrations characterizing the hormesis at the cellular level as will become more clear in Chapter 14.

9.2 MOLECULAR DESCRIPTORS FOR SCREENING COMBINATORIAL LIBRARIES

Hitherto construction of new topological indices has been primarily aiming at improving the structure–property–activity correlations. Occasionally, there have been reports on molecular descriptors of unusually high discriminatory power. One such descriptor is known as the molecular ID number, which has been constructed by adding the connectivity index and all higher-order connectivity numbers of a molecule [1]. In the case of alkanes (acyclic graphs with maximal degree or valence of four) Balaban's index J has six pairs of duplicates among 335 dodecanes $C_{12}H_{26}$ [14], and molecular ID has one pair of duplicates among 4347 $C_{15}H_{32}$ alkane isomers. This is quite impressive, but molecular ID numbers are not suitable for screening combinatorial libraries in view that they are computationally intensive for polycyclic molecules. Namely, they require computing all paths in a molecule, which, in the case of cyclic graphs, can be time-consuming. Combinatorial libraries can have 100,000 or more virtual compounds. Molecular descriptors for screening such libraries have to satisfy the following conditions:

1. They should have high discriminatory power.
2. They have to be computationally simple.
3. They have to be conceptually simple.

Computationally simple means that such descriptors should be readily calculated. Conceptually simple means using elementary mathematical operations and interpretation of derived molecular descriptors in terms of elementary structural concepts, such as vertices, edges, and paths. One expects that structurally closely related molecules should have numerical indices of similar magnitudes. Descriptors that satisfy our constraints will be promising for (i) screening combinatorial libraries and (ii) as a molecular "similarity" index.

9.3 GENERALIZED CONNECTIVITY INDICES

We continue to elaborate on the already mentioned generalized connectivity indices, construction of which has formal similarity with the "classical" connectivity index χ. Recall that the connectivity index is bond additive, the bond contributions being $1/\sqrt{(m \times n)}$, where m, n are the vertex degrees of the bond end. Because the row sums of the adjacency matrix are the vertex degrees, formally one can construct the connectivity index χ by summing the bond contributions written as $1/\sqrt{(R_i \times R_j)}$, where R_i and R_j are the row sums of the corresponding rows of the adjacency matrix. Thus, as already mentioned, for the connectivity index χ, one can use the symbol $\chi(A)$, where A stands for the adjacency matrix. Similarly, Balaban's J index is bond additive, the bond contributions being $1/\sqrt{(R_i \times R_j)}$, where R_i and R_j are the row sums of the rows in the distance matrix of graph. Using the same formalism, one can represent Balaban's J index as $\chi(D)$, where D stands for the distance matrix.

Both $\chi(A)$ and $\chi(D)$ can now be viewed as special cases of the generalized connectivity indices $\chi(X)$, which are derived using different matrices X. In the next section, we will consider as matrix X the distance matrix between column vectors of adjacent vertices in a graph, using the Hamming distance as metrics [15,16]. Hamming distance is defined for strings of numbers of the same length. It is given by the number of places at which the corresponding elements in two strings (sequences) are different. Strictly speaking the Hamming distance is pseudo-distance because it does not always satisfy the three axioms on distance (1–3) which are the following:

1. Distance is is non-negative, $D \geq 0$.
2. $D(x, y) = 0$ if and only if $x = y$.
3. Distance does not depend on the direction of measuring, that is, $D(x, y) = D(y, x)$.
4. Distance satisfies the triangular inequality: $D(x, y) + D(y, z) \geq D(x, z)$.

It has been proven that the triangular inequality (4) follows from the three axioms. In the case of Hamming distance, the second axiom on distance allows that distance be zero even for cases that $x \neq y$. Hence, as in the case of pseudo-distance, the second axiom is relaxed and replaced by the requirement:

2. $D(x, x) = 0$ (but possibly $D(x, y) = 0$ for some distinct values $x \neq y$).

9.3.1 HAMMING DISTANCE MATRIX (HD)

For illustration of construction of the n-dimensional distance matrix, we have selected a pair of nonane isomers 3,4-dimethylheptane and 3-ethyl-2methylhexane

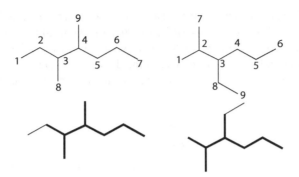

FIGURE 9.1 The most similar nonane isomers: 3,4-dimethylheptane and 3-ethyl-2-methylhexane. To better see their similarity, below the second isomers has been reflected, and the common substructure has been emphasized in bold.

illustrated in Figure 9.1, the adjacency matrices of which are listed at the top in Matrix 9.3. In construction of the n-dimensional distance matrix, one views the columns in adjacency matrices as vectors. The Hamming distance matrix may have zero elements off the main diagonal. This will happen whenever the corresponding columns of the adjacency matrix are identical.

In the case of the isomer 3,4-dimethylheptane, for the element (1, 2), the Hamming distance between columns (or rows)

$$(0, 1, 0, 0, 0, 0, 0, 0, 0) \text{ and } (1, 0, 1, 0, 0, 0, 0, 0, 0)$$

is equal to 3 because the two sequences have different elements only in three locations.

Similarly, the element (2, 3) is 5, the two sequences have different elements in five locations, and so on. In this way, one obtains the HD matrix of 3,4-dimethylhexane shown at the left middle part of Matrix 9.2. In this matrix, only Hamming distances between columns corresponding to adjacent vertices were considered. At left bottom part is shown HD matrix for 3,4-dymethylhexane when one considers all pairs of vertices, whether adjacent or not. Similarly follows the construction of the HD matrix of the 3-ethyl-2-metylhexane isomer, shown at the bottom right of Matrix 9.2. In this matrix, Hamming distances between all pairs of vertices were considered. Observe that the off-diagonal element (1, 7) in the case of the ^nD matrix of 3-ethyl-2-metylhexane is zero, illustrating that the Hamming distance is pseudo-distance. At the right of each matrix, we have listed the corresponding row sums (RS).

Observe an interesting property of the HD matrix elements. Although their structural interpretation is the distance (or pseudo-distance in the case that two columns are identical) between columns of the adjacency matrix, matrix elements are given as the sum of degrees of edge vertices. Thus, to bond (m, n), where m, n are the vertex degrees of each bond end point, belongs the entry $(m + n)$. This observation allows one to construct the HD matrix directly from inspection of a graph. This is true for acyclic graphs and all triangle-free cyclic graphs (graphs in which adjacent vertices have no common vertex). This simple rule holds also for nonadjacent matrix elements of the Hamming distance matrices for vertices that have no common adjacent vertex. For every common adjacent vertex the sum $(m + n)$ should be decreased by 2.

Top: the adjacency matrices for the two nonane isomers illustrated in Figure 9.1 and the corresponding row sums (RS); middle: the Hamming distance matrices (HD) obtained for the columns vectors corresponding to the adjacent vertices of nonane isomers graphs; bottom: the Hamming distance matrices (HD) obtained for the column vectors corresponding to the adjacent and nonadjacent vertices of nonane isomer graphs

Adjacent Matrices

									RS											RS
0	1	0	0	0	0	0	0	0	1		0	1	0	0	0	0	0	0	0	1
1	0	1	0	0	0	0	0	0	2		1	0	1	0	0	0	1	0	0	3
0	1	0	1	0	0	0	1	0	3		0	1	0	1	0	0	0	1	0	3
0	0	1	0	1	0	0	0	1	3		0	0	1	0	1	0	0	0	0	2
0	0	0	1	0	1	0	0	0	2		0	0	0	1	0	1	0	0	0	2
0	0	0	0	1	0	1	0	0	2		0	0	0	0	1	0	0	0	0	1
0	0	0	0	0	1	0	0	0	1		0	1	0	0	0	0	0	0	0	1
0	0	1	0	0	0	0	0	0	1		0	0	1	0	0	0	0	0	1	2
0	0	0	1	0	0	0	0	0	1		0	0	0	0	0	0	0	1	0	1

Hamming Distances Matrices

									RS											RS
0	3	0	0	0	0	0	0	0	3		0	4	0	0	0	0	0	0	0	4
3	0	5	0	0	0	0	0	0	8		4	0	6	0	0	0	4	0	0	14
0	5	0	6	0	0	0	4	0	15		0	6	0	5	0	0	0	5	0	16
0	0	6	0	5	0	0	0	4	15		0	0	5	0	4	0	0	0	0	9
0	0	0	5	0	4	0	0	0	9		0	0	0	4	0	3	0	0	0	7
0	0	0	0	4	0	3	0	0	7		0	0	0	0	3	0	0	0	0	3
0	0	0	0	0	3	0	0	0	3		0	4	0	0	0	0	0	0	0	4
0	0	4	0	0	0	0	0	0	4		0	0	5	0	0	0	0	0	3	8
0	0	0	4	0	0	0	0	0	4		0	0	0	0	0	0	0	3	0	3

Hamming Distances Matrices

									RS											RS
0	3	2	4	3	3	2	2	2	21		0	4	2	3	3	2	0	3	2	19
3	0	5	3	4	4	3	1	3	26		4	0	6	3	5	4	4	3	4	33
2	5	0	6	3	5	4	4	2	31		2	6	0	5	3	4	2	5	2	29
4	3	6	0	5	3	4	2	4	31		3	3	5	0	4	1	3	2	3	24
3	4	3	5	0	4	1	3	1	24		3	5	3	4	0	3	3	4	5	30
3	4	4	3	4	0	3	3	3	27		3	4	4	1	3	0	2	3	2	22
2	3	4	4	1	3	0	2	2	21		0	4	2	3	3	2	0	3	2	19
2	1	4	2	3	3	2	0	2	19		3	3	5	2	4	3	3	0	3	26
2	3	2	4	1	3	2	2	0	19		2	4	2	3	3	2	2	3	0	21

$$(9.3)$$

9.4 CONNECTIVITY INDEX χ(HD)

Construction of the generalized connectivity index χ(HD) for 3,4-dimethylheptane is illustrated in Table 9.2 at left. After making a list of bonds (the first column of

TABLE 9.2

List of Bonds and Bond Contribution to the Connectivity Index χ(HD) for 3,4-Dimethylheptane and 3-Ethyl-2-Methylhexane

	3,4-Dimethylheptane				3-Ethyl-2-Methylhexane		
Bond	$m \times n$	$= R_iR_j$	$1/\sqrt{R_iR_j}$	Bond	$m \times n$	$= R_iR_j$	$1/\sqrt{R_iR_j}$
1, 2	3×8	24	0.2041	1, 2	4×14	56	0.1336
2, 3	8×15	120	0.0913	2, 3	14×16	224	0.0668
3, 4	15×15	225	0.0667	2, 7	14×4	56	0.1336
3, 8	15×4	60	0.1291	3, 4	16×9	144	0.0833
4, 5	15×9	135	0.0867	3, 8	16×8	128	0.0884
4, 9	15×4	60	0.1291	4, 5	9×7	63	0.1260
5, 6	9×7	63	0.1260	5, 6	7×3	21	0.2182
6, 7	7×3	21	0.2182	8, 9	8×3	24	0.2041
χ(HD)			1.0505				1.0541

Table 9.2), one finds the row sum factors (the second column of Table 9.2), and the products (third column of Table 9.2), and finally calculates the reciprocal square roots (fourth column of Table 9.2). At the right part of Table 9.2 is shown the corresponding calculations for the 3-ethyl-2-metylhexane isomer.

The χ(HD) for the two nonane isomers are obtained by summing bond contributions and are listed in the last row of Table 9.2. As one can see, the χ(HD) are very similar, differing by less then 1%, which suggest that the χ(HD) index may be a suitable measure of molecular similarity. Based on the χ(HD) index, the two most similar pairs among the 11 pairs of nonane isomers are 3,4-dimethylheptane and 3-ethyl-2-methylhexane. This can be seen from the last column in Table 9.3, in which

TABLE 9.3

The Relative Similarity of a Selection of More Similar Nonane Isomers Based on the Difference in Their χ(HD) Values

I	II	χ(HD)$_I$	χ(HD)$_{II}$	Absolute Difference
3,4-MM C$_7$	3-E, 2-M C$_6$	1.0505	1.0541	0.0036
3-E C$_7$	4-E C$_7$	1.1466	1.1556	0.0094
3-M C$_8$	4-M C$_8$	1.1160	1.1257	0.0097
3,3-MM C$_7$	4,4-MM C$_7$	0.9853	1.0010	0.0157
2,3-MM C$_7$	3,4-MM C$_7$	1.0211	1.1560	0.0294
2-M C$_8$	3-M C$_8$	1.0860	1.1160	0.0301
4-M C$_8$	4-E C$_7$	1.1257	1.1556	0.0302
3-M C$_8$	3-E C$_7$	1.1160	1.1466	0.0306
2,3-MM C$_7$	3-E 2-M C$_6$	1.0211	1.0541	0.0330
2,6-MM C$_7$	2,5-MM C$_7$	0.9753	1.0142	0.0389
3,5-MM C$_7$	4-E, 2-M C$_6$	1.0826	1.0375	0.0451

are given the differences in χ(HD) indices for each pair. All the calculations reported in this section were made using Microsoft Excel.

9.4.1 SIMILARITY AMONG NONANE ISOMERS

The generalized connectivity index χ(HD) has been found to have high discriminating power in the case of two dozen or more acyclic graphs, some of which have several identical graph invariants, including the same Balaban's index J. Here we would like to show the potential use of χ(HD) for finding *highly similar structures*. Highly similar structures would be structures that overlap in local and global structural characteristics. The elements of HD matrix carry *local* bond information based on the additivity of the vertex valences. This information is augmented with information on more distant bonds through construction of row sums. By summing the row sums, one approaches *global* characterization of the molecule.

We will use χ(HD) as a similarity index on the 35 C_9H_{20} nonane isomers. In Table 9.4, we show listed lexicographically the ordered row sums of the Hamming distance matrices for all 35 nonane isomers, which can be viewed as nine-component vectors. A way to characterize similarities is to consider the absolute differences of the corresponding terms in two ordered row sums. In this way one finds that among 595 pairs of nonanes the most similar are 2,6-dimethylheptane and 2,5-dimethylheptane. Their nine-component vectors differ only in the last component by one. In

TABLE 9.4
The Ordered Row Sums of HD Matrix for 35 Nonane Constitutional Isomers Listed Lexicographically

C_9H_{20}	Ordered Row Sums	Ordered Row Sums	C_9H_{20}
n-nonane	8, 8, 8, 8, 8, 7, 7, 3, 3	17, 16, 14, 8, 4, 4, 4, 4, 3	2,4-MM,3-E C_5
2-M C_8	13, 9, 8, 8, 8, 7, 4, 4, 3	20, 14, 11, 8, 5, 5, 4, 4, 3	2,2,4-MMM C_6
2,5-MM C_7	13, 13, 9, 9, 8, 4, 4, 4, 3	21,10, 8, 8, 7, 5, 5, 5, 3	2,2-MM C_7
2,6-MM C_7	13, 13, 9, 9, 8, 4, 4, 4, 4	21, 11, 10, 9, 5, 5, 5, 3, 3	2,2,5-MMM C_6
3-M C_8	14, 9, 8, 8, 8, 7, 4, 3, 3	21, 21, 12, 5, 5, 5, 5, 5, 5	2,2,4,4-MMMM C_5
4-M C_8	14, 9, 9, 8, 7, 7, 4, 3, 3	22, 10, 9, 8, 7, 5, 5, 3, 3	3,3-MM C_7
2,4-MM C_7	14, 13, 10, 9, 7, 4, 4, 4, 3	22, 10, 10, 7, 7, 5, 5, 3, 3	4,4-MM C_7
3,5-MM C_7	14, 14, 10, 8, 8, 4, 4, 3, 3	22, 13, 11, 9, 5, 5, 4, 4, 3	2,4,4-MMM C_6
3-E C_7	15, 9, 8, 8, 8, 7, 3, 3, 3	22, 16, 9, 7, 5, 5, 5, 4, 3	2,2,3-MMM C_6
4-E C_7	15, 9, 9, 8, 7, 7, 3, 3, 3	22, 17, 8, 8, 5, 5, 5, 3, 3	2,2-MM,3-E C_5
2-M,4-E C_6	15, 13, 10, 8, 8, 4, 4, 3, 3	22, 17, 13, 5, 5, 5, 4, 4, 3	2,2,3,4-MMMM C_5
2,3-MM C_7	15, 14, 9, 8, 7, 4, 4, 4, 3	23, 15, 10, 7, 5, 5, 4, 4, 3	2,3,3-MMM C_6
2,3,5-MMM C_7	15, 14, 13, 10, 4, 4, 4, 4, 4	23, 16, 9, 8, 5, 5, 4, 3, 3	3,3-MM,4-E C_5
2,3,4-MMM C_6	15, 14, 14, 8, 4, 4, 4, 4, 3	24, 9, 9, 9, 9, 3, 3, 3,3	3,3-EE C_5
3-M,4-E C_6	16, 15, 8, 8, 8, 4, 3, 3, 3	24, 15, 9, 9, 5, 4, 4, 3, 3	2,3-MM,3-E C_5
3,4-MM C_7	15, 15, 9, 8, 7, 4, 4, 3, 3	24, 15, 15, 5, 5, 4, 4, 4, 4	2,3,3,4-MMMM C_5
2-M,3-E C_6	16, 14, 9, 8, 7, 4, 4, 3, 3	24, 23, 9, 5, 5, 5, 5, 5, 3	2,2,3,3-MMMM C_5
3-M,3-E C_6	17, 16, 9, 9, 7, 5, 3, 3, 3		

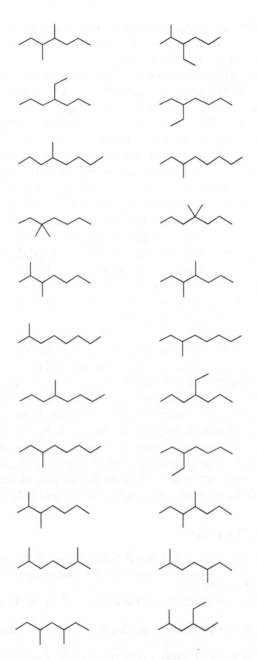

FIGURE 9.2 Eleven pairs of the most similar nonane isomers based on the index χ(DH).

addition, one finds 10 pairs of nonanes in which the ordered row sums differ by two units. All these 11 pairs are illustrated in Figure 9.2, but they have been ordered with respect to differences in their $\chi(HD)$ values.

If we consider the $\chi(HD)$ index as a *single* number characterization of molecules, we can use the absolute difference in $\chi(HD)$ values of two molecules as the measure of their similarity. Interestingly, now the pair 2,6-dimethylheptane and 2,5-dimethylheptane is not the most similar. The smallest difference in $\chi(HD)$ of 0.003579 belongs to the pair 3,4-dimethylheptane and 3-ethyl-2-methylhexane, the first pair in Figure 9.2. The lower part of Figure 9.1 illuminates why this is so. Observe that when suitably drawn the two molecules show considerable skeletal overlap. The CC bond in which they differ is attached to the same kind of carbon atom, so both molecules have identical (m, n) bond types. This is not the case with 2,6-dimethylheptane and 2,5-dimethylheptane.

The pair 2,6-dimethylheptane and 2,5-dimethylheptane, found the most similar using ordered row sums, is not even near the top in Table 9.3, but is close to the bottom of the list of most similar nonane pairs using the $\chi(HD)$ index! Clearly, the ordered row sums and the connectivity $\chi(HD)$ index, not surprisingly, have captured different structural features of graphs and do not carry the same information. But which of the two similarity measures would be better and more suitable when considering similarity of molecular properties?

Let us look at bond types (m, n) of the most similar pairs of nonane isomers of Figure 9.2. Looking in Figure 9.2 at the first pair, one find that both isomers have two bonds of type (1, 2), (1, 3), and (2, 3), and one bond of type (2, 2) and (3, 3). The second pair has the three bonds (1, 2) and (2, 3) and two bonds of type (2, 2). Actually the same is true for the first four pairs of isomers of Figure 9.2. However, the remaining pairs, including 2,6-dimethylheptane and 2,5-dimethylheptane, have bonds of a different kind. Thus 2,6-dimethylheptane has four (1, 3) bonds, two (2, 3) and (2, 2) bonds, and 2,5-dimethylheptane has three (1, 3) and (2, 3) bonds, one (2, 2) and (1, 2) bond. Hence, although these remaining pairs of isomers may globally be similar, locally they show visible differences. More on this subject, including use of $\chi(HD)$ to search for most different pairs of molecules one can find in Ref. [10].

REFERENCES AND NOTES

1. G. Grassy, B. Calas, A. Yasri, R. Lahana, J. Woo, S. Iyer, M. Kaczorek, R. Floc'h, and R. Buelow, Computer-assisted rational design of immunosuppressive compounds, *Nat. Biotechnol.* 16 (1998) 748–752.
2. M. Randić, On molecular identification numbers, *J. Chem. Inf. Comput. Sci.* 24 (1984) 164–175.
3. M. Randić, Molecular ID numbers: By design, *J. Chem. Inf. Comput. Sci.* 26 (1986) 134–136.
4. K. Szymanski, W. Müller, J. Knop, and N. Trinajstić, On Randić's molecular identification numbers, *Int. J. Quantum Chem.* 25 (1985) 413–415.
5. M. Randić, Ring ID numbers, *J. Chem. Inf. Comput. Sci.* 28 (1988) 142–147.
6. M. Dehmer and M. Grabner, The discrimination power of molecular identification numbers revisited, *MATCH Commun. Math. Comput. Chem.* 69 (2013) 785–794.
7. M. Dehmer, M. Grabner, and K. Varmuza, Information indices with high discriminative power for graphs, *PLoS One* 7 (2012) e31214.

8. M. Dehmer, L. Sivakumar, and K. Varmuza, Uniquely discriminating molecular structures using novel eigenvalue-based descriptors, *MATCH Commun. Math. Comput. Chem.* 67 (2012) 147–172.

9. M. Dehmer, M. Emmert-Streib, and M. Grabner, A computational approach to construct a multivariate complete graph invariant, *Inf. Sci.* 260 (2014) 200–208.

10. M. Randić, On molecular similarity based on a single molecular descriptor *Chem. Phys. Lett.* 599 (2014) 1–6.

11. M. Randić, Characterization of molecular branching, *J. Am. Chem. Soc.* 97 (1975) 6609–6615.

12. A. T. Balaban, Highly discriminating distance–based topological index, *Chem. Phys. Lett.* 89 (1982) 399–404.

13. M. Randić, Reported at the 10th annual meeting of the Int. Acad. Math. Chem. held in Split, Croatia June 7–10, 2014.

14. A. T. Balaban and L. V. Quintas, The smallest graphs, trees and 4-trees with degenerate topological index J, *MATCH Commun. Math. Comput. Chem.* 14 (1983) 213–233.

15. I. J. Schoenberg, Remarks to Maurice Frechet's article: Sur la definition axiomatique d'une classe d'espace distancies vectoriellement applicable sur l'espace de Hilbert. *Ann. Math.* 36 (1935) 724–732.

16. R. W. Hamming, Error detecting and error correcting codes, *Bell Syst. Tech. J.* 29 (1950) 147–160.

10 Aromaticity Revisited

Aromaticity has been one of the central topics of organic chemistry ever since 1865 when August Kekulé proposed his ring structure for benzene [1,2]. With E. Hückel's explaining the difference between the chemical structure of benzene C_6H_6 and cyclooctatetraene C_8H_8 using his MO approach, interest in structural understanding of aromaticity was initiated. In this chapter, we will review the theoretical approach to aromaticity, primarily in polycyclic conjugated hydrocarbons, leaving out the topic of aromaticity in heterocyclic systems. The reason for not considering heterocyclic systems is not because they are not important; on the contrary, most of organic chemistry is concerned with heterocyclic systems, which are highly important. However, unless we first *fully* understand the origin and the nature of aromaticity in hydrocarbon systems, it is not likely that we will understand the aromaticity of heterocyclic molecules. For those readers anxious to hear about aromaticity in molecules having heteroatoms, we recommend the review "Aromaticity as a Cornerstone of Heterocyclic Chemistry" by Balaban, Oniciu, and Katritzky [3].

10.1 CONJUGATED CIRCUITS

Although Kekulé was interested in the structural origin of aromaticity, Erlenmayer, who a year later proposed Kekulé valence structures for naphthalene, advocated that the characterization of aromaticity be based on *properties* of compounds. However, such a point of view has inherent difficulties when the approach is extended to molecules that show only partial aromatic properties. The structural approach to aromaticity, which was advocated by Kekulé, on one hand was crowned with the success of the Hückel's $4n + 2$ rule in resolving the fundamental difference between benzene and cyclooctatetraene. At the same time, the structural approach to aromaticity was frustrated by a lack of progress in generalizing this simple and elegant solution for monocyclic systems from monocyclic systems to polycyclic conjugated hydrocarbons. It is not that the problem did not attract the attention of a number of theoreticians, as it did, but the generalization appeared elusive for so long a time that, eventually, theoreticians, around the 1960s, apparently gave up in searching for a potentially nonexistant solution—a solution that was, however, found to be good in the mid-1970s but remained unrecognized by most chemists until the present time.

It is not that theoretical chemists of the mid-1960s were not good enough in trying to solve the problem, but as we will see, a *tool* to solve the problem was missing. The tool for solving the mystery of aromaticity is conjugated circuits [4]. Conjugated circuits were discovered in 1976 when one of the present authors was wondering how *different* were the three Kekulé valence structures of naphthalene:

In the case of benzene, both Kekulé valence structures are equivalent, being identical except for their orientation in space. In naphthalene, two of the three structures are symmetry-related, but the third structure is *different*. The question is: How different is the symmetry non-equivalent structure? With a close look at the naphthalene Kekulé valence structures it is easy to see that the unique Kekulé structure of naphthalene has two rings each having three C=C bonds just as have Kekulé structure of benzene. The other two Kekulé valence structures of naphthalene have but a single benzene ring with three C=C bonds.

From a theoretical point of view, there is here some disharmony. Naphthalene has two kinds of Kekulé valence structures, which contain different structural elements: One structure has *two* rings with alternating C=C and C–C bonds, and two structures have only a *single* ring with alternating C=C and C–C bonds. Looking a bit longer at the three Kekulé valence structures of naphthalene, one may discover *conjugated circuits*. Observe that the two Kekulé structures with a single ring with alternating C=C and C–C bonds have larger circuits on their *periphery* with regular alternation of the C=C and C–C bonds. These circuits involve five bonds having alternating C=C and C–C rather than three C=C and C–C bonds in each benzene ring. Rings of naphthalene with two C=C and four C–C bonds are of no interest, but a *circuit* having alternating C=C and C–C bonds may be of interest. Circuits involving alternating C=C and C–C bonds have been named *conjugated circuits*. Thus, in contrast to the benzene ring, which can be viewed as conjugated circuits of size 6, in naphthalene we have, in addition, a Kekulé valence structure with conjugated circuits of size 10. In larger molecules, there will occur larger conjugated circuits. The size of conjugated circuits is given by the number of carbon atoms involved; thus a conjugated circuit of size R_n contains $4n + 2$ π electrons. Hence, in naphthalene, there is one Kekulé valence structure having $2R_1$ and two Kekulé valence structures having $R_1 + R_2$ conjugated circuit, illustrated in Figure 10.1. It is not difficult to construct all Kekulé valence structures for smaller benzenoid hydrocarbons and then find and enumerate all conjugated circuits.

Construction of all Kekulé valence structures, and finding and enumerating all conjugated circuits in molecules of intermediate size, not to speak of buckminsterfullerene C_{60}, is neither straightforward nor simple. As we will see later, coronone, $C_{24}H_{12}$, having six benzene rings fused around a central benzene, has 20 Kekulé valence structures, and ovalene, $C_{32}H_{18}$, having eight benzene rings fused around a

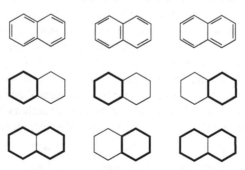

FIGURE 10.1   Three Kekulé valence structures of naphthalene and their conjugated circuits.

central naphthalene, has 50 Kekulé valence structures. But that is nothing comparing with C_{60}, found to have 12,500 Kekulé valence structures [5].

In Figure 10.2, we have illustrated the six Kekulé valence structures of pyrene. Below each Kekulé valence structure, we show their conjugated circuits. As one can see, each Kekulé valence structure of pyrene has five conjugated circuits. In the count, we include separate combinations of *disjointed* conjugated circuits occurring simultaneously in a single Kekulé valence structure as additional conjugated circuits. This is the case with the four Kekulé valence structures of pyrene in which the central CC bond is single. Also we count the *linearly dependent* conjugated circuit, for example, R_3 in the first and the last Kekulé valence structures, as an additional conjugated circuit despite that they can be obtained by superposition of two R_2 conjugated circuits present having common central R_1 conjugated circuits in the same structure and represented as $2R_2 - R_1$.

FIGURE 10.2 Illustration of conjugated circuits of pyrene.

According to a theorem by Gutman [6], polycyclic conjugated hydrocarbons—
be they benzenoid or non-benzenoid, alternant or non-alternant—having K Kekulé
valence each of these K structures has $(K - 1)$ conjugated circuit. The count includes
all conjugated circuits: linearly independent, linearly dependent, and disjoint, as
already illustrated in Figure 10.2 on pyrene. This is an important theorem as it tells in
advance how many conjugated circuits there are in each Kekulé valence structure, so
one can always be sure that all conjugated circuits have been found and counted. The
above publication from Gutman and Randić [6] reports also an additional important
result: It illustrates how from a *single* Kekulé valence structure (any Kekulé valence
structure) one can construct *all* the remaining Kekulé valence structures of a molecule. The construction is very simple and is illustrated in Figure 10.3 on one of the
Kekulé valence structures of pyrene. In the left column are shown five conjugated
circuits of the selected Kekulé valence structure. In the middle column are shown
conjugated circuits obtained by exchanging CC single and CC double bonds. At the
right is shown the completed Kekulé valence structures for the valence structures in
the middle column, which are the remaining Kekulé valence structures of pyrene.

Alternant molecules are all acyclic molecules and molecules having only rings
with an even number of atoms. Non-alternant molecules have rings with an odd number of atoms. Thus, in the case of an alternant molecule, all vertices of its molecular graph can be colored as black and white so that no vertices of the same color
are adjacent. The division of molecules into alternant and non-alternant is desirable
because of their distinct mathematical properties. Thus, in the case of alternant systems, all eigenvalues come in pairs, positive and negative. This is not the case with
non-alternant systems.

As already mentioned, it is not difficult to find conjugated circuits in smaller
benzenoid hydrocarbons. In benzenoid hydrocarbons of intermediate size, searching for conjugated circuits requires more attention. In Figure 10.4, we have illustrated six symmetry nonequivalent Kekulé valence structures of coronene. Because
coronene has 20 Kekulé valence structures, according to the theorem by Gutman,
each Kekulé valence structure should have 19 conjugated circuits. In Figure 10.5,
we have illustrated the 19 conjugated circuits for the last Kekulé valence structures
of coronene of Figure 10.4. Readers may try to find 19 conjugated circuits for any
of the remaining Kekulé valence structures of coronene as an exercise. Such an
exercise is likely to convince most people that it is advisable to delegate searching for conjugated circuits of benzenoids in larger hydrocarbons to computers. In
Appendix 14 in Figures 14.9 through 14.13, we have illustrated all conjugated circuits for the remaining five symmetry nonequivalent Kekulé valence structures of
coronene so that those who were trying to construct them can verify the correctness
of their results. One benefit of this exercise is to appreciate the significance of the
theorem by Gutman about the total number of conjugated circuits being $K(K - 1)$,
without which it is likely that some conjugated circuits, in the case of systems of
intermediate size, could be overlooked.

Observe that coronene, which is not so big molecule, has 380 conjugated circuits.
In the case of buckminsterfullerene C_{60}, the number conjugated circuits is well over
150 million (to be precise, 156,237,500, easily calculated from Gutman's theorem
stating that conjugated hydrocarbon having K Kekulé valence structures has $(K - 1)$

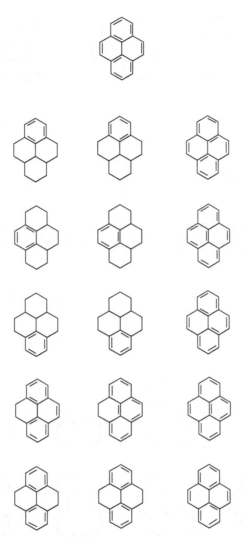

FIGURE 10.3 Construction of all Kekulé valence structures from conjugated circuits of a single Kekulé valence structure (shown at the top of the figure). At left is shown corresponding conjugated circuits. In the middle is shown conjugated circuits obtained by exchanging CC single and CC double bonds. At right is shown completed Kekulé valence structures.

conjugated circuits. There have been a number of publications dealing with enumerations of conjugated circuits in molecules of intermediate size [7–15].

It is well known that the number of Kekulé valence structures (K) is very useful information because, among molecules of equal size, those having larger K have larger resonance energy. So among $C_{24}H_{12}$ isomers, coronene has the largest K and is the most stable. But, as we will see later, generally the situation is not so simple. Examination of Fries structures, Clar sextets, and analysis of conjugated circuits will be more informative. Consider, for example, buckminsterfullerene, which has

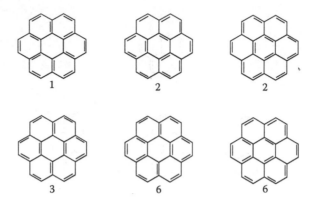

FIGURE 10.4   All symmetry nonequivalent Kekulé valence structures of coronene. Under each structure is shown the multiplicity of the structure, that is, how many times the structure occurs among 20 Kekulé valence structures of coronene.

FIGURE 10.5   All conjugated circuits of the last of Kekulé valence structures of Figure 10.4.

1812 isomers according to Austin and collaborators [16]; some of the 1812 isomers have more than 12,500 Kekulé valence structures but are less stable than buckminsterfullerene. Let us quote the abstract of the publication of Austin et al.:

A count of Kekulé structures for all 1812 distinct fullerene isomers of C_{60} shows that 20 isomers surpass the count of 12500 for icosahedral C_{60}, and demonstrates the lack of correlation between molecular-orbital indices of stability and raw Kekulé counts for fullerenes. Analysis of Kekulé structures in terms of benzenoid, cyclopentenoid and cyclopentadienoid rings reveals the source of the stability of icosahedral C_{60} in a localised model to be the fact that uniquely amongst the 1812 structural isomers it has a Fries Kekulé structure where all hexagons contain three double bonds and all pentagons none.

Conjugated circuits play the central role in three rather important topics of polycyclic conjugated hydrocarbons: (i) definition and expressions of resonance energy, (ii) generalization of the Hückel $4n + 2$ rule of aromaticity, and (iii) calculation of π electron ring currents. In the next three sections, we will address these three topics.

10.2   RESONANCE ENERGIES

It is generally understood that resonance energy (RE) is additional stabilization of conjugated cyclic and polycyclic systems due to delocalization of π electrons. Simple MO calculations offer some insight into the role of π electrons in monocyclic systems as reflected in the Hückel $4n + 2$ rule, which we can also view as coming from contributions from conjugated circuits of size $4n + 2$. It is natural then to assume that the same "monocyclic" contributions play a similar role in polycyclic systems and that the resonance energy can be mathematically expressed as a sum of the contribution of all conjugated circuits in a system. Thus, in the case of naphthalene, RE will be proportional to $(4R_1 + 2R_2)$ and is given by $(4R_1 + 2R_2)/3$, the average from the contributing three Kekulé valence structures.

In Table 10.1, we have listed *expressions* for RE for a number of smaller benzenoid hydrocarbons. Observe that, for the first time, one has *formulas, expressions*, for RE, which are fully determined by Kekulé valence structures of the molecule and do not depend on comparison of a molecule with a real or hypothetical standard having fewer delocalized π electrons. This is a significant conceptual advantage even though the choice of the numeric values for R_1, R_2, R_3, and R_4 may vary, may change from time to time, as estimates of experimental RE of some molecules change or computations offer alternative reliable values. We will not elaborate here on this theme because there is enough literature on RE based on conjugated circuits and theoretical calculations elaborating on conjugated circuits. We do recommend the sizable review article "Aromaticity" in *Chemical Reviews* for more details and literature [17]. We should also add that, in calculation of the resonance energies using the conjugated circuits model, only the linearly *independent* conjugated circuits have been considered, and the potential contributions from disjointed and linearly dependent conjugate circuits have been, for now, ignored, and it seems that they are minor.

TABLE 10.1

Resonance Energy (RE) Expressions in Terms of Conjugated Circuits of Size R_1, R_2, and R_3 for Benzene and 40 Smaller Benzenoid Hydrocarbons of Figure 10.6

1	$(2R_1)/2$	21	$(12R_1 + 10R_2 + 8R_3)/7$
2	$(4R_1 + 2R_2)/3$	22	$(28R_1 + 16R_2 + 11R_3)/11$
3	$(6R_1 + 4R_2 + 2R_3)/4$	23	$(40R_1 + 26R_2 + 12R_3)/14$
4	$(10R_1 + 4R_2 + R_3)/5$	24	$(44R_1 + 30R_2 + 14R_3)/15$
5	$(8R_1 + 6R_2 + 4R_3)/5$	25	$(52R_1 + 24R_2 + 12R_3)/16$
6	$(16R_1 + 8R_2 + 3R_3)/7$	26	$(58R_1 + 22R_2 + 13R_3)/17$
7	$(20R_1 + 10R_2 + 2R_3)/8$	27	$(86R_1 + 28R_2 + 12R_3)/22$
8	$(26R_1 + 6R_2 + 3R_3)/9$	28	$(90R_1 + 34R_2 + 12R_3)/23$
9	$(12R_1 + 8R_2 + 4R_3)/6$	29	$(24R_1 + 17R_2 + 12R_3)/10$
10	$(10R_1 + 8R_2 + 6R_3)/6$	30	$(96R_1 + 34R_2 + 12R_3)/24$
11	$(22R_1 + 12R_2 + 7R_3)/9$	31	$(42R_1 + 26R_2 + 12R_3)/14$
12	$(26R_1 + 16R_2 + 5R_3)/10$	32	$(76R_1 + 24R_2 + 14R_3)/20$
13	$(36R_1 + 16R_2 + 6R_3)/12$	33	$(52R_1 + 28R_2 + 10R_3)/13$
14	$(30R_1 + 18R_2 + 6R_3)/11$	34	$(36R_1 + 24R_2 + 14R_3)/13$
15	$(40R_1 + 20R_2 + 5R_3)/13$	35	$(192R_1 + 48R_2 + 28R_3)/40$
16	$(42R_1 + 14R_2 + 5R_3)/13$	36	$(64R_1 + 48R_2 + 27R_3)/20$
17	$(46R_1 + 18R_2 + 5R_3)/14$	37	$(48R_1 + 32R_2 + 16R_3)/16$
18	$(32R_1 + 14R_2 + 7R_3)/11$	38	$(56R_1 + 40R_2 + 22R_3)/18$
19	$(22R_1 + 14R_2 + 7R_3)/9$	39	$(140R_1 + 90R_2 + 46R_3)/35$
20	$(24R_1 + 12R_2)/9$	40	$(108R_1 + 80R_2 + 45R_3)/30$
		41	$(200R_1 + 160R_2 + 110R_3)/50$

10.3   IN SEARCH OF A GENERALIZED HÜCKEL $4n + 2$ RULE

There is no doubt that the famous Hückel $4n + 2$ rule is considered one of *the most important* results of early quantum chemistry today, just as it was 85 years ago. This is one of few very significant theoretical achievements that were immediately recognized as a great discovery, a contribution equally fascinating experimental and theoretical chemists. It explained, in a simple conceptual model, the dramatic difference in the chemistry of benzene and cyclooctatetraene, six-member and eight-member unsaturated rings. In addition, it also explained the stabilities of the cyclopentadienyl anion $C_5H_5^-$ and the tropylium ion (cycloheptatrienyl cation) $C_7H_7^+$. However, the $4n + 2$ rule has an important limitation: that it strictly applies only to *monocyclic* systems. This is but a fraction of a fraction of the chemistry of conjugated hydrocarbons.

It is not surprising that a number of theoretical chemists of early quantum chemistry tried to extend the rule to polycyclic conjugated systems, but they have not succeeded. This must have been frustrating because the goal is clear, but a solution was not in sight. The solution, as mentioned before, was found some 40 years later. It was published in one of the leading journals of chemistry in the world, the *Journal of the American Chemical Society*. Nevertheless, apparently, most organic chemists and theoretical organic chemists *are not aware* of the existing solution, at least as is reflected by the fact that when citing the Hückel $4n + 2$ rule, which is often seen in the chemical literature even today, there is no mention of the generalized Hückel $4n + 2$ rule for polycyclic systems. Let's recall:

>   *The Hückel $4n + 2$ rule*: If a cyclic, planar molecule has $4n + 2$ π electrons, it is considered aromatic.
>   The general $4n + 2$ rule is expressed in terms of the conjugated circuits and is also very simply worded.
>   *The general $4n + 2$ rule*: If a planar conjugated molecule has only $4n + 2$ π conjugated circuits, it is considered aromatic.

This, and much more was stated almost 40 years ago, but how many chemists have noticed it? One of the reasons may be because many chemists may not be familiar with the *tool* used to solve the problem: the conjugated circuits. Observe the fundamental difference between the Hückel $4n + 2$ rule and the general $4n + 2$ rule: The Hückel $4n + 2$ rule is *counting π electrons*; the general $4n + 2$ rule is *examining conjugated circuits*. In a way, expression of the Hückel $4n + 2$ rule in terms of the counting of π electrons, although correct, was unfortunate in misleading those trying to find a generalization to the rule in suggesting that the *count* of π electrons may be the critical factor, which it is not! Pyrene is the smallest benzenoid system that is aromatic, but it has 16, that is, $4n$ π electrons, and so is coronene with 24 π electrons and many other benzenoid hydrocarbons.

Instead of counting π electrons, chemists should pay attention to *conjugated circuits* (which were not around in the 1960s). But even just paying attention to cycle in molecules, the difficulties of pyrene and coronene would have been resolved. We can restate the the Hückel $4n + 2$ rule and the general $4n + 2$ rule in terms of cycles as the following:

>   *The Hückel $4n + 2$ rule*: If a cyclic, planar molecule has only $4n + 2$ π cycles, it is considered aromatic.
>   *The general $4n + 2$ rule*: If a polycyclic, planar molecule has only $4n + 2$ π cycles, it is considered aromatic.

It is not difficult to see that although pyrene has 16 π electrons it has 6, 10, and 14 cycles or circuits and has no $4n$ conjugated circuits. Similarly, coronene, another $4n$ π electron system, has only 6, 10, 14, 18, and 22 conjugated circuits, all $4n + 2$, and has no $4n$ conjugated circuits. In fact, all benzenoid hydrocarbons, regardless of how many rings they have, have *necessarily* only $4n + 2$ conjugated circuits [18]. As we will see soon, when it comes to non-benzenoid hydrocarbons in the above generalization of the Hückel $4n + 2$ rule, one has to replace cycles (circuits) with "conjugated circuits," which not only eliminates isolated odd cycles, but shows that circuits can be of size $4n + 2$ and $4n$, resulting in molecules having different degrees of aromaticity.

In Figure 10.5, we have illustrated all conjugated circuits in a single Kekulé valence structure of coronene, $C_{24}H_{12}$. All conjugated circuits of coronene are *peripheries* of smaller benzenoid systems and are subgraphs of the coronene graph and combinations of them. In Figure 10.6, we show 40 smaller benzenoid hydrocarbons for which,

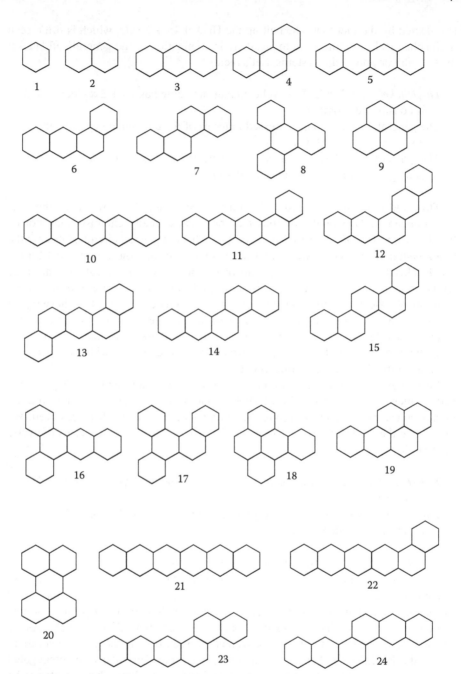

FIGURE 10.6 Benzene and 40 smaller benzenoid hydrocarbons whose RE expressions are listed in Table 10.1.   (*Continued*)

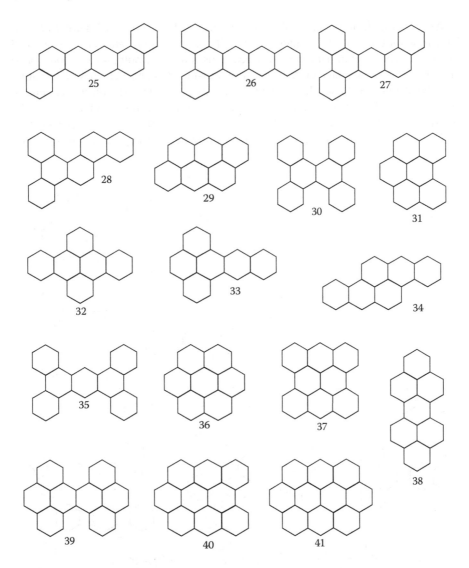

FIGURE 10.6 (CONTINUED)   Benzene and 40 smaller benzenoid hydrocarbons whose RE expressions are listed in Table 10.1.

in Table 10.1, we have listed the expression for molecular resonance energy (RE). We have limited the count of conjugated circuits to R_1, R_2, and R_3, which make the largest contributions to molecular RE. The count of conjugated circuits is necessarily even because all conjugated circuits come in pairs of Kekulé valence structures. However, because larger conjugated circuits, such as R_3, may be linearly dependent—that is, they can be decomposed in smaller linearly independent conjugated circuits—when considering expressions molecular RE, they are not counted, so occasionally one comes across odd counts of larger conjugated circuits in the expression for molecular RE. The smallest such case is the count of R_3 in phenanthrene (#4 in Figure 10.6).

As mentioned, non-benzenoid hydrocarbons, such as those in Figure 10.7, will have in addition to $4n + 2$ conjugated circuits also $4n$ conjugated circuits denoted by Q_n. In general, non-benzenoid hydrocarbons may have both, $4n + 2$ conjugated circuits and $4n$ conjugated circuits, because four-member rings can have alternating C=C bond and C–C bonds. It is easy to see that azulene, the first molecules in Figure 10.8, has only $4n + 2$ conjugated circuits because the central CC bond between the fused pentagon and heptagon in both Kekulé valence structures is necessarily a C–C single bond. This makes the periphery of azulene conjugated circuits R_2 with

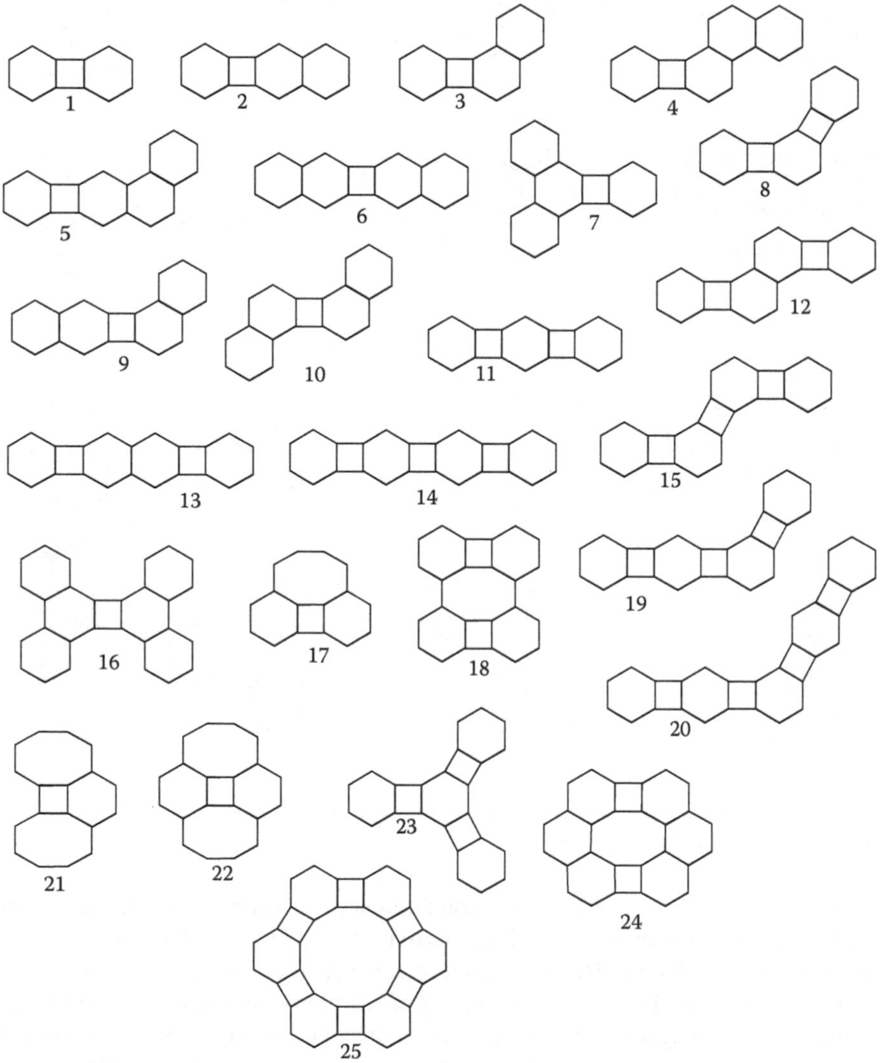

FIGURE 10.7 25 non-benzenoid alternant hydrocarbons whose RE expressions are listed in Table 10.2.

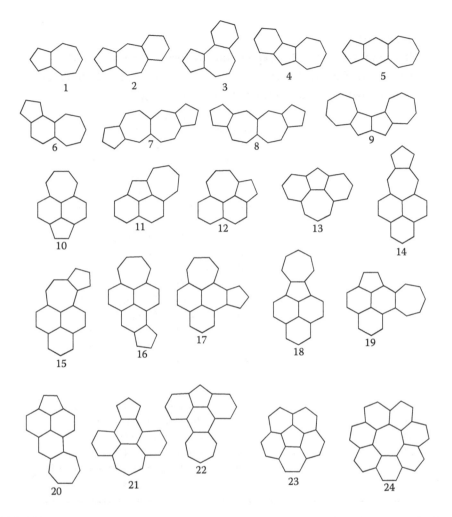

FIGURE 10.8 Two dozen non-benzenoid non-alternant polycyclic hydrocarbons having only $4n + 2$ conjugated circuits whose RE expressions are listed in Table 10.3.

five C=C bonds. In order to find out how many conjugated circuits and of which size appear in each structure, one has to examine all their Kekulé valence structures.

We will now very briefly examine in more detail Tables 10.2 through 10.5. Let us start by first comparing linear and angular biphenylene analogues consisting of several fused hexagons and squares. From Table 10.2 and Figure 10.7, one can see that RE of linear systems have higher RE than analog systems with angularly fused hexagons and squares. This is in contrast to the relative RE of linear and angular acenes. This information is sufficient to conclude that anti-kekulene (the last compound in Figure 10.7) is not expected to show much aromatic properties. In contrast, Kekulene, which has 12 benzene rings fused in a larger ring, is expected to show good stability, which is lacking in linearly fused eight benzene rings.

TABLE 10.2

Resonance Energy (RE) Expressions in Terms of Conjugated Circuits of Size R_1, R_2, and R_3 and Q_1, Q_2, and Q_3 for 25 Smaller Non-Benzenoid Alternant Hydrocarbons of Figure 10.7

1	$(8R_1 + 2Q_1 + 4Q_2 + Q_3)/2$
2	$(14R_1 + 5R_2 + 2Q_1 + 4Q_2 + 3Q_3)/7$
3	$(16R_1 + 4R_2 + 4Q_1 + 6Q_2 + 2Q_3)/8$
4	$(34R_1 + 10R_2 + 2R_3 + 6Q_1 + 10Q_2 + 3Q_3)/13$
5	$(32R_1 + 8R_2 + 2R_4 + 4Q_1 + 8Q_2 + 4Q_3)/12$
6	$(24R_1 + 12R_2 + 2Q_1 + 4Q_2 + 3Q_3)/12$
7	$(38R_1 + 8R_2 + R_3 + 8Q_1 + 10Q_2 + 3Q_3)/14$
8	$(28R_1 + R_2 + 12Q_1 + 20Q_2 + 4Q_3)/13$
9	$(26R_1 + 12R_2 + 4Q_1 + 6Q_2 + 6Q_3)/11$
10	$(32R_1 + 12R_2 + 8Q_1 + 5Q_2 + 2Q_3)/13$
11	$(28R_1 + 2R_2 + 2R_3 + 8Q_1 + 16Q_2 + 4Q_3)/12$
12	$(56R_1 + 8R_2 + 20Q_1 + 32Q_2 + 19Q_3)/21$
13	$(44R_1 + 8R_2 + 2R_3 + 8Q_1 + 16Q_2 + 12Q_3)/16$
14	$(88R_1 + 6R_2 + 2R_3 + 28Q_1 + 55Q_2 + 13Q_3)/29$
15	$(92R_1 + 6R_2 + 49Q_1 + 73Q_2 + 14Q_3)/34$
16	$(164R_1 + 40R_2 + 12R_3 + 32Q_1 + 16Q_2 + 16Q_3)/41$
17	$(8R_1 + 2R_3 + 2Q_1 + 6Q_2 + 5Q_3)/6$
18	$(80R_1 + 2R_2 + 20Q_1 + 41Q_2 + 14Q_3)/8$
19	$(90R_1 + 9R_2 + 2R_3 + 36Q_1 + 64Q_2 + 156Q_3)/31$
20	$(156R_1 + 68R_2 + 18R_3 + 52Q_1 + 67Q_2 + 40Q_3)/37$
21	$(2R_1 + 4R_2 + R_3 + 2Q_1 + 8Q_2 + 4Q_3)/6$
22	$(10R_1 + 4R_2 + 4Q_1 + 12Q_2 + 9Q_3)/9$
23	$(94R_1 + 6R_2 + 54Q_1 + 78Q_2 + 13Q_3)/35$
24	$(159R_1 + 68R_2 + 16R_3 + 52Q_1 + 65Q_2 + 23Q_3)/41$
25	Try to solve

The non-benzenoid non-alternant compounds of Table 10.3 and Figure 10.8 are quite interesting because they have only $4n + 2$ conjugated circuits and thus classify as *pure* aromatic compounds, just as do all benzenoid hydrocarbons. The first compound, azulene, is well known to be aromatic, and we now see that this should not be surprising. The same is also the case with the last compound of Table 10.3 and Figure 10.8, 7-circulene, which was synthesized 30 years ago [19]. Observe from Table 10.3 that although all conjugated circuits in these molecules are of size $4n + 2$, these molecules have a relatively small number of Kekulé valence structures and a small number of conjugated circuits R_1, with the exception of corannulene (5-circulene) and 7-circulene, which have five and seven benzene rings fused to the central ring, respectively.

Thanks to Klaus Haffner, George F. Thiele, and Carsten Mink [20], here we may add a short story on dicyclopenta[a,e]cyclooctene, an interesting 14π electron system.

TABLE 10.3
Resonance Energy (RE) Expressions in Terms of Conjugated Circuits of Size R_1, R_2, R_3, and R_4 for 24 Non-Benzenoid Non-Alternant Hydrocarbons of Figure 10.8

1	$(2R_2)/2$	13	$(2R_3)/2$
2	$(2R_1 + 2R_2 + 2R_3)/3$	14	$(8R_1 + 8R_2 + 4R_3 + 4R_4)/6$
3	$(2R_1 + 2R_2 + 2R_3)/3$	15	$(8R_1 + 8R_2 + 4R_3 + 4R_3)/6$
4	$(2R_1 + 2R_2 + 2R_3)/3$	16	$(4R_1 + 2R_2 + 6R_4)/4$
5	$(2R_3)/2$	17	$(4R_1 + 2R_2 + 2R_3 + 4R_4)/4$
6	$(16R_1 + 8R_2 + 3R_3)/7$	18	$(8R_1 + 8R_2 + 4R_3 + 4R_3)/6$
7	$(20R_1 + 10R_2 + 2R_3)/8$	19	$(4R_1 + 2R_2 + 2R_3 + 4R_4)/4$
8	$(4R_2 + 2R_4)/3$	20	$(4R_1 + 2R_2 + 6R_4)/4$
9	$(4R_2 + 2R_4)/3$	21	$(2R_4)/2$
10	$(4R_1 + 2R_2 + 6R_3)/4$	22	$(2R_4)/2$
11	$(4R_1 + 4R_2 + 4R_3)/4$	23	$(30R_1 + 20R_2 + 5R_3)/11$
12	$(4R_1 + 4R_2 + 4R_3)/4$	24	$(112R_1 + 70R_2 + 35R_3)/29$

In 1988, they synthesized 2,7-di-*tert*-butyldicyclopenta[*a,e*]cyclooctene, shown in Figure 10.9 at right and found that the molecule has a planar eight member ring. Let Haffner et al. speak for themselves. Here is the abstract of their paper titled "Synthesis and Structure of 2,7-Di-*tert*-butyldicyclopenta[*a,e*]cyclooctene," of which Figure 10.9 is a part:

> **A completely planar, delocalized 14 π-electron system** which does not contain cyclically conjugated substructures characterizes the title compound **2**. A productive synthesis of the 1,2-bis(*tert*-butylcyclopentadienyl)ethane **1** enabled the construction of the novel non-benzenoid aromatic hydrocarbon **2** which also contains one of the few planar eight-membered rings.

We are also reproducing concluding remarks of their article, in which the title compound is referred to as compound 10:

> 10 not only belongs to the few hydrocarbons with a planar eight-membered ring, but in addition it is also a completely planar 14 π-electron system which has no further cyclic conjugated subunits apart from the 14 π-perimeter. The molecular structure and spectroscopic properties justify the classification of 10 as a non-benzenoid aromatic hydrocarbon.

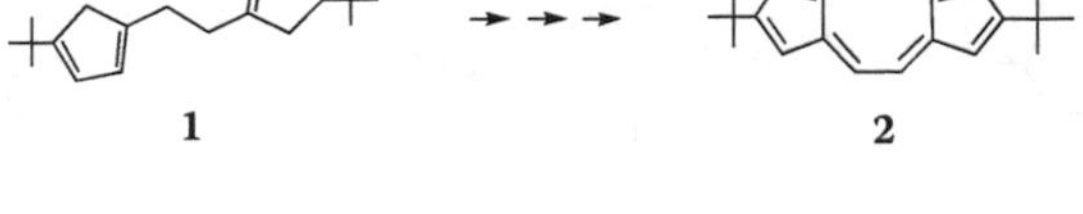

FIGURE 10.9 The starting and the final compounds in the synthesis of derivative of dicyclopenta[*a,e*]cyclooctene (10).

The molecule has only two Kekulé valence structures, like cyclooctene, but in contrast to cyclooctatetraene, which has four C=C bonds in its valence structures, cyclooctene in 2,7-di-*tert*-butyldicyclopenta[*a,e*]cyclooctene has only three C=C bonds in its Kekulé valence structures, which are part of the 14 π electron conjugated circuit on the molecule's periphery. The two CC bonds bridging the periphery are essentially single bonds, like the central bond in azulene. If in azulene and dicyclopenta[*a,e*]cyclooctene one is to ignore π electron interactions across the essentially single CC bond, the molecules would transform in monocyclic 10 and 14 π electron systems for which the Hückel $4n + 2$ rule would hold. One could therefore say that the Hückel $4n + 2$ rule holds for monocyclic conjugated systems and polycyclic systems that reduce to monocyclic systems when the essentially single CC bonds are ignored.

The compounds of Figure 10.10 and Table 10.4 are non-alternant non-benzenoids, but in contrast to the compounds of Figure 10.8, they have, in addition to $4n + 2$ conjugated circuits, also destabilizing $4n$ conjugated circuits. The smallest $4n$ conjugated circuits are Q_2 (having eight π electrons), which occur in compounds 3, 4, 6, 8, and 9, which have two adjacent fused pentagons. In other compounds of Figure 10.10, the smallest destabilizing $4n$ conjugated circuits are of size Q_3 (having 12 π electrons). The structures 1 (1:10 pentenoheptalene), 3 (aceheptylene or cyclopent-[*cd*]azulene), and 5 (azuleno[8,8a,1,2-*def*]heptalene) are known as Hafner's hydrocarbons. Their derivatives were synthesized by Hafner and coworkers between 1958 and 1970, respectively [20–24]. For a theoretical discussion of Hafner's hydrocarbons, see Ref. [25].

Finally, in Figure 10.11 and Table 10.5 have been grouped compounds having *only* $4n$ conjugated circuits, the compounds that can be classified as anti-aromatic. Observe that in the cata-condensed systems of Figure 10.11, all internal CC bonds are the so-called "essentially single CC bonds," which means that these CC bonds are a single bond in all Kekulé valence structures. In the conjugated circuit model, such bonds do

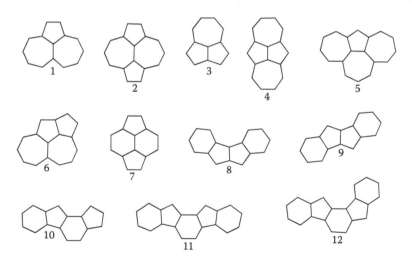

FIGURE 10.10 A dozen non-benzenoid non-alternant polycyclic hydrocarbons having $4n + 2$ and $4n$ conjugated circuits whose RE expressions are listed in Table 10.4.

TABLE 10.4
Resonance Energy (RE) Expressions in Terms of $4n + 2$ and $4n$ Conjugated Circuits for a Dozen Smaller Non-Benzenoid Non-Alternant Hydrocarbons of Figure 10.10

1	$(4R_2 + 2Q_3)/3$
2	$(8R_2 + 2R_3 + 2Q_3)/4$
3	$(4R_2 + 2Q_2)/3$
4	$(6R_2 + 2R_3 + 2Q_2)/4$
5	$(6R_2 + 4Q_3 + 2Q_4)/4$
6	$(6R_2 + 2R_3 + 2Q_2 + 2Q_3)/4$
7	$(4R_1 + 2R_2 + 6Q_3)/4$
8	$(4R_1 + 2Q_2 + 4Q_3 + 2Q_4)/4$
9	$(8R_1 + Q_2 + 4Q_3 + 2Q_4)/5$
10	$(2R_1 + Q_3 + 2Q_4)/3$
11	$(8R_1 + 2Q_3 + 4Q_4 + 2Q_5)/5$
12	$(4R_1 + 2Q_3 + 4Q_4 + 2Q_5)/4$

not play any role and can be disregarded, erased. This transforms such a molecule to a $4n$ monocyclic structure, which has but two Kekulé valence structures.

As we have seen, the Hückel $4n + 2$ rule classifies all monocyclic systems in aromatics and anti-aromatics. By extension the generalized Hückel $4n + 2$ rule classifies all polycyclic conjugated hydrocarbons also into aromatic and anti-aromatic, but in addition, there are compounds "in between." Consider Figure 10.12 showing a collection of structures in between the two extreme cases of aromaticity, benzene and cyclooctatetraene. The structures in between suggest a gradual transition between the two. The selection of polycyclic compounds in between clearly must have a variable degree of both aromatic and anti-aromatic structural elements. Below benzene, the prototypes of aromaticity, at the top of the figure, and cyclooctatetraene, one of the prototypes of anti-aromaticity, at the bottom, and in between, there are compounds whose aromaticity gradually decreases. These intermediate cases cannot be classified as aromatic or anti-aromatic but as partially aromatic or partially anti-aromatic.

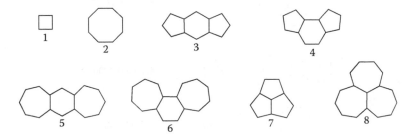

FIGURE 10.11   Several non-benzenoid polycyclic hydrocarbons having only $4n$ conjugated circuits whose RE expressions are listed in Table 10.5.

TABLE 10.5

Resonance Energy (RE) Expressions in Terms of $4n$ Conjugated Circuits for Several Alternant and Non-Alternant Non-Benzenoid Hydrocarbons of Figure 10.11

1	$(2Q_2)/2$
2	$(2Q_2)/2$
3	$(2Q_3)/2$
4	$(2Q_3)/2$
5	$(2Q_4)/2$
6	$(2Q_4)/2$
7	$(6Q_2)/3$
8	$(6Q_3)/3$

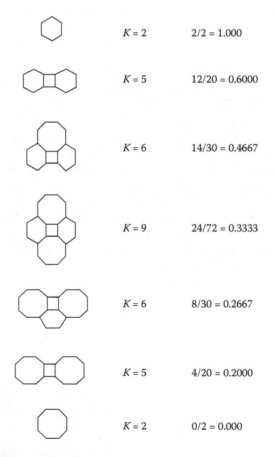

	$K = 2$	$2/2 = 1.000$
	$K = 5$	$12/20 = 0.6000$
	$K = 6$	$14/30 = 0.4667$
	$K = 9$	$24/72 = 0.3333$
	$K = 6$	$8/30 = 0.2667$
	$K = 5$	$4/20 = 0.2000$
	$K = 2$	$0/2 = 0.000$

FIGURE 10.12 Set of compounds starting with benzene and ending with cyclooctatetrene ordered in decreasing degree of aromaticity.

TABLE 10.6

Resonance Energy (RE) Expressions in Terms of $4n + 2$ and $4n$ Conjugated Circuits for Seven Compounds of Figure 10.12

1	$(2R_1)/2$
2	$(8R_1 + 2Q_1 + 4Q_2 + Q_3)/5$
3	$(8R_1 + 2R_3 + 2Q_1 + 6Q_2 + 5Q_3 + Q_4)/6$
4	$(8R_1 + 4R_2 + 4R_3 + 2R_4 + 4Q_1 + 12Q_2 + 10Q_3 + 15Q_4)/9$
5	$(2R_1 + 4R_2 + 2R_4 + 2Q_1 + 8Q_2 + 4Q_3 + 4Q_4)/6$
6	$(4R_2 + 2Q_1 + 8Q_2 + 2Q_4)/5$
7	$(2\,Q_2)/2$

The missing *conjugated circuits*, the "tool" for such classification, has been the reason for failure of numerous attempts to generalize the Hückel $4n + 2$ rule.

If one wants to go back and discuss on a quantitative level aromaticity of compounds illustrated in the gradual changes from aromatic benzene to anti-aromatic cyclooctatetraene, one should examine the RE for the seven compounds of Figure 10.12. In Table 10.6, we have listed the RE expressions for these molecules from which one can see quantitatively how going down from benzene to cyclooctatetraene, the number of $4n$ conjugated circuits increases and their anti-aromatic contributions decrease the aromaticity of these molecules.

In Figure 10.12, in addition to the count of Kekulé valence structures, we show also fractions in which the numerator gives the count of $4n + 2$ conjugated circuits in each molecule and the denominator gives the total number of conjugated circuits, which equals $K(K - 1)$. The fraction thus indicates the percentage of $4n + 2$ conjugated circuits in each molecule, and as one can see from Figure 10.12, it goes from 1 (100%) for benzene to 0 for cyclooctatetraene. This quotient appears to be a useful structural index of aromatic character for molecules that do not qualify as fully aromatic or fully anti-aromatic! Observe that the proposed index of aromaticity is fully structural. It will be interesting to see how well it parallels the reduced aromatic characteristics of these compounds.

10.4 ON CLASSIFICATION OF POLYCYCLIC CONJUGATED COMPOUNDS

In summary, one may classify polycyclic conjugated hydrocarbons as *fully aromatic*, *partially* aromatic, *partially* anti-aromatic, and *fully* anti-aromatic. This classification allows grouping together compounds that are expected to exhibit aromatic properties, limited aromatic properties, limited anti-aromatic properties, and anti-aromatic properties, respectively. Clearly, fully aromatic are the compounds of Figures 10.6 and 10.8, which have only $4n + 2$ conjugated circuits. Observe that aromaticity of benzene, coronene, and benzenoid hydrocarbons will be more pronounced than those of azulene, dicyclopenta[a,e]cyclooctene, corannulene, and related non-benzenoid compounds, which should be a sufficiently strong indication to dismiss the arguments of those who prefer to characterize aromaticity in terms of molecular properties. Even when one puts azulene, dicyclopenta[*a,e*]cyclooctene, corannulene, and structurally related molecules aside, there is enough variation in aromatic properties among the benzenoid molecules of Figure 10.6 to hint that aromaticity should not be characterized, least to be defined, by molecular properties, because these aromatic compounds already show considerable variations in some of their properties and biological activities.

Let us consider but a single property, or lack of it, among benzenoid hydrocarbons: that of carcinogenity. It was proposed by Jerina and coworkers that the presence of the so-called *bay regions* in benzenoid hydrocarbons contribute to their carcinogenicity [26]. Here we have illustrated the bay regions in benzanthracene (one of the highly carcinogenic benzenoid hydrocarbons) and related benzo[*a*]pyrene (there being evidence to link to the formation of lung cancer [27]):

As one can see, bay regions involve three benzene rings, one of which is terminal, and the central ring is a "kink" ring. The presence of a bay region may not be sufficient to induce cancer because benzo[*e*]pyrene

has bay regions but is non-carcinogenic [28]. For more on topological correlation and cancer [29] and a quantitative approach to carcinogenesis inhibition [30], one may consult chapters four and six on the quantitative approach to carcinogenesis inhibition in the book on modifying cancer genesis [31]. We should mention that there are also benzenoid hydrocarbons that have no bay regions but which have been found to be carcinogenic.

Graph theoretical characterization and computer generation of certain carcinogenic benzenoid hydrocarbons and identification of bay regions were reported by K. Balasubramanian, J. J. Kaufman, and W. S. Kosky in 1980, almost 35 years ago [28].

Our book, being about solved and unsolved problems in structural chemistry, induced us to select carcinogenicity as a property of benzenoid hydrocarbons to illustrate considerable variations of some of the properties of this group of compounds because there are several issues that have not been clarified and understood concerning benzenoids and their carcinogenicity. Consider a question, which we consider to be unsolved:

Problem 12

What are, if any, the structural factors that may answer why some benzenoids do not follow the regularity that holds for most benzenoid hydrocarbons that associates the presence of the bay regions with carcinogenicity?

Here is another speculation that could be examined more closely: Are highly stable benzenoids, among which is benzo[e]pyrene, less or not carcinogenic? If one writes the structural formula of benzo[e]pyrene as

which is its Clar structure, we see that benzo[e]pyrene has three Clar's π aromatic sextets. This would suggest it to be fairly stable (unreactive). As we said, this is a speculation, and more thorough examination of a large group of benzenoids needs to be investigated. Only then one will find the "exceptions," relative to the majority, with respect to the presence or absence of bay regions and carcinogenicity and whether they are false positive or false negative.

In the next chapter on Clar's aromatic sextets, we elaborate on Clar structures and discuss differences in properties among "fully aromatic" compounds, which should further support the notions that aromaticity ought to be characterized on structural factors and not properties.

The borderline between partially aromatic and partially anti-aromatic compounds may be a matter of convention and convenience because truly, from the point of view of structural chemistry, these two groups differ only in degree and not substantively. The fully anti-aromatic compounds, again, are well defined as the compounds that have *only* $4n$ conjugated circuits, in contrast to fully aromatic compounds that have been defined as the compounds that have *only* $4n + 2$ conjugated circuits.

In the next chapter, we describe Clar's aromatic sextet theory, which clarifies differences between fully aromatic systems because one can definitely say, "All benzenoids are aromatic, but some are more aromatic than others," to paraphrase George Orwell's (1903–1950) *Animal Farm*, "All animals are equal, but some are more equal than others" [32], which alluded to privileges of party members in communist countries installed in Europe after World War II. Clar's aromatic sextet theory offers clues to how can one recognize more aromatic benzenoids from less aromatic. Past efforts to arrive at structural factors that are crucial for understanding aromaticity (at least in the case of hydrocarbons) have been marred by emphasis on the *count* of π electrons and giving prominence to molecular periphery instead of considering π electron (conjugated) circuits. Observe also that molecular periphery is often associated with the largest conjugated circuits in a molecule, and we know well that the larger conjugated circuits contribute much less to molecular aromaticity than the smaller conjugated circuits R_1, R_2, and R_3. So molecular periphery does not play any special role when it come to aromaticity. For example, molecules of Figure 10.13 have $4n$ π-electrons, which is irrelevant for their aromatic properties, but sometimes it takes long time to forget the old habits.

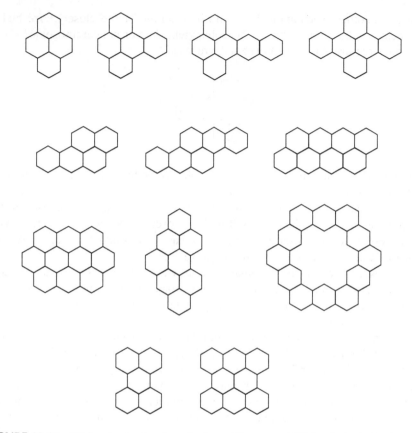

FIGURE 10.13 Molecules having $4n\ \pi$ electrons. The central CC bonds in the last two benzenoid hydrocarbons are C–C single bonds in all Kekulé valence structures.

REFERENCES AND NOTES

1. A. Kekulé, Sur la constitution des substances aromatiques, *Bull. Soc. Chim.* 3 (1865) 98–110.
2. A. Kekulé, Untersuchungen uber aromatische Verbindungen, *Ann. Chem. Pharm.* 137 (1866) 129–196.
3. A. T. Balaban, D. C. Oniciu, and A. R. Katritzky, Aromaticity as a cornerstone of heterocycic chemistry, *Chem. Rev.* 194 (2004) 2777–2812.
4. M. Randić, Conjugated circuits and resonance energies of benzenoid hydrocarbons, *Chem. Phys. Lett.* 38 (1976) 68–70.
5. D. J. Klein, T. G. Schmalz, G. E. Hite, and W. A. Seitz, Resonance in C_{60}—Buckminsterfullerene, *J. Chem. Soc.* 108 (1986) 1301.
6. I. Gutman and M. Randić, A correlation between Kekulé valence structures and conjugated circuits, *Chem. Phys.* 41 (1979) 265–270.
7. M. Randić, V. Solomon, S. C. Grossman, D. J. Klein, and N. Trinajstić, Resonance energies of large conjugated hydrocarbons by a statistical method, *Int. J. Quantum Chem.* 32 (1987) 35–59.
8. I. Gutman, Counting independent conjugated circuits in benzenoid hydrocarbons, *Z. Phys. Chem. (Leipzig)* 271 (1990) 793–798.

9. X. Guo and M. Randić, Recursive formulae for enumeration of L-M conjugated circuits in structurally related benzenoid hydrocarbons, *J. Math. Chem.* 30 (2001) 325–342.

10. X. Guo and M. Randić, Recursive method for enumeration of linearly independent and minimal conjugated circuits of benzenoid hydrocarbons, *J. Chem. Inf. Comput. Sci.* 34 (1994) 339–348.

11. X. F. Guo, M. Randić, and D. J. Klein, Analytical expressions for the count of LM-conjugated circuits in benzenoid hydrocarbons, *Int. J. Quantum Chem.* 60 (1996) 943–958.

12. C.-D. Lin, Efficient method for calculating the resonance energy expression of benzenoid hydrocarbons based on the enumeration of conjugated circuits, *J. Chem. Inf. Comput. Sci.* 40 (2000) 778–783.

13. C.-D. Lin and G.-Q. Fan, Algorithms for the count of linearly independent and minimal conjugated circuits in benzenoid hydrocarbons, *J. Chem. Inf. Comp. Sci.* 39 (1999) 782–787.

14. I. Lukovits and D. Janežič, Enumeration of conjugated circuits in nanotubes, *J. Chem. Inf. Comput. Sci.* 44 (2004) 410–414.

15. I. Lukovits, M. Diudea, and E. Kalman, Enumeration of conjugated circuits in graphite, in: *Semiconductor Conference, 2004. CAS 2004 Proceedings*, Vol. 1, 2004.

16. S. J. Austin, P. W. Fowler, P. Hansen, D. E. Monolopoulis, and M. Zheng, Fullerene isomers of C_{60}. Kekulé counts versus stability, *Chem. Phys. Lett.* 228 (1994) 478–484.

17. M. Randić, Aromaticity of polycyclic conjugated hydrocarbons, *Chem. Rev.* 103 (2003) 3449–3605.

18. I. Gutman and S. J. Cyvin, Conjugated circuits in benzenoid hydrocarbons, *J. Mol. Struct. (Theochem)* 184 (1880) 159–163.

19. K. Yamamoto, T. Harada, M. Nakazaki, T. Naka, Y. Kai, S. Harada, and N. Kasai, Synthesis and characterization of [7]circulene, *J. Am. Chem. Soc.* 105 (1983) 7171–7172.

20. K. Haffner, G. F. Thiele, and C. Mink, Synthesis and structure of 2,7-Di-*tert*-butyldicyclopenta[*a,e*]cyclooctene, *Angew. Chem. Int. Ed.* 27 (1988) 1191–1192.

21. K. Hafner and J. Schneider, *Angew. Chem.* 70 (1958) 702.

22. K. Hafner and J. Schneider, Darstellung und Eigenschaften von Derivaten des Pentalens und Heptalens, *Liebigs Ann. Chem.* 624 (1959) 37–47.

23. K. Hafner and R. Fleischer, New Synthesis of Aceheptylene, *Angew. Chem. Int. Ed.* 9 (1970) 247.

24. K. Hafner, G. Hafner-Schneider, and F. Bauer, Azuleno[8,8a,1,2-*def*]heptalene, *Angew. Chem. Int. Ed.* 7 (1968) 808–809.

25. M. Asgar Ali and C. A. Coulson, Properties of Hafner's heptalene and pentalene derivatives, *Mol. Phys.* 4 (1961) 65–79.

26. D. M. Jerina, P. E. Lehr, H. Yagi, O. Hernandez, P. Dansette, P. G. Wislocki, G. Wislocki, A. W. Wood, R. I. Chang, W. Levin, and A. H. Conney, Mutagenicity of benzo(a)-pyrene derivatives and the description of a quantum mechanical model which predicts the phase of carbonium ion formation from diol epoxides, in: *In-Vitro Activation in Mutagenesis Testing*, F. J. De-Seres, J. R. Bond, and R. M. Philpot (Eds.), Elsevier, Amsterdam, 1976.

27. M. F. Denissenko, A. Pao, M. Tang, and G. P. Pfeifer, Preferential formation of benzo[a] pyrene adducts at lung cancer mutational hotspots in P53, *Science* 274 (1996) 430–432.

28. J. J. Kaufman, P. C. Hariharan, W. S. Kosky, and K. Balasubramanian, Quantum chemical and other theoretical studies of carcinogens, their metabolic activation and attack on DNA constituents, in: *Molecular Basis of Cancer, Part A: Macromolecular Structure, Carcinogens, and Oncogens*, R. Rein (Ed.), Alan R. Liss, Inc., New York, 1985, pp. 275–363.

29. A. T. Balaban, Topological correlation and cancer, Chapter 4, in: *Modeling of Cancer Genesis and Prevention*, N. Volculetz, A. T. Balaban, I. Niculescu-Duvǎz, and Z. Simon (Eds.), CRC Press, Boca Raton, FL, 1991.

30. N. Volzuletz, A. T. Balaban, and I. Niculescu-Duvăz, Quantitative approach to carcino-genesis inhibition, chapter six, in: N. Volculetz, A. T. Balaban, I. Niculescu-Duvăz, and Z. Simon (Eds.), *Modeling of Cancer Genesis and Prevention*, CRC Press, Boca Raton, FL, 1991.

31. N. Volzuletz, A. T. Balaban, I. Niculescu-Duvăz, and Z. Simon, *Modeling of Cancer Genesis and Prevention*, CRC Press, Boca Raton, FL, 1991.

32. G. Orwell, *Animal Farm*, Penguin Group, London, 1946.

11 Clar Aromatic Sextet

11.1 INTRODUCTION

Eric Clar (1902–1987) was an outstanding organic chemist who has synthesized more than 100 benzenoid hydrocarbons [1]. He has been considered, deservingly, the "father" of benzenoid chemistry. In addition to synthesis of numerous benzenoid hydrocarbons, Clar also initiated a novel approach to rationalize collected experimental data, mainly UV spectra and early NMR chemical couplings. In doing this, he developed a new theoretical model for benzenoid molecules, the so-called the "aromatic sextet" theory [2]. Armit and Robinson [3] introduced the notion of the aromatic sextet in chemistry a few years before the historic paper by Heitler and London in 1927 on the H_2 molecule [4] that opened the new field of theoretical chemistry, quantum chemistry, which grew over the decades into a new branch of chemistry with its own problems, concepts, insights, solutions, and enlightenments that were previously unavailable.

But important insights were made in chemistry and chemical bonding even before 1927 when quantum mechanics initiated its present quantitative molecular modeling. Two outstanding illustrations are (i) Gilbert N. Lewis and his octet rule [5,6] and (ii) Eric Clar and his aromatic sextets. We will here elaborate on the latter. According to Armit and Robinson, the six-member rings in unsaturated compounds to which they refer as the π electron aromatic sextet and for which they have introduced the "circle" notation, have unusual stability. The idea of π sextets has been traced to Crocker and Armstrong [7–9], but it was Clar, not Robinson, Armit, Crocker, and Armstrong, who *extended* the notion of the "aromatic sextet" beyond *single* six-member rings to *polycyclic* benzenoid hydrocarbons. Clar's approach has been lucidly described in his booklet *The Aromatic Sextet* [2], in which, as experimental support for his approach, he discusses the UV spectra and NMR spectra of selected benzenoid compounds and the relative stability and the reactivity of polycyclic conjugated systems.

In the next section of this chapter, we first outlined Clar's aromatic sextet theory and follow with a very recent development relating to *quantitative* characterization of Clar's aromatic sextet theory. It appears that over the past 40 years and more, since the publication of Clar's aromatic sextet booklet in 1972, Clar's theory has been mostly ignored by quantum chemists, who appear to be preoccupied with *ab initio* calculations. An exception have been theoretical chemists working in chemical graph theory, most of whom, one may say, have been also equally ignored, overlooked, or marginalized by most quantum chemists. The first theoretical approach to characterization of Clar's theory of local benzenoid regions appeared in 1967 in the publication by Polansky and Derflinger in the *International Journal of Quantum Chemistry* [10]. They used the coefficients of molecular orbitals to characterize individual benzene rings in benzenoid hydrocarbons.

11.2 QUALITATIVE CLAR AROMATIC SEXTET THEORY

Armit and Robinson introduced the notion of six-member rings in unsaturated compounds associated with the π electron aromatic sextet that bestows on them unusual stability, for which they have introduced the "circle" notation. Robinson later abandoned this novel model to characterize benzenoid compounds. It was Clar who not only adopted the notion of aromatic sextets, but significantly developed the model into a "theory," which could explain several regularities of selected properties of these compounds. Clar's aromatic sextet theory assigns to individual benzenoid hydrocarbons novel structural formulas, which are obtained by following the rules summarized in Table 11.1.

In Figure 11.1, we have illustrated Clar's structural formulas for benzene and a number of smaller benzenoid systems. According to Clar, benzenoid hydrocarbons can be divided into three groups: (i) the fully benzenoid hydrocarbons, the rings of which are either the sextet rings or "empty" rings (Figure 11.2); (ii) the benzenoid systems with a *single* Clar structure, which may have, in addition to aromatic sextet rings, also empty rings and rings with one C=C bond (Figure 11.3); and (iii) the benzenoid systems that have two or more Clar structures, which are summarily represented graphically as single structures with a sextet to which are added "arrows," which signify the so-called "migrating" sextets (Figure 11.4).

The arrows indicate alternative locations for the aromatic sextets. Observe, however, that if the migrating sextet "arrows" end in adjacent rings, two sextets cannot *simultaneously* occupy the adjacent rings. For instance, this happens in chrysene shown at the end of the second row in Figure 11.1, benzo[b]chrysene in the middle, and benzo[a]pyrene, at the end of Figure 11.1.

As experimental support for his model, Clar considered regularity in shifts of certain bands toward the violet and beyond in the UV spectra of structurally related benzenoids. As one moves from green linear hexacene toward colorless, the most branched dibenzo[g,p]chrysene (Figure 11.5), the shifts reflect the increasing stability of these benzenoids. Moving from linear hexacene to dibenzochrysene also parallels the increase in the number of aromatic sextets in these compounds.

The Clar sextet theory has received fair attention in the literature and continues to remain a subject of study as is reflected in several recent publications [11–16]. Unfortunately, Clar's sextets have not yet attracted the visible attention of quantum chemists, who could investigate if the roots of Clar's theory have some hidden relationship with quantum mechanics. As the relationship of Clar's sextet theory to quantum mechanics is unexplained and unknown, we are tempted to pose this

---

TABLE 11.1

Rules for Construction of Clar's Formulas for Benzenoid Hydrocarbons

(i) Circles representing aromatic sextets should not be placed in adjacent rings.
(ii) Clar structure has the maximal number of inscribed circles possible such that all carbon atoms not part of sextets are incident to one C=C bond.
(iii) The arrows indicate the adjacent alternative sites of aromatic sextets.
(iv) Rings having adjacent arrow ends cannot be simultaneously occupied by sextets.

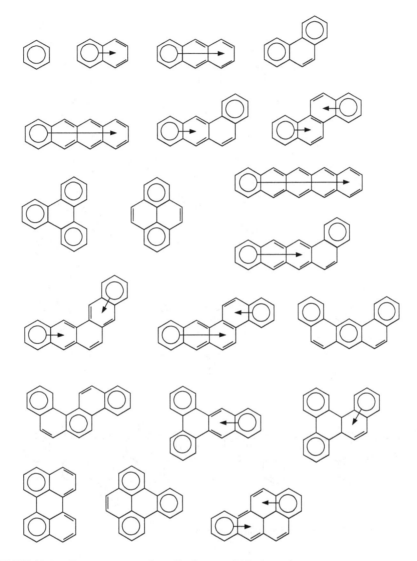

FIGURE 11.1 Clar structures of smaller benzenoid hydrocarbons.

as one of the important unsolved problems of structural chemistry, which can be formulated as

Problem 13

Using the standard quantum chemical approaches and general methodology of quantum mechanics, seek non-empirical justifications of Clar's sextet theory.

Interestingly, the MO theory and VB theory have apparently different conceptual and computational formalism, so in the early days of quantum chemistry,

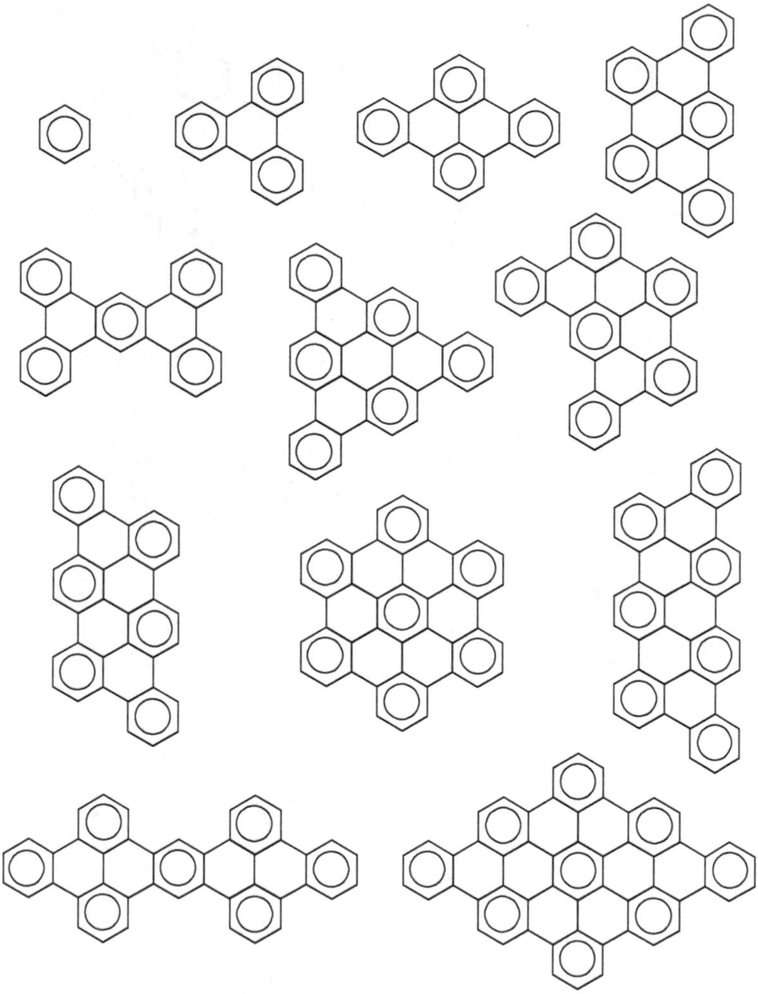

FIGURE 11.2 Fully benzenoid aromatic compounds.

theoreticians were looking to find their equivalence. The density functional theory (DFT) of Walter Kohn [17] had a harder time being accepted, but it is now almost on an equal level in quantum chemical calculations. DFT may even have some advantage over the Gaussian *ab initio* calculations [18] concerning applications to larger molecules but at the cost of lower precision. The more recent great and fast expansion of DFT undoubtedly was catalyzed by the award of the Nobel Prize in chemistry in 1998 to Walter Kohn *"for his development of the density-functional theory,"* which he shared with John A. Pople, the advocate of Gaussian *ab initio* calculations *"for his development of computational methods in quantum chemistry."*

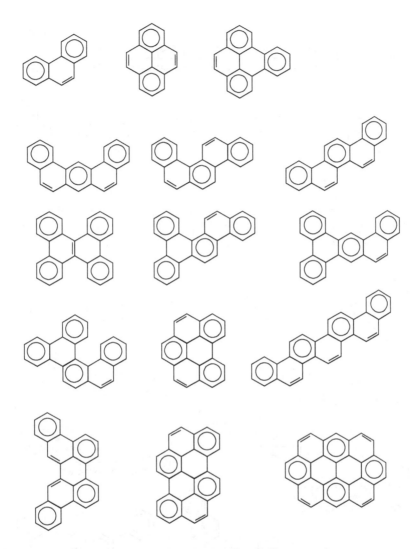

FIGURE 11.3   Beznenoid hydrocarbons having "fixed" Clar aromatic sextets.

11.3   CLAR AROMATIC SEXTET REVISITED

The famous Hückel $4n + 2$ rule [19], which explained the unusual stability of mono-cyclic systems, is widely and well known. In contrast, Clar's rule on the great stability of benzenoid systems having $6n$ π electrons, which is much broader as it applies to polycyclic conjugate systems, appears not to be so widely known. According to Clar [2],

> The fully benzenoid hydrocarbons have $6n$ π electrons, n being an integer. It becomes obvious that the most stable hydrocarbons do not follow Hückel's rule: $(2 + 4n)$ π electrons [19]. Therefore the latter must be strictly limited to monocyclic systems.

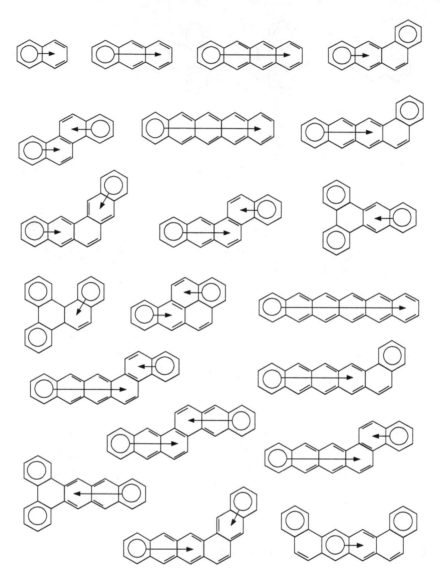

FIGURE 11.4   Clar formulas for smaller benzenoid hydrocarbons having migrating sextets.
(*Continued*)

Fully benzenoid hydrocarbons are those for which Clar's formulas contain only sextet rings and "empty" rings and no CC double bonds. The smallest fully benzenoid hydrocarbon is triphenylene.

It is unclear how such a profound empirical prediction, which characterizes the most stable benzenoid systems, could have been overlooked or ignored for such a long time by so many theoretical chemists. Clar's approach to novel representation of benzenoid systems may have appeared to some as a phantasm, but Clar, as we have already mentioned, has reported on a number of experimental results that support his

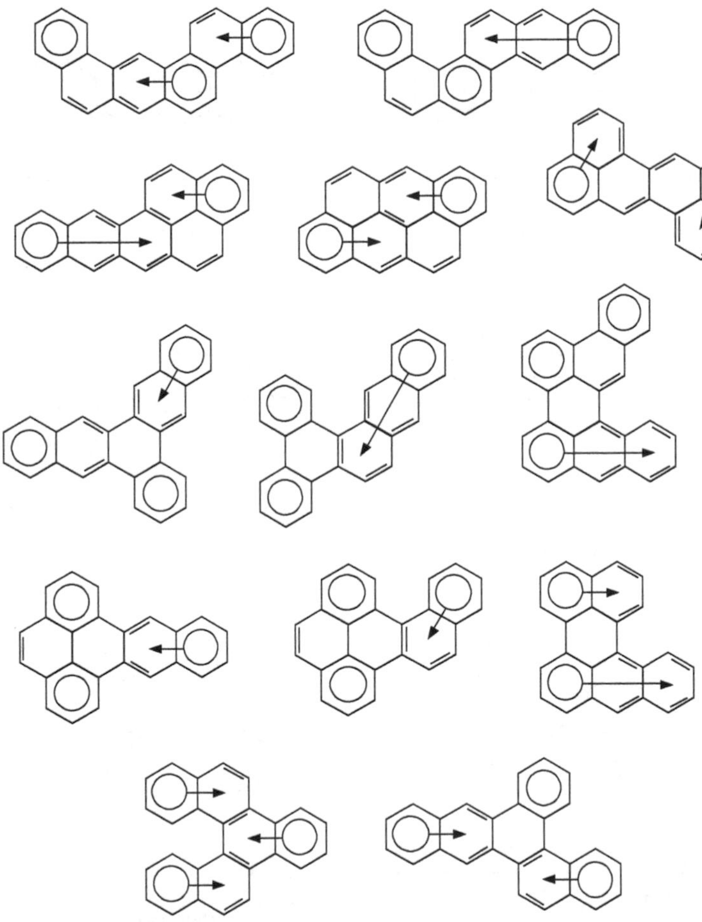

FIGURE 11.4 (CONTINUED)   Clar formulas for smaller benzenoid hydrocarbons having migrating sextets.

views on benzenoid systems and his theory of aromatic sextets. Support for his views is based on relative shifts of UV spectra along a series of benzenoids having increasing numbers of π electron sextets, already mentioned and illustrated in Figures 11.5 and 11.6. The Clar aromatic sextet theory as of today is empirical and mostly qualitative. We, in continuation of this section, offer a numerical model for Clar's structures, which bear some relationship to the properties underlying Kekulé valence structures of the molecule considered. This qualifies as a *quantitative* model of Clar's sextets, but for this, we have first to consider Pauling bond orders of benzenoid hydrocarbons.

11.4   PAULING BOND ORDERS

In 1935, Linus Pauling introduced his bond orders [20], a simple and elegant concept for numerical characterization of individual CC bonds in benzenoid hydrocarbons. The Pauling bond order of a CC bond is defined for benzenoid and non-benzenoid

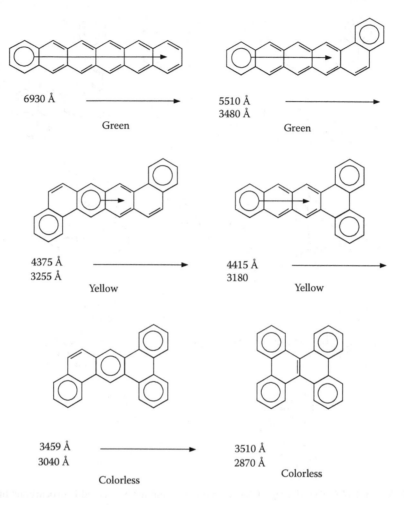

6930 Å　⟶　Green

5510 Å
3480 Å　⟶　Green

4375 Å
3255 Å　⟶　Yellow

4415 Å
3180　⟶　Yellow

3459 Å
3040 Å　⟶　Colorless

3510 Å
2870 Å　⟶　Colorless

FIGURE 11.5 Shift of UV and visible spectral peaks as the number of aromatic sextets increases from hexacene to dibenzo[g,p]chrysene.

conjugated hydrocarbons as the relative frequency that a selected CC bond occurs as a C=C bond in all Kekulé valence structures of the molecule.

Pauling bond orders were correlated with experimental CC bond lengths in numerous benzenoid hydrocarbons and have offered insights into the variations in CC bonds in these molecules. Pauling bond orders found additional application in quantum chemistry whether used alone or combined with their counterpart in HMO, the Coulson bond orders [21]. One such illustration, outlined in Chapter 3 in the section *"The Difference in HMO and PPP Bond Orders,"* showed that the differences between the simple Hückel MO calculations and more elaborate self-consistent-field MO calculations, known as the Pariser-Parr-Pople MO method [22], points to so-called Fries valence structures. These are Kekulé valence structures that have the largest number of benzene rings with three C=C bonds [23]. Pauling bond orders continue to be of interest in structural chemistry. For example, recently S. Narita and

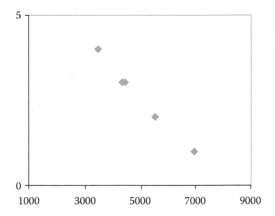

FIGURE 11.6 Plot of the number of Clar's aromatics sextets in benzenoid hydrocarbons of Figure 11.5 against wavelength (in Å) of their absorption peaks.

coworkers [24] found a good linear relationship between Pauling bond orders and bond lengths for fullerene molecules C_{60} and C_{70}.

11.5 RING BOND ORDERS

Looking back to the early days of quantum chemistry, there are two scientists who made the most important contributions that helped acceptance of the novelty of quantum mechanics in chemistry: Erich Hückel and Linus Pauling. Hückel extended the atomic orbital model to molecules by considering π electrons of benzene and arrived at his $4n + 2$ rule for aromaticity of monocyclic systems. Pauling used the atomic orbital model in constructing hybrid orbitals: sp^3 for tetrahederal carbons in alkanes and saturated compounds involving CC single bonds; sp^2 for trigonal carbons in alkenes, benzenoid hydrocarbons, and structurally related compounds involving CC double bonds; and sp for diagonal carbons in alkynes and compounds involving CC triple bonds. In addition, Pauling also introduced the CC bond orders that characterize variations of CC bonds in molecules having more than one Kekulé valence structure.

The notion of sp^3, sp^2, and sp hybrids has been essential for qualitative understanding of the geometry of most molecular structures although it was clear that in strained cyclic structures the hybridizations and bonds would be affected. Thus, Coulson speaks of "banana" bonds (that is, "bent" bonds) in cyclopropane C_3H_6 and similar compounds, and the notion of maximum overlap of hybrids forming bonds was invoked for a qualitative characterization of such systems. However, overlaps of orbitals forming bonds could be calculated, which made it possible to generalize the sp^3, sp^2, and sp to sp^n, where n need not be an integer [25–27]. It's been 30 years since Pauling in 1935 introduced hybrids to the emergence of this particularly simple generalization, suited for characterization strained molecules.

There have been few modifications of the Pauling bond order, but in contrast to hybrid orbitals, it took 80 years for the arrival in 2014 of the ring bond order, which is an important and simple generalization of the Pauling bond order. As we will see, ring bond orders (RBOs) form the basis for a theoretical explanation of the Clar aromatic sextet theory, which was around for more than 40 years—justified hitherto only on Clar's empirical arguments.

The ring bond orders could have been introduced in 1935 as they do not require any modifications of the Pauling bond order concept but simply an extension of its application from considering *individual* CC bonds to the six CC bonds of *individual* benzene rings of benzenoid hydrocarbons. One can calculate the ring bond order of a benzene ring in a benzenoid hydrocarbon by first finding Pauling bond orders for the six CC bonds making the ring up and then summing them, or by counting C=C bonds in the benzene ring in all the Kekulé valence structures of the benzenoid hydrocarbon and dividing the count by K, the number of Kekulé valence structures.

In Figure 11.7, we show the ring bond orders for 20 smaller benzenoid hydrocarbons. From Figure 11.7, one can observe a number of regularities for the calculated ring bond orders. Observe that all benzene rings in linear systems have the same RBO, which gradually decreases from 8/3 (which is approximately equal to 2.67) in naphthalene to 10/4 (which is equal to 2.50) in anthracene, to 12/5 (which is equal to 2.40) in tetracene, to 14/6 (which is approximately equal to 2.33) in pentacene, and to 16/7 (which is approximately equal to 2.29) in hexacene. In the case of infinite linearly fused benzene rings RBO of a benzene ring is 2.00. However, this is of little interest in chemistry because the longest known stable linear chains have eight benzene rings with the ring bond order of 20/9 (which is approximately equal to 2.22). The highest possible value of RBO is 3.00, which belongs to a ring in a benzenoid hydrocarbon if the ring has three C=C bonds in all Kekulé valence structures of the benzenoid hydrocarbon. An illustration is biphenyl (diphenyl), besides the case of benzene. Strictly speaking biphenyl and benzene are non-benzenoids, because benzenoids are defined as built of fused benzene rings. So the question is: Is it possible to have a benzenoid hydrocarbon with a ring having bond order 3.000. We expect the answer to be: No. We also expect that our mathematical colleagues will find a proof to that. The value 22/10 may be the smallest value belonging to stable linear nonacene detected in a matrix isolation.

As already mentioned, by adding the Pauling bond orders of a ring, one gets the average count of C=C in individual rings of a benzenoid molecule. In naphthalene, one finds for the average C=C 8/3. In anthracene, in two out of four Kekulé valence structures, every ring has three C=C bonds, and in the remaining two structures, every ring has only two C=C bonds, resulting in 10 C=C bonds for four Kekulé structures or on the average 10/4. In tetracene, as illustrated in Figure 11.8, one obtains the average of 12/5, and so on, as shown in Figure 11.8.

From Figure 11.7, one can see that in nonlinear cata-condensed benzenoids, terminal rings and adjacent rings linearly fused have the largest ring values. On the other side, the "kink" rings (the rings at which fusion of benzene rings changes direction) have the smallest ring bond orders. The largest ring bond orders in Figure 11.7 after benzene have two terminal benzene rings of dibenzoanthracene (2.92) and terminal benzene rings of fully benzenoid triphenylene (2.89). The high RBO for triphenylene could have been expected, in the view that it is a fully benzenoid system. It appears somewhat surprising that the two rings of dibenzoanthracene, which is not fully a benzenoid system, has even higher values of RBO. However, observe that

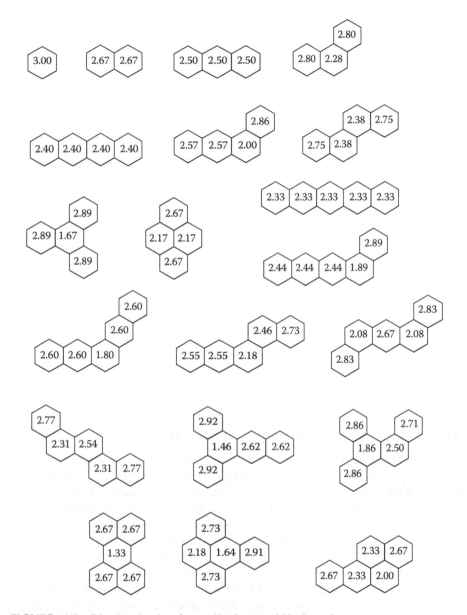

FIGURE 11.7   Ring bond orders for smaller benzenoid hydrocarbons.

in dibenzoanthracene the ring adjacent to the maximal RBO rings has a ring bond order of 1.46, and the "empty" ring of triplenylene has a higher value of 1.67. With exception of the central ring of perylene, which has "essentially" single CC bonds (the bonds that are CC single bonds in all Kekulé valence structures of a molecule), the "empty" rings of dibenzoanthracene and triphenylene have the smallest values of the ring bond orders among the ring of benzenoids of Figure 11.7.

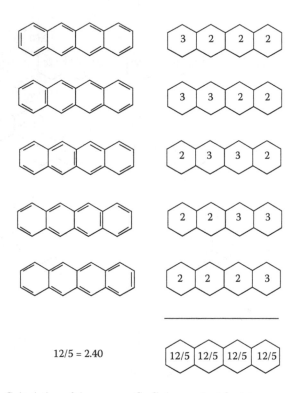

12/5 = 2.40

FIGURE 11.8   Calculation of the average C=C ring content for tetracene.

11.6   QUANTITATIVE AROMATIC SEXTET THEORY

In Figures 11.1 through 11.4, we have illustrated Clar structures showing sextet rings, empty rings, C=C bonds, and migrating rings for benzene and 60 smaller benzenoid hydrocarbons. These figures should be compared to the numerical Clar structures for the same benzenoids shown in Figure 11.7. By making such a comparison, one is comparing results of two hitherto *unrelated* models: the Clar model, based on the concept of aromatic sextets, migrating sextets, and "empty" rings, and the ring bond order model, which is based on Pauling bond orders. A remarkable parallelism between the *qualitative* characterization of the Clar model and the *quantitative* characterization of rings based on bond orders is overwhelming. The ring bond orders clearly discriminate rings having "fixed" aromatic sextets from rings having two CC double bonds and rings having one CC double bond. In the case of rings of "migrating" sextets, the length of the arrows indicates the range of their migration. As one can see, the "fixed" sextets have the largest RBO, the ring belonging to "migrating" sextets have intermediate RBO, and the "empty" rings and rings with "fixed" CC single bonds, such as the central bonds in perylene and zethrene, have the smallest RBO. Observe also that naphthalene rings in perylene and zethrene have the *same* magnitude for ring bond orders as have the rings in naphthalene (2.67). This is expected and is due to a lack of conjugation between the π electrons in two naphthalene moieties.

TABLE 11.2

Regularities in RBO Values in the Phenanthrene (C₁₄H₁₀)–Benzanthracene Family as the Number of Rings Increases

Ring	$C_{14}H_{10}$	$C_{18}H_{12}$	$C_{22}H_{14}$	$C_{26}H_{16}$	Difference in Numerator
Linear	14/5	18/7	22/9	26/11	+4
Kink	11/5	14/7	17/9	20/11	+3
Terminal	14/5	20/7	26/9	32/11	+6

Looking closer at the numerical values for RBOs in structurally related benzenoids, one observes the RBO of migrating sextets decreases with the length of the "arrow," that is, with the increase in the domain that the migrating rings cover. This has been already noticed for linear acenes, with which we had for RBO the sequence 8/3, 10/4, 12/5, 14/6. A similar regularity can be found for the sequence of phenanthrene, benzanthracene, benztetracene, benzpentace: 14/5, 18/7, 22/9, 26/11, and so on for similar cases. In Table 11.2, we have collected results for other rings of the phenanthrene–benzanthracene family. As one can see from the Table 11.2, there is very simple regularity in the gradual increase of the magnitudes of RBOs in this family as the number of ring increases. This regularity allows one to construct ring bond orders for large members of the family without knowing the Pauling bond orders. Clearly, being numerical, the RBO approach for characterization of Clar structures has more information than the Clar structures that remain qualitative. Thus RBO transformed the *qualitative* aromatic sextet model to the *quantitative* model.

We can now pose the following

Problem 14

Find regularities in RBO analogous to those in phenathrene–benzanthracene in other benzenoid families, including fibonacene, if possible.

Pauling bond orders, which have been generally accepted, just as have been accepted their counterpart in MO theory, should have some tangible connection with quantum chemistry. This may be indicative that Clar's aromatic π electron sextet theory might also have tangible connection with quantum mechanics and be one of its reflections when considering benzenoid hydrocarbons. So we dare to propose

Problem 15

Find the hidden connection of the Clar sextet theory with quantum mechanics or prove that there is no such.

We added "*or prove that there is no such*" not to hint that this may be the case, but just to formulate the problem along the usual parlance of mathematicians. A good starting point for this research could be revisiting the paper by Polanski and Derflinger [10] and seeing if it can be upgraded to an *ab initio* level. If the above problem may appear to some too vague for rigorous computational interest of quantum chemists, we may add that, in fact, this problem in its nature is not much different from finding the hidden connection of the Pauling bond orders and quantum mechanics or proving that there are no such. Although most chemists expect that Pauling bond orders belong to quantum chemistry, just as Coulson's bond orders do, which are the result of quantum chemical calculations (regardless of the level of approximations used), the Pauling bond orders are a pure combinatorial result.

The parallelism between Clar's intuitive notion of aromatic sextets, migrating sextets, and "empty" (of sextets) rings and the relative magnitudes of the numerical ring bond orders is so complete that there is not a single exception. So, in this respect, this theory is comparable with the Woodward–Hoffmann rules [28–31] initially formulated to explain the impressive stereospecificity of electrocyclic reactions under thermal and photochemical control. One lengthy review article on Woodward–Hoffmann rules ends with *"Exceptions: None."* The same is the case with our numerical characterization of π aromatic sextets with RBO.

One can interpret the numerical characterization of rings based on Pauling bond orders in benzenoid systems as a quantitative representation of Clar's qualitative aromatic sextet theory. Characterization of benzene rings in benzenoid hydrocarbons, which is of interest for discussion of local aromaticity of these compounds, has received some attention in the past [32] and more recently [33]. Of the more recent work, we should mention here (i) the partition of π electrons of C=C bonds in separate rings [34] and (ii) the partition of Pauling bond orders to individual benzene rings [35]. Both approaches are related to the ring bond orders outlined here with one important distinction: In this latest approach, summarized in Figure 11.7, the contributions from π electrons of C=C bonds shared by two rings *have not been partitioned between the two rings*, but have been assigned *separately* to *both* rings.

Calculating the average C=C count for rings may be simpler than summing Pauling bond orders for individual rings because instead of finding the frequency of individual bonds appearing as C=C in all Kekulé valence structures one enumerates the number of C=C bonds in individual rings in all Kekulé valence structures. The sum rules for Pauling bond orders can also shorten the time to find Pauling bond orders as it does allow one to find Pauling bond orders of neighboring CC bonds without counting.

We will end this outline of numerical Clar structures by applying it to benzene rings of buckminsterfullerene C_{60} [36]. Buckminsterfullerene has only two kinds of CC bonds: the CC bonds between pentagons and hexagons and the CC bonds between hexagons (Figure 11.9). Klein et al. have reported the Pauling bond orders for these two CC bond types to be 7/25 and 11/25, respectively [37]. It follows that the ring bond order for pentagons of C_{60} are $5 \times 7/25 = 1.40$, and the ring bond orders for hexagons are $3 \times 7/25 + 3 \times 11/25 = 2.16$. Because all hexagons have necessarily the same RBO, it follows that all hexagons are the migrating aromatic sextets.

In Figure 11.10, we show a few benzenoids having benzene rings with RBOs of similar magnitude (2.16 ± 0.01). If one is to characterize the *fully benzenoid* systems, which have a *single* Clar structure, as *fully aromatic* and benzenoids having

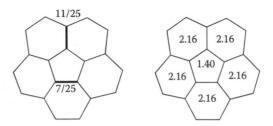

FIGURE 11.9 Pauling bond orders for the two non-symmetry-related CC bonds of buckminsterfullerene and the ring bond orders for a fragment of C_{60}.

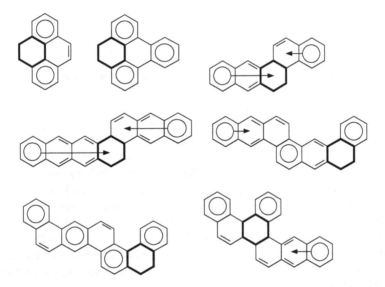

FIGURE 11.10 Several benzenoid hydrocarbons having benzene rings of moderate local aromaticity (shown in bold) with average C=C count of around 2.16, similar to benzene rings in buckminsterfullerene.

migrating sextets as *moderately aromatic*, this would suggest that buckminsterfullerene has moderate aromatic character.

11.7 RESONANCE GRAPHS

Kekulé valence structures continue to offer an alternative basis for description of benzenoid and non-benzenoid conjugated hydrocarbons, which complements characterization of molecules by quantum chemistry, which tends almost exclusively to consider molecular orbital approaches as the frame for discussion of molecular properties. We have already seen applications of conjugated circuits (which are extracted from individual Kekulé valence structures) for clarification of the notion of aromaticity in polycyclic conjugated hydrocarbons and characterization of the relative magnitudes of molecular resonance energies and will also later consider the use of conjugated

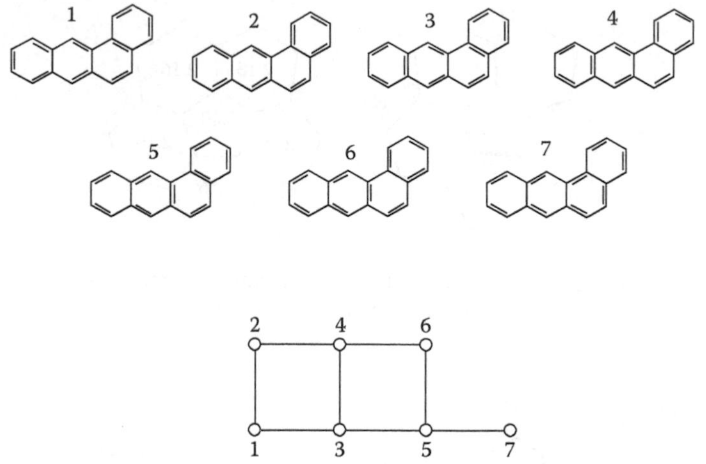

FIGURE 11.11   Seven Kekulé valence structures of benzanthracene and their resonance graph.

circuits in calculations of the ring currents in benzenoid and non-benzenoid conjugated hydrocarbons. In this section, we outline yet another aspect of Kekulé valence structures, which deserves better visibility than it is currently receiving in organic and theoretical chemistry. This is the notion of *resonance graphs*, which relate to the interrelationship between individual Kekulé valence structures in benzenoid molecules.

Let us examine the seven Kekulé valence structures of benzanthracene illustrated in Figure 11.11, which have been arbitrarily labeled 1–7. We now introduce the notion of "adjacency" of Kekulé valence structures, which depends on the distributions of C=C bonds in individual benzene rings of two structures following the rules: two Kekulé structures K_i and K_j are adjacent, $a_{ij} = 1$, if C=C bonds in Kekulé structures K_i and K_j are in the same location in all rings except one, where they exchange positions with C–C bonds; $a_{ij} = 0$ otherwise.

The resulting binary matrix can be informally referred to as the "interaction" matrix of the "resonance" matrix **R**, and so the defined graph as the resonance graph.

A close look at Figure 11.11 shows that, in the first row of the **R** matrix for K_1, the first Kekulé structure, the nonzero elements are only a_{12} and a_{13}. For K_2, the second Kekulé structure, the nonzero elements are only a_{21} and a_{24}, and so on. In this way, one finds the "adjacency," "interaction," or "resonance" matrix for Kekulé valence structures of benzanthracene to be

$$\mathbf{R} = \begin{bmatrix} 0 & 1 & 1 & 0 & 0 & 0 & 0 \\ 1 & 0 & 0 & 1 & 0 & 0 & 0 \\ 1 & 0 & 0 & 1 & 1 & 0 & 0 \\ 0 & 1 & 1 & 0 & 0 & 1 & 0 \\ 0 & 0 & 1 & 0 & 0 & 1 & 1 \\ 0 & 0 & 0 & 1 & 1 & 0 & 0 \\ 0 & 0 & 0 & 0 & 1 & 0 & 0 \end{bmatrix}$$

The corresponding graph is illustrated at the bottom of Figure 11.11. Observe that the "resonance" graph has seven vertices because benzantracene has seven Kekulé valence structures. The graph has eight edges, there being 16, twice as many, R_1 conjugated circuits in the seven Kekulé valence structures of benzanthracene. The degrees (valences) of the vertices of the resonance graph are determined by the number of R_1 conjugated circuits in the Kekulé valence structures. Thus, in the case of benzanthracene, the Kekulé valence structures 1, 2, and 6 have two R_1 conjugated circuits and constitute the three bridging vertices of the resonance graph. Structures 3, 4, and 5 have three R_1 conjugated circuits and constitute the three branching vertices of the resonance graph. Last, structure 7 has one R_1 conjugated circuit and is represented by the terminal vertex of the resonance graph. As drawn in Figure 11.11, the resonance graph of benzanthracene has a simple geometrical form; it consists of two fused squares and an additional exocyclic bond.

In Figure 11.12, we have illustrated a number of resonance graphs for a collection of smaller benzenoid hydrocarbons. Because all resonance graphs are a combination of n cubes, one can construct for resonance graphs a simple code, which consists of a list of the number of cubes of different dimension they contain. Thus, the code of the resonance graph of benzanthracene is 2, 1, the graph having two two-dimensional cubes (squares) and one one-dimensional cube (the line segment). In Table 11.3 in the last column labeled Euler code, we have listed the codes for all the graphs of Figure 11.12. We thus honor Euler and his famous "polyhedral formula," which relates the number of vertices, the number of edges, and the number of faces for any polyhedron:

$$F + V = E + 2.$$

Here V stands for the number of vertices, E for the number of edges, and F for the number of faces of a polyhedron.

Observe that Euler codes listed in Table 11.3 allow classification of benzenoid hydrocarbons. Clearly, benzenoid hydrocarbons that are iso-Kekuléan, such as chrysene and benzphenanthrene, will have the same Euler code because their Kekulé valence structures are in one-to-one correspondence. However, it is possible that two different benzenoid hydrocarbons that have different resonance graphs may still have the same Euler codes. It may be of interest to find conditions in which different benzenoids will have the same Euler codes, which is currently an open problem. In Figure 11.13, we have grouped several pairs of resonance graphs of structurally related benzenoid hydrocarbons, in order to show that there are regularities between resonance graphs of structurally similar benzenoids.

Resonance graphs were independently discovered first by W. Gründler in then East Germany, who published his work in 1983 in a rather obscure journal of a local university [38]. Ten years later, S. El-Basil apparently introduced them in two publications considering Kekulé structures as graph generators and lattice generators [39,40]. In 1996, Randić and coworkers addressed the topic of resonance in cata-condensed benzenoid hydrocarbons in which resonance graphs were introduced and illustrated on a number of benzenoid molecules [41,42]. Not long after, I. Gutman, in a letter to M. Randić, pointed out that W. Gründler had published a paper in the

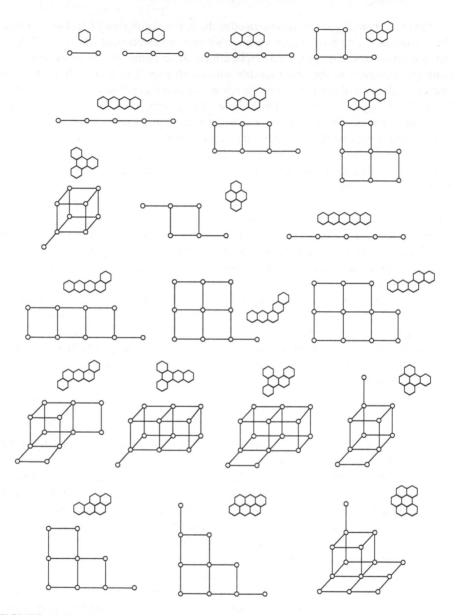

FIGURE 11.12   Resonance graphs of smaller benzenoid hydrocarbons.

science journal of the University of Halle, Germany, an article on resonance graphs
in 1983. Thus, the work of W. Gründler was resurrected and saved from oblivion.

Publications by Randić et al. apparently have attracted the attention of mathematicians, and thus, resonance graphs became subjects of a number of mathematical
studies. As noticed by T. Pisanski (see Ref. [43]), resonance graphs are a special
case of median graphs, which are well known in mathematical literature [44–50].

TABLE 11.3
The Leading Eigenvalue, Resonance Energy, and Euler Code of the Resonance Graphs of Smaller Benzenoid Hydrocarbons

	Benzenoid	K	Leading Eigenvalue	RE	Euler Code
1	Benzene	2	1.0000	0.869	1
2	Naphthalene	3	1.4142	1.323	2
3	Anthracene	4	1.6180	1.600	3
4	Phenanthrene	5	2.1358	1.933	1, 1
5	Tetracene	5	1.7321	1.822	4
6	Benzanthracene	7	2.4737	2.291	2, 1
7	Chrysene	8	2.6762	2.483	3, 0
8	Triphenylene	9	3.0455	2.654	1, 0, 1
9	Pyrene	6	2.2361	2.098	2, 2
10	Pentaphene	6	1.8019	1.876	5
11	Benztetracene	9	2.6491	2.531	3, 1
12	Pentene	10	2.8536	2.705	4, 1
13	Benzo[b]chrysene	11	2.9476	2.828	5, 0
14	Dibenzo[ah]anthracene	12	3.2415	2.986	1, 2, 0
15	Dibenzo[ac]anthracene	13	3.4343	2.948	2, 0, 1
16	Benzo[g]chrysene	14	3.4987	3.209	2, 1, 0
17	Benzo[e]pyrene	11	3.1690	2.906	1, 1, 1
18	Benzo[a]pyrene	9	2.7058	2.594	3, 1
19	Anthanthene	10	2.7321	2.650	3, 2
20	Benzo[ghi]perylene	14	3.3383	3.128	1, 3, 1

According to Klavžar et al. [43], median graphs are not fully understood and are obviously therefore of continuing interest in mathematics. It was H. Zhang, P. C. B. Lam, and W. C. Shiu [51–53], who established as one of the most important properties of the resonance graph that they are median graphs. Median graphs are graphs all the vertices of which have the property that every triplet of vertices (x, y, z) has a median. Median is defined as a vertex at the cross-sections of intervals between vertices (x, y), (x, z), and (y, z), and intervals between a pair of vertices consists of all vertices on the shortest paths between the selected pair. In less formal language, a median is a vertex that belongs to the shortest paths between each pair of vertices x, y, z. For illustration and clarification, in Figure 11.14, we have illustrated median for a set of three vertices of a simple graph built from two fused cubes.

It is clear that the concept of medians may be of considerable importance in mathematics but may remain a curiosity in chemistry. However, the fact that resonance graphs consist of combinations of fused n-dimensional cubes is of considerable importance and may even find some practical relevance. For $n = 1$, $n = 2$, and $n = 3$, graphs of the n-dimensional cubes are illustrated at Figure 11.15a. In Figure 11.15b are illustrated two alternative representations of three-dimensional cubes in a plane.

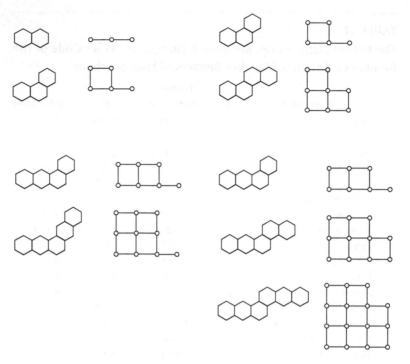

FIGURE 11.13 Regularities between resonance graphs of structurally similar benzenoid hydrocarbons.

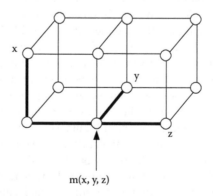

FIGURE 11.14 Illustration of the median vertex for the triplet (x, y, z). The edges in bold represent the set of shortest paths between vertices x, y, and z having a single common vertex (median).

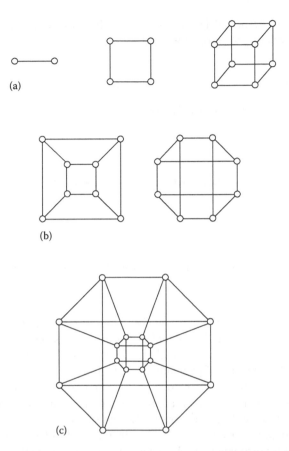

FIGURE 11.15 (a) One-dimensional cube, two-dimensional cube, and three-dimensional cube. (b) Two two-dimensional representations of three-dimensional cube. (c) A two-dimensional representation of a four-dimensional cube obtained by connecting two corresponding vertices of two regular octagons.

At Figure 11.15c is shown a graph of a four-dimensional cube ($n = 4$), which can be drawn by connecting two copies of a 3-D cube, just as a 3-D cube can be obtained by connecting two 2-D cubes (squares) and just as a 2-D cube can be obtained by connecting two 1-D cubes (edges).

We should add that resonance graphs were independently discovered in mathematics by F. Zhang and coworkers [51], who reported that the resonance graphs of catacondensed benzenoid hydrocarbons have a Hamiltonian path [52]. A Hamiltonian path is path that includes all vertices of a graph. A graph is Hamiltonian if there is a cyclic path that covers all vertices of a graph. A graph is semi-Hamiltonian if there is a path covering all vertices of a graph. It is not difficult to see that the finding of F. Zhang and coworkers is not true for the resonance graphs of pericondensed benzenoid systems as is illustrated by the resonance graphs of anthanthrene and benzo[*ghi*]perylene shown in the middle of the last row in Figure 11.12.

The remaining 19 graphs of Figure 11.12 have Hamiltonian paths. Let us propose three problems relating to resonance graphs:

Problem 16

Try to characterize peri-condensed benzenoid graphs that have a Hamiltonian path.

Problem 17

Try to characterize benzenoid hydrocarbons whose resonance graphs have only 1-D and 2-D cubes.

Problem 18

Try to characterize benzenoid hydrocarbons whose resonance graphs have only 3-D cubes.

Returning to Figure 11.12, can you detect additional regularities between the resonance graphs of different benzenoid hydrocarbons besides the obvious cases of benzene, naphthalene, anthracene, and tetracene, which are given as the paths of increasing length? The case of phenantherene, benzantracene, and benztetracene, is not difficult to detect, but more interesting is a comparison of resonance graphs of naphthalene and phenathrene, then anthracene and benzanthrace, or tetracene and benztetracene (in the middle of Figure 11.12), and so on.

We have illustrated several regularities in Figure 11.13 that point to possible extensions to other resonance graphs. From the case of naphthalene–phenanthrene (top left part of Figure 11.12), one can see that one can obtain a graph of phenathrene by combining the resonance graphs of naphthalene and benzene so that one connects the smaller graph with two "vertical edges" of the bigger graph. The same is true for the case of anthracene–benzanthracene, and one also expects that the same will be the case for tetracene and benzantracene. That this is the case one can verify from Figure 11.12. A similar relationship can also be seen in the case of benzanthracene–pentaphene (bottom left part of Figure 11.13).

The situation is somewhat different in the case phenanthrene–chrysene. Let us use the following notation: $[G]$ is the resonance graph of benzenoid hydrocarbons, $[G - g]$ is the resonance graph of G when the terminal vertex in linear positions is removed, and $[G - gg]$ is the resonance graph of G when the terminal vertex and adjacent vertex in linear positions are removed. On the other hand, $[G - g^*]$ is the resonance graph of G when the terminal vertex in a "kink" position is removed, and $[G - gg^*]$ is the resonance graph of G when the terminal vertex and adjacent vertex in linear positions are removed. With this notation, the earlier outlined two cases belong to the case $[G - g] \ddagger [G - gg]$, where $\ddagger$ represents the process of "linking" the two components $[G - g]$ and $[G - gg]$ with "vertical" edges as outlined earlier. In the case of phenanthrene–chrysene, one combines the resonance graph of the smaller molecule [G] with the graph [G − gg*], forming the union [G] ‡ [G − gg*]. So observe that the difference is whether one augments a graph G by a linear or a "kink" benzene ring.

The last case (bottom right part of Figure 11.13), illustrates a combination of both "linking" processes. One starts with benzanthracene (G₁) and adds to it a "kink" ring, obtaining a new graph (G₂) having five fused squares placed in two rows (bottom right middle part of Figure 11.13). Now to the graph (G₂) one adds linearly a benzene ring, which calls for the operation [G₂ − gg*], obtaining as the resonance graph the graph shown at the bottom right part of Figure 11.13 of eight fused squares. We claim this to be the right process to arrive at the resonance graph of dibenzo(b,k) chrysene, the last benzenoid of Figure 11.13.

REFERENCES AND NOTES

1. E. Clar, *Aromatische Kohlenwasserstoffe, Polycyclische Systeme*, Springer Verlag, Berlin, Heidelberg, 1941.

2. E. Clar, *The Aromatic Sextet*, John Wiley & Sons, London, 1972.

3. T. W. Armit and R. Robinson, Polynuclear heterocyclic aromatic types, Part II. Some anhydronium bases, *J. Chem. Soc.* 127 (1925) 1604–1618.

4. W. Heitler and F. London, Wechelwirkung neutraler Atome und homöopolare Bindung nach der Quantenmechanik, *Z. Phys.* 44 (1927) 619.

5. G. N. Lewis, *Valence and the Structure of Atoms and Molecules*, Dover Publications, Inc., New York, 1966.

6. G. N. Lewis, The atom and the molecule, *J. Am. Chem. Soc.* 38 (1916) 762–785.

7. E. C. Crocker, Application of the octet theory to single-ring aromatic compounds, *J. Am. Chem. Soc.* 44 (1922) 1618–1630.

8. H. E. Armstrong, The structure of cycloid hydrocarbons, *Proc. Camb. Soc. (Lond)* 6 (1890) 95–106.

9. A. T. Balaban, P. v. R. Schlyer, and H. S. Rzepa, Crocker, not Armit and Robinson, begat the six aromatic electrons, *Chem. Rev.* 105 (2005) 3436.

10. O. E. Polansky and G. Derflinger, Zur Clar'schen Theorie Lokaler Benzoider Gebiete in Kondenzirten Aromaten, *Int. J. Quantum Chem.* 1 (1967) 379–401.

11. G. Portella, J. Poater, and M. Solà. Assessment of Clar's aromatic p-sextet rule by means of PDI, NICS and HOMA indicators of local aromaticity, *J. Phys. Org. Chem.* 18 (2005) 785–791.

12. K. P. Vijayalakshmi and C. H. Suresh, Pictorial representation and validation of Clar's aromatic sextet theory using molecular electrostatic potentials, *New J. Chem.* 34 (2010) 2132–2138.

13. F. Zhang, X. Guo, and H. Zhang, Advances of Clar's aromatic sextet theory and Randic's conjugated circuit model, *Open Org. Chem. J.* (Suppl. 1-M6) (2006) 87–111.

14. M. Solà, Forty years of Clar's aromatic π-sextet rule, *Front. Chem., Theor. Comput. Chem.* 1 (2013) 22.

15. Z. Sun, S. Lee, K. H. Park, X. Zhu, W. Zhang, B. Zheng, P. Hu, Z. Zeng, S. Das, Y. Li, C. Chi, R.-W. Li, K.-W. Huang, J. Ding, D. Kim, and J. Wu, Indolo[2,3-b]carbazoles with tunable ground states: How Clar's aromatic sextet determines the singlet biradical character, *J. Am. Chem. Soc.* 135 (2013) 18229.

16. D. J. Klein and A. T. Balaban, Clarology for conjugated carbon nanostructures: Molecules, polymers, graphene, defected graphene, fractal benzenoids, fullerenes, nanotubes, nanocones, nanotori, etc., *Open Org. Chem. J.* (Suppl. 1-M3) (2011) 27–61.

17. Density Functional Theory (DFT) is a continually developing quantum chemical model initiated on two theorems by Pierre Hohenberg and Walter Kohn published in: P. Hohenberg and W. Kohn, Inhomogeneous electron gas, *Phys. Rev.* 136 (1964) B864–B871.

18. Gaussian is computer program initiated by J. A. Pople and his research group W. J. Hehre, W. A. Lathan, R. Ditchfield, and M. D. Newton in 1970 at Carnegie-Mellon University in Pittsburgh, Pennsylvania.

19. E. Hückel, *Grundzuge der Theorie ungesattiger und aromatischer Verbindugen*, Verlag Chemie, Berlin, 1938.

20. L. Pauling, *The Nature of the Chemical Bond and the Structure of Molecules and Crystals*, Cornell University Press, Ithaca, NY, 1939.

21. C. A. Coulson, *Valence*, 2nd ed., Oxford University Press, Oxford, 1961.

22. J. N. Murrell, S. F. A. Kettle, and J. M. Tedder, *Valence Theory*, John Wiley & Sons Ltd, New York, 1970.

23. K. Fries, Über bicyclische Verbindungen und ihren Vergleich mit dem Naphtalin, *Justus Liebigs Ann. Chem.* 454 (1927) 121–324.

24. S. Narita, T. Morikawa, and T. Shibuya, *J. Mol. Struct. THEOCHEM* 532 (2000) 37.

25. M. Randić, Hybridization by the maximum overlap method, *Int. J. Quantum Chem.* 8 (1974) 643–676.

26. M. Randić and Z. B. Maksić, Hybridization by the maximum overlap method, *Chem. Rev.* 72 (1972) 43–53.

27. M. Randić and Z. Maksić, Maximum overlap hybridization in cyclopropane and some related molecules, *Theor. Chim. Acta* 3 (1965) 59–68.

28. R. B. Woodward and R. Hoffmann, Stereochemistry of electrocyclic reactions, *J. Am. Chem. Soc.* 87 (1965) 395.

29. R. Hoffmann and R. B. Woodward, Selection rules for concerted cycloaddition Reactions, *J. Am. Chem. Soc.* 87 (1965) 2046.

30. R. B. Woodward and R. Hoffmann, The conservation of orbital symmetry, *Acc. Chem. Res.* 1 (1968) 17–22.

31. R. B. Woodward and R. Hoffmann, The conservation of orbital symmetry, *Angew. Chem. Int. Ed.* 8 (1969) 781–853.

32. M. Randić, Graph theoretical approach to local and overall aromaticity of benzenoid hydrocarbons, *Tetrahedron* 31 (1975) 1477–1481.

33. M. Randić, A. T. Balaban, and D. Plavšić, Applying the conjugated circuits method to Clar structures of [n]phenylenes for determining resonance energies, *Phys. Chem. Chem. Phys.* 13 (2011) 20644–20648.

34. M. Randić, Algebraic Kekulé formulas for benzenoid hydrocarbons, *J. Chem. Inf. Comput. Sci.* 44 (2004) 365–372.

35. I. Gutman, T. Morikawa, and S. Narita, On π-electron content of bonds and rings in benzenoid hydrocarbons, *Z. Naturforsch.* 59a (2004) 295–298.

36. H. W. Kroto, J. R. Heath, S. C. O'Brien, R. F. Curl, and R. E. Smalley, Buckminsterfullerene, *Nature* 318 (1985) 162–163.

37. D. J. Klein, T. G. Schmalz, G. E. Hite, and W. A. Seitz, Resonance in C_{60} Buckminsterfullerene, *J. Chem. Soc.* 108 (1986) 1301–1302.

38. W. Gründler, Signifikante Elektronenstrukturen fur benzenoide Kohlenwasserstose, *Wiss. Z. Univ. Halle* 31 (1982) 97–116.

39. S. El-Basil, Generation of lattice graphs. An equivalence relation on Kekulé counts of catacondensed benzenoid hydrocarbons, *J. Mol. Struct. (Theochem.)* 288 (1993) 67–84.

40. S. El-Basil, Kekulé structures as graph generators, *J. Math. Chem.* 14 (1993) 305–318.

41. M. Randić, D. J. Klein, S. El-Basil, and P. Calkins, Resonance in large benzenoid hydrocarbons, *Croat. Chem. Acta* 69 (1996) 1639–1660.

42. M. Randić, Resonance in catacondensed benzenoid hydrocarbons, *Int. J. Quantum Chem.* 63 (1997) 585–600.

43. S. Klavžar, P. Žigert, and G. Brinkmann, Resonance graphs of catacondensed even ring, systems are median, *Discrete Math.* 253 (2002) 35–43.

44. W. Imrich and S. Klavžar, Recognizing median graphs is subquadratic time, *Theor. Comput. Sci.* 215 (1999) 123–136.

45. W. Imrich, S. Klavžar, and H. M. Mulder, Median graphs and triangle-free graphs, *SIAM J. Discrete Math.* 12 (1999) 111–118.

46. S. Klavžar and H. M. Mulder, Median graphs: Characterizations, location theory and related structures, *J. Combin. Math. Combin. Comput.* 30 (1999) 103–127.

47. H. M. Mulder, The structure of median graphs, *Discrete Math.* 24 (1978) 197–204.

48. F. Zhang, X. Guo, and R. Chen, Z-transformation graphs of perfect matching of hexagonal systems, *Discrete Math.* 72 (1988) 405–415.

49. H. Zhang, Z-transformation graphs of a perfect matchings of plane bipartite graphs: A Survey, *MATCH Commun. Math. Comput. Chem.* 56 (2006) 457–476.

50. S. Klavžar, P.·Zigert, and G. Brinkmann, Resonance graphs of catacondensed even ring systems are median, *Discrete Math.* 253 (2002) 35.

51. R. Chen and F. Zhang, Hamilton paths in Z-transformation graphs of perfect matching of hexagonal systems, *Discrete Appl. Math.* 74 (1997) 191–196.

52. P. C. B. Lam, W. C. Shiu, and H. Zhang, Elementary blocks of plane bipartite graphs, *MATCH Commun. Math. Comput. Chem.* 49 (2003) 127–137.

53. H. Zhang, P. C. B. Lam, and W. C. Shiu, Resonance graphs and a binary coding for the 1-factors of benzenoid systems, *SIAM J. Discrete Math.* 22 (2008) 971–984.

12 Renormalization in Chemistry

We introduce the topic of renormalization in chemistry by considering calculation of ring currents in polycyclic conjugated hydrocarbons based on the conjugated circuit model.

12.1 CONTINUUM VERSUS DISCRETE MODELS

The discussion of conjugated circuits in previous chapters of this book have focused attention on resonance energies and aromaticity in general. In this section, we focus on use of conjugated circuits for calculations of ring currents. Ring currents have been at the center of attention of a number of theoretical chemists, starting with Linus Pauling's work in 1936 on the diamagnetic anisotropy of aromatic molecules [1] through the most recent *ab initio* calculations of induced current densities by Patrick Fowler and others covering a span of some 60 years. Those interested in the long history of ring current calculations may consult two recent reviews on this subject, the paper by Lazzereti on ring currents [2], and the paper by Mallion and coworkers on topological ring currents [3–8]. We focus in this section on the latest, conceptually and structurally different, approach to ring currents in cyclic and polycyclic conjugated hydrocarbons. It is so different and unique because it is based on *enumerations* as a method of calculation rather than *calculus* as a tool for computation of ring currents. In view of this dramatic distinction in the character of the calculations, it is worth digressing and recalling some thoughts of G. N. Lewis, the father of the pre-quantum chemistry octet rule of covalent bonding. The octet rule was mostly applied to first row atoms: carbon, oxygen, and nitrogen. The octet rule states that atoms of *main-group elements* tend to combine in a molecule in such a way that each atom involves eight *electrons* in its *valence shell*.

Excluding enumeration of isomers, Lewis was the first in chemistry to recognize the importance of *counting*, which has been one of the main tools of graph theory. Graph theory was unknown to chemists during the early growth of quantum chemistry in early 1930, and remained unknown to chemists almost to the present time. The early attention to the importance of enumerations in structural chemistry is reflected in G. N. Lewis writing in 1923 (more than 90 years ago) [9]:

> Two quantitative methods have been available to scientists. One consists in counting and the other consists in measuring. The former has been the basis of the theory of numbers; the latter has lead to the development of geometry. The first of these sciences has been the mere plaything of abstruse mathematicians; the second has become the working tool of the scientist and engineer. Geometry is based on the theory of the continuum, and so also is the closely related science of calculus. We have been taught that an integration of the infinitesimal elements of a continuum may be approximately

replaced by a summation of finite terms, but that the former method is exact and absolute while the second gives but an approximation. Are we not now going to be obliged to reverse this decision and to recognize that the branch of mathematics which will come nearest to meeting the needs of science will be the theory of numbers, rather than a theory of extension, and that measuring must be replaced by counting?

12.2 RING CURRENTS VIA THE CONJUGATED CIRCUITS MODEL

The novel approach to calculation of ring currents in benzenoid and non-benzenoid cyclic and polycyclic conjugated structures is based on *enumeration* of contributions from currents in individual conjugated circuits in a compound. In this graph theoretical model, Kekulé valence structure are partitioned into a set of conjugated circuits, for which individual currents are calculated. Conjugated circuits are the circuits within individual Kekulé valence structures in which there is a regular alternation of CC single and CC double bonds [10–12]. Calculations start by enumeration of bond contributions to current in all conjugate circuits. The calculated CC bond currents for a number of molecules, including coronene and its numerous non-benzenoid isomers [13,14], have been compared with *ab initio* MO computed current densities in the same compounds [15], and an impressive qualitative agreement, illustrated in Figure 12.1 on one of the coronene isomers, was found.

One should recall that *ab initio* MO computation of π electron current densities and the graph theoretical approach to bond currents are two as disparate theoretical models as one can possibly imagine. The former is based on quantum chemical *computations* and the latter on graph theoretical *enumerations*. The graph theoretical approach starts with construction of all conjugated circuits for a molecule, the number of which is $K(K - 1)$, where K is the number of Kekulé valence or resonance structures of the molecule [16]. In Figure 12.2, we have illustrated all conjugated circuits of the coronene isomer of Figure 12.1.

In the next step, conjugated circuits of size $4n + 2$ are assigned counterclockwise current circulation, and conjugated circuits of size $4n$ are assigned clockwise current circulation. When all such currents are *enumerated* and added for all Kekulé valence structures, one obtains the current pattern shown at Figure 12.1b.

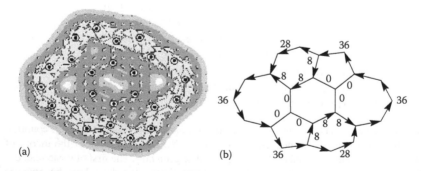

(a) (b)

FIGURE 12.1 π Electron current densities calculated by *ab initio* MO approach (a) and as obtained from graph theoretical approach based on calculations of contributing currents in $4n + 2$ and $4n$ conjugated circuits (b) for an isomer of coronene.

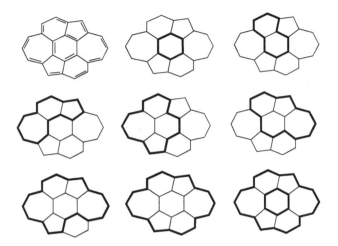

FIGURE 12.2 Conjugated circuits for one of the Kekulé valence structures of the coronene isomer of Figure 12.1.

12.2.1 ILLUSTRATION: RING CURRENTS IN PYRENE

We will illustrate calculation of ring currents in pyrene, which we have selected for two reasons: It is a simple peri-condensed benzenoid system with only six Kekulé valences structures (each having five conjugated circuits); thus, calculations can be reported fully and will not be difficult to follow. Another reason for picking pyrene is that it has 16 π electrons, which is a $4n$ number, and has been mentioned and caused some confusion relating to the expectation that it does not follow the Hückel $4n + 2$ rule of aromaticity, yet pyrene is very aromatic! Confusion here is due to a lack of awareness of the fact that the Hückel $4n + 2$ rule holds strictly only for monocyclic systems. Statements that naphthalene, anthracene, and other cata-condensed systems are aromatics because they are $4n + 2$ systems is equally wrong, being based on misinformation about the applicability of the Hückel rule. The basic misconception is thinking that the *number of π electrons* has fundamental influence on aromaticity although, as we have already pointed out, it is the *number of π electrons in conjugated circuits* that counts!

When speaking of the Hückel rule, one should continually remind readers that it holds for monocyclic conjugated systems only with the caveat that it also holds for molecules such as bicyclic azulene because the central CC bond in azulene is a single CC bond in both Kekulé valence structures of azulene, the conjugated circuits of which have therefore 10 π electrons. The aromaticity in azulene is strictly speaking not due to the Hückel rule and HMO but due to conjugated circuits, which do not involve the central CC single bond and which can thus be erased, or modified HMO in which π–π interactions across the essentially single CC bonds are ignored.

In Figure 10.2 in Chapter 10 we have illustrated conjugated circuits for the six Kekulé valence structures of pyrene. In Figure 12.3, we show the ring currents for the conjugated circuits of the six Kekulé valence structures of pyrene. All ring currents are counterclockwise in the view of the fact that all conjugated circuits are

FIGURE 12.3   Bond currents for conjugated circuits of pyrene.

the $4n + 2$ type. By adding contributions for individual CC bonds of the first three Kekulé valence structures, one obtains the results shown at the top left side of Figure 12.4. The CC bond currents for the remaining three Kekulé valence structures give similar results, which are shown at the top right of Figure 12.4. Observe, these bond currents do not represent the mirror images of the respective symmetry-equivalent structures, because the current directions remain anti-clockwise. By adding the contributions of the two sets of Kekulé valence structures, one obtains the result shown at the bottom of Figure 12.4. As one can see, the resulting CC currents can be interpreted as ring currents of magnitude 18 for the top and the bottom C_6 rings and ring currents of magnitude 10 for the central C_6 rings. The bond currents of the central CC bond cancel one another, and the magnitudes of the remaining interior CC bonds are equal to the difference of the ring currents of terminal and central benzene rings.

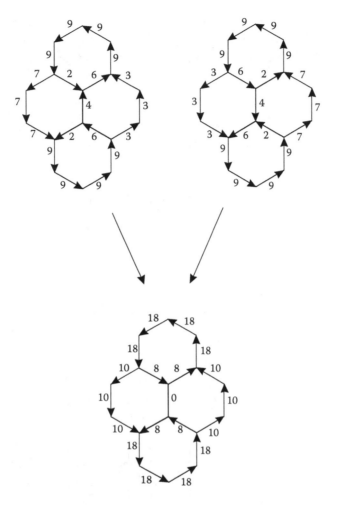

FIGURE 12.4 Count of CC bond currents for pyrene.

If one is interested in finding magnetic flux, one should take into account the size of the area of circuits contributing to the magnetic properties of the rings. In Figure 12.5, we show the conjugated circuits of the six Kekulé valence structures, but this time, we have inserted in the benzene rings labels 1, 2, 3, 4, which indicate the number of benzene rings encompassed, and thus indicate the relative areas of individual circuit.

In the case of conjugated circuits of size R_1 and R_2, the areas of conjugated circuits agree with the number of fused rings, but conjugated circuits of size R_3 can not only be of different shape (as is the case with R_3 in anthracene and phenanthrene), but can also have different areas. This is the case with the two R_3 conjugated circuits of the two central Kekulé valence structures in the last two rows in Figures 12.3 and 12.5. The revised CC bond current of a given bond is obtained by multiplying the

FIGURE 12.5 Ring currents for different conjugated circuits of pyrene when areas of circuits are taken into account.

CC bond current of that bond by the magnitude of the area of the conjugated circuit producing that CC bond current. The top left part of Figure 12.6 shows the revised CC bond currents for the first three Kekulé valence structures of pyrene. The revised CC bonds currents for three remaining symmetry-equivalent Kekulé valence structures are shown at the top right side of Figure 12.6. When adding the contributions for individual CC bonds, one obtains the results shown at the bottom of Figure 12.6. Now the resulting CC currents give rise to ring currents of magnitude 36 for the top and the bottom C_6 rings and ring currents of magnitude 24 for the central C_6 rings. The results are quantitatively different but qualitatively similar in confirming the dominant role of the terminal benzene rings and lesser contributions for the central rings of pyrene.

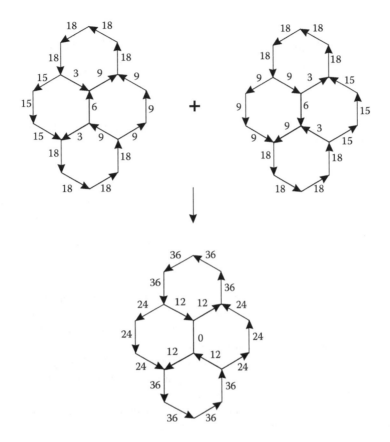

FIGURE 12.6 Count of CC bond currents in pyrene based on magnetic flux.

12.3 NORMALIZATION OF RING CURRENTS

If one is to compare CC bond currents in different molecules, except for comparisons of currents in molecules having the same number of Kekulé resonance structures, then one has to normalize CC bond currents with $1/K(K - 1)$, the number of conjugated circuits used in their calculation, because CC bond currents are additive to contributing conjugated circuits. Such a procedure is straightforward. However, in the case of molecules having essentially single CC bonds, such as perylene (Figure 12.7), it leads to results that are not consistent with the model. Namely, perylene on one side can be viewed as a single molecule having $K = 9$ Kekulé structures, but because there are no conjugations involving both naphthalene moieties, it can also be viewed, in this model, as two *independent* naphthalene molecules, each having $K = 3$ Kekulé structures.

If perylene is viewed as a single molecule having nine Kekulé valence structures (Figure 12.8) then normalized currents in its benzene rings are $36/(9 \times 8) = 0.50000$ as compared to normalized currents in benzene rings of naphthalene, which are $4/(3 \times 2) = 0.66667$. If one is to weigh contributing currents from conjugated circuits

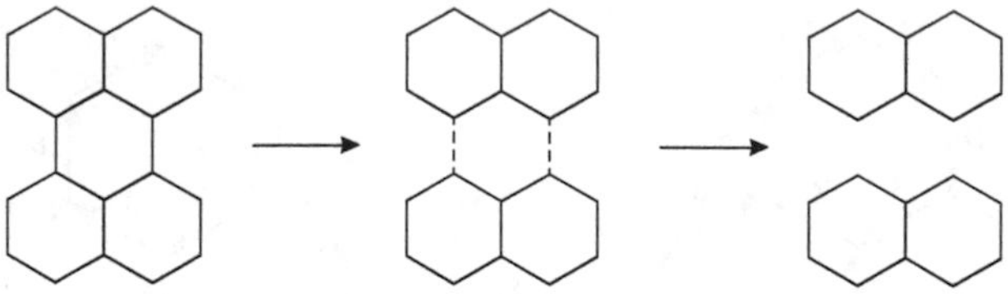

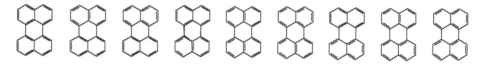

FIGURE 12.7 In view that perylene has non-interacting naphthalene subunits, it can, on this model, be viewed as two naphthalene molecules.

FIGURE 12.8 Nine Kekulé valence structures of perylene, in all of which the central bonds connecting naphthalene subunits are devoid of conjugation.

of different areas and make them proportional to the areas of conjugated circuits to account for the magnetic flux rather than magnetic field, then we have $54/(9 \times 8) = 0.75000$ and $6/(3 \times 2) = 1.00000$ for perylene viewed as a single molecule and as two non-interacting naphthalene subunits, respectively.

As we can see, naphthalene subunits in perylene, instead of having bond currents the same as in naphthalene, have significantly lower bond currents. It is not important how small or how great the difference is; in this model, there should be *no difference* in bond currents between benzene rings of naphthalene and perylene, clearly pointing to an inherent inconsistency of the model.

12.4   RENORMALIZATION OF RING CURRENTS

This kind of internal inconsistency when considering a computational model for systems, parts of which consist of self-similar subunits, have been recognized to occur in physics, in quantum electrodynamics, and statistical mechanics. In problems in which one is considering the same object simultaneously at different scales, internal inconsistencies may emerge because at different scales the same components of a system may be characterized by different parameters. Renormalization is a procedure that adjusts for changes of scale. It was first developed in quantum electrodynamics (QED) in the mid-1950s to deal with infinite integrals in perturbation theory and ill-defined differences of infinite quantities. When proposed, it was viewed with suspicion but eventually was accepted as an important procedure that makes theory *self-consistent*. Keneth G. Wilson (1936–2013) got the Nobel Prize in 1982 for systematic resolution of these problems in physics, formulating the fundamental work on the renormalization group. We may also mention, what is not so well known, that K. G. Wison has considerable interest in theoretical chemistry.

In chemistry, an internal inconsistency was for the first time recognized very recently in calculations of CC bond currents in modeling some benzenoid hydrocarbons. Non-interacting molecular subunits (e.g., naphthalene units in perylene)

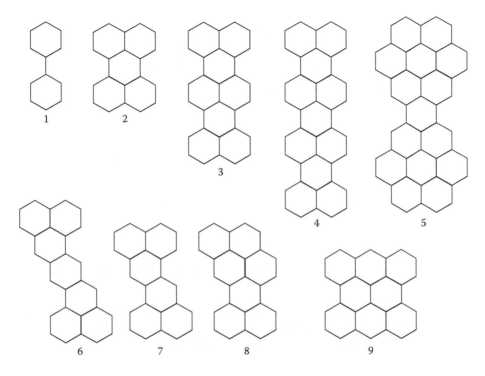

FIGURE 12.9 Benzenoid systems with essentially single CC bonds.

have to be scaled as to reproduce ring currents of the same magnitude as in naphthalene if one wishes to obtain internally consistent results. In this chapter, we discuss and outline renormalization for benzenoid systems such as biphenyl **1**, being the smallest member of the class, and perylene **2**, the most well-known illustration. Other related systems include terrylene **3**, quaterrylene **4**, dicoronylene **5**, dibenzo[*de,op*]pentacene **6**, zethrene **7**, naphtho[*8,1,2-bcd*]perylene **8**, and phenanthro-[*1,10,9,8-opqra*]perylene **9** of Figure 12.9.

The same problem also arises in a number of non-benzenoid molecules having essentially single CC bonds, molecules such as benzo[*a*]fluoranthene **1**, benzo[*b*]-fluoranthene **2**, benzo[*j*]fluoranthene **3**, benzo[*k*]fluoranthene **4**, pleiadene **5**, and rubicene **6**, which is molecular fragment of C_{70}, all illustrated in Figure 12.10. The essentially single bonds in benzofluoranthenes are CC bonds of the pentagonal rings and exocyclic CC bonds of naphthalene and anthracene subunits in pleiadene and rubicene, respectively.

12.4.1 NAPHTHALENE

In Figure 10.1 in Chapter 10, we have shown conjugated circuits in naphthalene, from which it is not difficult to calculate that the ring currents in CC bonds of naphthalene are of magnitude 4 units if each conjugated circuit contributes current of unit amplitude, and of magnitude 6 units if each conjugated circuit contributes current proportional to the area of the circuit, which is taken to be 1 for single benzene and

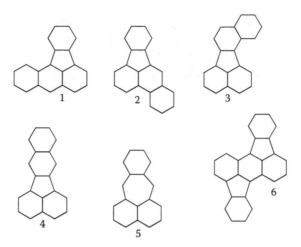

FIGURE 12.10 A selection of non-benzenoid hydrocarbons having essentially single CC bonds.

2 for the area of naphthalene. Because there are six conjugated circuits in naphthalene when normalized, the currents are $4/6 = 0.66667$ and $6/6 = 1$, respectively. The bond currents of the central CC bond cancel one another.

12.4.2 PERYLENE

Perylene consists of two naphthalene units connected by two single bonds, which remain C–C single bonds in all nine Kekulé valence structures of perylene shown in Figure 12.8. Thus, in this model, based on the count of conjugated circuits, perylene becomes equivalent to two isolated naphthalenes and is expected to have the same ring currents as naphthalene. The count of conjugated circuits of perylene gives for the bond currents in CC bonds of naphthalene periphery the magnitude 36 units if each conjugated circuit contributes current of unit amplitude and of magnitude 54 units if each conjugated circuit contributes current proportional to the area of the circuit. In the view that perylene has $K = 9$, the normalization is given by $K(K - 1) = 72$, which gives for the CC bond currents $36/72 = 0.50000$ instead of the expected 0.66667 and the CC bond currents of $54/72 = 0.7500$ when areas of conjugated circuits are taken into account instead of the expected 1.00000.

12.5 CONCLUDING REMARKS

Hence, there is a discrepancy in the calculated currents in naphthalene units, which have in perylene decreased visibly in magnitude. This inconsistency can be corrected by renormalization. Instead of the normalization factor $1/K(K - 1)$ one has to multiply it with the factor of $4/3$, which will increase the calculated currents of 0.50000 and 0.75000 to 0.66667 and 1.0000, respectively.

As mentioned earlier, in physics the introduction of renormalization was viewed for about a year with great skepticism and suspicion before being accepted because it involved novelties unheard of before, such as the implication that a charge of electron may be different in different situations. Renormalization in chemistry is conceptually as simple as it can be: In models in which two (or more) naphthalenes are non-interacting, they are identical to an isolated naphthalene. As far as we know, renormalization has not been reported previously in chemistry although it appears that it was considered by Ken Wilson, but without success. Hence, in the case of ring current calculations, at least those based on the conjugated circuits model, the earlier calculations involving molecules having essentially single bonds, such as perylene, should be revised and corrected. It is interesting that early HMO ring current calculations based on conjugated circuits of Gomes and Mallion [17] give the same ring currents in napthalene and perylene benzene rings. Unfortunately, in that paper, only the overall results were given without details from which one could see the results for contributions of individual conjugated circuits of perylene.

Observe that the renormalization occurs in the model based on conjugated circuits and may be of potential interest in some VB calculations, but whether and how, if relevant, it is for MO calculations remains yet to be seen. For example, the London–Pople–McWeeny ring current intensities [18–20], expressed as a ratio to the ring current intensity calculated by the same method for benzene, are for the benzene rings of naphthalene 1.09 and for the naphthalene moieties rings of perylene 0.97, but the ring current for the central ring in perylene is 0.24. Similar values are found for the naphthalene moieties rings and central rings in zethrene (0.87 and 0.37, respectively). The numerical intensities shown above have been reproduced as reported in the publication by Gomes and Mallion [17], who calculated ring currents using the conjugated circuits approach. They do not give many details on their computations but significantly have zero ring currents for the central benzene rings in perylene and zethrene. In addition, they also have zero ring currents for the five membered rings in acenaphthylene and flouranthene, which involve "essentially" single CC bonds. In contrast, in the London–Pople–McWeeny ring currents model, these rings have 0.11 and 0.05 intensities in acenaphthylene and flouranthene, respectively. Clearly MO methods give to central rings in perylene and zethrene and the five membered rings in acenaphthylene and flouranthene small ring currents, which is not consistent with expectations based on conjugated circuits. It remains to be seen if the so-calculated MO ring current intensities for all rings have to be further revised due to the notion of self-similarity of interacting molecular fragments, which is related to the notion of renormalization. Such explorations will require a high level of precision in ring current calculations, which have not been available in the past, but may be in the near future.

Those interested in pursuing the problem of renormalization associated with ring currents in systems with essentially single CC bonds may consider the following:

Problem 19

Consider systems having essentially single CC bonds connecting smaller benzenoid (and possibly non-benzenoid) fragments and see if there is a way that allows finding renormalization factors of such systems *without* examination individually of all Kekulé valence structures.

12.6 PLAGIARISM THAT IS NOT

We will end this chapter with a story, which has some similarity with anecdotes, but cannot be considered as a typical anecdote, which are characterized as stories about something interesting or funny, or short amusing true story, a short account of an interesting or humorous incident. Our story, as will be seen, although true and deserving attention, is neither funny nor humorous. On the contrary, in a way, it is sad. If one considers, as some do, that an anecdote also relates to secret or hitherto unknown details of history or biography, then the story to follow could be counted as an anecdote of this chapter. This is a story about the fate of the paper [14] *π-Electron Currents in Polycyclic Conjugated Hydrocarbons: Coronene and Its Isomers Having Five and Seven Member Rings*, by the authors of this book and their colleagues M. Vračko and D. Vukičević. The paper was published in the *International Journal of Quantum Chemistry* in the year 2012. Figure 12.1a shows π electron current densities calculated by the *ab initio* quantum chemical method by Balaban, Bean, and Fowler [15] and on the right obtained by *enumeration* of CC bond current contributions for an isomer of coronene having five- and seven-member rings. Although this figure illustrates the results reported, it also illustrates the *fundamental novelty* in theoretical chemistry: that the model based on calculus (i.e., the quantum chemical *ab initio* computation) and the model based on discrete mathematics (i.e., the enumerations of contributions from conjugated circuits) show *remarkable parallelism*!

This fascinating continuum-discrete duality in ring current calculations is a *fundamental novelty* in theoretical chemistry that deserves greater attention and better understanding, which adds to the great significance of this work. After the manuscript was submitted to the *Journal of Computational Chemistry*, unexpectedly in a letter one of the editors sent the following short report from one of the two anonymous reviewers:

> The paper can not be accepted in any journal because of plagiarism. This is evident—see Ciesielski, A., Krygovski, T. M., Cyranski, M. K., Dobrowolski, M. A., Aihara, J., *Phys. Chem. Chem. Phys.* 2009, 11, 11447–55.

We were unaware of this particular publication of Ciesielski et al. (ref. [21] in this chapter) and, while waiting to get a copy of that paper, were utterly puzzled at how our work, using our model that we originated, could be plagiarism. Finally, a copy of the paper by Ciesielski et al. arrived. A close look at the paper by Ciesielski et al. showed considerable parallelism and apparent similarities and dissimilarities. The title of their paper was "Graph Topological Approach to Magnetic Properties of Benzenoid Hydrocarbons," but it could have been equally descriptive if it was: *π-Electron Currents in Benzenoid Hydrocarbons*. We considered coronene and its isomers having five- and seven-member rings, and Ciesielski et al. considered benzenoid hydrocarbons. Their calculations used conjugated circuits (to which they referred as symmetric differences), and their illustrations of conjugated circuits of benzenoids had six-member rings. The coronene isomers that we considered have five-, six-, and seven-member rings, but putting aside that the two papers considered unrelated structures, the approach was similar enough to convince an unfamiliar anonymous reviewer that the two publications involved plagiarism, because Ciesielski et al. failed to cite our earlier work on the use of conjugated circuits for calculation of ring currents.

Despite that there are some apparent similarities of the two papers, the apparent similarities are, in our view, not sufficient for evoking plagiarism charges. One should examine similarities and differences in depth and detail before making criminal accusations. This is not the place to elaborate on the misleading charge of plagiarism, which only shows amazing irresponsibility of the reviewer in question. After seeing the publication of Ciesielski et al. authors of the "plagiarized" paper found no reason why should they cite the paper, which considered a different set of molecules. The only reason to cite the paper would be to indicate that authors of the paper failed to cite our earlier work on the calculation of ring currents using conjugated circuits—but that may not be a sufficiently strong reason to cite anyone. If the anonymous reviewer had closely looked at the paper by Ciesielski et al., he or she should have seen that Ciesielski et al. cited a book by Gutman as reference to symmetric difference (the term they use for conjugated circuits) in which Gutman wrote, *"...it was Milan Randić (1976) who first recognized the great significance of alternating cycles for the chemical behavior of conjugated molecules. He named them 'conjugated circuits'..."*

The above should have sufficed to show not only that our paper was not plagiarism, but that this information was available for the reviewer, who was incompetent and superficial, yet arrogant and irresponsible. Chemists who have not heard of conjugated circuits should not evaluate publications on chemical graph theory whether considering benzenoid or non-benzenoid conjugated hydrocarbons. Period.

Use of synonyms to cover plagiarism has been known, particularly in social sciences [22]. In mathematics and natural sciences, use of synonyms is not so infrequent as it may offer alternative labels that may have distinct descriptive features. Thus, for example, Clar referred to benzenoid systems having only π electron aromatic sextets and "empty" rings as the "fully benzenoid systems" [23]. Polansky and Gutman have referred to the same as the "all benzenoid systems" [24], and Dias refers to these most stable benzenoids as the "total resonant sextet benzenoids [25]. There are no problems with synonyms in chemistry *for informed persons*, but *uninformed chemists*, as was the anonymous referee of our paper on ring currents, *can be confused*, which is, of course, their problem!

REFERENCES AND NOTES

1. L. Pauling, The diamagnetic anisotropy of aromatic molecules, *J. Chem. Phys.* 4 (1936) 673–677.
2. P. Lazzaretti, Ring currents, *Progr. Nucl. Magn. Res. Spect.* 36 (2000) 1–88.
3. J. A. N. F. Gomez and R. B. Mallion, Aromaticity and ring currents, *Chem. Rev.* 101 (2001) 1349–1383.
4. R. B. Mallion, Topological ring-currents in condensed benzenoid hydrocarbons, *Croat. Chem. Acta* 86 (2013) 387–406.

5. T. K. Dickens and R. B. Mallion, An analysis of topological ring-currents and their use in assessing the annulene-within-an-annulene model for super-ring conjugated systems, *Croat. Chem. Acta* 86 (2013) 387–406.

6. T. K. Dickens and R. B. Mallion, π-electron ring-currents and bond-currents in some conjugated altan-structures, *J. Phys. Chem. A* 118 (2014) 3688–3697.

7. T. K. Dickens and R. B. Mallion, Topological ring-currents and bond-currents in the altan-[r,s]-coronenes, *Chem. Commun.* 51 (2015) 1819–1822.

8. T. K. Dickens and R. B. Mallion, Ring-currents assessment of the annulene-within-an-annulene model for some large coupled super-ring conjugated systems, *Croat. Chem. Acta* 87 (2013) 221–232.

9. G. N. Lewis, *Valence and the Structure of Atoms and Molecules*, Dover Publications, Inc., New York, 1966.

10. M. Randić, Aromaticity of polycyclic conjugated hydrocarbons, *Chem. Rev.* 103 (2003) 3449–3605.

11. M. Randić, Conjugated circuits and resonance energies of benzenoid hydrocarbons, *Chem. Phys. Lett.* 38 (1976) 68–70.

12. M. Randić, Aromaticity and conjugation, *J. Am. Chem. Soc.* 99 (1977) 444–450.

13. M. Randić, A graph theoretical approach to π-electron currents in polycyclic conjugated hydrocarbons, *Chem. Phys. Lett.* 500 (2010) 123–127.

14. M. Randić, M. Novič, M. Vračko, D. Vukičević, and D. Plavšić, π-electron currents in polycyclic conjugated hydrocarbons: Coronene and its isomers having five and seven member rings, *Int. J. Quantum Chem.* 112 (2012) 972–985.

15. A. T. Balaban, D. E. Bean, and P. W. Fowler, Patterns of ring current in coronene isomers, *Acta Chim. Slov.* 57 (2010) 507–512.

16. I. Gutman and M. Randić, A correlation between Kekulé valence structures and conjugated circuits, *Chem. Phys.* 41 (1979) 265–270.

17. J. A. N. F. Gomes and R. B. Mallion, A quasi-topological method for the calculation of relative «Ring-current» intensities in polycyclic, conjugated hydrocarbons, *Rev. Port. Quim.* 21 (1979) 82–89.

18. R. McWeeny, Ring currents and proton magnetic resonance in aromatic molecules, *Mol. Phys.* 1 (1958) 311–321.

19. R. B. Mallion, On the magnetic properties of conjugated molecules, *Mol. Phys.* 25 (1973) 1415–1432.

20. J. A. Pople, The molecular orbital theory of aromatic ring currents, *Mol. Phys.* 1 (1958) 175–180.

21. A. Ciesielski, T. M. Krygovski, M. K. Cyranski, and J.-I. Aihara, Graph topological approach to magnetic properties of benzenoid hydrocarbons, *Phys. Chem. Chem. Phys.* 11 (2009) 11447–11455.

22. J. Aziz, F. Hashim, and N. A. Razak, Anecdotes of plagiarism: Some pedagogical issues and considerations, *Asian Social Sci.* 8 (2012) 29–34.

23. E. Clar, *The Aromatic Sextet*, Wiley & Sons, London, 1972.

24. I. Gutman and O. E. Polansky, *Mathematical Concepts in Organic Chemistry*, Springer-Verlag, Berlin, 1986.

25. J. R. Dias, The most stable class of benzenoid hydrocarbons and their topological characteristics—Total resonant sextet benzenoids revisited, *J. Chem. Inf. Comput. Sci.* 39 (1999) 144–150.

13 Graphical Bioinformatics

Graphical bioinformatics is a relatively new branch of bioinformatics, which was initiated 30 years ago with a publication by Hamori and Ruskin [1]. It had a slow beginning and only around the year 2000 started to flourish. There are two recent reviews on graphical bioinformatics [2,3] in which those interested can find information on research in graphical informatics over the past dozen years. In Table 13.1, we have presented milestones in graphical bioinformatics that followed the work of Hamori and Ruskin. Almost half of the 30 years of graphical bioinformatics, since the pioneering work of Hamori, followed by Jeffrey's modification of the chaos game for representation of DNA, was essentially qualitative. Various graphical representations were mostly visually inspected.

It was in the year 2000 that qualitative graphical bioinformatics saw dramatic change—change to a quantitative discipline. One of the present authors (MR) was attending a seminar at the Natural Resources Research Institute of the University of Minnesota in Duluth by a visiting scientist from India, Dr. Ashesh Nandy. He was presenting his two-dimensional graphical representation of DNA. Nandy was showing and comparing *visually* graphical representations of various DNA. In Figure 13.1, we show a segment of DNA constructed by moving along the DNA sequence following the rule: For each A, move one step along –x-axis; for each C, move one step along +y-axis; for each G, move one step along +x-axis; for each T, move one step along –y-axis. So constructed 2-D graphical representations allowed Nandy to arrive at some insights into similarities of various DNA. Looking at these 2-D graphical representations of DNA, it was immediately clear to MR that such graphical representations of DNA can be *numerically* characterized by D/D matrices, just as one can numerically characterize molecular structures having fixed geometry. Over the following days in Duluth, a manuscript on *numerical* characterization of graphical representations of DNA was prepared [4].

This publication opened a novel direction in graphical bioinformatics by offering elaborate *quantitative* characterization of qualitative graphical data on DNA. It was during the seminar by Nandy that MR realized the existing tool of chemical graph theory, which is mostly concerned with *numerical* characterization of structure of chemical compounds for use in structure–property–activity relationship studies, can be extended to *quantitative* characterizations of graphical representations of DNA. Significantly, in addition to showing (i) that numerical characterizations of virtual geometrical patterns are possible and practical, this made possible (ii) that the analysis and comparative studies of data on DNA and proteins could be delegated to computers; (iii) that it is also possible to obtain useful information in some cases *without* prior need for alignment of sequences; and (iv) last but not least, that using numerical characterization of DNA one could *recover* the lost information of some simplified 2-D graphical representations of DNA. For example, although the first exon of the β-globin gene of goat has 87 bases, the diagram of graphical representation in

TABLE 13.1
Milestones in Graphical Bioinformatics

	Year	Topic
1	1983	3-D graphical representation of DNA
2	1990	Chaos Game representation of DNA
3	2000	Numerical characterization of DNA graphical plots
4	2001	Numerical characterization of proteomics maps
5	2003	Spectral representation of DNA
6	2004	Graphical representation of DNA as a map
7	2004	Virtual genetic code
8	2005	Hormesis at the proteome level
9	2006	Graphical alignment of DNA
10	2008	Graphical alignment of proteins
11	2008	Amino acid adjacency matrix
12	2008	Representation of RNA without loss of information
13	2012	Exact solution to protein alignments
14	2014	Computer program for exact solution to protein alignments

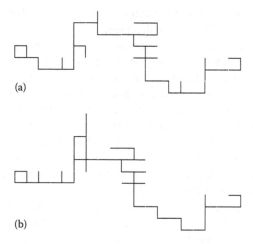

(a)

(b)

FIGURE 13.1 Graphical representation of Nandy of the first exon of β-globin gene of goat (a) and bovine (b).

Figure 13.1 has only 51 vertices. This is because each AG followed by A or GA followed by G and each CT followed by C or TC followed by T cancels one another. In addition to such cancellations, bases that follow may have fragments that repeat themselves. The diagram of DNA shown in Figure 13.1 is what in graph theory is called a *trail* (a path over which one walks), and DNA is a *walk*, which can repeat segments over the trail. Matrices used to represent graphical diagrams of DNA in fact record the walk, not the trail, and thus recover the lost information, which offers better comparative results of different DNA sequences.

13.1 CHARACTERIZATION VERSUS INDEXING

As we have mentioned, in our view, the publication on the numerical characterization of the 2-D graphical representation of DNA of Randić et al. has *opened* a novel direction in graphical bioinformatics. However, this view has been challenged by two anonymous referees of a recent paper, "On map representation of DNA" [5], which was initially submitted to *SAR and QSAR in Environmental Research* (see Appendix 14). The problem was with blurring the distinction between "characterization" and "indexing," insinuating that "indexing" DNA is tantamount to quantitative graphical bioinformatics. At best, indexing offers very limited quantitative characterization of individual DNA maps. A comparison between indexing and characterization can be illustrated on a person. If one has information on the height, weight, and eye color of a person, that is indexing. It can exclude persons who have different height, weight, and eye color, but there are too many individuals who have the same height, weight, and eye color. Characterization of a person corresponds to having fingerprints and a picture of a person. Such is the difference between having qualitative and quantitative information.

The paper [5], as its title says, is about *map* representations of DNA and elaborates on earlier publications in which four color *maps* for representing DNA were introduced; it is not a review on 2-D representations of DNA. The context of our statement refers to *quantitative* graphical bioinformatics and not to *numerical* indices for graphical DNA representation. Bioinformatics is not about a *single* paper, or even *several* papers, in which a few numerical parameters are introduced to serve indexing DNA. Bioinformatics is about *comparative* study of DNA, proteins, and the proteome, about *quantitative* characterization of the *similarity* of DNA and proteins. *Comparative study* and *indexing* are not the same. Even the paper by Randić, Vračko, Nandy, and Basak [4] on the numerical characterization of the 2-D graphical representation of DNA of Nandy, in which 2-D graphical representation of DNA is represented by a 92×92 (symmetrical) matrix that has more than 4000 numerical matrix elements, which as mentioned, has *opened* the way to *quantitative graphical bioinformatics*, strictly speaking, does not qualify as a contribution to *quantitative graphical bioinformatics*. The first paper in *quantitative graphical bioinformatics* is the paper titled "On the Similarity of DNA Primary Sequences" by Randić and Vračko [6], published in the year 2000, in which the 2-D graphical representation of the first exon of the β-globin gene of eight species were considered and the similarity/dissimilarity table reported. A few months earlier, also in the year 2000, Randić published a work on the "Condensed Representation of DNA Primary Sequences" [7,8], in which all DNA considered were represented by a 4×4 matrix in which rows and columns have labels A, C, G, and T, and the matrix element XY counts the occurrence of adjacency of nucleotide X being followed by nucleotide Y. In Table 13.2, we have listed the coding sequences of the first exon of β-globin gene of nine species.

In Matrix 13.1, we show the condensed matrices for the first exon of the nine species of Table 13.2. Observe from Matrix 13.1 great similarity of the condensed matrices of human and gorilla although differences with lemur are considerable. Similarly, there is visible similarity of the condensed matrices of cattle and goat and a lack of similarity between rabbit and rat.

	A	C	G	T
A	AA	AC	AG	AT
C	CA	CC	CG	CT
G	GA	GC	GG	GT
T	TA	TC	TG	TT

Cattle

	A	C	G	T
A	5	2	8	2
C	3	5	2	6
G	8	8	12	6
T	0	1	13	4

Chicken

	A	C	G	T
A	5	3	7	4
C	7	7	3	7
G	6	10	13	4
T	0	4	11	0

Goat

	A	C	G	T
A	5	2	8	2
C	3	4	2	8
G	8	9	12	5
T	0	2	13	2

Gorilla

	A	C	G	T
A	4	4	7	2
C	3	7	2	7
G	8	6	13	9
T	1	2	15	2

Human

	A	C	G	T
A	4	4	7	2
C	3	7	2	7
G	8	6	12	9
T	1	2	15	2

(13.1)

Lemur

	A	C	G	T
A	4	2	9	5
C	4	2	1	8
G	9	7	11	7
T	1	4	14	4

Opossum

	A	C	G	T
A	3	7	8	3
C	7	4	0	9
G	8	5	9	6
T	2	4	12	4

Rabbit

	A	C	G	T
A	5	1	8	3
C	4	5	1	5
G	7	6	13	11
T	0	4	15	1

Rat

	A	C	G	T
A	6	4	6	4
C	2	6	1	9
G	7	8	11	6
T	4	0	15	2

In Table 13.3, we show the similarity/dissimilarity table for the nine coding sequences of Table 13.2 based on viewing the 16 matrix elements of the condensed matrices as 16-component vectors and calculating the Euclidean distance between these vectors as a measure of the similarity/dissimilarity. As one can see, the smallest entries in Table 13.3, 1.00 and 3.46, belong human and gorilla and cattle and goat, respectively.

This work can be viewed as a contribution to bioinformatics, not graphical bioinformatics, because it is not based on *graphical* representations of DNA, but on combinatorics (the combinatorial properties of the prime DNA sequences). So, similarly, the works of Nandy, in which there have been introduced *numerical* parameters

TABLE 13.2
The Coding Sequences of the First Exon of β-Globin Gene of Nine Species
(for Easier Comparison of Bases, Each 10 Bases Are Grouped Together)

Cattle (*Bos taurus*)
ATGCTGACTG CTGAGGAGAA GGCTGCCGTC ACCGCCTTTT GGGGCAAGGT
GAAAGTGGAT GAAGTTGGTG GTGAGGCCCT GGGCAG

Chicken (*Gallus gallus*)
ATGGTGCACT GGACTGCTGA GGAGAAGCAG CTCATCACCG GCCTCTGGGG
CAAGGTCAAT GTGGCCGAAT GTGGGGCCGA AGCCCTGGCC AG

Goat (*Capra hircus*)
ATGCTGACTG CTGAGGAGAA GGCTGCCGTC ACCGGCTTCT GGGGCAAGGT
GAAAGTGGAT GAAGTTGGTG CTGAGGCCCT GGGCAG

Gorilla (*Gorilla gorilla*)
ATGGTGCACC TGACTCCTGA GGAGAAGTCT GCCGTTACTG CCCTGTGGGG
CAAGGTGAAC GTGGATGAAG TTGGTGGTGA GGCCCTGGGC AGG

Human (*Homo sapiens*)
ATGGTGCACC TGACTCCTGA GGAGAAGTCT GCCGTTACTG CCCTGTGGGG
CAAGGTGAAC GTGGATGAAG TTGGTGGTGA GGCCCTGGGC AG

Lemur (*Eulemur macaco*)
ATGACTTTGC TGAGTGCTGA GGAGAATGCT CATGTCACCT CTCTGTGGGG
CAAGGTGGAT GTAGAGAAAG TTGGTGGCGA GGCCTTGGGC AG

Opossum (*Didelphis virginiana*)
ATGGTGCACT TGACTTCTGA GGAGAAGAAC TGCATCACTA CCATCTGGTC TAAGGTGCAG
GTTGACCAGA CTGGTGGTGA GGCCCTTGGC AG

Rabbit (*Oryctolagus cuniculus*)
ATGGTGCATC TGTCCAGTGA GGAGAAGTCT GCGGTCACTG CCCTGTGGGG
CAAGGTGAAT GTGGAAGAAG TTGGTGGTGA GGCCCTGGGC

Rat (*Rattus norvegicus*)
ATGGTGCACC TAACTGATGC TGAGAAGGCT ACTGTTAGTG GCCTGTGGGG
AAAGGTGAAC CCTGATAATG TTGGCGCTGA GGCCCTGGGC AG

relating to his graphical representations of DNA, do not belong to bioinformatics but to molecular biology, or if one insists on using the word "graphical," it belongs to graphical molecular biology. Recall that the issue is not who was first to associate numerical parameters to DNA graphical representations but who upgraded *qualitative graphical bioinformatics* to *quantitative graphical bioinformatics*! Clearly, we are not to let two anonymous reviewers decide on the origin of quantitative graphical bioinformatics.

From Table 13.1, it should be clear that the "numerical characterizations" mentioned in Table 13.1 relate to *graphical bioinformatics*, to numerical characterizations that facilitate comparative studies of DNA and result in a *similarity/dissimilarity table*, which is the basis for construction of phylogenetic trees. None of the numerical

TABLE 13.3

Similarity–Dissimilarity Table for the Eight Exons of Table 13.2

	Cattle	Chicken	Goat	Gorilla	Human	Lemur	Opossum	Rabbit	Rat
Cattle	0	8.37	3.46	5.92	5.83	6.00	9.59	7.62	10.00
Chicken		0	7.21	9.64	9.70	10.10	10.68	10.49	10.10
Goat			0	6.86	6.78	5.66	9.49	8.37	8.94
Gorilla				0	1.00	7.68	9.43	5.74	8.89
Human					0	7.48	9.06	5.83	8.72
Lemur						0	7.62	7.62	8.60
Opossum							0	11.49	10.00
Rabbit								0	10.10
Rat									0

indexing prior to and after the year 2000 have been or could be used for construction of reliable similarity/dissimilarity tables. So we view the comments of the referees, shown below, relating to quantitative graphical bioinformatics, inappropriate. This referee is confusing the distinction *between* quantitative and qualitative graphical representation of DNA *and* quantitative and qualitative graphical bioinformatics.

Referee 1:
The paper ignores all seminal developments of 2D methods and mentions the 3D method of 2000 as if the 2D representations didn't exist. Their 2D four color maps were not the only 2D methods proposed by researchers!

Similarly, the authors claim that numerical characterization took place in 2000 only whereas it had actually taken place through geometrical means in 1998 and 1999. Ignoring such developments reflect poorly on the background research capabilities and fairness of the authors.

Referee 2, Comments to the Author:
The following comments are factually incorrect:
However, such graphical representations of DNA were qualitative, which was a considerable limitation. This all changed by the year 2000 when it was shown how one can accompany qualitative 2-D graphical representations of DNA with quantitative numerical characterizations.

In short, we strongly disagree with the above comments, and it suffices just to state that numerical characterization could be qualitative and quantitative. We maintain that before 2000 numerical characterizations of DNA were qualitative, not quantitative, in nature.

One of the "overlooked" contributions that Referee 2 mentioned is the "pioneering study" in the numerical characterization of DNA, the paper by Raychaudhury and Nandy, "Indexing Scheme and Similarity Measures for Macromolecular Sequences" [9]. We agree that this may be a "pioneering study" in the numerical qualitative characterization of DNA, but it does not represent the numerical *quantitative* characterization of DNA, a contribution to solving problems relevant for *quantitative* graphical bioinformatics. We claim that the first numerical quantitative characterization of DNA is the paper by Randić, Vračko, Nandy, and Basak [4].

13.2   MILESTONES OF GRAPHICAL BIOINFORMATICS

Let us return to the milestones of graphical bioinformatics. Table 13.1 ends with the most significant milestone of graphical bioinformatics: the exact solution to a pairwise alignment of proteins (and DNA). We briefly describe the exact solution, first to illustrate that it is conceptually and computationally simple enough that it could be understood and appreciated by graduate students and, second, to show that the solution to the alignment problem is unusually *efficient*. We emphasize the word *efficient* in order to draw attention to the distinction between having an exact solution, which we have, and having an *efficient* solution, which we also have. Recall the case of Karle and Hauptman, who found the solution to the X-ray diffraction inverse problem but had a difficult time for their result to be accepted—in part, because the computations were lengthy and tedious. According to Hauptman [10], it took them 30 days of calculations to solve one of the more complex crystal structures (this was, of course, in the time before computers), which today computers would do in a few minutes at most. For comparison, consider two proteins having approximately 170 amino acids. It takes around 10 minutes, at most 15 minutes, to align these two proteins *without* the use of computers or calculators—just pencil and paper! This illustrates the efficiency, which, in part, as we will show on the following pages, is due to the inherent nature of the approach, which eliminates from consideration adjacent pairs of amino acids that appear in one protein and not in the other. The number of such appearances can be a relatively high percentage of the total number of amino acid pairs present in proteins.

Before addressing the protein alignment problem, we will outline two of our distinct graphical representations of DNA, which are significantly different from those initiated by Hamori and Jeffrey. Hamori represented DNA as a zigzag curve in three-dimensional space, and Jeffrey represented DNA as a two-dimensional image. The novelty of our first new graphical representation, referred to as spectral, is that it has imbedded in it definite numerical content, which is a *novelty* [11]. Such representations are called spectral because they distantly resemble molecular spectra, but by having numerical content, one can arithmetically manipulate sequences and shift them and subtract them, arriving at graphical DNA alignment. The construction of the map representation of DNA is reminiscent of the graphical representation of prime numbers by Ulam in a spiral form. One writes a DNA sequence as a spiral, which, when the adjacent cells having labels of the same nucleobase are fused, results in a 2-D maps. In view of the fact that there are four nucleobases, A, C, G, T (adenine, cytosine, guanine, and thymine, respectively), and that one can associate with each type of nucleobase a color, the resulting map takes four colors.

13.3   SPECTRAL REPRESENTATION OF DNA

Among the dozen alternative graphical representations of DNA that emerged during the past 20 to 25 years, the spectral representations are unique among graphical representations of bio-sequences. Spectral representations of DNA are obtained by associating with the four bases, A, C, G, and T, in the case of DNA four horizontal lines assigning to them the numerical values from one to four. In the case of proteins,

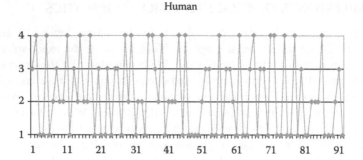

FIGURE 13.2 Spectral representation of the coding sequence of the first exon of β-globin gene of human.

one uses 20 horizontal lines, assigning to them the numerical values 1–20 for the 20 natural amino acids. Then one depicts in sequential order, in the case of DNA, the four bases and, in the case of proteins, the 20 amino acids as spots over the corresponding horizontal lines in the x, y plane. By connecting the adjacent spots by lines, one obtains graphical representations that are reminiscent of *spectra* in physics and chemistry. For illustration, in Figure 13.2, we illustrate spectral representation of the first exon (coding sequence of DNA) of the β-globin gene of human.

The first spectral representation of DNA appeared in 2003 [11,12]. Three years later, it was shown how one could arrive graphically at the alignment of DNA using spectral representations [13]. Similarly, spectral representations of proteins one can shift and subtract, and in this way, one could detect graphically the degree of alignment between two proteins [14] without the use of any approximations, which typify most computer-based alignment software.

13.4 FOUR-COLOR 2-D MAPS OF DNA

We will here elaborate on 2-D representations of DNA known as the "four-color" representation of DNA. The present approach is an extension of the approach that has been presented before by some of the present authors [15,16]. The complexity of DNA and proteins is such that one should welcome additional descriptors for comparative studies of DNA and proteins. Introduction of 2-D DNA *maps* allows construction of numerical descriptors, which go beyond local sequence characteristics. Also, 2-D map representations, just as is the case with spectral representations of DNA and proteins, are less sensitive to minor local changes in sequence composition. The basic idea behind the "four-color" representation of DNA is to transform a sequence into a map. For illustration, consider the short DNA sequence of the first exon of the β-globin gene of cattle (*Bos taurus*), listed in Table 13.2. One can select triangular, square, or hexagonal tessellation in a plane and write the selected DNA sequence in a spiral form. When adjacent cells having the same label are fused into a single region, one obtains a map in which regions having the same label are colored by the same color. This straightforward construction transforms a linear DNA sequence into geometrical object, similar to geographical maps.

Once a map is constructed, one searches for numerical characterizations of the so-constructed map. If two DNA sequences are very similar, then also the derived 2-D maps will be similar. Similarly, if two 2-D maps are widely different, then the corresponding DNA sequences will also be very different. Because of loss of information accompanying use of mathematical invariants as descriptors, it is possible that different sequences have similar maps and several of the same invariants. Hence, similar maps need not correspond to similar sequences. This is one of the reasons for considering alternative graphical representations of DNA sequences and construction of different sets of molecular descriptors, expecting that among such collections of descriptors there will be some sufficiently different for different DNA sequences.

13.4.1 FOUR-COLOR 2-D MAP OF A SEQUENCE OF CATTLE DNA

The 2-D map shown in Figure 13.3a belongs to the first exon of the β-globin gene of cattle (*Bos taurus*) and was obtained by representing the DNA sequence as a spiral over the square grid. After fusing adjacent cells belonging to the same base, as shown in Figure 13.3b, and coloring them by the same color, one obtains a four-color map of DNA. In the next step, one constructs the *inner dual* of the map. To obtain the inner dual, one replaces each map region by a vertex placed in its center and connects the vertices of adjacent regions by the edges. The dual graphs have definite geometry by being embedded in the x, y plane where all vertices have fixed (x, y) coordinates and edges have fixed lengths.

It is not widely known that the "squared" Euclidean distance matrix, which is the Euclidean distance matrix in which the elements are squared individually, allows using eigenvalues for comparison of maps of *different sizes* [17,18]. According to an intriguing theorem in linear algebra, when the individual elements of a distance matrix are squared, the resulting matrix, regardless of its size, has only *four eigenvalues different from zero*; of these, one is positive, and three are negative.

The above is a special case of a more general property of Euclidean distance matrices when their individual matrix elements are raised to higher powers. When entries

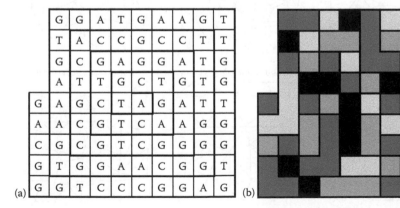

(a)

	G	G	A	T	G	A	A	G	T
	T	A	C	C	G	C	C	T	T
	G	C	G	A	G	G	A	T	G
	A	T	T	G	C	T	G	T	G
G	A	G	C	T	A	G	A	T	T
A	A	C	G	T	C	A	A	G	G
C	G	C	G	T	C	G	G	G	G
G	T	G	G	A	A	C	G	G	T
G	G	T	C	C	C	G	G	A	G

(b)

FIGURE 13.3 (a) Is a spiral representation of the first exon of bovine β-globin gene. (b) Is the 2-D map of the first exon of bovine β-globin gene obtained by fusing adjacent cells belonging to the same bases.

of the Euclidean distance matrix are raised to the second, third, fourth, and fifth power, etc., the numbers of nonzero eigenvalues of such a matrix are 4, 9, 16, 25, etc., respectively. Moreover, when one considers Euclidean matrices in 3-D, 4-D, and 5-D space, then the Hadamard squared distance matrix has not four, but five, six, and seven nonzero eigenvalues, respectively—one of which is always positive, and all the others are negative. The matrices obtained by raising individual matrix elements to higher powers (known as Hadamard multiplication) rather than using standard multiplication of matrices are thus of considerable interest in comparative studies of maps.

13.4.2 HADAMARD SQUARED EUCLIDEAN DISTANCE MATRICES

In Figure 13.4, we show dual graphs of the four-color maps representing (the first exon of β-globin gene of) for the nine species listed in Table 13.2. The four-color maps and their dual graphs are based on the square grid as a template. In Table 13.4,

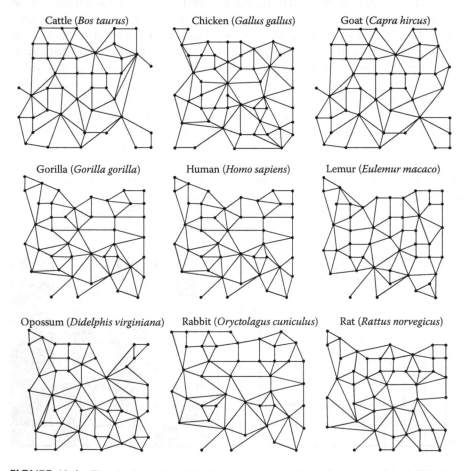

FIGURE 13.4 The dual graphs of the four-color maps of the first exon of β-globin gene of the nine species listed in Table 13.2. (Reproduced from On map representation of DNA, *Croat. Chem. Acta* © (2013) Croatian Chemical Society, Zagreb, Croatia.)

TABLE 13.4

Absolute Magnitudes of Four Nonzero Eigenvalues of the "Squared" Euclidean Distance Matrices Associated with the Dual Graphs Representing the Nine Species

	λ_1	λ_2	λ_3	λ_4
Cattle (*Bos taurus*)	1327.09	620.33	582.74	124.02
Chicken (*Gallus gallus*)	1682.70	872.92	645.76	164.02
Goat (*Capra hircus*)	1452.91	728.58	597.91	126.42
Gorilla (*Gorilla gorilla*)	1475.42	800.81	544.22	130.39
Human (*Homo sapiens*)	1479.83	805.21	543.38	131.23
Lemur (*Eulemur macaco*)	1561.82	819.60	595.66	146.55
Opossum (*Didelphis virginiana*)	1714.75	915.68	656.37	142.70
Rabbit (*Oryctolagus cuniculus*)	1339.35	719.67	514.05	105.63
Rat (*Rattus norvegicus*)	1563.6	802.362	612.397	148.843

we have collected for the nine species the four nonzero eigenvalues belonging to the "squared" (Hadamard) Euclidean distance matrix, the elements of which are the squared Euclidean distances between vertices of the corresponding dual graph. One can notice that the variations of the leading eigenvalue (λ_1) among the species are considerable, the λ_1 values are in the range 1327 to 1715.

13.5 SIMILARITIES AND DISSIMILARITIES AMONG THE DNA SEQUENCES

Different graphical representations generally involve different artifacts in the analysis and will be accompanied by different signal-to-noise ratios. The way to reduce the chance of including false similarity results is to use several 2-D graphical representations *simultaneously*. In this way, one may expect that "accidental" coincidences in the similarity–dissimilarity testing will be more likely to be detected, which will result in more reliable numerical data.

In Table 13.5, we have collected the data on similarity–dissimilarity between the nine species based on the comparison of the first exon of their β-globin gene. All the similarity–dissimilarity values have been calculated as the corresponding distance between vectors ($\lambda_1, \lambda_2, \lambda_3, \lambda_4$) of Table 13.4, and have been normalized so that the smallest entry equals 1, which occurs for the pair human–gorilla. If Table 13.5 is viewed in isolation, it may pass on some incorrect information. In order to extract more reliable information from such 2-D representations of DNA, one may consider using *simultaneous* similarity tables based on different grids.

It is of interest to see how this similarity–dissimilarity table parallels other similarity–dissimilarity tables. We have selected Table 13.6 derived using Clustal Omega, which can be taken as a standard for such a comparison. Table 13.6 was

TABLE 13.5

Similarity–Dissimilarity Table for the Nine Species with Spiral Representations of DNA Based on the Four Eigenvalues of Table 13.4

	Cattle	Chicken	Goat	Gorilla	Human	Lemur	Opossum	Rabbit	Rat
Cattle	0	69.8	26.3	37.4	38.4	48.7	77.8	19.4	47.5
Chicken		0	43.9	38.5	37.8	22.5	9.2	63.5	22.6
Goat			0	14.6	15.5	22.6	51.6	22.6	21.4
Gorilla				0	1.0	16.3	45.5	25.7	17.8
Human					0	15.7	44.7	26.7	17.4
Lemur						0	30.1	41.1	3.8
Opossum							0	70.7	30.6
Rabbit								0	41.3
Rat									0

TABLE 13.6

Similarity between the Nine Species Based on Clustal Omega Program

	Cattle	Chicken	Goat	Gorilla	Human	Lemur	Opossum	Rabbit	Rat
Cattle	0	25.6	3.5	8.1	8.1	18.6	30.6	13.1	24.4
Chicken		0	22.1	26.1	26.1	35.9	28.6	30.0	35.9
Goat			0	10.5	10.5	20.9	31.8	15.5	22.1
Gorilla				0	0.0	26.1	28.6	10.0	19.6
Human					0	26.1	28.6	10.0	19.6
Lemur						0	42.9	27.8	34.8
Opossum							0	33.7	40.7
Rabbit								0	
Rat									0

kindly supplied to us by an anonymous reviewer of our publication [5] and was derived using Clustal Omega program. It basically determines similarities between DNA sequences based on the number and size of gaps after DNA sequences have been aligned.

Notice that although qualitatively the two tables parallel each other, they also show differences. This is not alarming because the two tables measure similarities with respect to distinct properties of DNA sequences. If one compares the similarities of Table 13.5 with the "standards" of Table 13.6, one can see some agreement but also some disagreement, particularly when one considers less-similar species. The comparison of the two tables can be made easier by plotting the entries of the two tables one against the other.

Two modifications of the four-color map representation of DNA that may be of interest are offered in the following problems:

Problem 20

Make a table of similarity between pairs of DNA sequences by modifying the construction of the four color maps by considering exactly aligned DNA sequences.

This introduces gaps. Gaps should be left uncolored and ignored. In the view that DNA alignment has been very recently exactly solved [19], the exact solutions of DNA alignment should not be difficult to obtain.

Problem 21

Discuss the differences between the similarity/dissimilarity table based on the four color map and the table based on the four color map involving gaps.

13.6   IS A FOUR-COLOR MAP OF A DNA SEQUENCE A USEFUL NOVELTY?

Graphical representation of DNA as a four-color 2-D map is a novelty that was introduced 10 years ago [15] and has been revisited and further developed a few years later [16] and very recently [5]. Novelty as an attribute is not sufficient to justify an approach as a new scientific tool; it has to be shown to be *useful* in applications. There appears to be a widespread misunderstanding that toolmakers *ought* also to demonstrate use of a novel tool by solving some biological problems. Obviously, if somebody uses a particular tool, it has become useful; if nobody uses it, it will fade into oblivion. To demand that toolmakers, after demonstrating how to use a tool, should solve chemical or biological problems is as absurd as it is absurd to expect that chemists and biologists have to design the tools and instruments that they use. Toolmakers' expertise may be in applied discrete mathematics, graph theory, and combinatorics, not necessarily in problems of chemistry or biology, just as the expertise of chemists and biologists does not necessarily extend to topics of discrete mathematics, which may be desirable in building novel theoretical tools.

The recent manuscript on graphical representation of DNA as a 2-D map [5] elaborates on some further refinement of the approach. It has been submitted to the respectable *SAR and QSAR in Environmental Research* but has received unusually hostile reception by two reviewers, the short versions of which are shown below:

Reviewer 1:
I think the authors should stop proposing new models and try to apply a model to solve some real problems instead of proposing new models like homework. Such immature approach as evidenced by the authors only contributes to intellectual pollution and wastes time and effort of the real problem solvers.... I suggest that this manuscript be summarily rejected.

Reviewer 2:
...the authors suggest new methods but do not show any substantial and innovative biological applications. In that sense, it is an exercise in the useless proliferation of indices without showing their relevance in bioinformatics. So, I strongly suggest that the paper should be rejected.

For a full version of the above reports on four-color representation of DNA see Appendix 15. Both reports show not only an unrealistic expectation that tool *makers* have also to be tool *users* but also an unfamiliarity of the reviewers with the literature in the area covered by the manuscript (*vide infra*). Instead of arguing with anonymous "experts" on tool developing, let us quote the abstract and the summary of a recent publication, which has used the above-described tool, the 2-D map representation of DNA, for solving the problem of bioinformatics. G. Agüero-Chapin et al. explored the adenylation domain repertoire of nonribosomal peptide synthetases using an ensemble of sequence-search methods [20]. Here is their abstract in full:

The introduction of two dimensional (2D) graphs and their numerical characterization for comparative analyses of DNA/RNA and protein sequences without the need of sequence alignment is an active yet recent research topic in bioinformatics. Here we used a 2D artificial representation (four color maps) with a simple numerical characterization through topological indices (TIs) to aid the discovering of remote homologous of Adenylation domains (A-domains) from the Nonribosomal Peptide Synthetases (NRPS) class in the proteome of the cyanobacteria Microcystis aeruginosa.

The above publication, coming from four countries (Portugal, Cuba, Mexico, and Belgium), ends with the following conclusion:

The utility of graphical approaches in bioinformatics has been demonstrated by the introduction of the four-color maps and the TIs (topological indices) as a cooperative tool for detecting remote homologous of A-domains (Adenylation domains) in the proteome of Microcystis aeruginosa. Since each sequence search method extracted different features from the protein sequences, their integration allows a more exhaustive description of certain protein class and therefore provides a higher yield for the detection of remote protein homologous. The knowledge of the complete repertoire of A-domains in the proteome of cyanobacteria species may allow unraveling new NRPS (Nonribosomal Peptide Synthetases) clusters for the discovery of novel natural products with important biological activities.

This ought to suffice to clarify the situation concerning use of topological indices and graphical approaches in chemical graph theory and bioinformatics, including 2-D DNA map representations. In Table 6.3 in Chapter 6 we have summarized diverse areas of application of topological indices. Just as in mathematics one counterexample suffices to show that a conjecture is false, so we hold that in chemical literature one counterexample to the claims of reviewers of a scientific paper should be sufficient to show that the presumptions of the reviewers are false. In our case, this counterexample is the paper by G. Agüero-Chapin et al., which disqualifies the two reviewers of *SAR and QSAR in Environmental Research*, showing clearly that their assertions that the proposed new method does not solve some real problems (Reviewer 1) and does not have any substantial and innovative biological applications (Reviewer 2) show their immature approach and their exercise in useless foretelling what is relevant in bioinformatics. The question is who is competent and who is incompetent when it comes to evaluation of a new tool: those who use it or those who do not know how to use it. To use their own words and style, their evaluations should be *"summarily rejected"* or *"strongly rejected,"* not being worth the paper they are written on. In plain English, their reports are without authority, of no value, and hence meaningless.

The above discussion hopefully resolves the distinction between wishful thinking and wishful speculations of reviewers and chemists at large and the constructive efforts of authors to enrich the repertoire of tools available for solving the problems of physics, chemistry, or biology. The above reports illustrate that there is not only a lack of understanding between the "two cultures" (science and humanity) on which C. P. Snow lectured in 1959 [21], but that there are "two cultures" within the science community. According to Wikipedia [22], the "two cultures" of C. P. Snow are characterized by the following: *Its thesis was that "the intellectual life of the whole of western society" was split into the titular two cultures—namely the sciences and the humanities—and that this was a major hindrance to solving the world's problems.*

Based on the above comments and well-documented hostility toward chemical graph theory over several decades mentioned previously [23], one may say that in addition to the "two cultures" of science and humanities, there are, at least within the sciences, if not two cultures, then "two subcultures": the world of designers of new tools and the world of users solving problems in science. Hence, following Wikipedia, one can similarly state *that while this need not be a hindrance to solving many of science's problems, it has been a major hindrance to solving some of the science problems.* One may ask, for example, why was the exact solution to the problem of protein alignment waited for so long—well over 40 years? The answer is fairly obvious. This is not because there were no talented and imaginative scientists around, but because there was no suitable *tool* to solve the problem! And again, we may add that nobody is asking any real problem solvers to do anything, but letting those with vision decide for themselves to choose what tool they wish to use.

13.7   AMINO ACID ADJACENCY MATRIX

Let us introduce the amino acid adjacency matrix (AAA matrix) [24] as a tool of graphical bioinformatics. This is a 20×20 matrix, the rows and columns of which belong to 20 natural amino acids, which have been ordered as follows:

A R N D C Q E G H I L K M F P S T W Y V,

which corresponds to the alphabetical order of the three-letter codes for these amino acids:

ala-arg-asn-asp-asx-cys-glu-gln-glx-gly-his-ile-
leu-lys-met-phe-pro-ser-thr-trp-tyr-val

The matrix element (i, j) of the AAA matrix gives the count of the occurrence of the adjacent pair of amino acids (i, j) in the protein sequence. The element is zero if the sequence does not contain adjacent amino acids pair (i, j), the element is one, two, or three if the sequence contains adjacent amino acids pair (i, j) once, twice, or three times, and so on.

In Table 13.7, we have listed two proteins that we will continue to use for illustration. In Matrix 13.2, we show the AAA matrix for a short portion of 30 amino acids of the first protein of Table 13.7, the segment (61–90):

$$...LGPSSIGPDLKPIGNPYSWNSNATVIFLDQ...$$

TABLE 13.7

Two Proteins of *Saccharomyces cerevisiae*

Protein 1:

KILGIDPNVTQYTGYLDVEDEDKHFFFWTFESRNDPAKDPVILWLN
GGPGCSSLTGLFFELGPSSIGPDLKPIGNPYSWNSNATVIFLDQPVN
VGFSYSGSSGVSNTVAAGKDVYNFLELFFDQFPEYVNKGQDFHIAG
ESYAGHYIPVFASEILSHKDRNFNLTSVLIGNGLT

Protein 2:

PSKLGIDTVKQWSGYMDYKDSKHFFYWFFESRNDANDPIILWLNG
GPCSSFTGLLFELGPSSIGADMKPIHNPYSWNNNASMIFLEQPLGV
GFSYGDEKVSSTKLAGKDAYIFLELEFEAFPHLRSNDFHIAGESYAG
HYIPQIAHEIVVKNPERTFNLTSVMIGNGIT

	A	R	N	D	C	Q	E	G	H	I	L	K	M	F	P	S	T	W	Y	V
A	0	0	0	0	0	0	0	0	0	0	0	0	0	0	0	0	1	0	0	0
R	0	0	0	0	0	0	0	0	0	0	0	0	0	0	0	0	0	0	0	0
N	1	0	0	0	0	0	0	0	0	0	0	0	0	0	1	1	0	0	0	0
D	0	0	0	0	0	1	0	0	0	0	1	0	0	0	0	0	0	0	0	0
C	0	0	0	0	0	0	0	0	0	0	0	0	0	0	0	0	0	0	0	0
Q	0	0	0	0	0	0	0	0	0	0	0	0	0	0	0	0	0	0	0	0
E	0	0	0	0	0	0	0	0	0	0	0	0	0	0	0	0	0	0	0	0
G	0	0	1	0	0	0	0	0	0	0	0	0	0	0	2	0	0	0	0	0
H	0	0	0	0	0	0	0	0	0	0	0	0	0	0	0	0	0	0	0	0
I	0	0	0	0	0	0	0	0	2	0	0	0	0	1	0	0	0	0	0	0
L	0	0	0	1	0	0	0	1	0	0	0	1	0	0	0	0	0	0	0	0
K	0	0	0	0	0	0	0	0	0	0	0	0	0	0	1	0	0	0	0	0
M	0	0	0	0	0	0	0	0	0	0	0	0	0	0	0	0	0	0	0	0
F	0	0	0	0	0	0	0	0	0	0	1	0	0	0	0	0	0	0	0	0
P	0	0	0	1	0	0	0	0	0	1	0	0	0	0	0	1	0	0	1	0
S	0	0	1	0	0	0	0	0	0	1	0	0	0	0	0	1	0	1	0	0
T	0	0	0	0	0	0	0	0	0	0	0	0	0	0	0	0	0	0	0	1
W	0	0	1	0	0	0	0	0	0	0	0	0	0	0	0	0	0	0	0	0
Y	0	0	0	0	0	0	0	0	0	0	0	0	0	0	0	1	0	0	0	0
V	0	0	0	0	0	0	0	0	0	1	0	0	0	0	0	0	0	0	0	0

$$(13.2)$$

As one can see from Matrix 13.2, if one goes to row L and column G, there is an entry 1, which stands for LG, the first pair of amino acids. The next pair, GP, is registered in row G and column P. Going to row G and column P, the entry is 2 because the pair GP appears also in location 7. The considered AAA matrix has 29 entries—29 adjacent pairs for a list of 30 amino acids.

AAA matrices have been found of interest for characterization of segments of proteins and whole proteins in their comparative studies. Invariants of the AAA matrix have been used for characterization of proteins. As we will see in the next section, 20-component vectors based on the differences in rows and columns of the AAA matrix were used in neural network analysis for characterization of trans-membrane regions of ND6 proteins in order to discriminate between transmembrane and non-transmembrane regions of ND6 proteins [25]. In another study, El-Lakkani and El-Sherif [26] presented a 3-D amino acid adjacency matrix used on nine ND5 proteins of different species and found high correlation between their results in comparison with results obtained with ClustalW program [27]. Similarly, independently invariants from amino acid adjacency matrix and decagonal isometric matrix were also constructed and similarly used in neural network analysis.

13.8   ON TRANSMEMBRANE PROTEINS

In this section we report briefly on the work of Choudhury, Zhukov, and Novič on mathematical characterization of transmembrane proteins membrane and non-membrane regions [25]. To encode the transmembrane and non-transmembrane protein segments, these authors used the amino acid adjacency matrix and decagonal isometric matrix (DIM) independently. The selected matrix invariants mathematically used to characterize each of the membrane protein segments are implemented independently and are used to distinguish between the transmembrane and non-transmembrane segments of membrane spanning proteins.

For this purpose, first the transmembrane protein sequences are segmented into the transmembrane and non-transmembrane regions. The non-transmembrane regions are further divided into polypeptide segments of length 20 residues (Figure 13.5). It was essential to have the length of the non-transmembrane segments similar to that of the transmembrane segments in order to ensure better training of the classification models. For encoding, all the transmembrane and non-transmembrane regions are used independently in the AA adjacency matrix and DIM. The encoded transmembrane and non-transmembrane segments are divided into training and test sets as stipulated in Table 13.8.

The issue was to check if the numerical descriptors derived from the AA adjacency matrix are able to discriminate the transmembrane segments from the non-transmembrane segments using *principal component analysis* (PCA). As PCA is a projection of multidimensional data onto a coordinate system defined by the principal components, it gives an initial validation regarding the choice of descriptors. Two independent *counter propagation neural network* (CPNN) models were developed using the invariants derived from both matrices to distinguish between the transmembrane and non-transmembrane segments of the protein sequences.

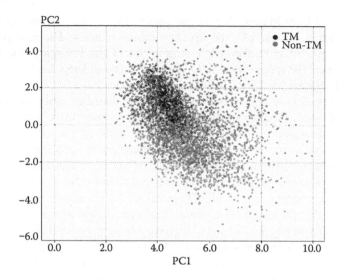

FIGURE 13.5 Principal component analysis. The transmembrane (black) and non-transmembrane (gray) segments form two different clusters.

TABLE 13.8

Training and Test Sets of Transmembrane and Non-Transmembrane Segments

		Number of Segments	
Sets	Total Segments	Transmembrane	Non-Transmembrane
Training	4204	1867	2337
Test	450	200	250

Figure 13.6 shows that the PCA analysis based on the row sum vector derived from the AA adjacency matrix well characterizes the transmembrane segments numerically. The transmembrane and non-transmembrane segments, represented in Figure 13.5 by the black and gray circles, respectively, are well-separated over the first and second principal components. The overlap between the two clusters is very small. Thus, it has been validated that the selected mathematical descriptors can bring out the characteristic features of the protein segments. The descriptors were able to distinguish the sequence characteristics of the two types of protein segments successfully.

The CPNN model was optimized for both the training and the test sets simultaneously varying different network parameters. The goal was to obtain the optimal network parameters that minimize misclassification. In the final step, the optimized network was tested for its prediction ability. The following network parameters are found to be optimal: the network size, 40 × 40; the number of epochs, 500; and the maximum correction factor, 0.9. In Figure 13.6 is shown the top map of the optimized network with the transmembrane and non-transmembrane segments in two distinct clusters. The network was able to correctly classify 95.67% of the transmembrane segments in the training set. For the test set, the error was 8.67%.

Finally, the 20-dimensional vectors based on the decagonal isometric matrix were used to represent the transmembrane and non-transmembrane segments. When used to train and optimize a CPNN network, the following configuration for the optimized network was found: the network size, 40×40; the number of epochs, 500; and the maximum correction factor, 0.5. The network shows 14.3% error in recall ability and 27.1% error in prediction ability with error threshold at 0.501.

With the growing number of proteins sequenced, there is a necessity for novel techniques to analyze protein sequences in order to determine their structure and function. The most commonly used protein sequence descriptors are based on evolutionary information and physicochemical properties. Even though these methods have proven to be efficient in most cases, in cases of transmembrane proteins, they may fall short. As the vast field of transmembrane proteins largely remains unexplored with many transmembrane proteins yet to be sequenced, it is possible to obtain new protein sequences without any known homolog. In such cases, traditional sequence analysis methods based on alignment profiles would not be sufficient. The evolutionary information-based descriptors appear inadequate, and indices based on physicochemical property can cause ambiguities. Therefore, it is of considerable interest to develop novel methods based on sequence information alone to represent protein sequences.

The two matrix representations of the protein segments, the amino acid adjacency matrix and the decagonal isometries matrix, are derived from the sequence information alone. As has been demonstrated, mathematical descriptors, dependent on the sequence information alone, have successfully revealed the underlying characteristics and patterns of given sequences. Their numerical nature also makes them easier to incorporate into a mathematical model. In addition, as has been well illustrated in chemical graph theory, when considering characterization of molecules, one can

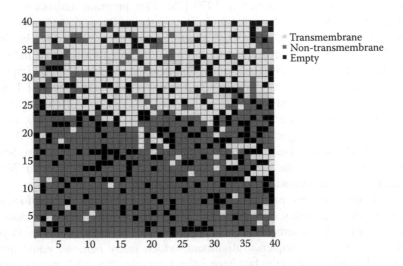

FIGURE 13.6 Top map of the optimized network. The transmembrane (light gray) and non-transmembrane (gray) segments form two different clusters.

derive different invariants to be used as descriptors from the same matrix representation, depending on the problem to be addressed.

In conclusion, characterization of proteins using mathematical descriptors to characterize transmembrane regions of proteins was successful. Both representations, the amino acid adjacency matrix and the decagonal isometries matrix, used independently to encode the protein segments have been successful in revealing the sequence characteristics particular to a specific group of protein segments in classifying them accordingly as transmembrane and non-transmembrane. The accuracy of the former method was better, which offers novel challenges for further development of the latter one.

13.9 ON THE SEQUENCE ALIGNMENT PROBLEM

One of the central problems of bioinformatics is DNA and protein alignment, which allows one to arrive at the degree of similarity between different DNA and proteins. Graphical bioinformatics [2,3,6] allows one to arrive at measures of similarity–dissimilarity of DNA and proteins without considering DNA or the protein alignment problem.

Although graphical representations of DNA and proteins allow quantitative comparisons of different sequences without prior alignment, we should add that spectral representations of DNA and proteins can also be used to search for DNA and protein alignments. Thus, this central problem of bioinformatics of determining the degree of sequence alignment is not beyond the reach of graphical approaches in bioinformatics. It appears that the potential of graphical representations of bio-sequences for alignment and comparisons has been hitherto overlooked.

Recall that the problem of sequence alignment, which has occupied computer scientists for about five decades, resulted in numerous computer programs and packages. Among the better known is the dynamic program for global alignment by Needleman and Wunsch, reported in 1970 [28]. This program outlines a general approach applicable to searching for similarities in the amino acid sequences of two proteins. A computer program for local alignment was reported in 1981 by Smith and Waterman [29]. These programs were followed by programs on rapid and sensitive protein similarity searches of Lipman and Pearson in 1985 [30] and improved tools for biological sequence comparison by the same authors in 1988 [31]. Finally, in 1990 came the basic local alignment search tool (BLAST) of Altschul et al. [32], one of the most widely used computer programs in bioinformatics today, considered to be the standard algorithm for similarity analysis. With its updated follow up, "Gapped BLAST and PSI-BLAST: A New Generation of Protein Database Search Programs" [33] of 1997, the two publications were cited 80,000 times in Web of Science, together with annual citations of around 5000.

There is no doubt that until recently the most that we knew in bioinformatics was due to the availability of very useful computer programs. But that does not mean that further improvements are not possible! Very recently, at least a 45-year-old problem of protein sequence alignment, which many believed could not be solved mathematically exactly, has been solved *exactly*. "Exactly" means *without* use of *approximations*, such as empirical parameters, statistical information, scoring based on penalties for gaps, insertions and deletions, and of course, *without* use of trial-and-error methodology. The trial-and-error methodology is abundantly used in most, if not all, current computer programs for protein and DNA alignments.

The article in which the exact solution was described was titled *"Very Efficient Search for Protein Alignment—VESPA"* [34] because while having an exact solution is important, even more important is that the algorithm leading to the solution is highly efficient. In the next section, we elaborate on the exact solution to the protein alignment problem.

Recall that the exact solution of the inverse problems of X-ray diffraction, the phase determination problem, by Hauptman and Karle [11], believed by many to be impossible to solve, involves heavy calculations. It took Hauptman and Karle one month to solve a single crystal structure having several heavy atoms in its unit cell. This required Fourier analyses of about 6000 diffraction spots, which, as Hauptman writes [10], would take less than three minutes on a computer today. In contrast, the VESPA algorithm for the exact solution to the protein alignment, takes about 15 minutes or less using a pen and pencil, no calculator or computer, to align a pair of proteins having about 170 amino acids. The high efficiency of the exact solution is due to the fact that the algorithm identifies pairs of adjacent amino acids that appear only in one sequence and not in the other and thus eliminates the need for their further examination. To find more about how the exact solution of protein alignment was extended to DNA alignment, consult ref. [19].

Let us add that there is an important distinction between computer-based alignment searches and a graphical approach to the same problem. It takes at least *two* sequences to make computer-based comparisons. In contrast, in the case of graphical representations of DNA and proteins one can characterize a *single* biological sequence! This is a very important distinction between graphical representations of bio-sequences and computer-based analyses of sequences, which one should keep in mind. This allows one to compile, characterize, and catalog DNA and protein sequences of interest for later visitation.

13.10   AMINO ACID SEQUENTIAL MATRIX

One of the present authors arrived at the exact solution of the protein alignment problem [34] by modifying the amino acid adjacency matrix [24]. The amino acid adjacency matrix (AAA), as mentioned before, is a 20×20 matrix, in which each row and column belong to one of the 20 natural amino acids. The matrix element $(AAA)_{ij}$ counts how many times in the primary sequence of a protein is amino acid i followed by amino acid j. The 20 amino acids have been ordered as follows:

A, R, N, D, C, Q, E, G, H, I, L, K, M, F, P, S, T, W, Y, V.

Amino acids of the ND6 proteins of human and mouse selected for comparison are shown in Table 13.9.

The AAA matrix for the ND6 human protein of Table 13.9 is shown in Matrix 13.3. The first row of the amino acid adjacency matrix for the ND6 human protein thus is

0 1 0 0 0 0 0 1 0 1 2 0 2 0 0 0 0 0 1 0 0.

	A	R	N	D	C	Q	E	G	H	I	L	K	M	F	P	S	T	W	Y	V
A	0	1	0	0	0	0	0	1	0	1	2	0	2	0	0	0	1	1	0	0
R	0	0	0	0	0	0	1	1	0	0	0	0	0	0	0	0	0	1	0	0
N	0	0	0	0	0	0	0	0	0	0	0	0	0	2	0	1	0	0	0	0
D	0	0	0	0	0	0	0	1	0	0	0	0	0	0	1	0	0	0	1	0
C	0	0	0	0	0	0	0	0	0	0	0	0	0	0	0	0	0	0	1	1
Q	0	0	0	0	0	0	0	0	0	0	0	0	0	0	0	0	0	0	0	0
E	1	0	0	1	0	0	1	2	0	1	0	0	0	0	0	0	0	0	2	2
G	2	1	1	0	1	0	1	4	0	0	4	0	1	4	0	3	0	1	1	4
H	0	0	0	0	0	0	0	0	0	0	0	0	0	0	0	0	0	0	0	0
I	1	1	0	0	0	0	2	1	0	1	1	0	0	0	0	0	0	0	2	3
L	1	0	1	0	0	0	0	1	0	2	1	0	1	2	0	1	0	1	1	5
K	0	0	0	0	0	1	0	0	0	0	0	0	0	1	0	0	0	0	0	0
M	1	0	0	0	0	0	1	2	0	1	0	0	2	0	0	0	0	0	1	2
F	0	0	1	0	0	0	0	2	0	1	1	0	0	0	0	1	0	0	0	3
P	0	0	0	0	0	0	1	0	0	2	2	0	0	0	0	1	0	0	0	0
S	0	0	0	0	0	0	0	3	0	0	0	1	0	0	1	1	0	1	0	3
T	1	0	0	0	0	0	0	1	0	0	0	0	0	0	0	0	1	0	0	0
W	0	0	0	0	0	0	0	1	0	0	1	0	1	0	1	0	0	0	0	1
Y	1	0	0	2	0	0	1	2	0	1	1	0	1	0	1	0	1	0	0	0
V	0	0	1	0	0	0	1	7	0	2	4	1	1	2	0	2	1	0	1	8

$$(13.3)$$

It registers the presence of the pairs AR, AG, AI, and AW occurring once and the pairs AL and AM occurring twice in the primary sequence of the ND6 human protein. The zeros indicate that in the sequence of this protein there are no adjacent amino acids AA, AN, AD, AC, AQ, and so on. Similarly, the first row of the amino acid adjacency matrix for the ND6 protein of mouse, the second protein of Table 13.9, is

---

TABLE 13.9
Amino Acids of Two ND6 Proteins Selected for Comparison

Human

MMYALFLLSV	**GLVMGFVGFS**	**SKPSPIYGGL**	**VLIVSGVVGC**	**VIILNFGGGY**
MGLMVFLIYL	GGMMVVFGYT	TAMAIEEYPE	AWGSGVEVLV	SVLVGLAMEV
GFVLWVKEYD	GVVVVVNFNS	VGSWMIYEGE	GSGFIREDPI	GAGALYDYGR
WLVVVTGWPL	FVGVYIVIEI	ARGN		

Mouse

MNNYIFVLSS	**LFLVGCLGLA**	**LKPSPIYGGL**	**GLIVSGFVGC**	**LMVLGFGGSF**
LGLMVFLIYL	GGMLVVFGYT	TAMATEEYPE	TWGSNWLILG	FLVLGVIMEV
FLICVLNYYD	EVGVINLDGL	GDWLMYEVDD	VGVMLEGGIG	VAAMYSCATW
MMVVAGWSLF	AGIFIIIEIT	RD		

Note: Each group of 10 amino acids is separated for easier count. The first 45 amino acids used for construction of the superposition AAA matrix are shown in bold.

1 0 0 0 0 0 0 2 0 0 1 0 2 0 0 0 2 0 0 0

By comparing only the two first rows of the ND6 proteins, one can already see that besides being similar they also show dissimilarity. Although the AAA matrix is accompanied with a loss of information on the sequential locations of pairs of adjacent amino acids, it nevertheless offers a useful characterization of proteins. The sequential labels, sites, or locations here indicate the exact position of an amino acid in the sequence of the protein. For example, the sequential labels for alanine are 4, 72, 74, 81, 97, 142, 144, and 171.

The sequence of the row sums of the AAA matrix gives the abundance of amino acids, which for human and mouse ND6 proteins are, respectively,

8 3 3 3 1 0 10 29 0 12 17 2 10 9 6 10 3 5 11 31

8 1 5 6 4 0 8 26 0 14 26 1 2 2 3 8 6 5 9 20

The similarity of the two sequences, although limited, again suggests limited similarity of the two proteins. Additional 20-component vectors can be constructed by considering the diagonal entries of the AAA matrices, which are for the human and mouse ND6 proteins, respectively,

0 0 0 0 0 0 1 4 0 1 1 0 2 0 0 1 1 0 0 8

1 0 1 1 0 0 1 4 0 2 0 0 1 0 0 1 0 0 1 2

These numbers indicate the presence of a succession of the same amino acids in the sequence. The above are just a few invariants of the AAA matrix, which may be of interest in a comparative study of proteins.

13.10.1 RECOVERY OF LOST INFORMATION

The problem of considerable interest is whether it is possible to recover some lost information accompanying the AAA matrix. Recall how quantitative graphical bioinformatics started by applying the D/D matrix approach developed earlier for characterization of 3-D structures of molecules [12], which was found to be applicable for characterization to 2-D graphical representations of DNA [35]. In doing this, not only that one arrived at a *numerical characterization* of 2-D graphical representations of DNA, but one was able to recover the lost information due to cancellation of A-G and C-T "moves" on the x, y coordinate grid.

In the case of the AAA matrix, in order to recover the lost information on the location of individual amino acids in the sequence of proteins, one should record the *sequential numbers* of adjacent amino acids rather than their *abundance*. This is possible by modifying the amino acid adjacency matrix and using the sequential labels of amino acids as input so that the resulting matrix has *full* information on protein sequences. The so-modified AAA matrix allows, if required, reconstruction of the protein sequence. Observe that now the matrix elements are not numbers, but sets of numbers, the special case of which are numbers that can also be viewed as sets, which occur when amino acid i is followed by amino acid j in the protein sequence only once.

Unexpectedly, by recovering the lost information of the AAA matrix, one arrives at a *tool* to solve the protein alignment problem exactly. This means finding the alignment of amino acids *without* the use of approximations, such as penalties for gaps, deletion and insertions of amino acids, use of statistical information, and *without* use of the trial and error search for alignment of individual amino acids. In order to align two proteins, it suffices to *superimpose* the sequential AAA matrices of two proteins. From such superposition, one can immediately identify amino acid entries in two proteins that differ by 0, ±1, ±2, ±3, ±4, etc., in their sequential labels. This procedure is not only exact but also highly efficient. Because the algorithm is very efficient, it was named VESPA, an acronym for "very efficient search for protein alignment."

Here we report on a computer program for the Randić algorithm, prepared by the second author [36] of the report on comparison of the exact versus approximate protein alignment [37]. The program VESPA was written in the computer language Python, which is an open source script language that runs on most platforms. This time, the acronym VESPA stands for "very exact solution of the protein alignment." The program VESPA can be freely downloaded from the Internet.

Amino acid adjacency matrices, which in addition to the information on abundance of individual amino acids also give the count of successive pairs of amino acids, are accompanied by some loss of information, in the view that information on the location of various pairs is not known. A remedy for this loss of information is to record the locations of pairs of adjacent amino acids rather than just their occurrence. In this way, the AAA matrix of Matrix 13.13 has a new form shown in Matrix 13.4, in which we show only the portion of the matrix involving elements G–V. The modified AAA matrix shown in Matrix 13.5 is associated with the sequence of the first 45 amino acids of the human ND6 protein. Observe the elements (G, L) and (G, F) that have in the corresponding AAA matrix the abundance count 2, are now replaced by two sequential numbers. Although many matrix elements of the modified AAA matrix may remain to be single numbers, in general, the elements of the modified AAA matrix are not numbers, but sets, that is, collections of numbers. In Matrix 13.5, we show the same portion of the AAA matrix for the second protein of Table 13.10.

	G	H	I	L	K	M	F	P	S	T	W	Y	V
G	28			11,29			15,18						36
H													
I			42	43								26	33
L			32	7		5			8				12,30
K								22					
M	14					1					2		
F				6					19				16
P			25						23				
S	35				21			24	20				9
T													
W													
Y	27												
V	10,17,38		41	31	13				34				37

(13.4)

$$(13.5)$$

	G	H	I	L	K	M	F	P	S	T	W	Y	V
G	28			18,11,29			36						
H													
I							5					26	33
L	17,30,44		32		21	41	11		8				13
K								22					
M													42
F				12									6,37
P			25						23				
S	35			10				24	9				
T													
W													
Y	27		4										
V	10,17,38			7,43		13			34				

By recording the locations of the adjacencies of amino acids it immediately became clear that a superposition of two such sequential AA matrices, shown in Matrix 13.6 (for regions G–V) will give information on locations of *adjacent pairs*, and thus, one will be able to align corresponding segments of protein. The solution to the protein alignment problem was found by an unexpected realization that the solution to another, much lesser problem, that of recovering the *lost* information of AAA matrices, suddenly offered a tool for finding the solution of a far more important problem, that of solving exactly the protein alignment problem.

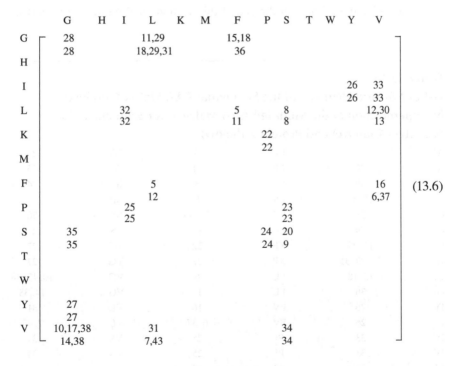

$$(13.6)$$

	G	H	I	L	K	M	F	P	S	T	W	Y	V
G	28 / 28			11,29 / 18,29,31			15,18 / 36						
H													
I												26 / 26	33 / 33
L			32 / 32				5 / 11		8 / 8				12,30 / 13
K								22 / 22					
M													
F				5 / 12									16 / 6,37
P			25 / 25						23 / 23				
S	35 / 35							24 / 24	20 / 9				
T													
W													
Y	27 / 27												
V	10,17,38 / 14,38			31 / 7,43					34 / 34				

Let us look more closely into Matrix 13.6. First, observe that of the more than 50 pairs that occur in this portion of the sequential AAA matrix there are only 16 amino acid pairs in which the sequential numbers of two proteins are very close numbers. In this illustration, the listed matrix elements below differ by zero. It is easy to list and order these pairs by inspection of the table:

(8, 8) (22, 22) (23, 23) (24, 24) (25, 25) (26, 26) (27, 27) (28, 28) (29, 29)
(32, 32) (33, 33) (34, 34) (35, 35) (38, 38).

This shows that one can align the segments (1–45), of two proteins with amino acid gaps at the locations 8, 29 and 35. The listed amino acids constitute just about half of the amino acids for the segments considered, and the other half are of no interest when searching for protein alignment. This significantly contributes to the efficiency of the search for the solution because no time was wasted on amino acids that cannot be paired. Simply stated, all amino acids not listed above as being paired can be ignored, which has been done in Matrix 13.6. In Table 13.10, we have listed all the remaining pairings of amino acids of the two proteins in the interval 1–45 merely to show that one need not manipulate the sequential amino acid matrices but only the corresponding nonzero matrix elements.

From Table 13.11, when extended to all 172 amino acids, one can order all pairings of amino acids into groups having differences −3, −1, 0, +1, +2, +3, and +4 as is shown in Table 13.12. This completes the search for the alignments of protein 1 and protein 2. In this particular case, we have not found any paired amino acids for the difference −4 and −2. Later, at the end of this section on alignment of proteins, we

---

TABLE 13.10

List of Nonzero Elements of the Sequential AAA Matrix Obtained by Superposition of the Modified AAA Matrices for Segment 1–45 of Protein 1 (above) and Protein 2 (below)

AL	4	LF	5	SG	35
AL	20	LF	11	SG	35
GC	39	LS	8	SP	24
GC	15, 39	LS	8	SP	24
GG	28	LV	12, 30	SS	20
GG	29	LV	13	SS	9
GL	11, 29	KP	22	YG	27
GL	18, 29, 31	KP	22	YG	27
GF	15, 18	FL	6	VG	10, 17, 38
GF	36	FL	12	VG	14, 38
IY	26	FV	16	VL	31
IY	26	FV	6, 37	VL	7, 43
IV	33	PI	25	VS	34
IV	33	PI	25	VS	34
LI	32	PS	23		
LI	32	PS	23		

TABLE 13.11

List of Sequential Labels of Adjacent Pairs of Amino Acids for Human Protein of Table 13.9 That Also Occur in the Protein of Mouse at Locations That Differ by 0, ±1, ±2, ±3, ±4, in All 64 Pairs

–4
–3 [17]
–2
–1 [48, 93, 127, 140, 154, 157, 160, 168, 169]
0 [8, 22, 23, 24, 25, 26, 27, 28, 29, 32, 33, 34, 35, 38, 39, 46, 47, 52, 53, 54, 55, 56, 57, 58, 59, 60, 61, 62, 65, 66, 67, 68, 69, 70, 71, 72, 73, 76, 77, 78, 79, 82, 83, 98, 99, 109, 153]
1 [12, 92]
2 [29, 103, 111]
3 [89]
4 [10]

TABLE 13.12

Output of Alignment of the Human and Mouse Proteins of Table 13.9 Based on Considering Common Pairs of Amino Acids in Two Proteins

```
MMYALFLLSVG LVMGFVGFSSKPSPIYGGLVLIVSGVVGCVI ILNFGGGYMGLMVFLIYLGGMMVVF

   ||    ||        ||||||||| ||||| |||      |||    |||||||||||| |||

MNNYIFVLSSLFLV GCLGLALKPSPIYGGLGLIVSGFVGCLMVLGFGGSFLGLMVFLIYLGGMLVVF


GYTTAMAIEEYPEAWGSGVEVLVSVLVG LAMEVGF  VLWVKEYD  GVVVVVNFNSVGSWMIYEGE

||||||| ||||| |||        ||   |||   ||   || ||            ||

GYTTAMATEEYPETWGSNWLILGF LVLGVIMEVFLICVLNY  YDEVGVI    NLDGLGDWLMYEVD


GSGFIREDPIGAGALYDYGRWLVVVTGWPLFVGVYIVIEIARGN

   ||        || || ||      |||

DVGVMLEGGIGVAAMYSCATWMMVVAGWSLFAGIFIIIEITRD
```

Source: Reproduced from Protein alignment: Exact versus approximate. An illustration, *J. Comput. Chem.* © (2015) Wiley, New York.

will, using spectral representations of proteins, show the pictorial representations of graphical alignment of the two proteins of Table 13.7.

The manuscript on the exact solution of the protein alignment problem [34] does not mention "the exact solution" neither in the title, nor in the abstract. In mathematics, a solution obtained without the use of empirical parameters, approximations, or approximate methods (including the trial-and-error method) is *exact*, regardless of the title or abstract outlining its construction. The title of the manuscript on the exact solution of the problem of protein alignment has already been mentioned: *Very efficient search for protein alignment—VESPA*. The acronym VESPA was added for two reasons: (i) In a hope that eventually somebody may be interested in preparing a computer program to facilitate use of VESPA among chemists and biologists and (ii) as a mnemonic for simplicity and elegance, which one may say on one side characterizes the outlined "very efficient search for protein alignment" but can also stand for "very elegant search for protein alignment." Vespa is a well-known "elegant" Italian scooter, which allows users to be well dressed when using it. Finally, *vespa*, in Italian language means *wasp*. This may be a reminder that if not using, or misusing, VESPA, one may be hurt! However, more seriously, the title *Very Efficient Search for Protein Alignment* was selected, knowing that it is more important to announce the *efficiency* of solving the protein alignment problem than to advertise that it represents the "exact" solution, which careful readers should recognize on their own.

We have referred to Matrices 13.4 and 13.5 as a sequential AAA matrix because their entries are the sequential labels of amino acids in proteins. These matrices have all the information that is contained in the corresponding AAA matrices, but in addition, they also include information on the locations of amino acids in the corresponding sequences. The elements of such matrices, as already mentioned, are sets, not numbers (sets, of course, can have a single element, including zero, which are numbers). Whether the sequential AAA matrices here outlined are the first matrices in mathematical literature whose elements are sets, we don't know. That is less important, in our view, than the fact that they are the *first* matrices that offered the exact solution to the over 45-year-old open problem on pairwise protein alignment. Observe that they have not only information on individual amino acids but on all *pairs of adjacent* amino acids.

In practical calculations, one need not be concerned with construction of AA sequential matrices as a matrix. It will suffice just to list only the nonzero matrix entries of 400 possible matrix elements, similar to the short list of Table 13.10.

In this outline on the exact solution of the alignment problem, we were deliberately dwelling on some details in order to illustrate the origin of the great efficiency of the algorithm. As we have seen, the solution focuses on *common pairs* of amino acids in two proteins. It is not difficult to modify the same approach for considering DNA alignments as has been outlined in a paper titled Very Efficient Search for Nucleotide Alignments [19].

13.11 COMPUTER PROGRAM FOR THE EXACT SOLUTION OF THE PROTEIN ALIGNMENT PROBLEM

The program by Pisanski [36,37], titled: *Protein Alignment: Implementation of Randic Algorithm VESPA*, is shown in Appendix 16. The program is written in computer language Python. Without the input information, it counts 42 lines (about one page) and an additional 12 lines for "print" commands. In the Appendix 17, we have listed an input sample of proteins having about 170 amino acids, and Table 13.11 shows the output of the program VESPA, which lists differences in sequential amino acids of the two proteins considered. The listed numbers are the amino acid sequential labels of the first protein and entries [ ], showing no aligned amino acid pairs (the case of differences −4 and −2) have been omitted in Table 13.12.

The exact solution has found 57 pairs of adjacent amino acids or 77 single amino acids, shown in Table 13.12. In Table 13.13 are 18 added alignments of *single* amino acids in two proteins to those of Table 13.12, obtaining total of 95 aligned amino acids.

In Table 13.14, we show the output of BLAST computer software for the same two proteins of human and mouse, which starts with amino acid 5 and ends with amino acid 172. The middle line shows the aligned amino acids, which give 51 aligned pairs, giving 66 single amino acids, and when 22 *single* aligned amino acids are added, one obtains the total number of aligned amino acids to be 89. The difference of six amino acids arose due to failure of BLAST to identify six adjacent pairs of amino acids.

Users may consider and select alternative versions of the output of the program that Pisanski has prepared:

1. *Full* program (Appendix 16), which prints as output (i) the sequential numbers of all amino acid adjacent pairs for the pair of proteins considered, and (ii) the sequential numbers for amino acids having the same sequential shift (differences in sequential locations).
2. *Basic* program, which prints the sequential numbers for amino acids having the same sequential shift.

TABLE 13.13

Alignment of the Human and Mouse Proteins of Table 13.12 to Which Are Added Alignments of Single Amino Acids in Two Proteins

```
MMYALFLLSVG LVMGFVGFSSKPSPIYGGLVLIVSGVVGCVI ILNFGGGYMGLMVFLIYLGGMMVVF

|     | ||   || |  |    |||||||||| ||||| |||   |  |||    |||||||||||| |||

MNNYIFVLSSLFLV GCLGLALKPSPIYGGLGLIVSGFVGCLMVLGFGGSFLGLMVFLIYLGGMLVVF


GYTTAMAIEEYPEAWGSGVEVLVSVLV GLAMEVGF  VLWVKEYD  GVVVVVNFNSVGSWMIYEGE

|||||||| |||||| |||    |   || |  |||    ||    || ||    |    | | ||

GYTTAMATEEYPETWGSNWLILGF LVLGVIMEVFLICVLNY  YDEVGVI   NLDGLGDWLMYEVD


GSGFIREDPIGAGALYDYGRWLVVVTGWPLFVGVYIVIEIARGN

|   |  ||  | |   |  | || || || |  | ||| |

DVGVMLEGGIGVAAMYSCATWMMVVAGWSLFAGIFIIIEITRD
```

Source: Reproduced from Protein alignment: Exact versus approximate. An illustration, *J. Comput. Chem.* © (2015) Wiley, New York.

TABLE 13.14
Output of BLAST for Pairwise Alignment of Human and Mouse Proteins of Table 13.9

```
Human   5   LFLLSVGLVMGFVGFSSKPSPIYGGLVLIVSGVVGCVIILNFGGGYMGLMVFLIYLGGMM   64
            +F+LS   ++G +G + KPSPIYGGL LIVSG VGC+++L FGG ++GLMVFLIYLGGM+
Mouse   5   IFVLSSLFLVGCLGLALKPSPIYGGLGLIVSGFVGCLMVLGFGGSFLGLMVFLIYLGGML   64

Human  65   VVFGYTTAMAIEEYPEAWGSGVEVLVSVLVGLAMEVGFVLWVKEYDGVVVVNFNSVGSW   124
            VVFGYTTAMA EEYPE WGS   +L  +++G+ MEV F++ V  Y   V V+N + +G W
Mouse  65   VVFGYTTAMATEEYPETWGSNWLILGFLVLGVIMEV-FLICVLNYYDEVGVINLDGLGDW   123

Human 125   MIYEGEGSGFIREDPIGAGALYDYGRWLVVVTGWPLFVGVYIVIEIAR   172
            ++YE +  G + E  IG  A+Y    W++VV GW LF G++I+IEI R
Mouse 124   LMYEVDDVGVMLEGGIGVAAMYSCATWMMVVAGWSLFAGIFIIIEITR   171
```

Source: Reproduced from Protein alignment: Exact versus approximate. An illustration, *J. Comput. Chem.* © (2015) Wiley, New York.

3. *Common Fragment* program, which searches for a common given fragment in two sequences.
4. *Single Protein* program, which considers in isolation a *single* protein and searches for repetition of the same fragments of amino acids in the protein.

Here we will briefly comment only on the *Basic* program, the output of which is the sequential numbers for amino acids having the same shift. For more information on the basic program and other versions, one should consult [37]. Let us return to Table 13.11, the *exact* solution for the alignment human–mouse proteins of Matrix 13.3. We would like to emphasize three things:

1. Table 13.11 is the *exact* solution, which means that no approximations of any kind and no empirical parameters or statistical information have been used.
2. The solution is *unique* in the sense that there are no additional (different) solutions.
3. There may be several different *representations* of the solution, one of which has been is illustrated in Tables 13.12 and 13.13.

Although it is easy to comprehend points (i) and (ii), point (iii) requires some attention because it appears that the *representations* of the *solution* of protein alignments have been often presented as *solutions*. This in itself need not be confusing when discussing similarities and differences among proteins, but from a *conceptual point of view*, there is this *significant* difference in presenting as the "solutions" of the protein alignments results reported by various computer programs and the results reported using the VESPA program (by Pisanski). The former as a solution offers *one* of the possible representations when there is more than one possibility or the *only* possible representation. The latter reports the solution and lets researchers consider and construct one or more representations (when possible) of the exact solution. One should be aware that for any pairwise protein alignment there is *only one solution* (in human–mouse, the alignment differences are shown in Table 13.11), but there could be more than one representation of the solution.

There is here some distant analogy with graphs. Graphs can be represented in different ways (such as the two representations of the Petersen graph of Figure 2.18 in Chapter 2), but this is the same graph (defined by the same list of neighbors). In the cases of graphs, the situation is further complicated by the fact that for a graph with *n* vertices there are *n!* ways to choose labels. In the case of proteins and DNA, the sequential labels are unique.

Generally, once the *solution* of alignment of two proteins (Table 13.11) is known, it is not difficult to construct the *representation* shown in Table 13.12. It may therefore be expected that the reverse will also be possible and that from a given representation one should be able to identify sequential labels of adjacent amino acid pairs and thus construct the Table of Differences (analogous to Table 14.18), which represents the genuine solution to protein alignments. Let us hope to see such tables of sequential labels having common differences for pairs of adjacent amino acids—but for the past 45 years, apparently users of computer programs have not been reporting the solutions to the alignment but only their representations. This may be, from a practical point of view, more desirable and more useful but may be accompanied with some loss of information and the possibility of not detecting the existence of possible alternative representations having the same number of aligned amino acids.

We are not aware of published results of the pairwise protein alignment problem that explicitly list shifts for amino acids pairs, such as we report in Table 13.11. It is not likely that in all situations from a single representation of the solution the complete list of shifts for all amino acids pairs in two proteins can be fully reconstructed. Thus, the "sleeping giant," the report on the exact solution of the pairwise protein alignment problem [34], does not only report the first exact solution to the protein alignment problem, but may also be the first *complete* report on protein alignments, which is not accompanied by any loss of information. The "loss of information" here relates to possibility of having additional very similar alignments, like the *two* alignments having the same number of pairings for the case of Table 13.11. They differ in two locations from the one shown in Table 13.12. Instead of a pair of LV amino acids at locations 93, 94, the other representation has the pair VL at locations 92, 93. Observe that the protein alignment problem has a unique solution (shown in Table 13.11) but two essentially distinct "user friendly" representations having the same number of aligned amino acids! We are not aware if similar cases have been detected, discussed, and reported separately earlier in published pairwise alignment reports because the literature on protein alignments is colossal, but it is possible that they have been seen even more often that one may expect.

Let us briefly mention a comparison of the exact approach to protein alignment and an approximate approach, for which we selected BLAST [32,33]. What follows has been presented in a paper: Protein Alignment: Exact versus Approximate. An Illustration [36]. The manuscript was submitted to the *Proceedings of the National Academy of Sciences (PNAS)*, on the recommendation of R. Hoffmann, but it was declined to be processed, just as its predecessor: "Very Efficient Search for Protein Alignments—VESPA," [34], submitted to *PNAS* also on the recommendation of R. Hoffmann, in which was reported the exact solution to the protein alignment problem—a problem solved exactly after 45 years waiting! *PNAS* has in the past published papers on the development of *approximate* approaches to protein alignment (e.g., see ref. [31]) but declined to consider the *exact* approach on unconvincing grounds that the paper is too technical.

The user-friendly representations of Table 13.13 have been constructed by first identifying pairs of amino acids having the same sequential numbers (zero shift), followed by locating shift −1 (at positions 93/92 LV, 127/126 YE, 140/139 IG, 157/156 GW, 160/159 LF, 168/167 IE, and 169/168 EI). Observe that shift 48/47 GG of Table 13.11 has not been considered because GG at place 47 has already been taken into account as zero shift. One continues with shifts +1 (12/13 LV, 92/93 VL, and 153/154 VV), and so on. When all alignments have been considered, one finds 57 alignments as shown in Table 13.12. After constructing a user-friendly representation of the alignments of two proteins based on the output of the VESPA algorithm, one can add overlapping single amino acids, starting with M at position 1 and ending with R at the terminal part of the proteins (position 172). In doing this, one should not introduce new gaps, but the existing gaps can be shifted in their neighborhood to optimize overlap of isolated amino acids. As one can see from Table 13.14, in this way, we have introduced additional 18 *single* amino acid alignments. The single amino acid alignments, if adjacent to multiple amino acid alignments, may be viewed as possibly significant; otherwise, they may be accidental and not biologically relevant unless adjacent to pairs of common amino acids for both proteins.

A close look at Table 13.11 shows that significant overlap of amino acids of human and mouse ND6 proteins occurs mostly in regions 21 to 82 for difference 0 (no shift of proteins). Spectral representations of proteins are useful for visualizing protein alignments. In Figure 13.7, we show spectral representation of the human (top) and mouse (bottom) ND6 proteins.

Looking very closely, one can find a few segments that show local similarity, but overall, spectral representations are not necessarily viewer-friendly. However, because they carry numerical values, such representations can be subtracted, and the difference can be plotted as is shown in Figure 13.8.

Observe in the regions around 20 to 80 many spots having the value y = 0, the points being on the x-axis, which means that at these sites are the same amino acids in both proteins.

The advantage of the spectral representations is even better illustrated by considering graphically the protein alignment for the two proteins of *Saccharomyces cerevisiae* shown in Table 13.7, selected for the outline of the VESPA algorithm. Pictorial representation of this table is shown in Figures 13.9 through 13.11. In Figure 13.9, we show spectral representations of the protein 1 and protein 2 of Table 13.7.

This figure is obtained by assigning numerical values 1–20 to the 20 amino acids A–V. Each amino acid is represented by a point with coordinates (x, y), where x is the sequential position of amino acids and y is a number between 1 and 20, assigning 1 to alanine A, 2 to arginine R, 3 to asparagine N, and so on, ending with 20 assigned to valine V. As one can see, spectral representation offers visual comparison of two proteins. A close look at Figure 13.9 shows some similarity between the two proteins, which are better seen in Figure 13.9 than Table 13.7. As mentioned before, the spectral graphical representations of proteins (and DNA) have an important advantage over other graphical and tabular representations of proteins and DNA. Here we will consider shifting two proteins and subtracting such representation and, in this way, find regions in which two proteins have the same or similar segments of amino acids.

In Figure 13.10, we show the difference plot of the spectral representations of Figure 13.8. Except for a few amino acids around position 15, which appear to be identical, there is no overlap in amino acid location for the remaining part of the proteins. However, if we shift the two proteins by one position and construct the difference, we obtain the top part of Figure 13.11, respectively. As one can see, the difference plot of the spectral representations of the upper part of Figure 13.11, when the two protein sequences have been shifted by one site, one relative to another, shows visible overlapping of numerous amino acid sequences at the interval 22–98. In the interval of 76 consecutive locations, 62 amino acids can be aligned. When the two sequences are shifted for two locations, one obtains the lower part of Figure 13.11 in which, at locations 103–120, there is full alignment of a fragment of amino acids of the two proteins. By increasing the relative shifts to three and four locations,

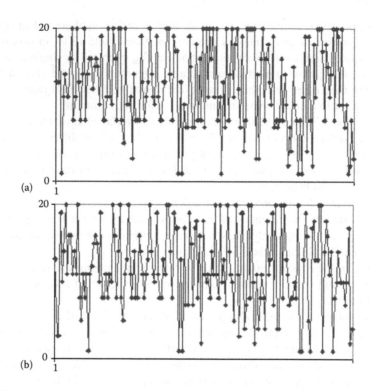

FIGURE 13.7   Spectral representation of the human (a) and mouse (b) ND6 proteins.

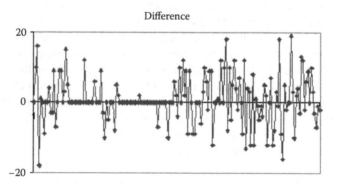

FIGURE 13.8   The difference plot of the spectral representations of human and mouse ND6 proteins.

respectively, we obtain similar graphs showing spectral differences of the two protein sequences around positions 160–170 and positions 130–145 shown in Figure 13.12.

We could continue but have so far found most of the overlapping segments of the two proteins. In addition to the above, one finds for zero difference five overlaps at sites 13–15, 34, 35, and 57, 58. This totals to alignment of 105 amino acids, not counting single amino acid *isolated* overlaps. The same two proteins when considered by

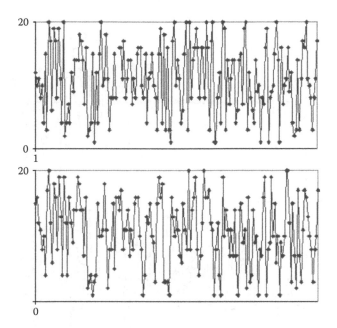

FIGURE 13.9 Spectral representation of the protein 1 and protein 2 of Table 13.7. (Reproduced from On a geometry-based approach to protein sequence alignment, *J. Math. Chem.* © (2008) Springer Verlag, Germany.)

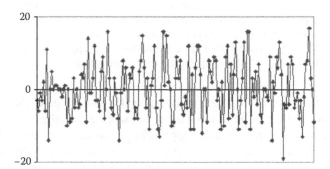

FIGURE 13.10 The difference plot of the spectral representations of Figure 13.9. Except for few amino acids at positions 13–15, 33, 34, and 57, 58, which are identical, there is no overlap in amino acid location for the remaining part of the proteins. (Reproduced from On a geometry-based approach to protein sequence alignment, *J. Math. Chem.* © (2008) Springer Verlag, Germany.)

BLAST have 115 amino acid overlaps, the count of which includes 13 single amino acid alignments, leaving a comparable count of 102 versus 105. Without more careful analysis, we cannot include single amino acid *isolated* overlaps because such may count the same amino acids several times as sequences are shifted. If one includes the count of single amino acid *isolated* overlaps, our count would give 145, suggesting 30 duplicate counts.

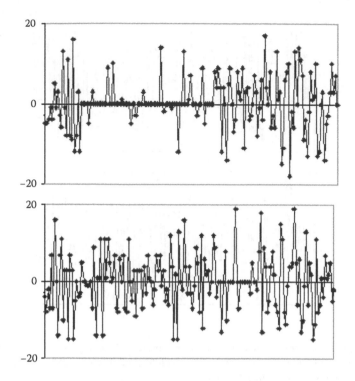

FIGURE 13.11   The difference plot of the spectral representations of Figure 13.9 when the two protein sequences have been shifted by one and two sites, respectively, one relative to another. Observe overlapping fragments of amino acid sequences around positions 22–98 and positions 103–120. (Reproduced from On a geometry-based approach to protein sequence alignment, *J. Math. Chem.* © (2008) Springer Verlag, Germany.)

   Our point here is to demonstrate that graphical alignment gives the results comparable to those of BLAST but without the use of scoring and other empirical parameters that are part of computer-based approaches to protein alignment. When using a graphical approach, if one would in advance know how much to shift two protein sequences, instead of using a trial-and-error approach (what many computer programs also do), one would illustrate graphical representations of one of the representations of the exact method. What the VESPA algorithm offers is finding in advance which relative shifts offer the exact solution. VESPA tells the user how much which amino acid has to be shifted in order to align two proteins. This is accomplished by examination of matrix elements of the sequential amino acid adjacency matrix.

   With this, we have ended our review of solved and unsolved problems in structural chemistry, which we hope has brought to the attention of readers not only a number of problems that deserve further consideration, but also a number of surprises, unexpected novel directions for exploring the structure and properties of molecules, including DNA, RNA, and proteins, and analysis of proteome. In the next chapter, we will summarize some aspects of this journey to solved and unsolved

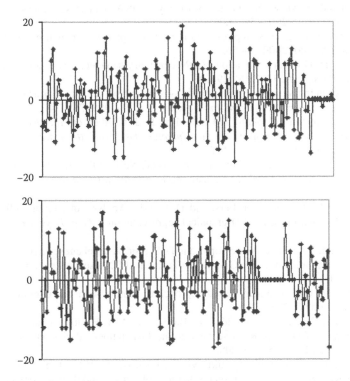

FIGURE 13.12 The difference plot of the spectral representations of Figure 13.9 when the two protein sequences have been shifted by three and four sites, respectively, one relative to another. Observe overlapping fragments of amino acid sequences around positions 160–170 and positions 130–145. (Reproduced from On a geometry-based approach to protein sequence alignment, *J. Math. Chem.* © (2008) Springer Verlag, Germany.)

problems with emphasis on the beauty and the overlooked beauty of the discussed problems in this book.

REFERENCES AND NOTES

1. E. Hamori and J. Ruskin, H curves, a novel method of representation of nucleotide series especially suited for long DNA sequences, *J. Biol. Chem.* 258 (1983) 1318–1327.
2. M. Randić, J. Zupan, A. T. Balaban, D. Vikić-Topić, and D. Plavšić, Graphical representation of proteins, *Chem. Rev.* 111 (2011) 790–862.
3. M. Randić, M. Novič, and D. Plavšić, Milestonmes in graphical bioinformatics, *Int. J. Quantum Chem.* 113 (2013) 2413–2446.
4. M. Randić, M. Vračko, A. Nandy, and S. C. Basak, On 3-D graphical representation of DNA primary sequences and their numerical characterization, *J. Chem. Inf. Comput. Sci.* 40 (2000) 1235–1244.
5. M. Randić, B. Horvat, T. Pisanski, G. Jaklič, and D. Plavšić, On map representation of DNA, *Croat. Chem. Acta* 86 (2013) 519–529.

6. M. Randić and M. Vračko, On the similarity of DNA primary sequences, *J. Chem. Inf. Comput. Sci.* 40 (2000) 599–606.
7. M. Randić, Condensed representation of DNA primary sequences, *J. Chem. Inf. Comput. Sci.* 40 (2000) 50–56.
8. M. Randić, On characterization of DNA primary sequences by a condensed matrix, *Chem. Phys. Lett.* 317 (2000) 29–34.
9. C. Raychaudhury and A. Nandy, Indexing scheme and similarity measures for macromolecular sequences, *J. Chem. Inf. Comput. Sci.* 39 (1999) 243–247.
10. H. A. Hauptman (as told to and Edited by D. J. Grothe), *On the Beauty of Science (A Nobel Laureate Reflects on the Universe, God, and the Nature of Discovery),* Prometheus Books, Amherst, NY.
11. M. Randić, M. Vračko, N. Lerš, and D. Plavšić, Novel 2-D graphical representation of DNA sequences and their numerical characterization, *Chem. Phys. Lett.* 371 (2003) 1–6.
12. M. Randić, M. Vračko, N. Lerš, and D. Plavšić, Analysis of similarity/dissimilarity of DNA based on novel 2-D graphical representation, *Chem. Phys. Lett.* 371 (2003) 202–207.
13. M. Randić, J. Zupan, D. Vikić-Topić, and D. Plavšić, A novel unexpected use of a graphical representation of DNA: Graphical alignment of DNA sequences, *Chem. Phys. Lett.* 431 (2006) 375–379.
14. M. Randić, On a geometry-based approach to protein sequence alignment, *J. Math. Chem.* 43 (2008) 756–772.
15. M. Randić, N. Lerš, D. Plavšić, S. C. Basak, and A. T. Balaban, Four-color map representation of DNA or RNA sequences and their numerical characterization, *Chem. Phys. Lett.* 407 (2005) 205–208.
16. M. Randić, K. Mehulić, D. Vukičević, T. Pisanski, D. Vikić-Topić, and D. Plavšić, Graphical representation of proteins as four-color maps and their numerical characterization, *J. Mol. Graph. Model.* 27 (2009) 637–641.
17. M. Kunz and Z. Rádl, Distribution of distances in information strings, *J. Chem. Inf. Comput. Sci.* 38 (1998) 374–378.
18. B. Horvat, G. Jaklič, I. Kavkler, and M. Randić, Rank of Hadamard powers of Euclidean distance matrices, *J. Math. Chem.* 52 (2014) 729–740.
19. M. Randić, Very efficient search for nucleotide alignments, *J. Comput. Chem.* 34 (2013) 77–82.
20. G. Agüero-Chapin, R. Molina-Ruiz, E. Maldonado, G. de la Riva, A. Sanchez-Rodrigues, V. Vasconcelos and A. Antunes, Exploring the Adenylation Domain Repertoire of Nonribosomal Peptide Synthetases Using an Ensemble of Sequence-Search Methods, *PLoS ONE* www.plosone.org, July 2013, vol. 8, issue 7, pp. 1–13, e65926.
21. C. P. Snow, *The Two Cultures*, Cambridge University Press, London, 2001.
22. http://en.wikipedia.org/wiki/cross-cultural_communication, August 2014.
23. M. Randić, On the history of the Randić index and emerging hostility towards chemical graph theory, *MATCH—Commun. Math. Comput. Chem.* 59 (2008) 5–124.
24. M. Randić, M. Novič, and M. Vračko, On novel representation of proteins based on amino acid adjacency matrix, *SAR QSAR Environ. Res.* 19 (2008) 339–349.
25. A. Roy Choudhury, N. Zhukov, and M. Novič, Mathematical characterization of protein transmembrane regions, *Sci. World J.* 2013 (2013). DOI.10.1155/2013/607830.
26. A. El-Lakkani and S. El-Sherif, Similarity analysis of protein sequences based on 2D and 3D amino acid adjacency matrices, *Chem. Phys. Lett.* 590 (2013) 192–195.
27. R. Chenna, H. Sugawara, T. Koike, R. Lopez, T. Gibson, D. Higgins, and D. Julie, Multiple sequence alignment with the Clustal series of programs, *Nucl. Acid Res.* 31 (2003) 3497–3500.

28. S. Needleman and C. D. Wunsch, A general method applicable to the search for similarities in the amino acid sequence of two proteins, *J. Mol. Biol.* 48 (1970) 443–453.
29. T. F. Smith and M. S. Waterman, Identification of common molecular subsequences, *J. Mol. Biol.* 147 (1981) 195–197.
30. D. J. Lipman and W. R. Pearson, Rapid and sensitive protein similarity searches, *Science* 227 (1985) 1435–1441.
31. W. R. Pearson and D. J. Lipman, Improved tools for biological sequence comparison, *Proc. Natl. Acad. Sci. U.S.A.* 85 (1988) 2444–2448.
32. S. Altschul, W. Gish, W. Miller, E. Myers, and D. Lipman, Basic local alignment search tool, *J. Mol. Biol.* 215 (1990) 403–410.
33. S. F. Altschul, T. L. Madden, A. A. Schaffer, J. Zhang, Z. Zhang, W. Miller, and D. J. Lipman, Gapped BLAST and PSI-BLAST: A new generation of protein database search programs, *Nucleic Acid Res.* 25 (1997) 3389–3402.
34. M. Randić, Very efficient search for protein alignments—VESPA, *J. Comput. Chem.* 33 (2012) 702–707.
35. M. Randić, A. F. Kleiner and L. M. DeAlba, Distance/Distance matrices, *J. Chem. Inf. Comput. Sci.* 34 (1994) 277–286.
36. T. Pisanski, University of Ljubljana, Ljubljana, and University Primorska, Kopar, Slovenia.
37. M. Randić and T. Pisanski, Proteins alignment: Exact versus approximate. An illustration *J. Comput. Chem.* 36 (2015) 1069–1074.

14 Beauties and Sleeping Beauties in Science

There have been in science known cases of important findings and results, which, after being published, received little or no attention and have been forgotten, sometimes for quite a long time. There have been in science also known cases of important findings and results, which, after being published, have not been noticed and have been since rediscovered—until eventually it was found that the results are not new. A famous illustration of the former is Gregor Mendel (1822–1884) who was a monk in the St. Augustine Monastery in Brno, Czech Republic (then Brünn, Austria-Hungary). There he first obtained a diploma in mathematics and continued his education in physics, chemistry, mathematics, zoology, and botany. Some 10 years later, in 1865, he started his work on the crossing of peas, which gave birth to genetics. He formulated his genetics laws, illustrated in Figure 14.1, which clearly illustrate the combinatorial properties of genetic laws.

Mendel presented his lecture *Versuche über Pflanzen-Hybriden* (*Experiments on Plant Hybridization*), at two meetings of the Natural History Society of Brünn in 1865, which were locally well received. In 1866, Mendel wrote his report, which he sent to one of the leading botanists of the time, Karl Wilhem von Nägeli, in Switzerland, who showed no interest. He published his paper *Versuche über Pflanzen-Hybriden* in the society's journal, *Verhandlungen des naturforschenden Vereines in Brünn*, in 1866 and sent 40 reprints of his work to various scholars in the world. By 1900, 16 years after Mendel died and 46 years after he started his work, which is the birth of genetics, Hugo de Vries of the Netherlands noticed *three* variations in primroses, and during the next 10 years, grew more than 50,000 plants, from which he was able to identify *eight* different mutation types. Again an illustration of combinatorics, a strong support for genetics, there being eight possible combinations of three symbols:

AAA AAB AAC ABB ABC BBC BCC CCC.

At the same time, Karl Correns in Germany and Erich von Tschermak in Austria also investigated mutations. In 1900, all three published their work, and all three papers cited the work of Mendel, which thus became fully recognized.

Not all unrecognized works are as fundamental as were the laws of Mendel and not all initially unrecognized works that are fundamental have been recognized by leading authorities as was the case with the tetrahedral carbon of Jacobus Henricus van't Hoff. There are enough unrecognized works of some significance and importance that remain overlooked and unrecognized for some time by most scientists even though such works may have received limited citations! When one comes across such papers that deserve more attention one should try to draw the attention of other chemists to such papers. We have been doing this in this book on several occasions and will continue to do so in this chapter.

Parents

AA - BB

↓

1st generation

AB AB AB AB

↓

2nd generation

AA AB BA BB

↓

3rd generation

↓

AA AA AA AA   AA AB BA BB   BB BA AB AA   BB BB BB BB

FIGURE 14.1 Mendel's genetic laws (illustration of combinatorics).

14.1 SLEEPING BEAUTIES REVISITED

> The man who does not read good books has no advantage over the man who cannot read them.
>
> **Mark Twain**

Citations of scientific papers may be useful in different ways to different people. Important as it may be to see some evidence that a particular publication of an author has been noticed, perhaps even beyond the author's expectation, it may be also of considerable interest to see that particular publications may have not been sufficiently recognized for their excellence as expected by the authors and a few who have been well aware of their merits. Many years ago, Eugene Garfield characterized papers reaching 1000 citations as "Classic" papers. He has also invited readers to inform him about influential work that, for a variety of reasons, may not be highly cited but deserve greater visibility. This amounts to a request to search for "Sleeping Beauties," which could be interpreted as papers that could have 1000 citations and more but have been far behind in the actual count of citations. Perhaps papers reaching 2500 citations, more than twice that of "Classic" papers, deserve a label of their own, and it appears to us that the label "Masterpiece" appears plausible. Classification and characterization of scientific papers as "classic," "masterpiece," important, significant, great, outstanding, fundamental, and so on, is not a necessity, but some publications deserve some distinction, and the mentioned attributes may serve some use, be they "Sleeping Beauties" or well recognized "Beauties."

Occasionally, one may come across a work that represents a giant step forward. "Giant," as a label for distinction of a scientific paper, we feel, should signify a contribution of great potential, such as papers that *open new directions* in research and work that opens new disciplines. A "Giant" work that has been hitherto overlooked accordingly can be referred to as a "Sleeping Giant." That Giant works that open new wide areas of scientific disciplines have occurred in recent times can be seen from the list in Table 14.1, in which we have collected works of science in general, which also have some potential interest in chemistry.

To be specific (and we speak of chemistry here), let us refer to papers having, or deserving to have, 5000 citations as "great" and papers having, or deserving to have, 10,000 citations as "outstanding." Presumably, there will be a few such publications that one can consider as giants among publications in any discipline. According to this, "Sleeping Giants" are publications that should, in different circumstances, be expected to reach such high number of citations but actually may have at best 50–100 or even fewer, which is 100 times fewer than they deserve. One such giant publication is BLAST [1,2], an acronym for "basic local alignment search tool," which may have earned the title of being one of the fundamental publications in biology and bioinformatics, not so much for its conceptual value but for its practical value.

Of course, one can speculate, be uncritical and not objective, when considering one's own work and contemplate whether this or that publication deserves the exquisite label of "Sleeping Giant." One may be right or wrong, and only time can tell if one was right or wrong! One should also keep in mind that "Giants" come in various sizes. Not all are as colossal as the Colossus of Rhodes!

Let's end this section with a comment on the two papers by D. Lipman and coworkers [1,2] already mentioned (BLAST), in which are described packages for computer search for protein alignments. There is no doubt that these two papers should be viewed as "Giant" among "Giants." The moment they appeared, they have been recognized by most, if not all, interested in protein alignments. This is attested

---

TABLE 14.1
Topics of a Selection of "Giants," Novelties That Opened New Directions of Theoretical Research

Topic	Year	Author
Statistics based on least residuals	1757	Joseph Boscovic
Partial ordering	1901	R. F. Muirhead
Hückel molecular orbitals	1930	E. Hückel
Linear programming	1947	G. B. Dantzig
Information theory	1948	C. E. Shannon
Gaussian calculations	1950	S. F. Boys
Quantitative structure–activity relationship	1961	Corwin Hansch
Density functional theory	1964	W. Kohn
		P. Hohenberg
Renormalization	1969	K. Wilson
Aromatic sextet theory	1970	E. Clar
Chaos game	1988	M. F. Barnsley

by the fact that jointly they are approaching 100,000 citations and have an annual increase in citations of approximately 5000!

We continue by considering Eugene Garfield's interest in finding publications that deserve more attention than they have been apparently receiving. Numbers may be misleading! Scientific papers that have received a dozen or fewer citations may be of little interest to many chemists and need not be unimportant on their own. Equally, if a paper has 100 or more citations, it does not mean necessarily that it has been noticed if a majority of chemists working in that area of chemistry are not aware of it, have not recognized its merits, or do not understand its significance. We will elaborate on several "Sleeping Beauties," as we will refer to such, which should deserve more attention from chemists, particularly among those working in structure–property and structure–activity studies.

14.2   ON BEAUTY IN SCIENCE

There have been numerous articles in scientific literature on beauty in science. We have mentioned in this book the book by Hauptman with such a title and a flock of papers on molecules considered to be beautiful, including dodecahedrane and buckminster-fullerene. There have been comments in scientific literature on "beautiful" molecules, which often have high symmetry, such as Platonic and Archimidean solids, which include the dodecahedrane (the Schlögl diagram of which is shown in Figure 14.2), and buckminsterfullerene (the Schlögl diagram of which is shown in Figure 14.3).

A beautiful illustration of buckminsterfullerene that displays cubic symmetry, constructed by T. Pisanski and M. Zaveršnik in Figure 14.4 can be found in Ref. [3]. In addition we also show a somewhat unusual representation of C_{60} displaying a Hamiltonian circuit in Figure 14.5. Observe that Schlögl diagrams can be viewed as graphs, so in addition to the beauty of molecules, one can also speak of the beauty of graphs. In Figure 14.6, we show a beautiful representation of "My Graph" from a recent publication by Coxeter [4], the leading mathematical figure in geometry. Beauty is more than high symmetry, and chemistry is more than the shape of molecules. As has been said, beauty is in the eyes of the beholder, and different beholders are likely to have different tastes, different preferences, different feelings, different

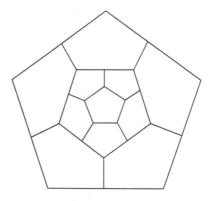

FIGURE 14.2   Schlögl diagram of dodecahedrane.

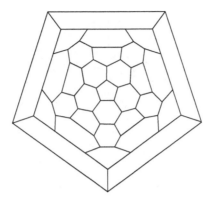

FIGURE 14.3 Schlögl diagram of Buckminsterfullerene C_{60}.

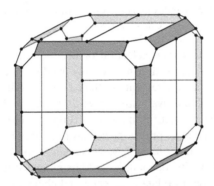

FIGURE 14.4 Buckminsterfullerene C_{60} displaying cubic symmetry as drawn by Pisanski and Zakrajšek (kindly sent by Professor T. Pisanski, April 2015).

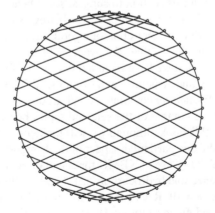

FIGURE 14.5 Buckminsterfullerene C_{60} displaying Hamiltonian circuit as drawn by Randić. (Reproduced from "On history of the Randić index and emerging hostility toward chemical graph theory," *MATCH Commun. Math. Comput. Chem.*, 2008.)

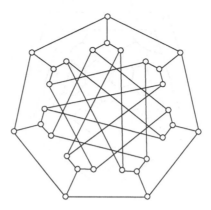

FIGURE 14.6 "The Coxeter graph as drawn by Milan Randić"—figure caption as published by H. S. M. Coxeter, *Congressus Numerantium* (Winipeg, Canada) 1981. (Reproduced from "On history of the Randić index and emerging hostility toward chemical graph theory," *MATCH Commun. Math. Comput. Chem.*, 2008.)

backgrounds, to expect that their selection of beauties in science and beauties in chemistry will be similar even if they may visibly overlap. Recently, accidentally, we came across a beautiful "advertisement" in the cafeteria in the famous Mayo Clinic, Rochester, Minnesota, which says, *"Love of beauty is taste. The creation of beauty is art."* The author of these beautiful thoughts is Ralph Waldo Emerson (1803–1882), an American essayist, lecturer, philosopher, and poet, one of the central figures of American intellectual life in 19th century. This sounds good and although it is hinting at food for the body by its location, it could be extended to include also food for the soul and, by extension, to relate to science and chemistry, especially if slightly modified to *"Love of beauty is good taste. The creation of beauty is great art."*

In this spirit, we wanted to be more specific when speaking of beauty in science by implying that beauty should have three main attributes: elegance, simplicity, and content. As we continue in this section to suggest some beautiful Sleeping Beauties and beautiful Sleeping Giants, we elaborate by listing two dozen illustrations of Beauties in Chemistry, which reflect upon our good taste—or lack of good taste—depending on who is judging. In Table 14.2, we have listed two dozen illustrations of beauty in chemistry, in which we have included two out of five "Sleeping Beauties" of the structure–property and structure–activity topic of Table 14.3, which are considered in this section. We neither elaborate on our arguments of the selections made nor are we to describe the content of the selected topics because most of the topics have been discussed in this book although we will make one exception.

Let us briefly describe one particular aspect of the Longuet-Higgins non-bonding molecular orbitals (NBMO)—use of NBMO for calculation of the number of Kekulé valence structures in benzenoid hydrocarbons [5]. The NBMO are molecular orbitals having zero eigenvalue ($\lambda = 0$). Recall the algorithm of Longuet-Higgins for determining the unnormalized eigenvector coefficients for a given eigenvalue λ_i, which has been mentioned before:

$$- \lambda_i c_i(r, s, t) + c_i(r) + c_i(s) + c_i(t) = 0.$$

TABLE 14.2
The Beauties of Chemistry

	Subject	Year	Author
1	Petersen graph	1898	J. Petersen
2	Octet rule	1925	G. N. Lewis
3	$4n + 2$ Hückel aromaticity rule	1930	E. Hückel
4	HMO	1930	E. Hückel
5	Pauling hybridization sp^n orbitals	1930	L. Pauling
6	GF matrix method	1941	E. Bright Wilson
7	Pauling bond orders	1947	L. Pauling
8	Longuet-Higgins non-bonding MO	1950	H. C. Longuet-Higgins
9	Frost-Musulin diagram	1953	A. A. Frost
			B. Musulin
10	Crystal field theory	1957	L. E. Orgel
			J. S. Griffith
11	Doering bullvalene	1963	W. Doering
12	Sachs theorem	1964	Sachs
13	Woodward-Hoffmann rules	1969	R. B. Woodward
			R. Hoffmann
14	Clar aromatic sextet	1970	E. Clar
15	Clarke's theorem	1971	F. H. Clarke
16	Conjugated circuits	1976	M. Randić
17	$4n + 2$ conjugated circuits aromaticity rule	1976	M. Randić
18	Gutman's theorem	1979	I. Gutman
19	Buckminsterfullerene	1985	R. F. Curl, Jr.
			H. Kroto
			R. E. Smalley
20	Simplex algorithm	1987	Dantzig
21	Orthogonal regression	1991	M. Randić
22	Matrix form for the connectivity index	2010	E. Estrada
23	Exact solution to protein alignment	2012	M. Randić
24	Ring bond order	2014	M. Randić
25	Numerical Clar sextet theory	2014	M. Randić

If $\lambda = 0$, the above reduces to

$$c_i(r) + c_i(s) + c_i(t) = 0,$$

that is, the sum of all adjacent coefficients for any carbon atoms is zero. Let us illustrate this by benzo[*ghi*]perylene, which is shown at Figure 14.7a, and in Figure 14.7b is shown its derivative obtained by adding an unsaturated exocyclic CC bond, which introduces into the system the zero eigenvalue $\lambda = 0$. The carbon atoms having zero coefficients in the NBMO are illustrated by adding small circles above them. We are interested in the nonzero coefficients of the accompanied NBMO, which are shown

TABLE 14.3

The "Sleeping Beauties" of Structure–Property and Structure–Activity Worlds

	Topic	Citations
1	$4n + 2$ and $4n$ rule for polycyclic conjugated systems	429
	Conjugated circuits	355
2	Orthogonal molecular descriptors	228
	Retro-regression	5
3	Variable connectivity index	96
	Search for active substructure	
4	Quantitative characterization of proteome maps	47
	Order from chaos	31
5	Graphical alignment of DNA	37
	Geometry based protein alignment	
	Beauty, but Not Sleeping	
1	The connectivity index $^1\chi$	2250
	Higher order connectivity indices	232

Note: In the last column are citations as of July 20, 2014.

in Figure 14.7c structure. In Figure 14.7d, we show how, with the help of the Longuet-Higgins sum rule: *The sum of all adjacent coefficients for any carbon atoms is zero*, the nonzero coefficients of NBMO have been found.

We start by assigning the value 1 to the most right carbon atom in the right bottom benzene ring. The sum rule allows one to find the coefficient of the remaining carbon atoms at the bottom row of carbon atoms. Now we consider carbon atoms in the row above. One can immediately find −1 for the most right carbon atoms, which, with the help of the sum rule, gives the values +2 and −3 for the remaining carbon atoms in this row. One can now continue and find the coefficients in the row above starting at left and getting +3 for the most left carbon. This leads to −5 and +6 for the remaining two carbon atoms, which allows one to find −3, +8, and 14 for the three top carbon atoms as can be seen at the top right of the diagram. As we have seen, finding the coefficients was simple, reflecting the simplicity and elegance of the procedure. But what about the content of this procedure?

Well, the absolute magnitude of the resulting coefficient, which is 14, gives the number of Kekulé valence structures of the parent structure, benzo[*ghi*]perylene. Thus, this procedure by Longuet-Higgins, according to our guidelines, qualifies as beautiful work. Even more, it qualifies as "good art" if not "great art" because it relates two "unconnected" theories, the molecular orbital theory, which is based on calculus, and the valence bond theory, which is based on enumeration of Kekulé valence structures.

Before continuing on the theme of beauty in science, we would like to encourage readers to search for their memory and chemical literature and make their own list of a dozen or two dozen beautiful papers, including their own, that, because of their beauty, they feel they deserve such recognition and wider attention. Recall the basic ingredients of beauty: elegance, simplicity, and content, not necessary in that order.

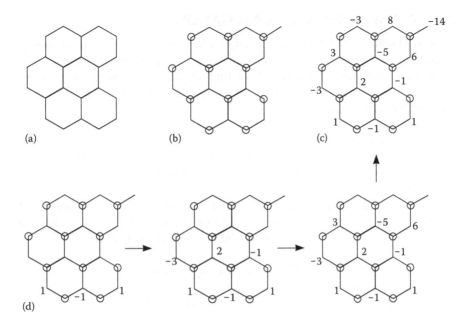

FIGURE 14.7 Approach of Longuet-Higgins for finding the number of Kekulé valence structures for benzenoid hydrocarbons using the coefficients of NMBO: (a) molecular graph of benzo[*ghi*]perylene; (b) locations of NBMO zero coefficients in derivative of benzo[*ghi*] perylene having acyclic C=C bond; (c) the non-zero coefficients of the NMBO of the derivative of benzo[*ghi*]perylene; (d) stepwise construction of the non-zero coefficients of the NMBO of the derivative of benzo[*ghi*]perylene using the zero sum rule.

Beauties deserve our attention, as has been well reflecting in beautiful thoughts of Frank Lloyd Wright (1867–1959), a well-known American architect, who wrote

If you ignore beauty you will soon find yourself without it. But if you invest in beauty, it will remain with you all the days of your life.

14.3 FIVE SLEEPING BEAUTIES

In Table 14.3, we have listed five topics that we will briefly review. The first is closely related to the notion of *aromaticity*, one of the central concepts of chemistry that appears to be elusive even though, as we will see, its conceptual clarification for the case of hydrocarbons was outlined over 35 years ago [6,7]. The second topic relates to the oldest statistical method, the method that has remained even today one of the most widely used tools for analysis of experimental data: the multivariate regression analysis (MRA) [8]. The third topic also relates to MRA, but rather than dealing with the method itself, it is concerned with construction and selection of molecular descriptors [9]. Next, we will consider one of the topics of bioinformatics: How to extract useful quantitative information from qualitative proteome maps [7]. Finally, we will address the topic of protein and DNA sequence alignments by describing, in contrast to the current computer manipulations of bio-sequences, an alternative: a non-empirical graphical approach to protein and DNA sequence alignments [8–10].

14.4 AROMATICITY AND CONJUGATION

Let us start with a comment on aromaticity—one of the most widely used concepts in chemistry and one that continues to cause confusion despite continuous efforts toward its "clarification." There is no doubt that clarification of aromaticity continues to attract chemists as is illustrated by numerous publications, including the following relatively recent publications:

What Is Aromaticity? [11]
Quantitative Measures of Aromaticity…and Their Interpretations [12]
To What Extent Can Aromaticity Be Defined Uniquely? [13]

Despite expectations that publications with such explicit references to aromaticity may clarify the situation, the concept of aromaticity remains, at best, ambiguous. Here are three dictionary explanations of the very word "ambiguous:"

1. Open to or having several possible meanings or interpretations
2. Doubtful or uncertain
3. Lacking clearness or definiteness; obscure; indistinct

What is not open to more than one possible meaning or interpretation, what is not doubtful or uncertain, what is not lacking clearness or definiteness, and is not obscure is that all three explanations of "ambiguous" apply to aromaticity. Why this is so?

Schleyer and Jiao [11] start their review on aromaticity by listing selected milestones that considered definitions and criteria for characterizing aromaticity the following:

Milestones on Aromaticity	
1865	Benzene structure (Kekulé)
1866	Substitution is more favorable than addition (Erlenmayer)
1910	Aromatic compounds have exalted diamagnetic susceptibility (Pascal)
1925	Electron sextet and heteroaromaticity (Armit-Robinson)
1931	Theory of cyclic $(4n + 2)$ π systems (Hückel)
1936	Ring current theory—free electron circulation around the benzene ring (Pauling)
1937	London diamagnetism—π electron current contribution to magnetic susceptibility
1956	Ring current effects on NMR chemical shifts (Pople)
1969	Modern study of diamagnetic susceptibility exaltation (Dauben)
1970	Magnetic susceptibility anisotropy (Flygare)
1980	Quantum chemical calculation of magnetic properties (Kutzelnigg)

Observe that it was only Kekulé who proposed that the definition of aromaticity should be of a structural nature, but the next year, Erlenmayer proposed that instead of a *structural* definition of aromaticity, the definition be based on characteristic *properties* of aromatic compounds, which, as we see from the above table, the rest of the researchers, with the exception of Hückel, followed. Observe also that, after Pauling, all the remaining listed milestones have focused on *magnetic* properties of delocalized π electrons of conjugated systems.

What one should observe is that fundamental contributions to the clarification of aromaticity originated with theoretical considerations of *structures*, not *properties* of benzenoid and non-benzenoid conjugated hydrocarbons, which apparently have not been included in the *Milestones on Aromaticity* by Schleyer and Jiao. We agree with the list by Schleyer and Jiao, but their list is just half of the truth because there are additional Milestones on Aromaticity, which also deserve attention. These omissions call for an *alternative* listing of the *Milestones on Aromaticity*. Here we use the word *alternative* as defined in the Merriam-Webster dictionary as offering and expressing a choice, not usual and traditional, existing or functioning outside the established society.

After examining our list of Milestones on Aromaticity, readers may judge for themselves whether the leaning toward the position of Erlenmayer rather than Kekulé is justified.

Milestones on Aromaticity

1865	Benzene structure (Kekulé) [14]
1927	Fries rule suggesting that Kekulé valence structures having more rings with three C=C bonds appear to be more important than others [15]
1931	Theory of cyclic $(4n + 2)$ π systems (Hückel) [16]
1962	Relating Fries Kekulé valence structures to the importance of electron repulsion (Randić) [17]
1972	Clar aromatic sextet—extension of the notion of the aromatic sextet to polycyclic systems (Clar) [18]
1972	The 6n π electron rule—necessary condition for identification of most stable benzenoid hydrocarbons (Clar) [18]
1976	Conjugated circuits—discovery of circuits of alternating C=C and C–C bonds within individual Kekulé valence structures (Randić) [19]
1977	Generalization of the Hückel $4n + 2$ rule. Use of $4n + 2$ and $4n$ conjugated circuits for definition of aromaticity (Randić) [20]
1979	The total count of conjugated circuits in a structure having K Kekulé valence structures is $K(K - 1)$, and any *single* Kekulé valence structure has information on all the remaining Kekulé valence structures (Gutman and Randić) [21]
2014	Ring bond orders (RBO)—quantitative representation of Clar's aromatic sextet theory (Randić) [22]

In Table 14.2 in which the beauties of chemistry are listed, one can find that half that list already contains half of the milestones on aromaticity listed above. However, if one consider elegance, content, and brevity of the remaining milestones, one may agree with us that all of them deserve to be characterized as beautiful.

The above contributed to recognition of structural factors important for aromaticity of polycyclic conjugated hydrocarbons. Of course, aromaticity is as important, if not even more important, in heterocyclic compounds, but the reason for restricting attention to conjugated cyclic and polycyclic *hydrocarbons only* is that unless we understand the origin of aromaticity in hydrocarbons, we are not likely to understand it in heterocyclic compounds.

We will not elaborate on the above list as several of the listed topics have been covered in this book and in the review article on aromaticity [23], in which readers can look for details. Let us briefly comment on ring bond orders (RBOs) obtained by averaging the Pauling bond orders for six CC bonds of each benzene ring in benzenoid hydrocarbons. In a way, it is quite surprising that it took almost 80 years for the concept of Pauling bond orders to be generalized to ring bond orders (RBOs)! And it was the RBOs that put the Clar aromatic sextet theory of 1972 to a quantitative (numerical) frame in 2014, more than 40 years after Clar introduced aromatic sextet theory! Amazing in a way and not amazing in the view that Kekulé valence structures, although never proclaimed dead, as was the case with HMO, were abandoned by most theoretical chemists. The two exceptions are a few theoretical chemists with continued interest in VB method, and most theoretical chemists interested in chemical graph theory.

It has been known for a long time that all benzenoid hydrocarbons are aromatic, but some are more aromatic than others. The challenge is to find what makes some more aromatic than others. Eric Clar came to an ingenious idea that aromatic sextets play here the crucial role and formulated his ideas by proposing his empirical aromatic sextet theory. It has been disappointing that quantum chemists have shown no interest in the theory of aromatic sextet, which remained qualitative for over 40 years. Clearly scientists can choose what they consider important and not important—and for most theoretical chemists, the topic of why some benzenoid molecules are more aromatic than others has not been important. In contrast, in social sciences and general public, the phenomenon that there has been an animal farm—"all animals are equal but some animals are more equal than others"—received considerable attention, as reflected in interest in the book *Animal Farm* of George Orwell [24], pen name of English novelist Eric Arthur Blair (1903–1950), which was an allegory on the communist political systems in Russia at that time.

In Figure 14.8, we have shown for the collection of smaller benzenoid hydrocarbons of Figure 11.7 in Chapter 11 the rings of the maximal RBO for all molecules. The rings with the maximal RBOs have been emphasized with the exception of linear acenes in which all rings, symmetry equivalent or not, have the same RBO value. One can observe some regularity for maximal molecular RBO. Thus the maximal molecular RBO increases from phenanthrene to benzanthracene to benztetracene as the linear fragment increases in its length. The same again occurs for thriphenylene and dibenzanthracene, the two molecules that approach the RBO of benzene. The last RBO ring in Figure 14.8 belongs to buckminsterfullerene C_{60}, and as one can see, it appears that the "spherical" aromaticity of buckminsterfullerene is, at best, moderate. In contrast, the count of Kekulé valence structures in comparison with benzenoid hydrocarbons having the same number of fused benzene rings in C_{60} is unusually high ($K = 12,500$). This is a sign of warning about simple comparison and extension of properties of planar systems to spherical. For planar systems, the number of Kekulé structures for molecules having the same number of fused benzene rings is an indicator of relative aromatic character, but very large K for spherical

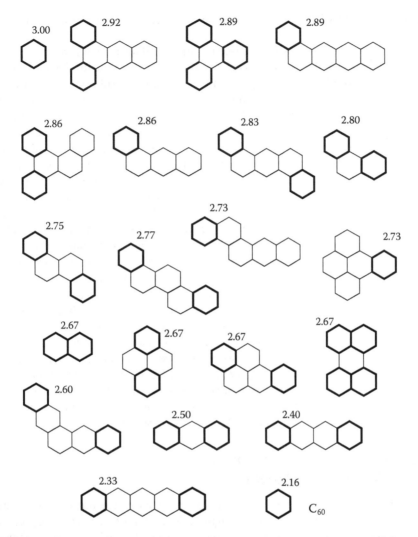

FIGURE 14.8 One of the symmetry nonequivalent rings with the maximal RBO is shown, except for linear acenes in which only one ring is shown (in linear acenes all rings, equivalent or not, have the same RBO).

systems need not parallel system aromaticity. As we have seen even among spherical systems of the same size (such as isomers of C_{60}), K is not a good indicator of relative stability.

In summary, the existence of two alternative routes to characterization of aromaticity is reminiscent of the existence of two alternative routes to characterization of QSAR. The "older" group has used *properties* as descriptors in structure–property–activity regressions, following Hansch, the "father" figure of early QSAR. The "younger" group has used *structure* descriptors in QSAR, which, despite vehement opposition by a few influential individuals, have been gaining a stronghold. However, we ought to mention that Hansch himself did not object to the use of mathematical descriptors in QSAR. Is the history to be repeated with the characterization of aromaticity?

Let us return to Kekulé, who proposed the structural formula of benzene and who considered that the definition of aromaticity should be of a structural nature, and to Erlenmayer, who followed by proposing the structural (Kekulé) formulas of naphthalene but suggested that the definition of aromaticity be based on characteristic *properties* of aromatic compounds. Erlenmayer was apparently followed by the rest of chemists who considered aromaticity (with exception of Hückel, who incidentally was not a chemist but a physicist). What apparently has been overlooked is that, from the stand of logic, to attempt to define aromaticity by considering properties of aromatic compounds for guidance is a case of "catch 22." Experimental properties, including bond lengths and magnetic properties, should not be used for *definition* of aromaticity, but can be used to argue that a particular compound is aromatic, less aromatic, or not aromatic. Calculated properties, including bond lengths and magnetic properties, can be used for *definition* of aromaticity, particularly if the model on which they are based is simple enough and captures important structural elements of considered compounds. Thus, use of Pauling bond orders and Coulson bond orders can be legitimately used for trying to arrive at a characterization of aromaticity. *Conjugated circuits* are such a simple structural concept, which, in addition to allowing a *simple* structural definition of aromaticity, allow one to find expressions for resonance energy (RE) and recently have been found useful in calculations of ring currents in benzenoid and non-benzenoid polycyclic conjugated hydrocarbons.

Before leaving the arena of ring bond orders, let us mention that it may be of considerable interest to see how ring bond orders based on Coulson's bond orders (using HMO [16], then PPP [25], and other more advanced computations, including *ab initio* calculations) would parallel Clar's aromatic sextet theory.

14.5   4n + 2 AND 4n RULE FOR POLYCYCLIC CONJUGATED SYSTEMS

Conjugated circuits have been hidden in Kekulé valence structures for more than 100 years and apparently continue to remain hidden to most chemists. The message that aromaticity of conjugated hydrocarbons has been clarified and that the famous Hückel 4n + 2 rule has been generalized for polycyclic systems, which has been eloquently expressed in a lengthy review article on aromaticity of polycyclic conjugated hydrocarbons [23], apparently has not attracted due attention despite that it was published in *Chemical Reviews*, the leading chemical journal. Despite that the review article has appeared in the most cited and widespread journal in chemistry with an impact factor over 40 and despite that it has nearly 400 citations and that the three seminal publications introducing conjugated circuits [19,20,26] together have more than 1000 citations, clearly conjugated circuits remain, for many chemists, an unrecognized basic *structural* concept, which today may well be as important as are the Kekulé valence structures. Conjugated circuits received some visibility among researchers interested in aromaticity as is reflected, for example, in the title of one of the papers in this area: *"Defining Rules of Aromaticity: A Unified Approach to the Hückel, Clar and Randić Concepts"* [27]. However, at the same time, there has been a case of avoiding the term "conjugated circuits" and referring to them as "symmetric difference" [28] as if that is something different. The term "symmetric difference" is used in mathematical literature as has been pointed out by Ivan Gutman [29], one of rare chemists who has a PhD in chemistry and mathematics. This is what Gutman, in Chapter 6 of his book [29], wrote concerning the "symmetric difference" and the "conjugated circuits":

> ...Alternating cycles and symmetric differences of 1-factors/Kekulé structures have been investigated in theoretical chemistry of different occasions (see Graovac et al. 1972, Cvetkovic et al. 1974, Gutman and Herndon 1975 and the references therein). However, it was Milan Randić (1976) who first recognized the great significance of alternating cycles for the chemical behavior of conjugated molecules. He named them "conjugated circuits" and developed a theory which is known under the name "the conjugated circuit model."
>
> The theory outlined in the present chapter has to be associated with the name of Milan Randić who discovered it (Randić 1976) and eventually elaborated it (Randić 1977a,b) and applied to numerous classes of conjugated molecules (Randić 1980, 1982, Randić et al. 1987b, and the references cited therein). In what follows we expose only the conjugated circuit model for benzenoid hydrocarbons. One should, however, note that the model covers a much wider class of conjugated systems (Randić 1977a,b, 1982)...

14.5.1 THE CONJUGATED CIRCUIT MODEL

In more recent times, conjugated circuits have been used for calculation of ring currents in conjugated systems as reported in the *Chemical Reviews* article, *Aromaticity of Polycyclic Conjugated Hydrocarbons* [23], and outlined in a previous chapter of this book. Unwarranted use of mathematical terminology in chemical literature by Ciesielski et al., [27] using the mathematical term "symmetric difference" instead of referring to "conjugated circuits" can only confuse the less informed as indeed was the case with one of the reviewers of the *Journal of Computational Chemistry* (as mentioned in the previous chapter).

It takes some imagination, according to E. B. Wilson, to appreciate novelty, and some recognized the merits of conjugated circuits immediately as soon as they were discovered. Thus Professor D. Hellwinkel of Heidelberg in a letter to the author of the paper on aromaticity and conjugation wrote (in March of 1977):

> I just came from reading your most interesting paper on Aromaticity and Conjugation (JACS 99 (1977) 444). I think that your extension of the Hückel Rule provides a most important improvement in the understanding of the properties of polynuclear conjugated systems. This even more so because your method of dividing fused systems formally into cyclic conjugated sub-entities can be understood and performed by every student without sophisticated theoretical/mathematical background. Personally I think that always these theories are the best ones which allow simple quantifications of inherent chemical intuitions. And this is exactly the case with your elegant substructure counting and weighting procedure. I am sure that these ideas will find very broad (and grateful) acceptance in chemical world...

Similarly, after a copy of the paper on conjugated circuits was sent to L. Pauling (dated March 24, 1976), there was reply acknowledging conjugated circuits. Below is a part of the letter reproduced from a review article on aromaticity [23]: *Aromaticity of Polycyclic Conjugated Hydrocarbons, Chem. Rev.* 103 © 2003, after obtaining permission from Linus Pauling, Jr., son of Linus Pauling:

> Dear Professor Randić:
> I was pleased to receive your letter and your paper, which I have examined with interest. I agree with you that it is better to make rather simple calculations, such as yours, than the very complicated ones.
> Your work on conjugated circuits reminds me of a paper that I wrote on the diamagnetic anisotropy of aromatic molecules, Journal of Chemical Physics 4, 673 (1936)...
> Again let me thank you for writing to me.
> Sincerely,
> Linus Pauling

So apparent confusion about what is aromaticity is not due to a lack of explanation of aromaticity in terms of the presence, or lack of the presence, of conjugated circuits of $4n + 2$ and $4n$ size in the set of Kekulé valence structures of compounds, but due to (i) lack of attention given to articles describing conjugated circuits and their use for characterization of aromaticity and (ii) not recognizing that these explanations, in fact, represent a generalization of the famous Hückel rule to polycyclic systems, which Hellwinkel recognized just as the paper was published, almost 40 years ago.

The three publications on conjugated circuits [19,20,26], despite totaling more than 1000 citations, can be viewed as Sleeping Beauties in view of the widespread unawareness that they solved the problem of aromaticity, at least for hydrocarbons, so many years back and in a sufficiently simple way that, according to Hellwinkel, "*can be understood and performed by every student without sophisticated theoretical/ mathematical background.*" So the right question to ask today is not, "What is the aromaticity problem?" but "Why is the aromaticity problem?" when, in the case of aromatic hydrocarbons, the answer has been known for more than 40 years!—at least to a few.

Let's end the topic of aromaticity and conjugated circuits with a beautiful problem:

Problem 22

Find a mathematical rationale for the close parallelism for the ring current in benzenoid and non-benzenoid hydrocarbons calculated using classical physics (*ab initio* density currents) and ring currents obtained by adding contributions from conjugated circuits.

14.6 ORTHOGONAL MOLECULAR DESCRIPTORS

Here we have another subject, orthogonal molecular descriptors [30,31], the topic of MRA, which, despite some visibility (the two papers listed above have together about 450 citations), continue not to be implemented by most MRA users. This suggests that the merits of orthogonal descriptors are either not appreciated or not understood. These two papers [30,31] published in 1991 and approaching their 25th anniversary, despite some visibility, may, for all practical purposes, be considered "Sleeping Beauties." There have been a dozen related papers that elaborate further on the use of orthogonal descriptors, an indication that there are chemists who recognize the fundamental significance of orthogonality in MRA.

Part of the problem is that it is not generally recognized that the basic message of MRA is that descriptors used in regression only define the *structure–property space* and that MRA regression equations do not relate to the *role of individual descriptors* in the structure–property space. Some of those using regression analysis may not be aware that a *single* regression equation involving several descriptors only defines the *structure–space* but does not allow quantitative *interpretation* of the coefficients of the regression equation. Thus, again, we have a case of a solved MRA problem that remains recognized, at best, as only partly solved until information on the inter-relationship of descriptors used for the set of compounds considered is also supplied directly or through orthogonalization of descriptors if one is interested in an interpretation of the results.

The situation, however, is not hopeless. In the book *On the Beauty of Science*, H. Hauptman [32], who shared the Nobel Prize in chemistry with J. Karle for solving the inverse problem of X-ray diffractions on crystals, described how a "massive" misconception can be successfully eradicated. Even after Hauptman and Karle showed that the problem had been solved, many crystallographers did not believe this and, in fact, believed that the problem could not be solved—but it has been solved! Time was passing, and the opposition did not show decline, in part possibly because one very influential and leading crystallographer in England was a very vocal opponent to the notion that the problem could be solved. The tide of change followed after Hauptman and Karle decided to determine the coordinates for atoms of a rather complicated crystal. The calculations involved construction of some 6000 Fourier transformations, which took them about one month—work that current computers, according to Hauptman, would finish in about three minutes! After they reported the result, the leading English crystallographer immediately recognized that this result demonstrated clearly the correctness of the crystal structure and the potential of the available method for solving the X-ray diffractions of crystals. So overnight, from the *most stubborn and vicious enemy*, the leading English crystallographer became the *most ardent supporter* of Hauptman's and Karle's work. It took some time for computer programmers to recognize the importance of the work of Karle and Hauptman and make a suitable computer package—use of which ensured the Nobel Prize for Karle and Hauptman.

Well, this story may illustrate how the fallacy of a single equation will eventually be resolved. We have to wait until one of the outstanding authorities in QSAR recognizes the merits of the orthogonal descriptors and encourages MRA users at large to supplement their current calculations with information on correlations between descriptors used in their work. Sooner or later, this will happen, but it would be better if it happens sooner as orthogonalization of MRA descriptors has been waiting on wide use since 1991, some 25 years.

14.7   VARIABLE CONNECTIVITY INDEX

The variable connectivity index has also been waiting for greater attention since 1991, when it was first briefly outlined [6]. It has received some attention, having a few citations short of 100, but in our view, it deserves at least having a few citations short of 1000, if not 10,000. You decide whether this speculation is sound or unsound. What makes the variable connectivity index unique, and this holds for other variable indices, is the following:

1. It is not a fixed molecular index; its magnitude varies from property to property and from regression to regression.
2. It is computationally efficient as it could be constructed after examination of a few dozen trials instead of searching for the best fixed descriptors among a few hundred descriptors.

For example, as has been illustrated in Table 6.13 in Chapter 6, it took only 20 steps to find the parameter that determine the optimal "valence" for carbons and nitrogen in 16 amino-alkanes for regression of their boiling points. The resulting standard error found was respectable s = 1.92°C. The variable connectivity index allows simple interpretation of the relative roles of heteroatoms, carbon, and nitrogen in the case of amino-alkanes. As can be seen from Table 6.13, the graph theoretical valences of carbon 1, 2, 3, 4 (which stand for primary, secondary, tertiary, and quaternary carbons) have to be increased by +1.25, becoming 2.25, 3.25, 4.25, and 5.25 primary, secondary, tertiary, and quaternary carbons, which translates, through the expression $1/\sqrt{(m\,n)}$, to effective contributions to terminal carbon bonds:

(1,2) instead of $1/\sqrt{2} = 0.707107$, one finds contribution (1,2) = 0.369800

(1,3) instead of $1/\sqrt{3} = 0.577350$, one finds contribution (1,3) = 0.323381

(1,4) instead of $1/\sqrt{4} = 0.500000$, one finds contribution (1,4) = 0.290957

and so on for additional bond types. As one can see, the relative contributions of CC bonds have been slightly diminished. Similarly for nitrogen, the graph theoretical valence is decreased by −0.65, which introduces an increase for the relative roles of C–N bonds. Thus, the effective contribution of the terminal carbon–nitrogen bond becomes

$$1/\sqrt{(2.25 \times 0.35)} = 1.12687$$

This is almost four times more than the contributions of terminal CC bonds. The relative contributions of carbon and nitrogen in other compounds and other molecular properties will generally differ. Recall that the contribution of bond (m, n) = $1/\sqrt{(m \times n)}$ is one of the solutions of the set of inequalities based on the ordering of isomers with respect to their boiling points, and inequalities can have numerous solutions and may differ for different properties.

So far, optimal parameters have been determined by a search, such as illustrated in Table 6.13. In a few instances, the optimal parameters determined statistically to have the smallest standard deviation have been found to have also some *structural* interpretation. This is very significant because there is no known reason that there should be any structural interpretation of optimal variable parameters.

Let revisit one illustration [33] involving the search for the optimal variable connectivity index for a set of 80 halo-alkanes, including only chlorine, bromine, and iodine compounds. In addition to differentiating between chlorine, bromine, and iodine in this particular work were discriminated also aliphatic carbon atoms (C_a), carbons forming cycles (C_c), and carbons forming C=C bonds (C_d), in all six variable parameters. The following optimal parameters were found:

$$C_a = 1.27, C_c = 1.23, C_d = 1.57, Cl = -0.235, Br = -0.653, I = -0.805.$$

As one can see, there is a little difference between acyclic (aliphatic) and cyclic carbons, around 3%, so one could have considered five variables and would obtain essentially the same regression (illustrated in Figure 6.8 in Chapter 6):

$$bp = 0.9727 \chi(x, y) + 3.1429 \text{ with } r^2 = 0.9729.$$

Here the sign $\chi(x, y)$ is the symbol for the variable connectivity index. The slight increase of the variable parameters for carbon atoms suggests a lesser role of carbons in comparison with halogen atoms on the boiling points of these compounds.

The trend of the variable parameters for halogen atoms shows an increasing role as we move from chlorine to bromine and iodine. That is apparent, but why this is so is not obvious. The three halogens differ in size, in mass, in charge, and in a number of other atomic properties. The variation of atomic mass of halogens and carbon against the optimal weights obtained from the best regression of the boiling points of 80 halo-alkanes suggested a decrease of mass with an increase of charge from the negative values of halogens toward positive carbon. Because the dependence resembled hyperbolic variation, it was of interest to plot reciprocal values of mass against the optimal weights, which is illustrated in Figure 6.7 in Chapter 6. As one can see, the plot of 1/mass against the variable weights shows an outstanding linear regression ($r^2 = 0.9999$).

Whenever in science two totally *unrelated* data, as are the atomic mass and optimal parameters of a regression shown in Figure 6.7 in Chapter 6, are found to be related, there is a time for pause. This is not accidental and not likely to be an isolated case. What this result shows, and what has not been known and has not been expected, is that atomic mass and other properties that parallel and correlate with the atomic masses of C, Cl, Br, and I, relate to the boiling points of these compounds. This means that atomic mass may parallel the relative boiling points of halo-alkanes. It also shows that characterization of such dependence is beyond the capabilities of "simple" connectivity indices with weights chosen in advance. In addition it clearly shows that objections of skeptics to mathematical descriptors in structure-property-activity studies have no merit whatsoever.

In another study involving alcohols, aldehydes, ethers, ketones, carboxylic acid, and ester, it was found that when the oxygen in each group is differentiated, the optimal weights of the variable connectivity indices for different oxygen types show good correlation with charges as calculated using quantum chemical calculations. As Aristotle (384 BC–ca. 322 BC) said, *"One swallow does not make a Spring,"* but two "swallows" may suggest that spring is coming and that additional similar correlations between optimization parameters and physicochemical atomic and bond properties may be expected!

14.8 PROTEOME MAPS: ORDER OR CHAOS?

Let us return to Figure 7.6 in Chapter 7, a schematic figure of a proteome map based on considering only the 20 most abundant spots (listed in Table 7.2 in Chapter 7). Although individual proteins have always the same (x, y) coordinates (assuming that the experimental conditions allow reproduction of such maps), each such map represents a "fingerprint" of a cell at the time and under the conditions in which the experiment has been carried out. So, in a way, just as atomic spectra and molecular spectra represent "fingerprints" of atoms, the successful interpretation of which tells us about the electronic structure of atoms and molecules, respectively, so a collection of proteome maps of individual cells are a source of information on the state of cells considered. The problem is that variations in the abundance of proteins appear chaotic when one compares proteome maps of the same cells under different conditions.

Proteome maps as a tool for exploration of cells will be very promising once the experimental technique improves sufficiently to offer experimental data of valid reproducibility, but there have been some laboratories that have been reporting good quality experimental data, which have been worth investigating. For example, Anderson and coworkers [34] reported, in 1996, on the effect of peroxisome proliferator on protein abundances in mouse liver cells. In that study, they reported results of variations in proteome maps with variations in concentrations of the selected peroxisome proliferator LY1712883. As we outlined in the section on proteome maps in Chapter 7, by 2001, a novel tool for quantitative comparative study of related proteome maps had been developed (see [7,35–40]). This made it possible to revisit earlier work, and in a publication, "Order from Chaos: Observing Hormesis at the Proteome Level" by M. Randić and E. Estrada [41], have reported on a J-shape dose response curve for the overall changes in the abundance of some hundred protein spots considered by Anderson and coworkers [34], illustrated in Figure 14.9.

When the manuscript was submitted to the *Journal of Proteome Research*, the authors received a one-sentence report: *"This paper will be highly cited."* This very positive evaluation has probably been "awarded" to this manuscript because this was the first time that a J-shaped dose response curve was reported at the cellular level for data coming from a *single cell*. Hormesis, the nonlinear dose response curve, which shows that at very small concentrations even a toxic substance can potentially be beneficial, has been observed and known for some time to hold for *whole*

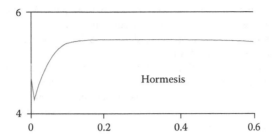

FIGURE 14.9 Illustration of hormesis on the cellular level.

organisms, animals and plants [41–49]. As an illustration, we may mention a paper in which a group of mice was first exposed to smaller amounts of X-ray radiation. Later, the same group of mice and the control group were both exposed to a lethal amount of X-ray radiation. The result was interesting: All mice in the control group were dead, but some mice in the group previously exposed to a smaller amount of radiation survived [50]. Clearly, mice previously exposed to a small amount of radiation developed some defensive mechanism, so a small amount of radiation appeared to be beneficial. Calabrese [41] was one of the pioneers in promoting hormesis at a time when the establishment was accepting only a linear dose response curve as viable. But accumulation of data on the *J*-shaped dose response curve could eventually not be ignored. That the *J*-shaped dose response curve was found on *cellular* data is therefore an import finding and was expected to be noticed and appreciated. It has now been 10 years since the paper appeared, and it has about 30 citations, which can only mean that the paper has either not been noticed or, if noticed, not appreciated. Perhaps this tells more about the "sleeping" community rather than about the "sleeping beauty," which is still waiting for her prince.

We may add that chaos, chaotic phenomena, and chaotic behavior are not so uncommon in science and have received the attention of mathematicians and scientists, people such as Henri Poincare, Jacques Hadamard, George David Birkoff, Andrei Nikolaevich Komogorov, John Edensor Littlewood, Stephen Smale, and Edward Lorenz. According to Lorenz, chaos can be defined as [51]

> Chaos: When the present determines the future, but the approximate present does not approximately determine the future.

The latest addition to the list of mathematicians contributing to clarifications on chaos is Yakov G. Sinai, who won the Abel Prize in 2014. The Abel Prize for Mathematical Sciences is an annual prize presented in Olso, Norway, to the winner by the king of Norway, and it is in the same amount as the Nobel Prize (which is approximately one million US dollars). It started in 2003 and was named after Niels Henrik Abel (1802–1829), an outstanding Norwegian mathematician, who, although he died very young (from tuberculosis), left an impressive mathematical legacy. Below we give short information on Yakov G. Sinai, under the title "Grappling with Chaos," written by M. Freiberger [52]:

The ability to see order in chaos has won the mathematician Yakov G. Sinai the 2014 Abel Prize. The Prize, named after the Norwegian mathematician Niels Henrik Abel, is awarded annually by the Norwegian Academy of Science and Letters and is viewed by many as on par with the Nobel Prize.

Chaos is something we see all around us every day. The weather, the fluctuations of the stock market, the eddies in our coffee as we stir in the milk, even the arrival times of buses display an unpredictability we would term chaotic. Yet, as mathematicians including Sinai have shown, it's possible to get some grip on chaotic systems by understanding their overall behavior.

To this, we may add that there are also numerous chaotic phenomena that are not dynamic. The "theoretical" illustrations include the chaos game by Barnsley [53]; the graphical representations of DNA by Jeffrey [54], which is illustrated in Figure 14.10 for the first exon of the lemur β globin gene; the modified chaos game representation of DNA [55,56]. Among experimental phenomena, we may mention the proteome maps, with which, even though the mass and charge of cellular proteins are not random, the abundance appears chaotic, that is, unpredictable from case to case. As is known, even small variations in the concentrations of toxic substances can produce unpredictable changes in the abundance of proteins in a cell, which is typical of chaotic phenomena.

In short, chaos phenomena, just as being part of mathematical physics, are also part of mathematical chemistry. So just as *"it's possible to get some grip on chaotic systems by understanding their overall behavior"* in physics, so *"it's possible to get*

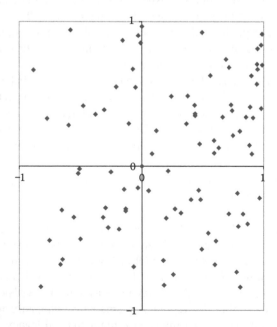

FIGURE 14.10 Graphical representation of 92 bases of the first exon of the β-globin gene for lemur according to the graphical approach by Jeffrey.

some grip on chaotic systems by understanding their overall behavior" in chemistry and biology and proteomics as shown by Randić and Estrada [41] and as has been shown by deducing a definite *J*-shaped dose response from the chaotic nature of proteome maps of mouse liver cells, reported by Anderson et al. [34], induced by a variation of concentrations of LY171883, a peroxisome proliferator, in mouse food.

14.9 GRAPHICAL ALIGNMENT OF PROTEINS

The problem of alignment of proteins is one of the central problems of bioinformatics. As has been mentioned, since 1970, computer science has offered a collection of programs that offer solutions to the protein alignment problem. What should not be overlooked is that all these are *approximate* solutions because they, in addition to involving a number of approximations, tend to use a trial-and-error approach. The outlined graphical alignment of proteins [10] is approximation-free; that is, it does not utilize any empirical parameters, such as various penalty schemes for gaps, deletions, and insertions, but it does not qualify as an exact solution because it uses trial-and-error methodology. The graphical approach to the protein alignment problem followed a graphical solution to the DNA alignment problem [57].

To arrive at a graphical alignment of proteins or DNA, the first step is to construct a spectral representation of proteins and DNA, respectively. In the case of DNA, one uses four equidistant horizontal lines that have been labeled as A, C, G, and T, which stand for the four nucleotide bases A = adenine, C = cytosine, G = guanine, and T = thymine. Then, going along the DNA sequence, one represents each base by a spot on the corresponding line, all spots are placed at equal distances along the *x* axis, so that, after connecting the adjacent spots by line segments, one obtains a pattern resembling atomic and molecular spectra. In Figure 13.2 in Chapter 13, we show the spectral representation of the first exon of the human β-globin gene:

ATGGTGCACCTGACTCCTGAGGAGAAGTCTGCCGTTACTGCCC
TGTGGGGCAAGGTGAACGTGGATGAAGTTGGTGGTGAGGCCC
TGGGCAG

Spectral representation of proteins uses 20 equidistant horizontal lines, each standing for one of the 20 natural amino acids. In Figure 13.7 in Chapter 13, we have illustrated spectral representations of two proteins of approximately similar lengths, listed in Table 13.10 in Chapter 13. The assignment of individual amino acids to individual horizontal lines could be arbitrary but, once adopted, should be followed in other similar graphical representations. We have selected alphabetical ordering of amino acids, based on three-letter codes for amino acids: ala (A), arg (R), asn (N), asp (D), for alanine, arginine, asparagine, aspartic acid, and so on. Already, visually from constructed spectra for the two proteins, one can see that some spectral regions show similarities. However, the real advantage of spectral representations of DNA and proteins is that one can easily search for a quantitative degree of similarity of bio-sequences, and one can detect the regions of protein or DNA in which they are aligned. To do this, one assigns to lines 1–20 numerical values 1–20 and constructs

388 Solved and Unsolved Problems of Structural Chemistry

the graphical representation in creating the difference of two spectral representations. Alternatively, one can arrange 20 amino acids on a periphery of a unit circle and assign to them radial angular coordinate values between 0 and 2π, which will limit all the differences in the range $(-2\pi, +2\pi)$.

For the two ND6 proteins of Figure 13.7, it is difficult to locate segments in which they are more similar; however, if one constructs the spectral difference shown in Figure 13.8, obtained by subtracting the two spectra, one can immediately see a sizable overlap of amino acids in location sites from 22 to 83. In this case, there are no additional segments of overlap of amino acid sections, which can be seen easily from Appendix 17, which lists sequential labels of adjacent pairs of amino acid in the two proteins as found from the exact solution to alignment of the two proteins considered. Hence, considerable time was thus saved, by knowing the exact solution, by not considering shifting the two proteins in trial-and-error search.

In contrast, in the case of the two proteins of Table 13.8, spectral representation of which was shown in Figure 13.9, the difference of their spectra, shown in Figure 13.10, does not show any region of overlap between the two proteins. There are few amino acids around position 15, which appear to be identical in two proteins; there are no additional overlaps of amino acid locations for the remaining part of the two proteins. However, as we have seen from the top part of Figure 13.11 when the two proteins have been shifted by one place, overlaps of sizable segments of the two proteins are found to be similar. When the two proteins have been shifted by two places, an additional region of similarity in proteins became visible (see Figure 13.11, bottom part). By continuing constructing additional differences of the spectra by shifting proteins by three and four sites, respectively, one finds two additional regions of shorter segments in which the two proteins show considerable similarity. The spectral differences, when the two proteins have been shifted by three and four places, are illustrated in Figure 13.12. In this particular application we found all the regions of alignments of two proteins in a few steps. All the computations can be easily made in a Microsoft Excel program. If one would in advance know that in this case, one should only consider differences from −1 to −4, then the outlined graphical approach would illustrate graphically the result of the *exact* solution of the protein alignment problem. However, use of the trial-and-error approach does not qualify the graphical alignment method as the exact solution.

Recall Kepler's conjecture on the density of packing uniform spheres outlined in the opening section of this book. The proof by Hales [58] was based on an *exhaustive* check by a computer of a very large, but finite, number of different arrangements. For a finite number of possibilities, there is little conceptual distinction between an exhaustive check of all possibilities and use of the trial-and-error approach to a check of all possibilities. Both could be very inefficient, in contrast to the exact solutions, which do not involve searches, which make them fairly efficient.

We will see in the ending section of this chapter, titled "Sleeping Giants," how the exact solution to the protein of pairwise alignment was found. The essence of the exact solution was *finding all the differences* in their labels between adjacent pairs of amino acids of two protein sequences. Having this information, graphical alignment of proteins is transformed into an illustration of the exact solution of protein alignments.

14.10 SLEEPING GIANTS

We will end this book on *Solved and Unsolved Problems in Structural Chemistry* with a brief description of a recent report on the *exact* solution of one of the central problems of bioinformatics: the problem of protein alignment.

As has been suggested [59], the characterization of some scientific publications as "Sleeping Beauties" relates to papers that have mostly been overlooked (not cited) for some time until another publication (the so-called "Prince") arrives that awakens the scientific community, just as the Prince awakened the Beauty by kissing her in fairy tales. In contrast to the characterization of "sleeping beauties" as publications hardly noticed upon their arrival, we would like to characterize scientific publications as "Sleeping Giants" that upon their arrival may have been noticed as being "giants" but noticed so by very few. To be quantitative, let us define the "very few" as some 1% of the scientific community that recognize the merits of a forthcoming publication. In other words "Sleeping Giants" are scientific publications that deserve recognition of at least a two orders of magnitude wider audience than they have been receiving.

Hence, the existence of Sleeping Giants does not mean that they have been totally overlooked (as is the case with the Sleeping Beauties), but that they are not receiving the attention that they deserve. The best illustration of a "Sleeping Giant" in theoretical chemistry is the paper by Walter Kohn on the density functional theory (DFT) [60]. DFT apparently has been overlooked by many quantum chemists until they have been awakened by the Swedish Academy of Sciences, which, in 1998, awarded the Nobel Prize to W. Kohn. However, the density functional theory was recognized by a few from its beginning and was gradually attracting a number of quantum chemists even before the Nobel Prize recognition of W. Kohn. Otherwise how would W. Kohn be nominated for the Nobel Prize in chemistry were it not for the few who recognized its merits early and made DFT visible enough.

Recall the statement of E. B. Wilson [61]: "...*every once in a while some new theory or new experimental method or apparatus makes it possible to enter a new domain. Sometimes it is obvious to all that this opportunity has arisen, but in other cases recognition of the opportunity requires more imagination.*" The "Sleeping Beauties" are papers that have been recognized by a few, who succeeded in awakening the sleeping communities of scientists. The "Sleeping Giants" are papers that have been recognized by a few, who may have succeeded in awakening their close neighbors, but the community of scientists at large apparently does not wish to be disturbed in their sleeping and dreaming.

In structure property and structure activity studies, we may mention as Sleeping Beauties the paper of Harry Wiener on W [62], and J. R. Platt on path numbers [63]. As the publications that may be viewed as Sleeping Giants, we may mention the papers on orthogonal molecular descriptors [30] and the papers on the variable connectivity indices [6]. In contrast, the paper on the connectivity index $^1\chi$ [64] does not qualify either as a sleeping beauty or a sleeping giant because it was never "sleeping" due to the alertness of Kier and Hall, who immediately recognized its merits and were able to alert others in the QSAR community about its virtues.

There is no doubt that the connectivity index has demonstrated its promise, but could one design a better connectivity index than the one proposed? Recollect that the connectivity index represents but one solution to the bond-type inequalities based on ordering of smaller alkane isomers relative to their boiling points. Alkanes are but a minor group of molecules and boiling points but a single property. To generalize the connectivity index, Kier and Hall made a significant step by introducing the valence connectivity indices, which as is well known, proved themselves to be very promising. But could one design better valence connectivity indices than those proposed?

The valence connectivity indices not only represent, strictly speaking, an ad hoc solution for discrimination of heteroatoms, even if plausible, but indirectly presume that there are *universal* valence weights characteristic for individual heteroatoms and valid for *all* their molecular properties. That the latter assumption is questionable follows from the already mentioned few regression equations of Kier and Hall in which the "classical" connectivity index χ produced better correlations than the corresponding valence connectivity index. Recognition that different molecular properties may require different heteroatom "weights" led to the idea of the *variable* connectivity index.

We will briefly outline the earliest publications on the variable connectivity index that have illustrated the novelty, exceptional properties, and the quality of the variable connectivity indices and variable indices in general. The variable indices have opened a novel chapter in QSAR studies, which may have not been recognized by some as a novelty worthy of closer attention. The seminal paper on the variable connectivity index, titled *"Novel Graph Theoretical Approach to Heteroatoms in Quantitative Structure Property—Structure Activity Relationships"* [6], the coming 25th anniversary of which is in 2016, is most likely to pass unobserved. More recent applications, as mentioned in the previous chapter, have demonstrated not only the exceptional potential of novel descriptors, but have also demonstrated that optimal parameters may, at least in a few cases so far studied, have important physicochemical interpretation. This, if nothing else, should suffice to terminate "complaints" heard over and over that mathematical descriptors are not relevant for the "real world" chemistry. The publication on the variable connectivity index, in our view, deserves to be characterized as "sleeping" despite limited recognition, whether as a "Sleeping Beauty" or even a "Sleeping Giant" we may eventually one day see— because, one day, its use may make hundreds of current molecular descriptors (topological indices) obsolete. Recall David Hilbert (1862–1943), a very distinguished German mathematician at the turn of the century, who has written [65], *"One can measure the importance of a scientific work by the number of earlier publications rendered superfluous by it."*

Let us end this diversion into the "sleeping" world of science by again mentioning the recent exact solution of the protein alignment problem [9], the paper that could be characterized as a "Beauty" or a "Giant," a "Sleeping Beauty" or a "Sleeping Giant." This will depend on whether it is immediately recognized as a great work, whether it will take years to be recognized for its importance, or whether only a small circle of scientists, a small percentage, say 1% of those in bioinformatics, find it to be a very significant contribution while others remain indifferent. Only time will tell.

14.11   SUMMARY

Coming to the end, an attentive and informed reader may have observed that the "solved problems" that we have selected belong to a special group of research papers that have received limited acknowledgment, at least as reflected in their limited citations. This does not mean necessarily that a limited number of researchers have seen them and have been aware of some of them, which could be the case had they been published in some obscure scientific journals. Because they have been published in outstanding journals, such as the *Journal of the American Chemical Society, Tetrahedron, Journal of Physical Chemistry, Journal of Chemical Information and Computer Science, International Journal of Quantum Chemistry, Chemical Physics Letters, Journal of Computational Chemistry, New Journal of Chemistry, Journal of Mathematical Chemistry, SAR and QSAR in Environmental Research, Journal of Molecular Graphics and Modelling*, and such, it is fair to assume that these papers may have been seen but, in view of their limited citations, tend to be avoided.

Apparently a sizable part of the QSAR community is unwilling to accept certain scientific results. Examples discussed in this book include (1) the necessity to use orthogonal molecular descriptors if they want their results to be interpreted quantitatively; (2) unnecessarily avoiding the use of descriptors that visibly duplicate to a considerable extent one another, instead of focusing attention on parts in which such descriptors differ; and (3) viewing the results expressed by a *single equation* as quantitative without offering and using the information on mutual correlation of the so-called "independent variables."

Orthogonalization of molecular descriptors was outlined approximately 30 years ago. It was pointed out more than 15 years ago that highly interrelated descriptors need not be avoided in QSAR and that using a single regression equation for interpretation of regression results is at best only a "half-truth." The future cannot be predicted, but do not expect an overnight "acceptance" of "abandoned" discoveries, an overnight awakening, and overnight learning from the "mistakes of the past." Resistance to novelty is not as uncommon as one may think, regardless of the relative importance of such novelties. There have been numerous illustrations of such mistakes in science. In physics, in some circles, there has been initial resistance toward Einstein's relativity theory. In biology, Darwin's evolution theory has been questioned. In chemistry, acceptance of quantum chemistry was unnecessarily slow in some circles. These may be notorious cases that everyone has heard of, but there are many lesser-known and less spectacular cases in physics, biology, and chemistry, including the abovementioned illustrations in QSAR.

In science, eventually truth prevails, but this may take a long time in some cases even at the present time. Consider the case of the exact solution of the X-ray diffraction pattern for crystals by Hauptman and Karle in the mid-1950s. Even after being solved, this problem has been considered by many as not possible to solve. Once a problem has been reported solved, this should have been sufficient to convince everyone of this fact, whether one understands the problem or solution or does not— because there are some, even if only a few, who can verify the results. Surprisingly, in the case of inverse X-ray diffraction analysis, many people continued to believe that the problem cannot have an exact solution.

Is this to be the same with the "Sleeping Giant" mentioned, just because many never thought that the problem now solved could be solved exactly? Hauptman and Karle started to work on a few crystals in order to promote their approach, but it was only after they succeeded in solving the crystal structures of an "impossible" mineral, colemanite, the structure of which was unknown at the time, that their work received greater attention. They have been given all the data necessary for a couple of thousand intensities to determine the positions of 20 atoms in the crystal unit cell. As mentioned, it took Hauptman and Karle about a month to calculate the phases of some thousands of diffracted rays by hand, having no computer at that time. The hostility toward the "exact solution" continued for a while. However, not surprisingly, the wife of Jerry Karle, who was a crystallographer, decided to use this method for her research, and very surprisingly, one of the main academic critics, who claimed all the time that the exact solution was not possible and that it could not be done, the crystallographer M. Woolfson (from York University in Britain), not only completely reversed himself after seeing that the method works, but became friendly and an advocate of the exact approach.

"Orthogonal molecular descriptors" can be viewed as a "Sleeping Giant" of QSAR. Let us hope that one of the main academic figures in QSAR, and there are several, who avoided the use of orthogonalized descriptors thinking that they add nothing to the MRA solution, not only completely reverses himself or herself after seeing that the method works, but becomes a friendly advocate of the use of orthogonal MRA regression equations.

This book could be characterized as offering a provocative review on some solved and unsolved problems in structural chemistry and particularly in structure–property–activity studies. Due to space limitation, we selected a few problems with which we have been better acquainted, but we are sure that there are additional interesting problems, some of which may deserve even greater attention and may broaden our horizons, arouse our curiosity, and may challenge our intellect.

Let us end the list of problems in this book with a "Giant Problem":

Problem 23

Mathematical Treatment of the Axioms of Chemistry: To treat ... by means of axioms, those chemical sciences in which already today mathematics plays an important part; in the first rank are quantum chemistry, statistical mechanics, chemical graph theory, and graphical bioinformatics.

This problem parallels Hilbert's Problem 6, which is *"To treat in the same manner, by means of axioms, those physical sciences in which already today mathematics plays an important part; in the first rank are the theory of probabilities and mechanics."*

This is the only problem on which Hilbert himself spent many years working and, thus, deserves the label of being a "Giant Problem"!

Let us hear of additional "sleeping beauties" and "sleeping giants," solved and unsolved, because one overlooked significant paper in science is one too many. It is very encouraging that we have not been alone in the search for solved and unsolved problems in chemistry. In one of the latest issues of *SAR and QSAR in Environmental Research* (April–June 2014) there is an article by J. Gasteiger titled "Some Solved and Unsolved Problems of Chemoinformatics" [66] in which the author highlights many of the problems of chemoinformatics during 40 years of its history. We took the liberty of quoting the first paragraph of the introduction of this paper fully as this need not be the opening words only for his publication, but with a change of the word "chemoinformatics" to other chemistry disciplines, would speak well of the whole of chemistry and beyond. Let us hope to see more of such contributions in the near future. Gasteiger's paper starts with

> The aim of this paper is to highlight many of the problems that chemoinformatics has solved in its more than 40 years of history. This should raise the spirit and pride of those working in chemoinformatics by recognizing that they are working in a field that has made decisive contributions to the development of chemistry and related scientific disciplines. On the other hand, it should also be realized that there are still many problems to be solved in the area of chemoinformatics. In fact, this simply emphasizes that chemoinformatics has matured into a scientific field of its own that will endure into the future. There are many open questions and interesting challenges in chemoinformatics waiting for novel ideas and powerful solutions to be developed. Furthermore, it emphases that chemoinformatics is attractive to students to join and meet these challenges.

Let us end this book by saying that this book illustrates that the above introductory statement of J. Gasteiger, equally applies to significant parts of this book, if one is to replace "chemoinformatics" with "chemical graph theory," with "graphical bioinformatics," and with "partial ordering"—the three similarly novel disciplines of chemistry that strongly overlap with chemoinformatics and deserve more attention.

14.12 APOLOGY

We apologize for occasional repetitions, which are not to suggest that some people need to hear the same thing twice before noticing the message (although hearing twice cannot hurt), but because of the variety of topics and problems considered, it may be expected that some readers may skip some sections of the book and thus miss out on some subject that may have wider potential interest. We apologize for hinting at the bias of some chemists toward "narrow" points of view that they cherish, but by that, we make no claims that their opinions are not valid—just that, in our view, they are not broad enough. After all, we are also biased; our bias is toward the use of mathematical descriptors for chemical phenomena; toward the use of discrete mathematics when appropriate, which includes graph theory and even HMO when appropriate; and toward the use of abstract graphical approaches with preference to numerical characterizations and matrix representations, which have an obvious advantage of being susceptible to computer manipulations.

Concerning the 23 numbered problems (and other suggestions scattered and unnumbered), by sticking to 23 to honor Hilbert, in no way do the majority of our problems bear similarity to those of Hilbert. Our problems, if solved, will not bring glory to the solvers, but our criterion for selection of the problems listed is that they ought to at least bring a publication to those involved in solving them!

After examining and reading in two days the proof of the whole book, which was corrected for various typing, spelling, and factual errors, we found that it would be desirable in a form of additional appendices to give an illustration of construction of orthogonal MRA descriptors (Appendix 20); to introduce and illustrate weighted Kekulé valence structures (Appendix 21); to add a comment with updates on Fries Kekulé valence structures (Appendix 22); and finally to offer a commentary on prolonged hostility toward chemical graph theory in some circles (Appendix 23). In this way, not only we have 23 problems but also 23 appendices to further stress our respect for great mathematician, David Hilbert.

Finally, the ultimate purpose of this book has been to draw the attention of chemists, in particular those interested in molecular structure, not only to a number of unsolved and incompletely solved problems, but also to a number of solved problems, some of which by some may have been unintentionally or intentionally overlooked. A call for awakening the sleeping community should be welcome; Sleeping Beauties and Sleeping Giants should be events of the past. We apologize also for trying to cover a limited area of chemistry, the one that we are familiar with, and hope that the "torch of the light" we carried over the pages of this book will from our hands pass to others, not forgetting that *In absentia luci, tenebrae vincunt.*

REFERENCES AND NOTES

1. S. Altschul, W. Gish, W. Miller, E. Myers, and D. Lipman, Basic local alignment search tool, *J. Mol. Biol.* 215 (1990) 403–410.
2. S. F. Altschul, T. L. Madden, A. A. Schaffer, J. Zhang, Z. Zhang, W. Miller, and D. J. Lipman, Gapped BLAST and PSI-BLAST: A new generation of protein database search programs, *Nucleic Acid Res.* 25 (1997) 3389–3402.
3. T. Pisanski, private communication, April 2015.
4. H. S. M. Coxeter, Congressus Numerantum (Winipeg, Canada) 1981.
5. H. C. Longuet-Higgins, Some studies in molecular orbital theory I. Resonance structures and molecular orbitals in unsaturated hydrocarbons, *J. Chem. Phys.* 18 (1950) 265–274.
6. M. Randić, Novel graph theoretical approach to heteroatom in quantitative structure-activity relationship, *Chemom. Intell. Lab. Syst.* 10 (1991) 213–227.
7. M. Randić, Quantitative characterization of proteomics maps by matrix invariants, in: *Handbook of Proteomics Methods*, P. M. Conn (Ed.), Humana Press, Inc. Totowa, NJ, 2003, pp. 429–450.
8. M. Randić, Very efficient search for nucleotide alignments, *J. Comput. Chem.* 34 (2013) 77–82.
9. M. Randić, Very efficient search for protein alignments—VESPA, *J. Comput. Chem.* 33 (2012) 702–707.
10. M. Randić, On a geometry-based approach to protein sequence alignment, *J. Math. Chem.* 43 (2008) 756–772.
11. P. von Rague Schleyer and H. Jiao, What is aromaticity?, *Pure Appl. Chem.* 68 (1996) 209–218.
12. A. R. Katritzky, K. Jug, and D. C. Oniciu, Quantitative measures of aromaticity for mono-, bi-, and tricyclic penta- and hexaatomic hegteroaromatics ring systems and their interrelationships, *Chem. Rev.* 101 (2001) 1421–1449.
13. M. K. Cyranski, T. M. Krygovski, A. R. Katritzky, and P. von Rague Schleyer, To what extent can aromaticity be defined uniquely?, *J. Org. Chem.* 67 (2002) 1333–1338.
14. A. Kekulé, Sur la constitution des substances aromatiques, *Bull. Soc. Chim.* 3 (1865) 98–110.
15. K. Fries, Über bicyclische Verbindungen und ihren Vergleich mit dem Naphtalin, *Justus Liebigs Ann. Chem.* 454 (1927) 121–324.

16. E. Hückel, *Grundzuge der Theorie ungesattiger und aromatischer Verbindugen*, Verlag Chemie, Berlin, 1938.
17. M. Randić, Comment on difference between bond orders calculated by SCF MO and simple MO method, *J. Chem. Phys.* 34 (1962) 693–694.
18. E. Clar, *The Aromatic Sextet*, John Wiley & Sons, London, 1972.
19. M. Randić, Conjugated circuits and resonance energies of benzenoid hydrocarbons, *Chem. Phys. Lett.* 38 (1976) 68–70.
20. M. Randić, Aromaticity and conjugation, *J. Am. Chem. Soc.* 99 (1977) 444–450.
21. I. Gutman and M. Randić, A correlation between Kekulé valence structures and conjugated circuits, *Chem. Phys.* 41 (1979) 265–270.
22. M. Randić, Novel insight into Clar's aromatic π-sextets, *Chem. Phys. Lett.* 601 (2014) 1–5.
23. M. Randić, Aromaticity of polycyclic conjugated hydrocarbons, *Chem. Rev.* 103 (2003) 3449–3605.
24. G. Orwell, *Animal Farm*. London: Penguin Group, 1946.
25. J. N. Murrell, S. F. A. Kettle, and J. M. Tedder, *Valence Theory*, John Wiley & Sons Ltd, New York, 1970.
26. M. Randić, A graph theoretical approach to conjugation and resonance energies of hydrocarbons, *Tetrahedron* 33 (1977) 1905–1920.
27. A. Ciesielski, T. M. Krygowski, M. K. Cyranski, and J.-I. Aihara, Graph topological approach to magnetic properties of benzenoid hydrocarbons, *Phys. Chem. Chem. Phys.* 11 (2009) 11447–11455.
28. A. Ciesielski, T. M. Krygowski, M. K. Cyranski, and A. T. Balaban, Defining rules of aromaticity: A unified approach to the Hückel, Clar and Randić concepts, *Phys. Chem. Chem. Phys.* (2011).
29. I. Gutman, S. J. Cyvin, Introduction to the Theory of Benzenoid Hydrocarbons, Springer, Berlin, 1989.
30. M. Randić, Resolution of ambiguities in structure-property studies by use of orthogonal descriptors, *Chem. Inf. Comput. Sci.* 37 (1991) 311–320.
31. M. Randić, Orthogonal molecular descriptors, *New J. Chem.* 15 (1991) 517–525.
32. H. A. Hauptman (as told to and edited by D. J. Grothe), *On the Beauty of Science (A Nobel Laureate Reflects on the Universe, God, and the Nature of Discovery)*, Prometheus Books, Amherst, NY.
33. M. Pompe and M. Randić, The variable connectivity index revisited—Insight into structural interpretation, *J. Chem. Inf. Modeling* (submitted).
34. N. L. Anderson, R. Esquer-Blasco, F. Richardson, P. Foxworthy, and P. Eacho, The effect of peroxisome proliferation on protein abundances in mouse liver, *Toxicol. Appl. Pharmacol.* 137 (1996) 75–89.
35. M. Randić, On graphical and numerical characterization of proteomics maps, *J. Chem. Inf. Comput. Sci.* 41 (2001) 1330–1338.
36. M. Randić, J. Zupan, and M. Novič, On 3-D graphical representation on proteomics maps and their numerical characterization, *J. Chem. Inf. Comput. Sci.* 41 (2001) 1339–1344.
37. M. Randić, F. Witzmann, M. Vračko, and S. C. Basak, On characterization of proteomics maps and chemically induced changes in proteomes using matrix invariants: Application to peroxisome proliferators, *Med. Chem. Res.* 10 (2001) 456–479.
38. M. Randić and S. C. Basak, A comparative study of proteomics maps using graph theoretical biodescriptors, *J. Chem. Inf. Comput. Sci.* 42 (2002) 983–992.
39. M. Randić, M. Novič, and M. Vračko, On characterization of dose variations of 2-D proteomics maps by matrix invariants, *J. Proteome Res.* 1 (2002) 217–226.
40. M. Randić, M. Novič, M. Vračko, and D. Plavšić, Study of proteome maps using partial ordering, *J. Theor. Biol.* 266 (2010) 21–28.
41. M. Randić and E. Estrada, Order from chaos: Observing hormesis at the proteome level, *J. Proteome Res.* 4 (2005) 2133–2136.

42. E. J. Calabrese, Historical blunders: How toxicology got the dose-response relationship half right, *Cell. Mol. Biol.* 51 (2005) 643–654.
43. E. J. Calabrese and L. A. Baldwin, Toxicology rethinks its central belief. *Nature* 421 (2003) 691–321.
44. E. J. Calabrese, Hormesis: Changing view of the dose–response, a personal account of the history and current status, *Mutat. Res.* 511 (2002) 181–189.
45. E. J. Calabrese and L. A. Baldwin, Chemical hormesis: Its historical foundations as a biological hypothesis, *Hum. Exp. Toxicol.* 19 (2000) 2–31.
46. E. J. Calabrese and L. A. Baldwin, The marginalization of hormesis, *Hum. Exp. Toxicol.* 19 (2000) 32–40.
47. E. J. Calabrese and L. A. Baldwin, Hormesis: The dose–response revolution, *Annu. Rev. Pharmacol. Toxicol.* 43 (2003) 175–197.
48. E. J. Calabrese and L. A. Baldwin, Tales of two similar hypotheses: The rise and fall of chemical and radiation hormesis, *Hum. Exp. Toxicol.* 19 (2000) 85–97.
49. E. J. Calabrese and L. A. Baldwin, Radiation hormesis: The demise of a legitimate hypothesis, *Hum. Exp. Toxicol.* 19 (2000) 76–84.
50. M. Randić and Z. Supek.
51. P. Rosario, *Underdetermination of Science: Part I*, Lulu.com, Raleigh, 2006, p. 68.
52. M. Freiberger, Grappling with chaos: The Abel Prize, Plus, May 2014; plus.math.org.
53. M. Barnsley, Chaos game, in: *Fractals Everywhere*, Morgan Kaufmann, 1993.
54. H. I. Jeffrey, Chaos game representation of gene structure. *Nucleic Acid Res.* 18 (1990) 2163–2170.
55. M. Randić, Another look at the chaos game representation of DNA, *Chem. Phys. Lett.* 456 (2008) 84–88.
56. M. Randić, M. Novič, D. Vikić-Topić, and D. Plavšić, Novel numerical and graphical representation of DNA sequences and proteins, *SAR QSAR Environ. Res.* 17 (2006) 583–595.
57. M. Randić, J. Zupan, D. Vikić-Topić, and D. Plavšić, A novel unexpected use of a graphical representation of DNA: Graphical alignment of DNA sequences, *Chem. Phys. Lett.* 431 (2006) 375–379.
58. T. C. Hales, A computer verification of the Kepler conjecture, in: *Proc. Int. Congress Mathematicians, Vol. II. Invited Lectures, Held in Beijing, August 20–28, 2002*, T. Li (Ed.), Higher Education Press, Beijing, China, 2002, pp. 795–804.
59. A. F. J. van Raan, Sleeping Beauties in science, *Scientometrics* 59 (2004) 467–472.
60. W. Kohn, From a Personal Introduction (a foreword in a book *Reviews of Modern Quantum Chemistry; A Celebration of the Contributions of Robert G. Parr, Vol. 1*), editor K. D. Sen, World Scientific, Singapore, 2002).
61. E. B. Wilson, *Introduction to Scientific Research*, McGraw-Hill, New York, 1952.
62. H. Wiener, Structural determination of paraffin boiling points, *J. Am. Chem. Soc.* 69 (1947) 17–20.
63. J. R. Platt, Influence of neighbor bonds on additive bond properties in paraffins, *J. Chem. Phys.* 15 (1947) 419–420.
64. M. Randić, Characterization of molecular branching, *J. Am. Chem. Soc.* 97 (1975) 6609–6615.
65. H. W. Eves, *Mathematical Circles Revisited*, Prindle, Weber and Schmidt, Boston, 1988.
66. J. Gasteiger, Some solved and unsolved problems of chemoinformatics, *SAR QSAR Environ. Res.* 25 (2014) 443–455.
67. M. Randić, R. Orel, and A. T. Balaban, D_{MAX} matrix invariants as graph descriptors. Graphs having the same Balaban index *J. MATCH—Commun. Math. Comput. Chem.* 70 (2013) 239–258.

Appendix 1: Early Hostility toward Chemical Graph Theory

Report received from the *Journal of Chemical Physics* in 1979 (reproduced with permission from "On the history of the Randić index and emerging hostility towards chemical graph theory," M. Randić, *MATCH Commun. Math. & Comput. Chem.*, vol. 59, copyright © 2008):

REFEREES REPORT FILE: 1363A8

TITLE: A CORRELATION BETWEEN KEKULÉ VALENCE STRUCTURES AND CONJUGATED CIRCUITS

Authors: Ivan Gutman and Milan Randić

My colleague, _____, and I have examined the above paper. We have decided not to referee it in the usual sense of providing an analysis of strengths and weaknesses because we feel that, being about neither chemistry nor physics, it is not appropriate for the JCP. This paper is representative of a genre which has grown up over the past few years in which various numbers are associated with a molecular structure (usually derived by simply counting the number of members of some class of substructures) and then correlations are sought between these numbers and some other molecular property, frequently a number resulting from another exercise of the same kind. These studies are usually coached in the language of graph theory but they do not lead to any deeper insights into the mathematics of structure, in fact they tend to be mathematically quite unsophisticated.

Since papers of this class contain no physics, negligible chemistry and near-trivial mathematics it is hard to say where they might find a home in the scientific literature. (Unfortunately a few of them have found their way into JCP and few other journals of similarly high standards). Many of these papers have been published in *Croat. Chem. Acta*, *Rev. Roum. Chem.*, *Bull. Chem. Soc. Japan* and *Mathematical Chemistry*; while I am not personally familiar with any of these journals, perhaps one of them would be appropriate place. I do not believe that this paper should be published in *Journal of Chemical Physics*. The use of graph theory [to enumerate the numbers and types of polycyclic systems, to single out single and double bonds, and to predict resonance energies] has been Dr. Randic's main concern for the past few years. He has published a large number of papers and notes on this method, some of which have presented interesting insights into the concept of resonance. I am afraid I do not believe that the present paper lives up to that criterion. The present paper presents the proofs of three theorems, one of which, (g), is used to find all Kekulé structures for coronene from the examination of a single Kekulé structure; and then the idea of disjoint conjugated circuits is discussed in relation to Clar's sextet idea. While these are amusing, they do not, in my opinion, contain enough physics or chemistry to make them of interest to the chemical physics community (or even to a significant fraction of that community). I realize that this is an opinion, not a fact, and that Dr. Randic will not like it any better than the other referee's report; however I think he realizes that most of the scientists interested in chemical bonding in aromatic polycyclic molecules do not find this approach very fruitful, and it is up to the practitioners of this method (Dr. Randic, for example) to convince the rest of the error of their judgment. This paper does not do that.

What is depressing is not that there are chemists who have opinions that are not based on facts, but that there are editors who make decision on such speculative reports, before asking author for rebuttal, forgetting or not knowing what Lynn Margulis [1] has said:

> So I don't see how people can have strong opinions ... Let me put this way: opinions aren't science. There is no scientific basis! It is just opinion!

After being rejected by the *Journal of Chemical Physics* the paper was (unchanged) sent to *Chemical Physics*, the leading scientific journal on chemical physics in Europe. The following is the report received from *Chemical Physics* (dated January 5, 1979):

> Although I have not checked the mathematics carefully, I suspect it is correct. I find this to be a very interesting, stimulating, and a clear paper about a subject that few would have thought contained such plums. I recommend publication.

This evaluation can serve as a rebuttal and at the same time a warning to reviewers and editors that reviewers always speak for themselves and not to speak for chemical community at large, which is an arrogant presumption on their part. One can find in the article: on the history of Randić index and emerging hostility towards graph theory [2] additional illustration of misevaluation of papers on chemical graph theory. We may add that after this article was published, the secretary of Eugene Garfield wrote a letter asking that we send the abstract of the paper, which Dr. Garfield wanted to put on the web. Eugene Garfield is an American scientist, among others, one of the founders of bibliometrics and scientometrics; the founder of the Institute for Scientific Information, who initiated *Current Contents*, the *Science Citation Index*; and the founding editor and publisher of *The Scientist*. The article did not have an abstract, so a one-page abstract was made and sent to Dr. Garfield. Depressing as the two reports appear, which could be devastating for younger researchers with limited experience on unlimited arrogance not uncommon in science, we in chemical graph theory have been receiving also very favorable views, indicating that we are on a right road. We have already illustrated in Chapter 14 the letters of D. Hellwinkel from Heidelberg and L. Pauling, who has been called as one of the 20 greatest scientists of all time [3]. To this we may add a report by an anonymous reviewer of *Tetrahedron*, one of prestigious journals of organic chemistry, on the paper of M. Randić on graph theoretical approach to conjugation:

> This paper presents a novel and impressively revealing application of graph theory to chemistry. I am amazed how the simple concept of conjugated circuits is utilized to provide a wealth of qualitative and quantitative information about conjugated

hydrocarbons. The contribution is timely in view of the widespread interest in structure-stability-reactivity relationships. The manuscript is lucidly written ...

This review was accompanied with the following questions:

1. Does this article incorporate novel and original organic chemistry—theoretical or experimental?
2. Will the paper be read with interest and enjoyment by a sizable group of organic chemists?
3. Does the paper describe a substantial, definitive piece of work?

On the scale 1–7 (7 being the best), the anonymous reviewer answered on all three questions with 7.

REFERENCES

1. J. Horgan, End of Science, *The End of Science: Facing the Limits of Science in the Twilight of the Scientific Age*, New York: Broadway Books (1996).
2. M. Randić, On the history of the Randić index and emerging hostility towards graph theory, *MATCH—Commun. Math. & Comput. Chem.* 69 (2008) 5–124.
3. J. Horgan, Profile: Linus C. Pauling—Stubbornly ahead of his time, *Scientific American* 266 (3), (1993) 36–40.

Appendix 2: Editorial Alert to Reviewers on Graph Theoretical Manuscripts

(Reproduced with permission from "On the history of the Randić index and emerging hostility towards chemical graph theory," M. Randić, *MATCH Commun. Math. & Comput. Chem.*, vol. 59, copyright © 2008.)

THE JCICS POLICY ON GRAPH THEORY MANUSCRIPTS

The Journal of Chemical Information and Computer Sciences is adopting a new policy concerning papers in the area of graph theory. To be publishable, a manuscript must now have truly innovative and unique ideas that make major advances in graph theory applied to chemical systems, or, it must use graph theory in a way as to make a large scientific advancement.

Accordingly, if the manuscript you are reviewing: (1) is presenting a trivial advance in graph theoretical methodology and lack any application to a chemical problem of significance (the significance of the work needs to be clearly delineated in the author's manuscript) or if it is of purely graph theory that should be published in a mathematical journal, we are asking you to note this in your review or (2) if the manuscript has no clearly stated scientific hypothesis that are being proved or disproved, i.e., if the work is simply yet another GT study of some system that is narrow in scope, or if the manuscript simply compares its methodology to existing GT methods without attempting to show its general applicability to solving problems in chemistry and informatics, we ask that you also indicate this in your review.

(Anonymous)

P. S.
There is nothing wrong with this Editorial Alert to Reviewers—except that it is solely addressed to graph theoretical manuscripts. Thus, apparently manuscript on other aspects of theoretical chemistry, including quantum chemistry, all "have truly innovative and unique ideas that make major advances in ... theory applied to chemical systems, or, must use ... theory in a way as to make a large scientific advancement."

Appendix 3: Letter Relating to the Calculation of the Symmetry of a Graph

(Reproduced with permission from "On the history of the Randić index and emerging hostility towards chemical graph theory," M. Randić, *MATCH Commun. Math. & Comput. Chem.*, 2008.)

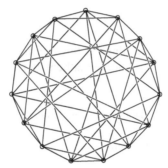

Dear Colleague,

Recently I analyzed the graph shown here (which arises in a study of isomerizations) and established to have automorphism group S_6. A referee of a distinguished journal finds my finding—obvious—but has not forwarded explanation. I determined the symmetry (automorphism) using canonical labels in a process which is neither obvious nor trivial. As you know, the determination of symmetry of graphs (as is also case with related problem of graph isomorphism) is NP-complete, generally viewed as intractable problems, not trivial. Do you find the problem and the solution obvious?

Please, send your brief comment to:

Milan Randić
3225 Kingman Rd.,
Ames, Iowa 50014
First Day of Summer 1982

P. S.
One refers to a problem as intractable if it is so hard that no polynomial time algorithm can possibly solve it.

With kind regards, Milan Randić

I contacted a number of leading mathematicians in graph theory and a number of outstanding chemists familiar with graph theory, including H. S. M. Coxeter, F. Harary,

R. Frucht, and E. K. Lloyd among mathematicians and K. Mislow, D. Herschbach, P. G. Mezey, A. T. Balaban, Hs. H. Günthard, H. Primas, M. Gielen, and A. Mead among chemists. From each of them, I received a supportive response. Professor Coxeter, at that time the president of the Mathematical Society of Canada, known to readers of *Scientific American* as "Mr. Geometry," and widely respected as the leading authority in combinatorial geometry sent a brief letter (dated July 25, 1982):

Dear Dr. Randić,

Thanks for your letter about the 15-point graph. I find it interesting and consider the referee to be incompetent; he should not have called your conclusion "obvious." It took me about ten minutes to see how to name the 15 vertices with the pairs of 6 symbols so that two of them are joined if they have no common symbol in common.

Best regards,
HSM Coxeter

This response was sent to the editor of the *Journal of Chemical Physics*, introducing Professor Coxeter as "the leading mathematician in this area." The response received follows by the editor:

It took only 15 minutes for the leading mathematician in this area to solve the problem.

To this the following reply was sent:

...No, I solved the problem, Professor Coxeter verified and found the solution correct. To see the difference here is cubic equation $x^3 - 2x^2 - x + 2 = 0$. Try to solve it using Cardano formulas and it may take you 10–15 minutes, but if I tell you that the solutions are $x = \pm 1$ and $x = 2$, you can verify that this is correct in a few seconds.

I did not elaborate that cubic equations with simple coefficients can be often solved quickly by other methods. This whole correspondence took some time, and meanwhile, I published this "simplistic" (or should I say "childish"?) paper (in collaboration with M. I. Davis) elsewhere [1]. The editor of the *Journal of Chemical Physics* apparently overlooked the difference between *solving a problem* and *verifying* that a solution is correct. Determining the symmetry of graphs is an NP problem (involving non-polynomial algorithms), and the verification of the same problem is P (polynomial in nature). I have, however, to express my respect for Professor Stout, the editor of the *Journal of Chemical Physics*, because he was not only kind enough to continue the correspondence, but was taking appropriate measures to get a better appraisal of my work.

My next paper in the symmetry of isomerization graphs, which I sent to him, was accepted by Professor Stout for publication. Unfortunately, soon after, before the paper was processed and printed, Professor Stout was replaced as the editor of the *Journal of Chemical Physics*, and Professor Light became the new editor. He sent the already *accepted* manuscript again for review. The accepted paper was rejected based on the letter shown below by a new referee:

Dear John,

Here are my comments on Randić's paper and a general assessment of this line of research. The reasons why the Journal has generally declined to publish this sort of things are pretty clear after even a quick run through this manuscript. Firstly there is precisely little here of any chemical substance at all. There is a fair bit of hand waving, dressed up for the occasion with plenty of graph theorist's jargon... Secondly, and this point transcends the first, graph theory at present does not have the mathematical tools to do many of the things we'd want it to do. So the discussion for example which starts on page 9 uses 'trial-and-error' or ' "educated guess" ' (his " ") type approaches to the problem of finding alternative graphical descriptions. This happens to be a fundamental part of the paper and in terms of pecking order of intellectual endeavors this must rank pretty near the bottom. It frankly reminds me of the geometrical puzzles my 10-year old does. To answer one of your questions, there is nothing incorrect in this paper, just not much of any depth at all...

REFERENCE

1. M. Randić and M. I. Davis, Symmetry properties of chemical graphs VI. Isomerization of octahedral complexes, *Int. J. Quantum Chem.* 26 (1984) 69–89.

Appendix 4: On Detection of the Illegal Deck of Cards for Graph Reconstruction

We will illustrate on the graph of 3-methylhexane in Figure A4.1 our (unpublished) method for detecting if a "deck of cards" having Ulam subgraphs has been tampered with.

In Figure A4.1, below the graph of 3-methylhexane at left we show in bold seven Ulam subgraphs the graph of 3-methylhexane. At the right part of the same figure are the same seven Ulam subgraphs with one of them being altered, hence representing an illegal set. The question is: Can one, once given a set of seven subgraphs determine whether the set is legal or illegal, that there is not some error in the subgraphs given, deliberate or accidental, and they do not represent Ulam subgraphs?

A way to find this is to construct for each given Ulam subgraph U(i) its Ulam subgraphs U(i, j). A graph having n vertices will have n Ulam subgraphs U(i), each of which will have ($n - 1$) Ulam subgraphs U(i, j). All these n ($n - 1$) subgraphs can be arranged as a symmetrical $n \times n$ matrix, such as is the matrix in Figure A4.2. It follows necessarily that the matrix is symmetrical because elements U(i, j) = U(j, i) only differ in the order in which vertices i and j have been deleted from the original graph G. As one can see, in the case of the Ulam subgraphs of Figure A4.1, there are eight different subgraphs when two vertices have been deleted: butane with an isolated atom and propane with two isolated carbons occurring 10 times; pentane and two ethanes with an isolated carbon occuring six times; isopentane occurring four times; while isobutane with an isolated carbon, ethane, and propane, and ethane with three isolated carbons occurring twice. This adds to 42, which is the number of nonzero matrix elements. One concludes that the Ulam subgraphs at the left side of Figure A4.1 are legitimate.

If one constructs all Ulam subgraphs for the set of subgraphs at the right side of Figure A4.2 and counts the so-obtained 42 subgraphs, one finds that a few of the subgraphs occur an odd number of times. This statement is easy to verify if one considers the subgraph of isopentane (shown at the top of Figure A4.3 and counts how many times it occurs in the Ulam subgraphs of the Ulam subgraphs in Figure A4.1. It is not difficult to see that, for the legal Ulam subgraphs at the left of Figure A4.1, the isopentane subgraph occurs four times, but in the case of the illegal set of subgraphs (at the right of Figure A4.1), the isopentane subgraphs occur only three times.

There are three additional subgraphs besides isobutane: isobutane with an isolated carbon, propane with two isolated carbons, and ethane with three isolated carbons, which occur an odd number of times. All this is strong evidence that the set of subgraphs at right in Figure A4.1 are not the Ulam subgraphs of the graph of 3-methylhexane.

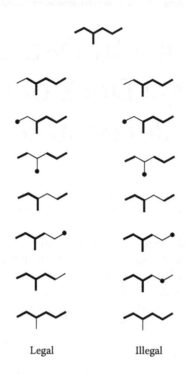

Legal   Illegal

FIGURE A4.1 At left, Ulam subgraphs of 3-methylhexane graph and, at right, one of the subgraphs has been changed.

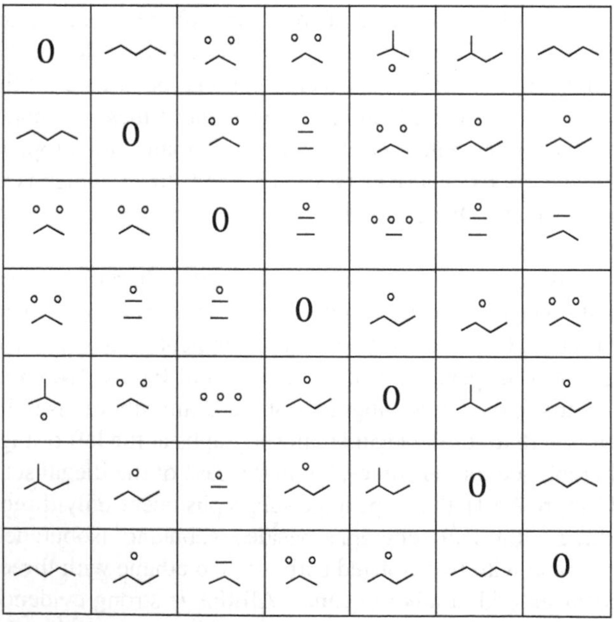

0	~~~	°° ⌢	°° ⌢	⊥ₒ	⅄	~~~
~~	0	°° ⌣	＝°	°° ⌢	° ~	° ~
°° ⌢	°° ⌢	0	°＝	° °° ＿	°＝	⁻⌢
°° ⌢	°＝	°＝	0	° ~	° ~	°° ⌢
⊥ₒ	°° ⌢	° °° ＿	° ~	0	⊥	° ~
⊥	° ~	°＝	° ~	⊥	0	~~~
~~	° ~	⁻⌢	°° ⌣	° ~	~~~	0

FIGURE A4.2 Ulam subgraphs of Ulam subgraphs of Figure A4.1 (left side).

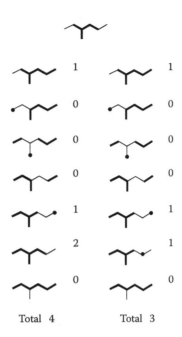

FIGURE A4.3 The count of occurrence of isopentane subgraphs in the two sets of subgraphs of Figure A4.1. A presence of an odd count indicates that the set does not represent the Ulam subgraphs of any graph.

Appendix 5: Solution to Halmos Handshake Problem

We have to account for the assignment of handshakes 0, 1, 2, 3, 4, 5, 6, 7, 8 to the nine persons giving an answer to Professor Halmos. In Figure A5.1, 10 persons are represented by 10 vertices, to which we will gradually assign labels. Let us start by assigning the letter A to one of the 10 vertices and assume that this is the person who made eight handshakes. We can represent each handshake between two persons as an edge connecting the corresponding vertices. Hence, we can draw eight edges emanating from vertex A as is illustrated in Figure A5.1a. As one can see, this leaves one unconnected vertex, to which we may give the label A*, which must stand for the spouse of person A because, according to the rule stated, one does not shake hands with one's spouse. With this, we assign labels 8 and 0 to A and A*, respectively, and are left with seven unassigned seven: 1, 2, 3, 4, 5, 6, 7. We continue and select one of the eight equivalent vertices, each already having one handshake, to which we give label B and attribute seven handshakes. Vertex B already has one edge (the handshake with A), so we draw an additional six edges coming out from vertex B as shown in Figure A5.1b. We are now left with the following five unidentified persons each already having two handshakes who had 2, 3, 4, 5, and 6 handshakes. The vertex with a single edge must be the spouse of B, labeled as B*. We select one of the five equivalent vertices (each having two edges) as person C, who had six handshakes, so we have to draw four additional handshakes coming from C as illustrated in Figure A5.1c. The vertex left with two edges must be representing the spouse of person C, indicated as C*. We are now left with four vertices each having three edges and three persons reporting 3, 4, and 5 handshakes. Let us assign to one of the four equivalent vertices label D and assume that this person had in total five handshakes. Because vertex (person) C already has three handshakes, we have to add two additional edges to vertex D, obtaining the graph shown in Figure A5.1d. The vertex with three edges must be spouse D* of person D. As one can see from the handshaking graph, the eight labeled vertices A–D and A*–D* have degrees 8, 7, 6, 5, 4, 3, 2, 1, 0, the answers given to Professor Halmos. Observe that pairs (A, A*), (B, B*), (C, C*), and (D, D*), which together shook eight hands, correspond to a wife and a husband. Hence, the last two vertices, each having four handshakes, represent Professor Halmos and his wife because if this were not the case, if Professor Halmos and his wife were any other couple, Professor Halmos would have heard the number four mentioned twice, which was not the case.

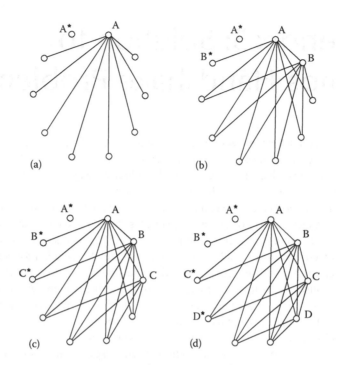

FIGURE A5.1   Ten persons shown as vertices and their handshakes shown as edges of the graph. (a) Person A having eight handshakes and spouse having zero handshakes. (b) Person B having seven handshakes and spouse having one handshake. (c) Person C having six handshakes and spouse having two handshakes. (d) Person D having five handshakes and spouse having three handshakes.

Answering the two additional questions: (1) What is the (here assumed) name of Mrs. Halmos? (2) How to identify the five couples by their names? is easy because each couple has eight handshakes together. The corresponding names are, say, Ann, Barbara, Carmen, Doris, Elisabeth, Francis, George, Henry, and Ivan (in short, the nine names A, B, C, D, E, F, G, H, and I). If we assume that A, B, C, D, and E stand for Ann, Barbara, Carmen, Doris, and Elisabeth, then F, G, H, and I, stand for D*, C*, B*, and A*, respectively, which identifies Mrs. Halmos as Elisabeth.

Appendix 6: The First Page of the Book of Euclid

Definition 1. A *point* is that which has no part.

Definition 2. A *line* is breadthless length.

Definition 3. The ends of a line are points.

Definition 4. A *straight line* is a line which lies evenly with the points on itself.

Definition 5. A *surface* is that which has length and breadth only.

Definition 6. The edges of a surface are lines.

Definition 7. A *plane surface* is a surface which lies evenly with the straight lines on itself.

Definition 8. A *plane angle* is the inclination to one another of two lines in a plane which meet one another and do not lie in a straight line.

Definition 9. And when the lines containing the angle are straight, the angle is called *rectilinear.*

Definition 10. When a straight line standing on a straight line makes the adjacent angles equal to one another, each of the equal angles is *right*, and the straight line standing on the other is called a *perpendicular* to that on which it stands.

Definition 11. An *obtuse angle* is an angle greater than a right angle.

Definition 12. An *acute angle* is an angle less than a right angle.

Definition 13. A *boundary* is that which is extremity of anything.

Definition 14. A *figure* is that which is contained by any boundary or boundaries.

Definition 15. A *circle* is a plane figure contained by one line such that all the straight lines falling upon it from one point among those lying within the figure equal one another.

Definition 16. And the point is called the *center* of the circle.

Definition 17. A *diameter* of the circle is any straight line drawn through the center and terminated in both directions by the circumference of the circle, and such a straight line also bisects the circle.

Definition 18. A *semicircle* is the figure contained by the diameter and the circumference cut off by it. And the center of the semicircle is the same as that of the circle.

Definition 19. *Rectilinear figures* are those which are contained by straight lines, *trilateral* figures being those contained by three, *quadrilateral* those contained by four, and *multilateral* those contained by more than four straight lines.

Definition 20. Of trilateral figures, an *equilateral triangle* is that which has its three sides equal, an *isosceles triangle* that which has two of its sides alone equal, and a *scalene triangle* that which has its three sides unequal.

Definition 21. Further, of trilateral figures, a *right-angled triangle* is that which has a right angle, an *obtuse-angled triangle* that which has an obtuse angle, and an *acute-angled triangle* that which has its three angles acute.

Definition 22. Of quadrilateral figures, a *square* is that which is both equilateral and right-angled; an *oblong* that which is right-angled but not equilateral; a *rhombus* that which is equilateral but not right-angled; and a *rhomboid* that which has its opposite sides and angles equal to one another but is neither equilateral nor right-angled. And let quadrilaterals other than these be called *trapezia*.

Definition 23. *Parallel* straight lines are straight lines which, being in the same plane and being produced indefinitely in both directions, do not meet one another in either direction.

Appendix 7: Full Quote of Immanuel Kant

Short quote of I. Kant (1786):

> In any special doctrine of nature there can only be as much proper science as there is mathematics therein. And since chemistry fails to satisfy this condition, chemistry can be nothing more than a systematic art or experimental doctrine, but never a proper science.

The above is a shortened quote of a longer and convoluted statement, which we have listed fully below:

> In any special doctrine of nature there can only be as much proper science as there is mathematics therein. And since chemistry fails to satisfy this condition, it is an improper science. So long, therefore, as there is still for chemical actions of matters on one another no concept to be discovered that can be constructed, that is, no law of approach or withdrawal of the parts of matter can be specified according to which, perhaps in proportion to their density or the like, their motions and all the consequences thereof can be made intuitive and presented a priori in space (a demand that will only with great difficulty ever be fulfilled), then chemistry can be nothing more than a systematic art or experimental doctrine, but never a proper science, because its principles are merely empirical, and allow of no a priori presentation in intuition. Consequently, they do not in the least make the principles of chemical appearances conceivable with respect to their possibility, for they are not receptive to the application of mathematics.

We thank Professor Ivan Gutman (University of Kragujevac, Serbia) for the correspondence on Immanuel Kant.

Appendix 8: Construction of Canonical Labeling for Diamantane

One has to assign labels 1–14 to the 14 carbon atoms of diamantane so that the smallest labels have as adjacent labels with the largest numbers. This requirement will ensure that the adjacency matrix, when its rows are read from left to right and from top to bottom, results in the smallest binary number. We show below the construction of the canonical labels for diamantane.

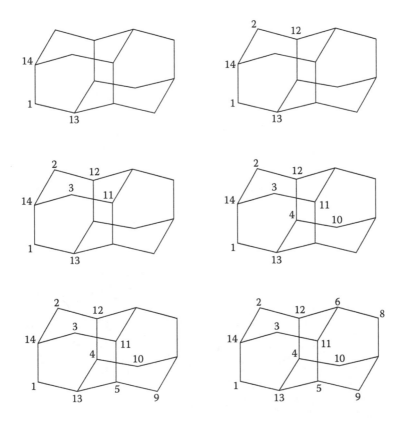

FIGURE A8.1 Construction of the canonical labeling for diamantane.

Appendix 9: Twelve Different Canonical Labeling of Diamantane

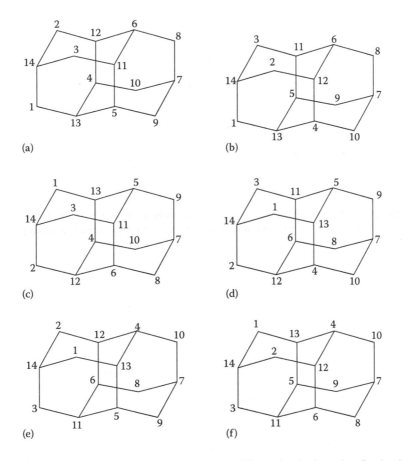

FIGURE A9.1 The six symmetry-related labeling of the molecule through reflection in the vertical plane passing through atoms 7 and 14 (structures from left to right and top to bottom have labels a–f).

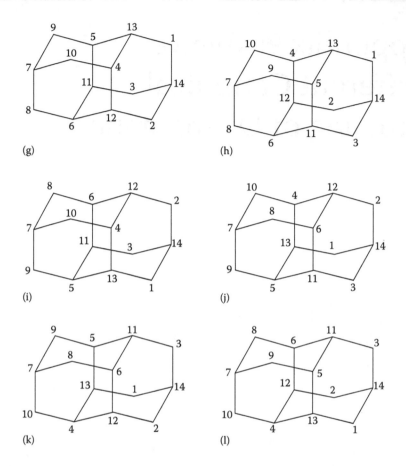

FIGURE A9.2 The six additional labeling obtained by reflection of vertex labels through the center of the molecule of the first six labels (structures from left to right and top to bottom have labels g–l).

Appendix 10: The Eigenvalues and Eigenvectors of Hexatriene

Table A10.1 shows the eigenvalues and eigenvectors of hexatriene. The coefficients a, b, c of Table A10.2 are proportional to the coefficients c, b, a respectively of Table A10.1, and thus also represent eigenvectors of hexatriene (with different normalization, and interchange of labels a and c).

TABLE A10.1
The Eigenvalues and Eigenvectors of Hexatriene

	$\lambda_1 = 1.8019$	$\lambda_2 = 1.2470$	$\lambda_3 = 0.4450$	$\lambda_4 = -0.4550$	$\lambda_5 = -1.2470$	$\lambda_6 = -1.8019$
a	0.2319	−0.4179	0.5211	−0.5211	−0.4179	0.2319
b	0.4179	−0.5211	0.2319	0.2319	0.5211	−0.4179
c	0.5211	−0.2319	−0.4179	0.4179	−0.2319	0.5211
c	0.5211	0.2319	−0.4179	−0.4179	−0.2319	−0.5211
b	0.4179	0.5211	0.2319	−0.2319	0.5211	0.4179
a	0.2319	0.4179	0.5211	0.5211	−0.4179	−0.2319

TABLE A10.2
The Sum Rule for the Coefficients of the Graph of Figure 4.6 in Chapter 4

	$\lambda_1 = 1.8019$	$\lambda_2 = 1.2470$	$\lambda_3 = 0.4450$
c	−0.2700	−0.1788	0.1026
b	−0.2165	−0.4018	−0.0569
a	−0.1202	−0.3222	−0.1279
−a	0.1202	−0.3222	0.1279
−a − b	0.3367	−0.0796	0.1849
a − c	0.1498	0.1434	−0.2305
2a + b − 2c	0.0831	0.1150	−0.5180
−a + c	−0.1498	0.1434	0.2305
c	−0.2700	0.1788	0.1026
a + b	−0.3367	0.0796	−0.1849

Appendix 10: The Eigenvalues and Eigenvectors of Hexatriene

Appendix 11: The Coefficients of All Antisymmetric Molecular Orbitals of the Tetracene Derivative $C_{20}H_{14}$

The coefficients a, b, c of antisymmetric molecular orbitals of the tetracene derivative $C_{20}H_{14}$ satisfy the same sum rules that hold for antisymmetric molecular orbitals of hexatriene in Table A10.1 in Appendix 10 that the magnitudes of the coefficients of symmetric and antisymmetric eigenvectors have the same magnitudes and only differ in their signs.

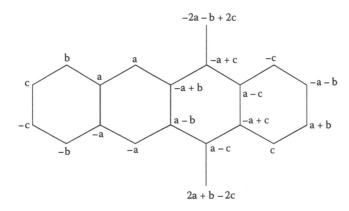

FIGURE A11.1   The coefficient of the antisymmetric eigenvectors.

Appendix 12: Halogen Compounds and Their Boiling Points

TABLE A12.1
Halogen Compounds and Their Boiling Points

Name	BP °C	Name	BP °C
1-Chlorobutane	78	Bromocyclopropane	68
2-Chlorobutane	65	1-Bromodecane	238
Chloroethane	12	Bromodichloromethane	88
1-Chlorodecane	223	Bromoethane	38
Chlorocyclopentane	113,5	Bromoform	150
Chlorocyclohexane	142	1-Bromoheptane	180
Chloroethylene	−14	1-Bromohexane	154
1-Chloro-2-butene	81	Bromomethane	4.2
3-Chloro-1-butene	64,5	1-Bromo-3-methylbutane	121
1-Chloroheptane	160,5	Bromomethyl-cyclopropane	111.5
1-Chlorohexane	134	1-Bromo-2-methylpropane	91
Chloroiodomethane	108,5	2-Bromo-2-methylpropane	73
Chloromethane	−24	1-Bromooctane	201.5
2-Chloro-2-methylbutane	85	1-Bromopentane	130
1-Chloro-2-methylpropane	67,5	1-Bromopropane	71
1-Chloro-2-methyl-1-propene	68	2-Bromopropane	60
3-Chloro-2-methyl-1-propene	73	3-Bromo-1-propene	71
1-Chlorononane	204	Bromotrichloromethane	105
1-Chlorooctane	182	1-Iodobutane	130
1-Chloropentane	108	2-Iodobutane	119.5
2-Chloropropane	35	Iodoethane	72
3-Chloro-1-propene	45	1-Iodoheptane	204
2-Chloro-2-methylpropane	51	1-Iodohexane	181
1-Bromobutane	100	Iodomethane	42
2-Bromo-2-butene	86	1-Iodo-3-methylbutane	147
4-Bromo-1-butene	99	1-Iodo-2-methylpropane	120.5
1-Bromo-2-Chloroethane	105	1-Iodooctane	225.5
2-Bromobutane	91	1-Iodopentane	154.5
Bromochloromethane	68	1-Iodopropane	103
1-Bromo-5-chloropentane	211	2-Iodopropane	89
1-Bromo-3-chloropropane	140	3-Iodo-1-propene	102
Bromocyclohexane	164	1-Bromopropane	71
Bromocyclopentane	137	Bromocyclopropane	68

TABLE A12.2
The Best Regressions Having from 1 to 4 Descriptors for the Boiling Points of 66 Halogenides

	Regression	Descriptors	r^2	s
	−58.582	Constant		
1	56.261	Kier and Hall (order 1)		
			0.9094	18.04
	−80.2720	Constant		
1	0.65011	Molecular weight		
2	38.420	Randić index (order 0)		
			0.9543	12.89
	−110.38	Constant		
1	56.747	Kier and Hall (order 1)		
2	−1.1635	Information content (order 0)		
3	20.232	Randić index (order 0)		
			0.9730	9.97
	−98.338	Constant		
1	48.939	Kier and Hall (order 1)		
2	−1.3182	Information content (order 1)		
3	34.606	Randić index (order 1)		
4	5.2473	Kier and Hall (order 2)		
			0.9792	8.80

Source: Pompe and Randić. *Acta Chim. Slov.* 54, 605–610, 2007.

REFERENCE

1. M. Pompe and M. Randić, Variable connectivity model for determination of pK(a) values of selected organic acids, *Acta Chim. Slov.* 54 (2007) 605–610.

Appendix 13: "News and Views" on Chemical Topology from *Nature*, November 15, 1974

Towards Chemical Topology

... Consider, for example, the following: can a coding system be devised which not only leads to the systematic reconstruction of an arbitrary carbon skeleton but which is essentially independent of the particular way adopted for labeling individual atoms or other distinguishing features? Alternatively, if a certain labeling system is adopted (usually, of course, that of attaching sequential numbers to the carbon atom present) can coded representation be found from which the topological identity of two differently numbered versions of the same N-carbon structure can be immediately inferred? (Other, that is, than by investigation of all possible N! permutations of labeling).

This problem, which shares its graph-theoretical content with several others in statistical mechanics the theory of disordered structures, has been taken up in a number of recent papers, most notably by M. Randić of the University of Zagreb writing in the Journal of Chemical Physics (60, 3920; 1974). *Professor Randić advances no rigorous proofs but suggests an ingenious algorithm by which a binary number code is assigned to each possible labeling of a given structure, the lowest number obtainable under systematic rearrangements providing both a canonical numbering scheme and an immediate comparison test for the identity of different networks of bonds. The method given seems to fall short of guaranteeing the discovery of the minimal labeling but seems to work well when tried on common systems of saturated and benzenoid hydrocarbons. Needless to say the canonical numbering systems come out differently from those traditionally adopted. (It is notable, and amusing, that the basic tool of the Randić method—the adjacency matrix for comparable structures—is now a familiar part of O-level mathematics (See* Schools Mathematics Project Book Y, 55; *Cambridge University Press, 1973) though likely to seem formidable to many research chemists educated in early era)....*

Appendix 14: Conjugated Circuits for the Remaining Five Symmetry Nonequivalent Kekulé Valence Structures of Coronene

FIGURE A14.1 Conjugated circuits for the first Kekulé structures of coronene of Figure 10.4 in Chapter 10.

FIGURE A14.2   Conjugated circuits for the second Kekulé structures of coronene of Figure 10.4.

FIGURE A14.3   Conjugated circuits for the third Kekulé structures of coronene of Figure 10.4.

FIGURE A14.4 Conjugated circuits for the fourth Kekulé structures of coronene of Figure 10.4.

FIGURE A14.5 Conjugated circuits for the fifth Kekulé structures of coronene of Figure 10.4.

Appendix 15: Hostile Reports of Anonymous Referees on Novelty in Graphical Bioinfomatics

Full reports of the two referees of the journal *SAR and QSAR in Environment Research* and the comments of the authors of a paper developing the four color map model for DNA [1].

15-May-2013

Dear Dr Randić:

I regret to inform you that our reviewers have now considered your paper but unfortunately feel it unsuitable for publication in SAR and QSAR in Environmental Research. For your information I attach the reviewer comments at the bottom of this email. I hope you will find them to be constructive and helpful. You are of course now free to submit the paper elsewhere should you choose to do so.

Thank you for considering SAR and QSAR in Environmental Research. I hope the outcome of this specific submission will not discourage you from the submission of future manuscripts.

Sincerely,

Dr James Devillers
Editor in Chief, SAR and QSAR in Environmental Research
j.devillers@ctis.fr
Reviewer(s)' Comments to Author:

Referee: 1
Comments to the Author
This paper is in an area of great current interest providing a new way of analyzing biomolecular sequences. The current authors have been instrumental in suggesting several new methods to view DNA sequences.

That said, the point of this kind of exercise is ostensibly to augment the tools of the biologist to unravel new features of DNA sequences and to assist in their work to understand the available sequences. The authors of this paper have already proposed several novel ways of viewing DNA sequences, but they do not seem to believe in the efficacy of their own work and therefore go on proposing new models every so often without analyzing the pros and cons of their earlier models or state why a new model is necessary. If any of these models are supposed to help the biologist, and then they come up with yet another model, what is the biologist to do? They mention (p. 7) that some map representations may have shortcomings: The question is how does a biologist know which representation to use in which instance? I think the authors should stop proposing new models and try to apply a model to solve some real problems instead of proposing new models like homework. Such immature approach as evidenced by the authors only contributes to intellectual pollution and wastes time and effort of the real problem solvers. The authors also do injustice to the existing literature and their paper have several drawbacks listed below. In view of these observations I suggest that this manuscript be summarily rejected.

1. Introduction: The paper ignores all seminal developments of 2D methods and mentions the 3D method of 2000 as if the 2D reps didn't exist. Their 2D four color maps were not the only 2D methods proposed by researchers!
2. Similarly, the authors claim that numerical characterization took place in 2000 only whereas it had actually taken place through geometrical means in 1998 and 1999. Ignoring such developments reflect poorly on the background research capabilities and fairness of the authors.
3. Page 7: [ref] left unfilled.
4. The authors harp on loss of information in 2D maps (pp. 5, 19, ...) without specifying where it arises. In several instances (pp. 5, 7) they mention that representation of two sequences could be similar when they are not, a vague generalization without actual examples. Possibly this arises from the misrepresentation that nothing happened before 2000!

Referee: 2
Comments to the Author
The following comments are factually incorrect.
However, such graphical representations of DNA were qualitative, which was considerable limitation. This all changed by the year 2000 when it was shown how one can accompany qualitative 2-D graphical representations of DNA with quantitative numerical characterizations [6,7]

One of the authors of Ref. [7], Dr. Ashesh Nandy, did "pioneering studies" in the numerical characterization of DNA sequences. Please see:

Raychaudhury, C.; Nandy, A. Indexation Schemes and Similarity Measures for Macromolecular Sequences. Paper presented at the Indo-US Workshop on Mathematical Chemistry, Shantiniketan, West Bengal, India, December 1998.
Raychaudhury, C.; Nandy, A. Indexing scheme and similarity measures for macromolecular sequences, J. Chem. Inf. Comput. Sci. 1999, 39, 243–247.

A simple numerical descriptor for quantifying effect of toxic substances on DNA sequences, A. Nandy and S.C. Basak, *J. Chem. Inf. Comput. Sci.*, 40, 915–919 (2000).

Although the paper is in an interesting area of bioinformatics, the authors suggest new methods but do not show any substantial and innovative biological applications. In that sense, it is an exercise in the useless proliferation of indices without showing their relevance in bioinformatics. So, I strongly suggest that the paper should be rejected.

Comment of authors to editors of scientific journals:
Be aware of false prophets. Have a respect for warning of Confucius: Those who do not know that they do not know are dangerous. Avoid them!

The above charges may appear to some uninformed readers plausible and therefore ask for response. We will be brief. Tool making is an art in itself, just as writing advanced computer programs is. Users should be instructed how to *use* new tools and new computer programs—and as a rule, that continues to be the case. Toolmakers and computer program designers should not be asked to solve the problems of chemistry or biology, which is what chemists and biologist should do. Tool making and computer programming requires familiarity with discrete mathematics. Use of novel tools and novel computer programs does not require familiarity and understanding of discrete mathematics. Anybody can use novel tools and novel computer programs. The *real problem solvers* should try novel tools and novel computer programs to find out which approach helps solving which particular problem that interests them. If they expect someone else to tell them *what* to do and *how* to do, they will be left behind, where they deserve to be, behind those who *know* what to do and how to do it. Attempts of both reviewers to block novel approaches is tantamount to making it easier for those of lesser capabilities to keep pace with the progress in science at the cost of *hindering* those in the forefront of research.

Both reviewers failed to recognize that finding a few numerical parameters for a graphical representation of DNA is not tantamount to numerical characterization, which allows making quantitative comparisons of DNA and proteins, which implies the possibility of constructing tables of similarity and construction of evolution trees, which are central to bioinformatics. In analogy with typical police work on suspects, indexing is like knowing the height and the weight of a suspect, which hardly suffices to identify suspects; characterization is like having fingerprints and DNA data, which tend to be unique to a person.

The comment that Referee 2 finds factually incorrect:
However, such graphical representations of DNA were qualitative, which was considerable limitation. This all changed by the year 2000 when it was shown how one can accompany qualitative 2-D graphical representations of DNA with quantitative numerical characterizations [6,7].

Is factually correct, while continuation of Referee 2:
One of the authors of Ref. [7], Dr. Ashesh Nandy, did "pioneering studies" in the numerical characterization of DNA sequences.

Is factually incorrect. Referee 2 apparently does not appreciate distinction between numerical *characterization* of DNA and (numerical) *indexing* of DNA. As anyone interested can find by examining few dictionaries, *characterization* is a description of qualities or peculiarities of somebody or something, the way in which someone or something is described as a particular type of person or thing. *Index* is something that serves to point out, or facilitate reference, while *indexing* is the act of clarifying and providing an index to make an item easier to retrieve.

By knowing the three numerical indices of Nandy for a DNA, you can say less about that DNA than by visually inspecting it, which is qualitative. By numerical characterization of DNA, introduced in 2000 by Randić and coworkers, not only that one can say more about the DNA considered than from visual inspections, but also one can quantify differences between considered DNA and delegate comparative study of DNA to computers.

The same explanation applies to the similar comments of Referee 1, who went even further with his statement: "The paper ignores all seminal developments of 2D methods...." Seminal development is development that opens novel directions in research. The paper of Hamory and Ruskin [2] was seminal development that Leong and Morgenthaler [3], Nandy [4], and others followed, which relates to visual, qualitative comparative study of DNA. Indexing of DNA of Nandy et al. was novelty; only time will tell how useful it will be and whether it can be characterized as seminal. It certainly is numerical, but it, in our view, neither represents characterization of DNA nor does it lead to matrix characterizations of DNA. If one looks for seminal publication that has led to numerical characterization of DNA, which opened a way to quantitative graphical bioinformatics, then that is the paper of D/D matrices of Randić and coworkers [5].

REFERENCES

1. M. Randić, B. Horvat, T. Pisanski, G. Jaklič, and D. Plavšić, On map representation of DNA, *Croat. Chem. Acta* 86 (2013) 519–529.
2. E. Hamori and J. Ruskin, H curves, a novel method of representation of nucleotide series especially suited for long range DNA sequences, *J. Biol. Chem.* 258 (1983) 1318.
3. P. M. Leong and S. Morgenthaler, Random walk and gap plots of DNA sequences, *Comput. Appl. Biosci.* 11 (1995) 503–507.
4. A. Nandy, A new graphical representation and analysis of DNA sequence structure: I. Methodology and application to globin genes, *Curr. Sci.* 66 (1994) 309–314.
5. M. Randić, A. F. Kleiner, and L. M. DeAlba, Distance/distance matrices, *J. Chem. Inf. Comput. Sci.* 34 (1994) 277–286.
6. M. Randić, On characterization of DNA primary sequences by a condensed matrix, *Chem. Phys. Lett.* 317 (2000) 29–34.
7. M. Randić, M. Vračko, A. Nandy, and S. C. Basak, On 3-D graphical representation of DNA primary sequences and their numerical characterization, *J. Chem. Inf. Comput. Sci.* 40 (2000) 1325–1244.

Appendix 16: The Computer Program for Exact Pairwise Protein Alignment

This computer program was written in Python by Professor Tomaž Pisanski of the Faculty of Mathematics and Physics of the University of Ljubljana, Slovenia and the University of Primorska, Koper, Slovenia. Python is a widely used high-level computer language.

```python
def protein(prot,ter):
    """ For each tuple of length ter record an entry composed
of lists of indices
where this tuple appears. """
    dic = {}
    for i in range(len(prot)-1):
        key = prot[i:i+ter]
        val = dic.get(key,[])
        val.append(i+1)
        dic[key] = val
    return dic

def compare_prot(prot1,prot2,ter=2,k = 4):
    """ """
    sprot1 = list(set(prot1))
    sprot2 = list(set(prot2))
    sprot = list(set(sprot1+sprot2))
    sprot.sort()
    dic1 = protein(prot1,ter)
    dic2 = protein(prot2,ter)
    res = {}
    for p in range(-k,k+1):
        res[p] = []
    for z in dic1:
#       print()
        if z in dic2:
            d1 = dic1[z]
            d2 = dic2[z]
            for j in d1:
                for p in range(-k,k+1):
                    if j+p in d2:
                        tmp = res.get(p,[])
                        tmp.append(j)
                        res[p] = tmp
            print("-----------------------------------------")
```

```
print("{0:3}{1:>30}\n{2:>33}".format(z,dic1.get(z,[]),dic2.get
(z,[])))
    print()
    print(prot1)
    print(prot2)
    print()
    for p in range(-k,k+1):
        rp = res[p]
        rp.sort()
        print(p,rp)

print()
```

Appendix 17: An Input and Output of the Program VESPA

We have selected ND6 proteins of human and mouse to illustrate an input and output of the program VESPA (very exact search for protein alignment).

```
# LIST OF SAMPLE PROTEINS FOR COMPARISON
prot_human ="MMYALFLLSVGLVMGFVGFSSKPSPIYGGLVLIVSGVVGCVIILNFGGG
YMGLMVFLIYLGGMMVVFGYTTAMAIEEYPEAWGSGVEVLVSVLVGLAMEVGFVLWVKEYDG
VVVVVNFNSVGSWMIYEGEGSGFIREDPIGAGALYDYGRWLVVVTGWPLFVGVYIVIEIARGN"
prot_mouse
="MNNYIFVLSSLFLVGCLGLALKPSPIYGGLGLIVSGFVGCLMVLGFGGSFLGLMVFLIYLGG
MLVVFGYTTAMATEEYPETWGSNWLILGFLVLGVIMEVFLICVLNYYDEVGVINLDGLGDWLMY
EVDDVGVMLEGGIGVAAMYSCATWMMVVAGWSLFAGIFIIIEITRD"
```

Program output

```
-4 []
-3 [17]
-2 []
-1 [48, 93, 127, 140, 154, 157, 160, 168, 169]
0 [8, 22, 23, 24, 25, 26, 27, 28, 29, 32, 33, 34, 35, 38, 39,
46, 47, 52, 53, 54, 55, 56, 57, 58, 59, 60, 61, 62, 65, 66,
67, 68, 69, 70, 71, 72, 73, 76, 77, 78, 79, 82, 83, 98, 99,
109, 153]
1 [12, 92]
2 [29, 103, 111]
3 [89]
4 [10]
```

Appendix 18: List of Problems of Structural Chemistry Presented in This Book

Problem 1

Looking at a list of smaller twin numbers, we noticed that prime number 5 appears to be the only prime number that appears in two twin prime pairs: (3, 5) and (5, 7). Is this the only prime number that is a member of two twin groups or, in other words, is the triplet (3, 5, 7) the only twin prime number triplet?

Problem 2

The question is whether there are some molecular fragments or sets of atoms and atomic groups, at least in the case of some physical properties of ensembles of molecules, which are critical structural factors for the considered molecular property and which could define such fragments as "physicophore."

Problem 3

Explore if the complexity algorithm mentioned above holds, that is, gives plausible results, for selected classes of molecules, such as (i) acyclic alkane graphs, (ii) cata-condensed benzenoid graphs, (iii) smaller polycyclic graphs.

Problem 4

Under what conditions will a pair of isospectral graphs g, h upon enlargement produce isospectral pairs G and H?

Problem 5

Is there isospectral folding? That is, are there two different protein foldings that will produce the same eigenvalues of the binary "folding" matrix?

Problem 6

How many different folding patterns exist for the $3 \times 3 \times 3$ lattice? This is of some interest if one contemplates the search for "isospectral" proteins.

Problem 7

Can one find the most "stable" folding pattern and/or find structural conditions required for the most stable folding patterns?

Problem 8

Calculate, using HMO, the plausibility of the extended edges on the reconstructed 5–7 zigzag model.

Problem 9

Find additional molecular structures containing the eigenvalues of hexatriene: ± 1.8019, ± 1.2470, ± 0.4450, but having no nodes on atoms.

Problem 10

As an exercise, write down the sum rules for antisymmetrical eigenvectors of hexatriene and show that the four-ring benzene derivative of Figure 4.6 in Chapter 4 satisfies such rules, which assures that this benzenoid also contains the eigenvalues of anti-symmetric eigenvectors of hexatriene. For the solution see Figure A11.1 in Appendix 11.

Problem 11

Select an arbitrary smaller graph. Find additional graphs that satisfy the same sum rules for their coefficients, which will then contain the eigenvalues of the smaller graph.

Problem 12

What are, if any, the structural factors that may answer why some benzenoids do not follow the regularity that holds for most benzenoid hydrocarbons that associates the presence of the bay regions with carcinogenicity?

Problem 13

Using the standard quantum chemical approaches and general methodology of quantum mechanics, seek non-empirical justifications of Clar's sextet theory.

Problem 14

Find regularities in RBO analogous to those in phenathrene–benzanthracene in other benzenoid families, including fibonacene, if possible.

Problem 15

Find hidden connection of the Clar sextet theory with quantum mechanics or prove that there is no such.

Problem 16

Try to characterize peri-condensed benzenoid graphs that have a Hamiltonian path.

Problem 17

Try to characterize benzenoid hydrocarbons whose resonance graphs have only 1-D and 2-D cubes.

Problem 18

Try to characterize benzenoid hydrocarbons whose resonance graphs have only 3-D cubes.

Problem 19

Consider systems having essentially single CC bonds connecting smaller benzenoid (and possibly non-benzenoid) fragments, and see if there is a way that allows finding renormalization factors of such systems *without* examination individually of all Kekulé valence structures.

Problem 20

Make a table of similarity between pairs of DNA sequences by modifying the construction of the four-color maps by considering exactly pairwise.

Problem 21

Discuss the differences between the similarity/dissimilarity table based on the four color map and the table involving gaps.

Problem 22

Find a mathematical rationale for the close parallelism for the ring currents in benzenoid and non-benzenoid systems calculated using classical physics and obtained by enumeration of contributions from conjugated circuits.

Problem 23

Mathematical Treatment of the Axioms of Chemistry:
To treat ... by means of axioms, those chemical sciences in which already today mathematics plays an important part; in the first rank are quantum chemistry, statistical mechanics, chemical graph theory, and graphical bioinformatics.

Observe that the parallelism of our Problem 23 and Hilbert's Problem 6 on mathematical treatment of the axioms of physics is about theoretical physics more than mathematics:

The investigations on the foundations of geometry suggest the problem: to treat in the same manner, by means of axioms, those physical sciences in which already today mathematics plays an important part; in the first rank are the theory of probabilities and mechanics.

So our Problem 23 on the mathematical treatment of the axioms of chemistry is about theoretical chemistry more than mathematics.

Appendix 19: Rigorous Definitions and Descriptions of a Selection of Mathematical Concepts of Discrete Mathematics

In this appendix, we have collected more rigorous definitions and descriptions of a selection of mathematical concepts of discrete mathematics.

DISCRETE MATHEMATICS

There are different ways in which one can define the term *discrete mathematics*. None of them is satisfactory on its own, but together they give some idea of what is discrete mathematics. Before we explain the term *discrete mathematics*, it is important to distinguish what is discrete from what is continuous. Roughly speaking, an object is discrete if it consists of distinct or isolated elements and continuous if one can move from one element to another without making sudden jumps. An example of a discrete object are electrons in an atom. They are discrete units, and one cannot take for instance half of an electron because it is not that unit, electron. An example of a continuous object is a line segment AB, which is a continuous whole. This means that if one moves from the point A to the point B, the segment *continues* without a break. A working definition of discrete mathematics is that it is the part of mathematics dealing with mathematical structures that are fundamentally discrete. More formally, discrete mathematics is concerned with finite or countably infinite sets. A set of objects able to be put into one-to-one correspondence with a subset of the set of natural numbers is countable.

PARTIAL ORDER

A set X is a finite or infinite unordered collection of distinct objects x called the elements of the set. We write $x \in X$ or $x \notin X$ to denote that x is an element of X or x is not an element of X, respectively. A set can be given by enumerating its elements in braces, e.g., $M = \{a,b,c,h,d\}$, or by defining the property or properties possessed exactly by the elements of the set, e.g., $S = \{x| \ x$ is an even positive integer less than 9$\}$. The number of elements of a finite set X, denoted by $|X|$, is called the cardinal number of X. Two sets X and Y are equal, written $X = Y$, if and only if they have the same elements. The set X is a subset of (or is included in) the set Y, denoted by

$X \subseteq Y$, if and only if every element of X is also an element of Y. If a set X is a subset of the set Y but $X \neq Y$, then $X \subset Y$, and it is said that X is a proper subset of Y. For any set X is $X \subseteq X$. The empty set $\emptyset$ contains no elements. There exists only one empty set and it holds for every set X. The union of the sets X and Y, denoted by $X \cup Y$, is the set of elements being in X, in Y, or in both X and Y. The intersection of sets X and Y, written $X \cap Y$, is the set containing all elements that belong to both X and Y. If $X \cap Y = \emptyset$, then X and Y are called disjoint.

Often the order of elements in a collection is important. As sets are unordered, a structure called *ordered m-tuple* was introduced to represent ordered collections. The ordered m-tuple $(x_1, x_2, x_3,..., x_m)$ is the ordered collection whose first element is X_1, second X_2, third X_3 and mth element X_m. Two ordered m-tuples $(x_1, x_2, x_3,..., x_m)$ and $(y_1, y_2, y_3, ... , y_m)$ are equal if and only if $x_i = y_i$, for $i = 1,2,..., m$. The ordered 2-tuples are called *ordered pairs*.

The Cartesian product of two sets X and Y, denoted by $X \times Y$, is the set of all ordered pairs (x, y), where $x \in X$ and $y \in Y$. A binary relation from set X to set Y is a subset of $X \times Y$. A binary relation R on a set X is a subset of $X \times X$. If $x, y \in X$ and $(x, y) \in R$, then we say that x is related to y by R and write xRy. A binary relation R on a set X is called reflexive if xRx for every $x \in X$; irreflexive if xRx for no $x \in X$; symmetric if xRy implies yRx; antisymmetric if xRy and yRx imply $x = y$; and transitive if xRy and yRz imply xRz. A binary relation R on a finite set can pictorially be represented by a directed graph in which each element of the set is represented by a small circle, called a vertex, and each ordered pair $(x, y) \in R$ by a directed arc or a directed line segment, called an edge, from x to y.

Binary relations are often used for ordering some or all of the elements of sets. A binary relation R on a set X is a partial order or partial ordering if R is reflexive, antisymmetric, and transitive. A set X together with a partial ordering R is called a partially ordered set or poset and is denoted by (X, R). In different posets, different symbols are used for a partial order. As a partial order generalizes the relation $\leq$, the symbol $\preccurlyeq$ is used to denote the partial order relation in any poset. The notation $x \preccurlyeq y$ means x precedes or equals y. If $x \preccurlyeq y$ and $x \neq y$, then we write $x \prec y$, meaning x is a predecessor of y (x is dominated by y) or y is a successor of x. Two elements x and y in a poset are comparable if either $x \preccurlyeq y$ or $y \preccurlyeq x$. Otherwise, x and y are incomparable. In other words, elements that are related are comparable. If any two elements in a poset $(X, \preccurlyeq)$ are comparable, then $\preccurlyeq$ is called a total order and the poset is a totally ordered set. A finite poset $(X, \preccurlyeq)$ can pictorially be represented by a directed graph, which can be simplified by deleting its edges representing reflexivity and edges implied by transitivity as well as by drawing the remaining edges upward and dropping all arrows. In other words, whenever $x \prec y$, place x at a lower level than y, and if x is an immediate predecessor of y, meaning there are no elements z satisfying $x \prec z \prec y$, connect x and y with an edge. The resulting diagram is the Hasse diagram of $(X, \preccurlyeq)$, named after the mathematician H. Hasse (1898–1979).

If y is an element of a poset such that $x \preccurlyeq y$ for all elements x of a nonempty subset S of the poset, then y is called an upper bound of S. An element y of the poset is called the least upper bound of S if y is an upper bound and $y \preccurlyeq z$ for all upper bounds of S. In a similar way, a lower bound and the greatest lower bound of a nonempty subset S of a poset are defined. An element k of a poset is called a lower bound of S if

$k \leqslant x$ for all elements x of S. An element k of the poset is the greatest lower bound of S if k is a lower bound and $z \leqslant k$ for all lower bounds z of S. A lattice is a poset with the property that every pair of elements has both a least upper bound and a greatest lower bound.

GRAPH

A graph is one of the very basic mathematical structures. It is defined as an ordered pair $G = (V,E)$ of a set $V = V(G) \neq \varnothing$ whose elements are called vertices and a (possibly empty) set $E = E(G)$ of unordered pairs of distinct elements of V called edges. The set $V(G)$ is called the vertex set of G and $E(G)$ its edge set. If $e_{ij} = \{i,j\} \in E$, then vertices i and j are said to be adjacent (to each other) and incident to the edge e_{ij}. Two edges are adjacent if they are incident with a common vertex. The degree of a vertex i is the number of edges incident with it. Alternatively, the degree of a vertex i is the number of vertices that are adjacent to i. Graphs are so named because they can be represented graphically (terms pictorially, geometrically, or diagramatically are also in use). It is customary to represent the vertices by small circles in a plane and to connect adjacent vertices by a line. The relative positions of small circles and the shapes of lines usually have no significance, and hence, there is no single correct way to diagramatically represent a graph. A drawing of a graph is also called graph.

A graph is widely used in chemistry in particular for representing the topology of a molecule, the totality of information on the mutual connectedness of all atoms in a molecule. The application of graphs in modeling many types of relations and processes in natural sciences, engineering sciences, and social sciences, as well as in many other areas of human activity required generalizations of the notion of the graph because the above definition of a graph is very restrictive. A graph defined in the above restrictive manner is called simple. A broader definition of a graph reads as follows: a graph is an ordered pair $G = (V,R)$ where V is the vertex set and R is a binary relation on V. As R can have different properties, this definition in a natural way introduces different types of graphs. For example, if R is symmetric and irreflexive, then G is a simple graph; if R is antisymmetric, then G is a directed graph; if R is not irreflexive, then G has loop(s); etc. A graph can be generally defined in the framework of different conception of the edge set as an ordered triple $G = (V,E,\varphi)$ of a set V of vertices, a (possibly empty) set E of edges ($V \cap E = \varnothing$), and an incidence function φ defined on E, which uniquely assigns to every element of E an ordered or unordered pair of (not necessary distinct) elements of V.

Appendix 20: QSAR Nightmare—No More

In Chapter 5, when discussing the use of co-linear descriptors in regression analysis, we mentioned the instability of regression equations. Here we will illustrate the *Nightmare of QSAR* on a regression using four connectivity indices: $^1\chi$, $^2\chi$, $^3\chi$, $^4\chi$, of which in particular $^1\chi$ and $^2\chi$ are highly intercorrelated. We conducted a multiple regression analysis (MRA) between the Hosoya index Z and the connectivity indices $^1\chi$, $^2\chi$, $^3\chi$, $^4\chi$ for 18 octane isomers, which is shown in Table A20.1.

The table shows the coefficients of the MRA stepwise regressions, which change drastically each time when an additional connectivity index is added. In the last two columns of the table we show the regression coefficient r and the standard error s are shown. Clearly, as the number of the descriptors increases, the standard error decreases and better regression equations follow. The unpredictable and chaotic changes of the coefficients in successive regression equations make the interpretation of the regression equation impossible—illustrating the QSAR nightmare.

In Table A20.2, we show a stepwise regression using orthogonalized $^1\chi$–$^4\chi$ connectivity indices. Here $^2\chi^*$ is the residual of a regression of $^2\chi$ against $^1\chi$, shortly to be denoted as R(2/1), which is the part of $^2\chi$ that cannot be calculated from $^1\chi$ via the regression equation. To obtain $^3\chi^*$, one first has to find the residual of a regression of $^3\chi$ against $^1\chi$, the R(3/1), because this is the part of $^3\chi$ that cannot be calculated from $^1\chi$, hence independent of $^1\chi$. However, the residual R(3/1) may correlate with $^2\chi^*$, that is, R(2/1). So one needs to consider their correlation in order to find $^3\chi^*$, which is obtained as the residual of a regression between the residuals R(3/1) and R(2/1). This residual can be denoted as R(3/2). The process of considering residuals of regressions involving residuals continues until all descriptors have been orthogonalized.

Observe that the statistical data, the coefficient of the regressions and the standard errors, are the same in Tables A20.1 and A20.2, but the regressions in Table A20.2 are stable, and the coefficients of descriptors already used do not change when a new orthogonal descriptor is added into the next stepwise regression. Hence, the relative contributions of individual descriptors in the regression are constant.

It is very interesting to compare the coefficients in Tables A20.1 and A20.2. Look at the entries along the diagonal in the two tables:

$$17.967, -3.471, 1.876, -0.561.$$

They are the same. The constant terms in Table A20.2 are all the same and equal to the constant term in the first equation. Because of this, from Table A20.1 alone (which is based on regressions using non-orthogonal connectivity indices $^1\chi$, $^2\chi$, $^3\chi$, $^4\chi$), one can construct the regression using orthogonalized connectivity indices

TABLE A20.1
Illustration of the QSAR Nightmare: Stepwise Regressions for the Hosoya Z Topological Index Using the $^1\chi$–$^4\chi$ Connectivity

$^1\chi$	$^2\chi$	$^3\chi$	$^4\chi$	Constant	r	s
17.967				– 40.4349	0.9869	0.472
6.233	– 3.471			6.4615	0.9951	0.313
28.631	3.016	1.876		– 85.3765	0.9987	0.175
22.020	0.935	1.079	– 0.561	– 57.1671	0.9991	0.168

TABLE A20.2
Stepwise Regression Using Orthogonalized $^1\chi$–$^4\chi$ Connectivity Indices

$^1\chi$	$^2\chi^*$	$^3\chi^*$	$^4\chi^*$	Constant	r	s
17.967				– 40.4349	0.9869	0.472
17.967	– 3.471			– 40.4349	0.9951	0.313
17.967	– 3.471	1.876		– 40.4349	0.9987	0.175
17.967	– 3.471	1.876	– 0.561	– 40.4349	0.9991	0.168

$^1\chi$, $^2\chi^*$, $^3\chi^*$, $^4\chi^*$ by using the coefficient along the main diagonal of Table A20.1, obtaining

$$Z = 17.967\ ^1\chi - 3.471\ ^2\chi^* \ 1.876\ ^3\chi^* - 0.561\ ^4\chi^* - 40.4349$$

The above outline holds only for stepwise regressions, which are MRA regressions, in which at every step one retains the descriptors previously used. For a more general case, when at each step previous descriptors can also be changed, interested readers should consult Refs. [1,2].

REFERENCES

1. M. Randić, Retro-regression—Another important multivariate regression improvement, *J. Chem. Inf. Comput. Sci.* 41 (2001) 602–606.
2. M. Randić and M. Pompe, Retro-regression—a way to resolve multivariate regression ambiguities, *Acta Chim. Slov.* 52 (2005) 408–416.

Appendix 21: Kekulé Valence Structures Weights

The Kekulé valence structures (or resonance structures) have been used in chemistry by assuming that all Kekulé valence structures have equal weight. This is the model, for example, used by Linus Pauling when he introduced his bond orders. However, in 1927, Fries came with the suggestion that Kekulé valence structures that have the largest number of benzene rings with three alternating double and single bonds are the most important [1]. In Figure 3.8 of Chapter 3, we have illustrated Fries valence structures for 10 smaller benzenoid hydrocarbons. As mentioned in Chapter 3, almost 35 years later, it was found that electron–electron repulsion may be a structural factor that may play some role by contributing to Fries Kekulé valence structures having a greater weight [2].

Occasionally, one comes across benzenoid hydrocarbons with several Kekulé valence structures that have the same number of benzene rings with alternating double and single bonds; however, some of those structures are distinct, that is, symmetry non-equivalent. Hence, such Kekulé valence structures need not be equally important for the aromaticity of these compounds. In order to find out how important individual Kekulé valence structures are, we propose a model in which the relative weight of Kekulé valence structures is given by their contribution to molecular resonance energy (RE).

We have examined all symmetry-non-equivalent Kekulé valence structures of all cata-condensed benzenoid hydrocarbons having at most four fused benzene rings, excluding benzo[g]phenanthene, which is iso-Kekuléan with chrysene (systematic name benzo[i]phenanthrene). Iso-Kekuléan benzenoid hydrocarbons have one-to-one correspondence between their Kekulé valence structures that not only result in the same expressions for the molecular RE but also involve the same contributions of the corresponding individual Kekulé valence structures. For all cases considered here, there are no two Kekulé valence structures in the same molecule that can be similarly decomposed into conjugated circuits, with the exception of two Kekulé valence structures in benzo[a]anthracene. In Figure A21.1, we show the seven Kekulé valence structures of benzanthracene, under each of which we show their decomposition into conjugated circuits. As one can see, among the seven structures there are two Kekulé valence structures having the same contribution $3R_1 + R_2$.

For these two Kekulé valence structures, we examined all conjugated circuits: (1) linearly independent circuits (shown in Figure A21.1), which contribute to the expression for RE, the number of which is equal to the number of fused benzene rings in a molecule; and (2) linearly dependent and disjoint conjugated circuits. The total number of all conjugated circuits is $K(K - 1)$. We adopted the following notation for disjoint conjugated circuits:

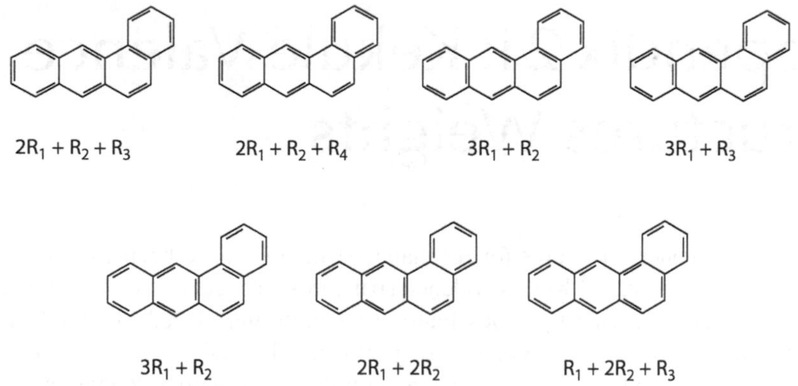

$2R_1 + R_2 + R_3$ $2R_1 + R_2 + R_4$ $3R_1 + R_2$ $3R_1 + R_3$

$3R_1 + R_2$ $2R_1 + 2R_2$ $R_1 + 2R_2 + R_3$

FIGURE A21.1 Contributions of linearly independent conjugated circuits of individual Kekulé valence structures of benzanthracene to molecular resonance energy.

R_1^2 = two disjoint R_1 conjugated circuits
R_1R_2 = disjoint R_1 and R_2 conjugated circuits
R_1R_3 = disjoint R_1 and R_3 conjugated circuits
R_1^3 = three disjoint R_1 conjugated circuits
$R_1^2R_2$ = two disjoint R_1 and one R_2 conjugated circuits, and so on

With this notation, the count of the conjugated circuits in the seven Kekulé valence structures of benzanthracene is as follows:

$$2R_1 + R_2 + R_3 + R_1^2 + R_1R_2$$
$$2R_1 + R_2 + R_4 + R_1^2 + R_1R_2$$
$$3R_1 + R_2 + 2R_1^2$$
$$3R_1 + R_3 + 2R_1^2$$
$$3R_1 + R_2 + R_1^2 + R_1R_2$$
$$2R_1 + 2R_2 + R_1^2 + R_1R_2$$
$$R_1 + 2R_2 + 2R_3 + R_4$$

As one can see, all seven expressions are different. The same is true for the eight cata-condensed non-iso-Kekuléan benzenoid built from five fused benzene rings.

We looked also at non-iso-Kekuléan cata-condensed benzenoids built from six fused benzene rings and found two pairs having the same count of all $K(K-1)$ conjugated circuits, but involving R_3 rings of different shape. As is well known, conjugated circuits R_3 are the smallest conjugated circuits of different shape, which in the case of cata-condensed benzenoids can be the periphery of anthracene and periphery of phenanthrene, which have different shape.

It occurred to us that we could propose Problem 24 and ask if, for still larger benzenoids, we will continue to see different contributions from different Kekulé valence structures, but decided that it is premature to propose such a problem without additional prior explorations. So we discarded Problem 24. It is interesting to mention that recent new evidence shows that Hilbert had initially 24 problems, but shortly before his lecture decided to discard one problem, announcing thus 23 problems. So we follow Hilbert, though our reasons for discarding Problem 24 are fairly obvious: We need a little bit more time to explore larger benzenoids and see if duplicate decomposition occurs among them, as we do not want that, within weeks of announcing the problem and speculating that there will be no duplicate, somebody comes with a counterexample.

REFERENCES

1. K. Fries, Uber bicyclische Verbindungen und ihren Vergleich mit dem Naphtalin, *Justus Liebigs Ann. Chem.* 454 (1927) 121–324.
2. M. Randić, Comment on difference between bond orders calculated by SCF MO and simple MO method, *J. Chem. Phys.* 34 (1962) 693–694.

Appendix 22: Fries Rule Revisited

In 1927, Fries suggested that among all Kekulé valence structures of a benzenoid hydrocarbon, the valence structures that have the largest number of benzene rings with three alternating CC double and single bonds are the most important (dominant) and could be viewed as being representative of a benzenoid hydrocarbon as, for instance, the single resonance structure of phenanthrene with all its three rings having alternating double bonds. Let us consider benzo[e]pyrene, which has 11 Kekulé valence structures. In Figure A22.1, we show the three Kekulé valence structures that have four benzene rings with three alternating CC double and single bonds, which thus might alone represent benzo[e]pyrene (ignoring the remaining eight).

Under each structure, we have shown the count of conjugated circuits, which contribute to molecular resonance energy (RE). Thus, the first two Kekulé valence structures have the same count: $4R_1 + R_3$, and for the last Kekulé valence structure, the count is $4R_1 + R_4$. For the relative contributions of conjugated circuits to molecular RE, the following inequalities hold:

$$R_1 > R_2 > R_3 > R_3 > \ldots$$

It follows therefore that only the first two Kekulé valence structures of Figure A22.1 make the largest contributions to molecular RE. Observe that the two Kekulé structures are symmetry-unrelated and thus, while having the same count of linearly independent conjugated circuits (which contribute to RE), may have different counts when all $K(K-1)$ conjugated are counted. Because benzo[e]pyrene has 11 Kekulé valence structures, each of them has 10 conjugated circuits. The count of all conjugated circuits for the first two Kekulé valence structure is

$$4R_1 + R_3 + 3R_1^2 + R_1R_3 + R_1^3$$

$$4R_1 + R_3 + 4R_1^2 + R_1^3$$

From the above inequalities, it follows that $R_1^2 > R_1R_3$, pointing thus to the central Kekulé valence structure of Figure A22.1 as the *most dominant* Kekulé valence structure of benzo[e]pyrene. Hence, we have been able to find a single most dominant Kekulé valence structure among the three Fries structures of benzo[e]pyrene.

This visitation of the topic of Fries Kekulé valence structure raises the question of whether this happens for all benzenoid hydrocarbons, and whether one will always find a *single* most dominant Kekulé valence structure. In Figure A22.2, we show the *single* most dominant Kekulé valence structure for 20 smaller benzenoid

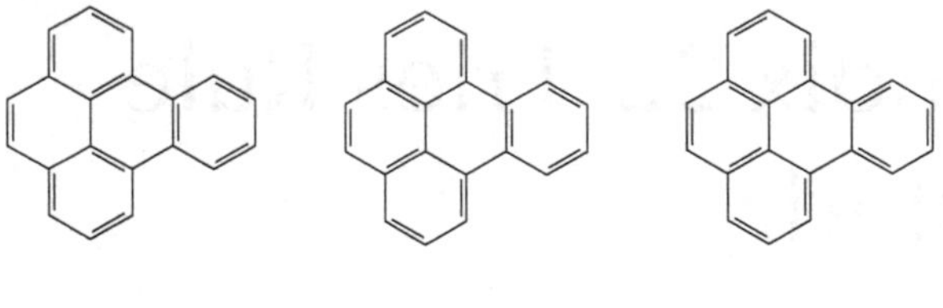

$4R_1 + R_3$ $4R_1 + R_3$ $4R_1 + R_4$

FIGURE A22.1   Fries Kekulé valence structures of benzo[*e*]pyrene.

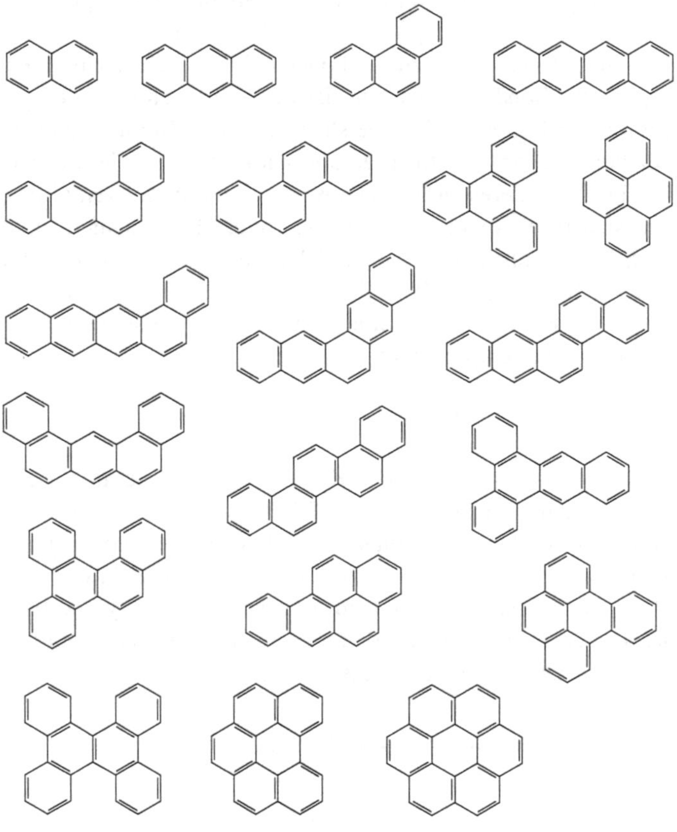

FIGURE A22.2   The most dominant Kekulé valence structure for 20 smaller benzenoid hydrocarbons.

hydrocarbons. All the cases considered have but a *single* most dominant Kekulé valence structure. It is premature to speculate whether this will hold for structures not so far examined. However, if it does for a sufficiently large representative sample, one could then consider to propose a conjecture to that effect. This problem is simpler than the *premature problem* of Appendix 21, because it concerns only the Fries structures of benzenoid hydrocarbons, but a solution of the problem in Appendix 21, if positive, would apply and solve the problem mentioned in this appendix.

Appendix 23: Late Hostility toward Chemical Graph Theory

The appendix section of this book started with Appendix 1: "Early Hostility toward Chemical Graph Theory," and, since in certain circles the hostility toward chemical graph theory has never ended, it seems appropriate to comment on one more recent incident in this area. The case that we have selected also illustrates (1) a lack of tolerance toward diversity and (2) cunning manipulations by part of establishment to protect the interest of a few scientists—a kind of bureaucracy that is emerging in science even in the twenty-first century.

Chemical graph theory as a visible discipline of chemistry started to grow in the mid-1970s, has undergone considerable evolution and expansion, and started to attract new researchers. One of the topics that received attention in the early days of chemical graph theory concerned isospectral graphs (unheard of, and unknown in chemistry in the 1950s, when the Hückel MO was receiving attention among chemists). However, by mid-1970, the time of HMO had passed, which in part accounts for the hostility of quantum chemists, many of whom identified (incorrectly) graph theory with HMO, possibly because the Hückel matrix was the same as the adjacency matrix of a graph.

The other area in which during these past 40 years chemical graph theory made visible contributions has been QSAR, as has been already mentioned and illustrated in the book. However, a number of influential researchers in QSAR, who followed the pioneering approach of Corvin Hansch to QSAR, objected to using mathematical descriptors in QSAR and continued to view the use of mathematical descriptors in QSAR as inappropriate. It appears that their objections reduce to the fact that mathematical descriptors do not contribute to their mechanistic modeling in QSAR. On the same grounds, one could object to using physicochemical properties as molecular descriptors as they do not contribute to a structural interpretation of QSAR modeling. Such opinions overlook the fact that molecular descriptors in *physicochemical models* should have *physicochemical interpretation* and that *molecular descriptors in graph theoretical models* should have *structural interpretation*. So we are neither trespassing to what is solely "their" territory, where only selected physicochemical properties are legitimate molecular descriptors, nor should they interfere to what is "our" territory: the application of discrete mathematics to chemistry—but they do.

In science, researchers can claim *monopoly* on methods, models, and algorithms and view them as their *property*, and those using them have to acknowledge the source of such. But all topics of science should be open to all. Thus, QSAR, which stands for quantitative structure–activity relationship, and not for the Hansch model of QSAR, should be a discipline open to all. Plainly speaking, QSAR is not synonym for *Hansch QSAR* model. Thus, QSAR should not exclude graph theoretical modeling and application of discrete mathematics.

In fact, linguistically speaking, the Hansch model of QSAR is based on regression between physicochemical properties and pharmacological activity, and is therefore a property–property regression. Most graph theoretical models consider regression between mathematical invariants (properties characterized by graph theoretical descriptors) and physicochemical properties or pharmacological activity, and are therefore similarly also property–property regressions. So strictly speaking, neither the Hansch model nor models using topological indices represent structure–activity relationships, but relate to property–activity relationships. However, there is an important difference between the two: many mathematical descriptors have direct *structural* interpretation, while this is not the case for the Hansch model of QSAR, which remains a regression between physicochemical properties and biological activity, not structure–activity.

Having been told the story about the background information of different approaches within QSAR, you are welcome to examine the editorial of the *Journal of Chemical Information and Modeling*, to be found in vol. 46, no. 3 (2006), p. 937, to which we object. In this editorial, among "Some guidelines to assist prospective Journal authors of manuscripts in the field of QSAR/QSPR analysis," you will find item #5:

> 5. **Only** *QSAR analyses that bring new insights on the mechanism of activity are encouraged....*

(Bold face was introduced by the senior author of this book to emphasize open **discrimination** in this editorial among *alternative* QSAR models, reminiscent of the time of "white only" labels.)

Clearly the editorial has selected a *single* model, the Hansch model of QSAR, as the only legitimate approach to QSAR that they will consider. We think that the role of editors and editorial boards is not to choose their favorite models and directions in science excluding others from consideration, but to review manuscripts sent to the journal and look for competent reviewers (even critical reviewers), and before deciding to wait for rebuttal of authors. We view the above decision that *only QSAR analyses that bring new insights on the mechanism of activity are encouraged* is scientifically unsound, unjustified, and hostile toward graph theoretical modeling using mathematical invariants (i.e., mathematical properties) of structures such as molecular descriptors, and other QSAR modeling, with the exception of Hansch QSAR approaches.

Editors and editorial boards of scientific papers may have their own opinions and preference on this or that model, but have no right to exclude models that they do not like or do not understand, from a journal the title of which includes chemical modeling. The graph theoretical manuscript ought to be sent to reviewers, who may express similar mistaken preference for this or that type of model, but authors should be *allowed* to respond to comments of reviewers. The level and degree of misconception of graph theory among chemists, and even physicists, is fairly widespread (as has been documented in several publications of several researchers in chemical graph theory), but how are those who don't know that they don't know going to find out about their bias when they ban papers that could open their eyes and from which they can learn about subjects on chemical modeling of which they may not be aware?

There is nothing wrong for an editorial to welcome papers on specific topics, such as Hansch QSAR. However, a journal covering chemical information and modeling ought to consider manuscript dealing with chemical information and modeling, which include graph theoretical models of structure–property–activity. Personal opinions and preferences of members of editorial boards and editors are irrelevant. Opinions, as Lynn Margulis had said, are opinions and not science.

We hope that our criticism of the policies of the *Journal of Chemical Information and Modeling* will not pass unnoticed, and that in the near future, item #5 of the "Guidelines to assist prospective journal authors of manuscripts in the field of QSAR/ QSPR analysis" will be eliminated so that the *Journal of Chemical Information and Modeling* will welcome all chemical modeling, including graph theoretical models, of structure–activity relationships. After all, *Journal of Chemical Information and Modeling* is a journal of the American Chemical Society, not a journal of Corwin Hansch Society, if there is such!

Subject Index

Name Index

9 780367 862268